In accordance with the latest syllabus prescribed by the Council for the Indian School Certificate Examinations, New Delhi.

MOST LIKELY

ISC QUESTION BANK

MATHEMATICS

CLASS XII

By

A PANEL OF AUTHORS

OSWAL PUBLISHERS
1/12, Sahitya Kunj, M.G. Road, Agra - 282002

EDITION
YEAR 2021

DISCLAIMER

With the ambition of providing standard academic resources, we have exercised extreme care in publishing the content. In case of any discrepancies in the matter, we request readers to excuse the unintentional lapse and not hold us liable for the same. Suggestions are always welcome.

ISBN : 978-93-91184-37-7

PUBLISHED BY

OSWAL PUBLISHERS

 1/12, Sahitya Kunj, M.G. Road, Agra - 282002

 (0562) 2527771-4, +91 7534077222

 info@oswalpublishers.in

 www.oswalpublishers.com

The cover of this book has been designed using resources from Freepik.com

Printed At Upkar Printing Unit, Agra

PREFACE

It is a matter of immense pride for us to present our 'MOST LIKELY ISC QUESTION BANK' series, especially prepared for students appearing for Board examinations in the oncoming year.

This book is a perfect capsule for building self-confidence during exam. preparation. Based on chunking strategy, the 'Categorywise–Chapterwise' format with its exhaustive set of questions allow the students to cover every category in a chapterwise manner.

Covering easy categories first boosts student's self-esteem and the ascending score braces them to take up challenging categories without fear. This prepares the student to see the exam paper as achievable at all times.

With its simple language and style, it is a one-stop solution for smart study. We are confident that the book will enable the candidates to develop a better understanding of the curriculum and help them organize their learning process. This book shall definitely prove to be a fruitful tool for the students and encourage them towards scholastic excellence.

Constructive suggestions for further improvement of the book are always welcome.

Note: Questions marked with '*' are frequently asked in previous years board examinations.

—**Publisher**

LET'S GO ORGANIC

Growing microgreens can be a good, easy and economical way for a student to start practicing plantation at home.

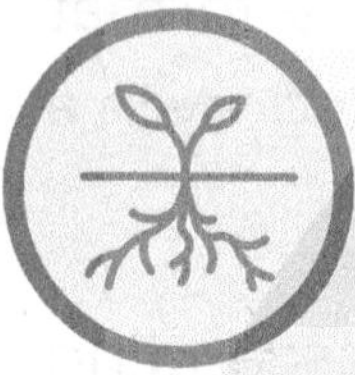

Microgreens are highly nutritious plants that are harvested after the sprout-stage and just before the maturity phase.

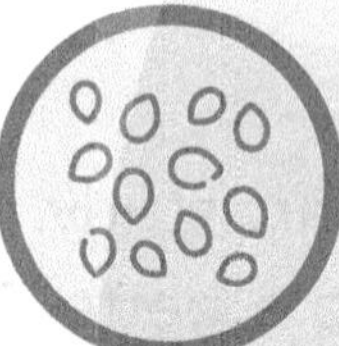

Microgreens may include mustard, coriander, basil seeds, dried peas, beans, broccoli, radish, spinach, beetroot, cauliflower and others.

It is more convenient to start growing microgreens that require low maintenance like mustard, coriander, peas, basil and others.

Start with finding a flat tray or a shoe-box or any tetra-pack carton.

Make holes in the box for easy drainage of water.

Spread two inches of soil in the box. You can get ready-made soil from the garden or nursery. Sprinkle your desired seeds over the soil.

You can spread a mixture of coco-peat and fertile topsoil or a stack of moist tissue papers over it. Remember to keep the tissue paper moist till the seeds start germinating.

Cover the box with newspaper and keep it in a place where there is no direct sunlight.

Spray water twice a day atleast.

On third and fourth day, once the leaves start to appear, start exposing the germinating microgreens to indirect sunlight.

You can try to expose them to direct sunlight for not more than 4 hours.

By ninth or tenth day, your microgreens will be ready, harvest them!

Microgreens can be eaten with sandwiches, wraps, salads, smoothies, juices or as garnishers on pizza, soup and curries.

CONTENTS

Word of Advice		7 - 12
1.	Chapter at a Glance	13 - 71
2.	Formulae Based Questions	72 - 124
3.	Find the Value Type Questions	125 - 155
4.	Prove the Following Type Questions	156 - 209
5.	Computational Questions	210 - 237
6.	Evaluate	238 - 261
7.	Data Based Questions	262 - 270
8.	Solve the Following Type Questions	271 - 296
9.	Graph Based Questions	297 - 306
10.	Practice Exercises	307 - 320
SC Solved Paper 2020 (With Comments of Examiners)		321 - 340

Note: Questions marked with '*' are frequently asked in previous years board examinations.

SUGGESTIONS FOR STUDENTS

1. Study and practice the entire syllabus keeping in view the pattern and the weightage of each chapter in the question paper.
2. Avoid selective study.
3. Stress on learning the concepts of topics like calculus and its applications as its weightage is maximum in the syllabus.
4. Make a list of all formulae and revise them frequently.
5. Understand the concepts of each topic and practice an adequate number of problems on a regular basis taking the assistance of the teacher, wherever required.
6. Identify the areas of weakness and try to resolve your problems
7. Revise the concepts of Class XI and integrate them with the Class XII syllabus (factorization, formulae of algebraic expressions, etc.)
8. Take part in all periodic tests conducted by the school sincerely.
9. Make use of the given reading time of 15 minutes, for reading the questions carefully and highlighting various key points in the statements of the questions
10. Develop logical and reasoning skills to have a clear understanding.
11. Revise all topics and formulae involved and make a chapter wise or topic-wise list of these.
12. Avoid calculation errors and increase accuracy to the highest levels while solving problems based on numerical values.
13. Learn the Formulae related to every topic after acquiring thorough understanding of each symbol used.
14. Manage your time effectively while attempting the question paper.
15. Practice mock/sample papers by strictly adhering to the stipulated time.
16. Make wise choices from the options available in the question paper and management time wisely.
17. Be methodical and neat in working.
18. All steps of calculation need to be simplified before proceeding to the next step.
19. Take sufficient rest before the examination.
20. Utilize the reading time properly.

Word of Advice

Chapter 2. Functions

Topics Found difficult By Students
- In proving the function "onto".
- Applying formulae of inverse trigonometric functions.
- Between the composite functions *fog* and *gof.*

1. Many students failed to prove the given binary operation using properties. They *used conventional operation instead of the binary operation* as defined in the question.
2. In problem involving *composite functions,* most of the students applied incorrect methods to solve it while many made errors in simplification of algebraic functions after applying the concept.
3. In proof involving one-one onto function, most students proved that the given function is one-one function, but some could not give any justification to say that it is onto function also.
4. In problem involving inverse function, many students do not read the question carefully and consequentially made error.

Chapter 3. Inverse Trigonometric Functions

Topics Found difficult By Students
- Applying formulae of inverse trigonometric functions.

1. Several students made errors while applying properties of inverse trigonometric functions, *converting one inverse trigonometric function into another equivalent inverse trigonometric function* and in simplification.
2. Many students made errors in applying the formula of $\sin^{-1} A + \sin^{-1} B$. Also, errors took place in simplifying and solving higher degree algebraic equations. Some students converted all terms into a particular inverse function form, for example $\tan^{-1}$ and could not handle the resulting equations.
3. A few students not only wrote incorrect formula for $\tan^{-1} x + \tan^{-1} y$ but also made errors while simplifying the expression.
4. Several students made errors while converting $\sec^{-1} x$ and $\text{cosec}^{-1} x$ to its correct form.

Chapter 4. Determinants

Topics Found difficult By Students
- Finding the area of a triangle using determinant.

1. Some students expanded the determinant directly to solve it, though it was clearly mentioned in the question *without expanding at any stage* or *using properties.*
2. Many students *applies properties of determinants in the incorrect order* to reduce further and *did not take common factors in a row/column* and hence could not simplify the determinant while others used incorrect co-factors after simplifying correctly.
3. Many students wrote incorrect formula of the area of a triangle using determinant.

Chapter 5. Matrices

Topics Found difficult By Students
- Multiplication of matrices.
- System for linear equations.

1. Many students made errors while finding product of a matrix by a matrix.
2. In solving linear equations using matrices, some students made simplification errors in the process of finding the values of x, y and z. A few students made errors while calculating cofactors and inverse matrix.
3. Many students did not know how to use the product of the given two matrices to solve the given equations. A few students incorrectly used Cramer's rule.
4. While proving the given matrix that is symmetric matrix which most students were unable to prove it as they *incorrectly used the concept of singular matrix instead of symmetric matrix.*
5. A few students could not solve the problem based on *symmetric matrix.*
6. In finding inverse of a matrix, a few candidates applied both row and column operations in the same matrix. In a number of cases, errors were committed at the later stages of simplification and also in using correct $X = A^{-1}B$ form.

Chapter 6. Differential Calculus

Topics Found difficult By Students

- Continuity.
- Proving differentiability.
- Mean value theorems, open and closed interval in Mean Value theorem.

1. Most of the students proved the continuity of a function but did not apply the conditions to check differentiability while many failed to calculate LHD and RHD.
2. Errors were made while finding the derivative of implicit function.
3. Majority of the students committed errors while *identifying open and closed intervals in the process of applying the properties of mean value theorems.* They also made errors while finding the value of 'x' in the open interval when $f'(x) = 0$.
4. Most of the students made errors while differentiating second time and in further simplification.
5. Majority of the students made errors in calculating limit of the given function.
6. A large number of students had an idea of solving the limit using L' Hospital's rule but had *no knowledge of differentiating the function* a^x.
7. In solving limits, mistakes were made while converting in 0/0 form while some applied L' Hospital's rule without observing the indefinite form 0/0.
8. A few students made mistakes while applying L' Hospital's rule successively.
9. Common errors made by students were in differentiation of exponential functions, elimination of the parameter used and in simplification especially while applying L' Hospital's rule and in successive differentiations.
10. Some students committed errors in solving Logarithmic equations as they could not make a difference between log function with base '10' and base 'e'.
11. Use of the rule for composite function as well as chain rule was forgotten by some students.

Chapter 7. Applications of Derivatives

Topics Found difficult By Students

- Rate measure.
- Increasing and decreasing functions.
- Maxima and Minima.

1. Some students could not test the *condition of increasing function* after finding the derivative of the function.
2. While finding the rate of increase or decrease using derivatives, several students could not form the function. Many made errors while differentiating the function with respect to time and ended up *differentiating with respect to 'x' instead of 't'.*

3. Several students failed to apply the correct formula for the surface area of a closed box with square base and express the same in terms of one variable. Some could not complete the 2nd order derivative test for maximization/minimization. A few found the value of the variable but could not find the dimensions of the box.
4. A number of students identified distance function in terms of one variable in the square root form where differentiation of it involves long calculations.
5. Many students did not have an idea of approximation concept as application of derivatives and a few students made mistakes while writing final answer in the correct form.
6. A few students made mistakes while finding the point of contact at which tangent to the curve is parallel to x-axis.

Chapter 8. Integral Calculus

Topics Found difficult By Students

- Integration by substitution and by parts.
- Definite integrals and their properties.
- Modulus function.

1. In solving definite integral involving modulus function, many students were unable to identify the correct intervals of absolute function for the given limits.
2. In integrals involving quadratic expression in square root, most students made mistakes while writing the quadratic expression in the form of difference of two perfect squares which lead to further simplification mistakes.
3. In integrals involving complex inverse trigonometric functions, most students attempted to the extent of simplifying the function by taking correct substitution but made errors while applying the concept of integration by parts to integrate the function.
4. In integrals involving trigonometric functions, most students could not complete the integration by using correct identity of trigonometric functions.
5. Majority of the students made simplification errors while applying the concept of standard integrals.
6. Majority of the students did not split the complex polynomial functions into individual functions while others applied incorrect substitution.
7. Common errors made were in substitution, simplification or applying integration by parts concept for further simplification.
8. Many students applied incorrect techniques for solution of different types of integrals.
9. Many students made errors in using properties of definite integrals.

Chapter 9. Differential Equations

Topics Found difficult By Students

- Identification of the types of differential equations and their framing.

1. While forming differential equations, most of the students differentiated the function correctly but did not *eliminate the arbitrary constant*. Some used the second order derivative method for eliminating constant.
2. In solving the given differential equation, most students solved the equation correctly but made errors while integrating a term at the end of the solution. *A few did not write constant C at the end of the solution.*
3. Majority of the students made mistakes in *identifying the type of given differential equation* for example, some identified the given equation as homogeneous equation whereas it was reducible to simple separable type equation.
4. Many students did not have an idea of formation of differential equation and made errors while solving it.
5. Most students made errors while transforming the given equation into linear differential equation and in finding the integration factor.
6. Some students did not have any knowledge of applications of differential equations.
7. Many students could not separate the variables to solve the differential equations.

Chapter 10. Probability

Topics Found difficult By Students

- ➢ Conditional Probability and their applications.
- ➢ Between mutually exclusive and independent events.
- ➢ Between dependent and independent events.
- ➢ Product and sum rule of probability.
- ➢ Probability distribution, Binomial Probability distribution.

1. A number of students made errors while finding the probability of drawing multiple balls one by one with or without replacement due to the lack of knowledge of the *basic idea of probability of drawing ball with or without replacement.*
2. Several students could not find the probability of mutually exclusive and independent events by applying the concept of *independent events.*
3. Most students committed simplification errors while finding the *probability of success and failure.*
4. A large number of students could not *identify all possible cases while transferring balls from one bag to another.* Some also made errors while writing the summation of product of probabilities of all the cases.
5. A few students got confused while using *Bayes' theorem* and the concept of total probability.
6. Many students used incorrect formula for Standard deviation and calculated incorrect values of n, p and q.
7. Some students could not understand the meaning of *at most one success.*
8. A large number of students could not find the probability of *neither of them* cases for example, *neither of them winning the race.*
9. Many students made errors in computing the values of p and q and *applying the binomial probability formula* $P(X = r) = {}^nC_r\, p^r q^{n-r}$. *Most were not clear in writing correct binomial distribution.*
10. Several students had difficulty in calculating the value of $P(A \cap B)$.

Chapter 11. Vectors

Topics Found difficult By Students

- ➢ Projection of a vector.
- ➢ Interchange of vector equation to Cartesian equation and vice-versa.
- ➢ Dot and Cross product of vectors.

1. A few students made simplification errors in applying dot and cross product.
2. In finding the area of a triangle using vectors, many students made errors in the process of finding any two sides of the triangle and some applied incorrect formula to find the area.
3. A few students applied incorrect formula for perpendicular vectors.
4. Many students failed to apply the concept of scalar triple product $a.(a \times b) = 0$ while a few made calculation errors while finding $a \times b$.
5. Many students had no idea of the concept of collinear vectors and coplanar vectors.
6. Some students made errors while comparing the components of collinear vectors.
7. In finding scalar projection, most students applied dot product instead of *applying projection formula.*
8. To find the *angle between vectors,* many students were unable to express the formula for $\cos\theta = (a.b)/(|a|\,|b|)$.
9. In co-planarity of vectors, some students expanded scalar triple product determinant incorrectly, many made errors while applying scalar triple product and a few failed to write the correct order of the same.
10. A few students did not know the *basic difference between position vector and vector.*

Chapter 12. Three-dimensional Geometry

Topics Found difficult By Students

- Family of planes and equation of straight line in 3D.

1. Many students applied the incorrect formula for *angle between the two lines* while others committed errors in simplification.
2. A number of students made errors in finding the *image of a point in the given plane*. A few applied incorrect method.
3. Most of the students did not apply the concept of perpendicular lines and the condition for the same incorrectly.
4. Some students made calculation errors while applying the cross-multiplication rule to find direction ratios.
5. Many students did not apply the *perpendicular distance formula from a given point to a plane.*
6. In finding the angle between the lines, some could not express the given equation into *desired form (symmetric form)* and hence could not calculate the direction ratios of the lines correctly.
7. Some students transformed Cartesian form of line into vector equation incorrectly.
8. Majority of the students could not find the correct values of intercepts on the respective axis.
9. Many students failed to prove lines are intersecting and used incorrect formula to find the shortest distance.
10. Many students made errors in applying the concept that *direction ratios of normal to the plane and direction ratios of perpendicular to the plane are proportional.*

Chapter 13. Application of Integrals

Topics Found difficult By Students

- Application of definite integrals: Sketching of curves, Identifying the area of the shaded region and finding the upper and lower limits from the graph.
- Area under the curve.

1. In sketching the given parabola curves and finding area closed between curves, many students were not able to associate the equations of the curves as a parabola and made mistakes while sketching the graph. The *concept of symmetry* was not used by a few students and hence, they could not find the *common portion* and *limits* correctly.
2. In most of the cases, rough sketch of the curve was not drawn correctly.
3. In sketching the graph of absolute value function, many students did not understand the concept of absolute value function and could not identify the limits of it by using the given conditions.
4. Common errors were made in finding the limits.

Chapter 14. Application of Calculus

Topics Found difficult By Students

- Application of calculus in Commerce and Economics.

1. A few students made mistakes while calculating the demand function and marginal revenue function.
2. Many students made errors in applying the concept of Revenue function $R(x)$ and differentiating the $R(x)$.
3. Several students used integrated marginal cost function $MC(x)$ to find cost function $C(x)$ without adding a constant to $C(x)$. This led to an incorrect solution.
4. Many students made errors while finding the profit function $P(x)$. Some made errors in the process of finding the value of 'x' at which the profit was maximum while applying the concept of maxima and minima. In a few cases, students failed to apply the second order derivative for maximisation.
5. In finding break-even point, many students misinterpreted the question and calculated incorrect revenue function $R(x)$, which led to incorrect calculation of break-even point.

6. While calculating the value of 'x' when average cost function AC was increasing, some students included the negative value of x.
7. Common errors were made in simplification.

Chapter 15. Linear Regression

Topics Found difficult By Students

➢ Lines of Regression of y on x and x on y.

1. Many students applied the incorrect formulae for calculation of regression coefficients b_{yx} and b_{xy}, thereby made errors in finding correlation coefficient and regression equations.
2. Many students made errors while applying the condition to identify correct regression coefficients of the given lines, thereby calculated incorrect correlation coefficient.
3. Some students considered incorrect regression equations due to lack of understanding of regression concept.
4. Many students made errors in simplification.

Chapter 16. Linear Programming

Topics Found difficult By Students

➢ Formation of LPP (Linear Programming Problem).

1. Many students could not express the given constraints in the form of linear inequalities correctly.
2. Many students did not write non-negative constraints $x \geq 0$ and $y \geq 0$.
3. A few students did not indicate the feasible region on the graph at all.
4. Some students noted incorrect corner points to find optimum value.
5. In some cases, all the constraints were not used and hence coordinates of all feasible points were not obtained.

❑❑

Chapter at a Glance

Set 1

Chapter 1. Relations

1. Any set of ordered pairs is called a **relation**, where the set of first components of ordered pairs is called **domain** and the set of second component is called **range.** It is generally denoted by R.

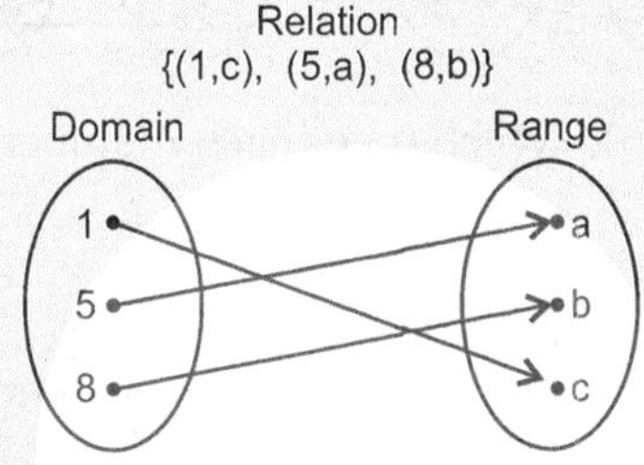

2. A relation in which every element of a set is related to itself only is called **identity** relation, *i.e.*

$$I_A = \{(a, a) : a \in A\}$$

3. A relation R on a set A is called the **empty** relation, if no element of A is related to another element of A, *i.e.*

$$R = \phi \subset A \times A.$$

It is the smallest relation on A.

4. A relation R on a set A is called the **universal** relation, if every element of A is related to each of the elements of A, *i.e.*

$$R = A \times A.$$

5. A relation in which every element is in relation with itself is called a **reflexive** relation, *i.e.* R is reflexive if $(a, a) \in R$ for every $a \in A$.
6. A relation R in a set is said to be **symmetric** if whenever we have $(a, b) \in R$, then we also have $(b, a) \in R$.

(a, b)	-------->	(b, a)
(1, 1)	-------->	(1, 1)
(2, 2)	-------->	(2, 2)
(3, 3)	-------->	(3, 3)
(1, 2)	-------->	(2, 1)

7. A relation in a set is said to be **transitive** if whenever we have $(a, b) \in R$ and $(b, c) \in R$, then we also have $(a, c) \in R$.

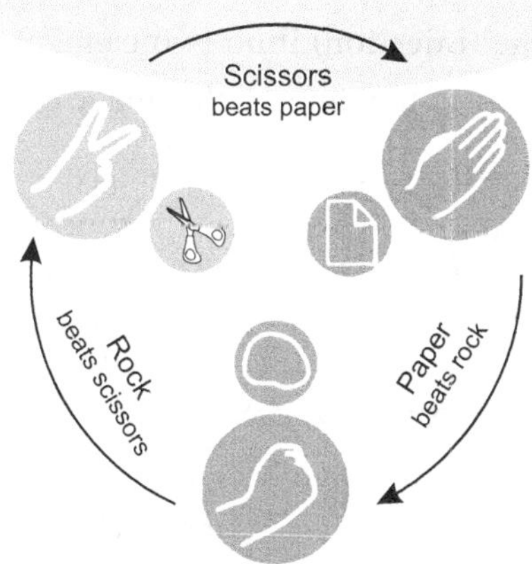

8. R is said to be an **equivalence relation** only if R is *reflexive, symmetric* and *transitive.* (all three condition must be satisfied)

Chapter 2. Functions

1. A **function** is a *special relation* in which no two different ordered pairs have the same first component, *i.e* set {(0, 1), (1, 1)} is a function but the set {(1, 0),(1, 1)} is not a function. It is generally denoted by *f*.

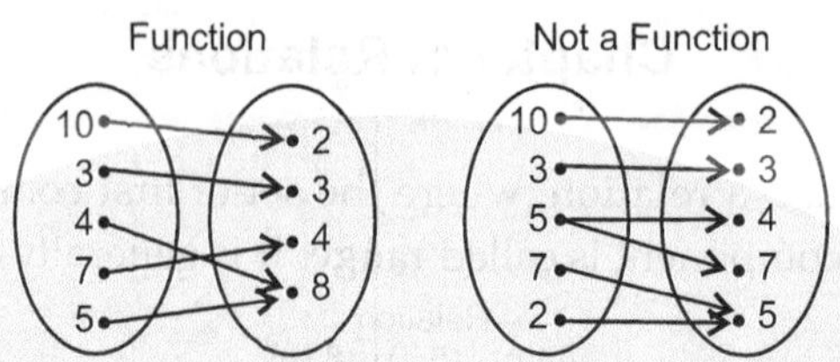

2. A **function** may be thought of as a rule which takes each member x of a set and assigns, or maps it to the same value y known at its image.

$$x \to \text{Function} \to y$$

3. The phrase "y is a function of x" means that the value of y depends upon the value of x, so :
   - y can be written in terms of x (*e.g.* $y = 3x$).
   - If $f(x) = 3x$, and y is a function of x, *i.e.* $y = f(x)$, then the value of y when x is 4 is $f(4)$, which is found by replacing x by 4.
4. What can go **into** a function is called the **Domain** (possible values of x)
5. What **may possibly come out** of a function is called the **Codomain.**
6. What **actually comes out** of a function is called the **Range** or **Image,** *i.e.* set of values of y corresponding to x. It is a subset of codomain.

Example :

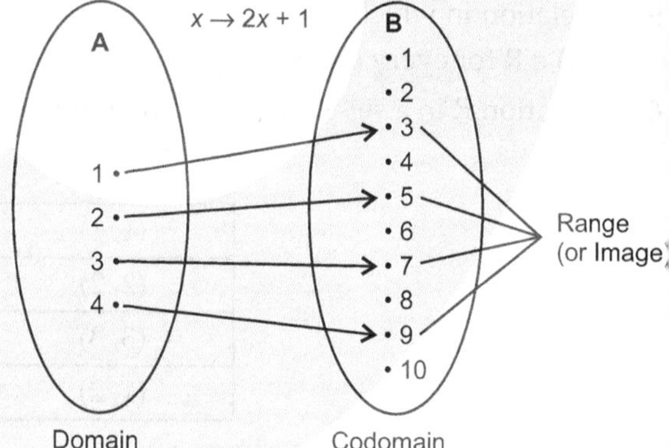

- The set "A" is the **Domain,**
- The set "B" is the **Codomain,**
- And the set of elements that get pointed to in B (the actual values produced by the function) are the **Range.**

  And we have :
- Domain : {1, 2, 3, 4}
- Codomain : {1, 2, 3, 4, 5, 6, 7, 8, 9, 10
- Range : {3, 5, 7, 9}

7. A function f is said to be **one-to-one (injection)** if no two elements in the domain of f correspond to the same element in the range of f, *i.e.* no y in the range is the image of more than one x in the domain.

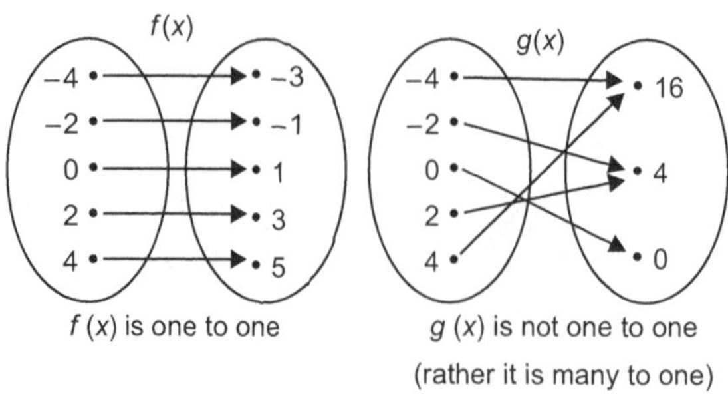

f (*x*) is one to one

g (*x*) is not one to one (rather it is many to one)

8. A function f is said to be **many-to-one** if two or more elements in the domain of f correspond to the same element in the range of f, *i.e.* atleast one y in the range is the image of more than one x in the domain. In the above example $g(x)$ is many-to-one function.

9. A function $f: A \to B$ is said to be an **into** function if there exists even a single element in B having no pre-image in A, *i.e.* range of f is a proper subset of (and not equal to) codomain of f. eg. In the figure below, element 4 in B has no pre-image in A, thus it is an into function.

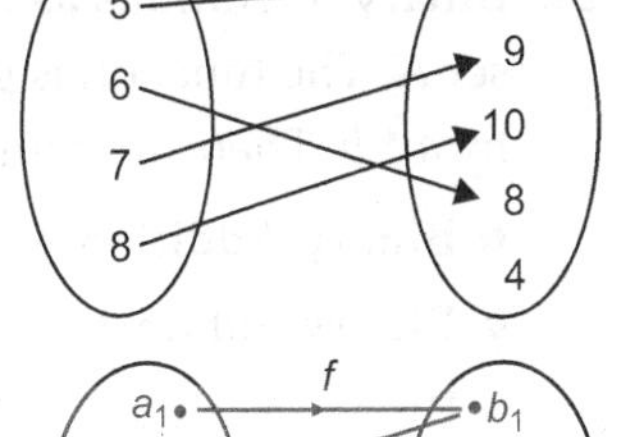

10. A function $f: A \to B$ is said to be an **onto** function if every element of B has a pre-image in A, *i.e.* $f(A) = B$, *i.e.* range of f = codomain of f. eg. In the figure below, it is an onto function because set B is entirely used up.

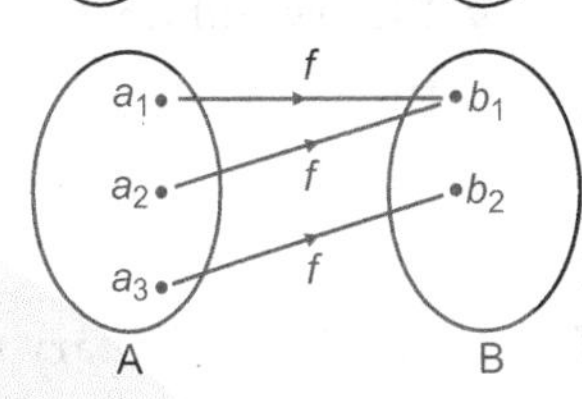

11. A function $f: A \to B$ is said to be a **real-valued function** if B (codomain) is a subset of **R** (set of all real numbers).

12. A function $f: A \to B$ is said to be a **real function** if both A (domain) and B (codomain) are subsets of **R** (set of all real numbers).

13. A function that reverses that another function is called **inverse** function. It takes us back to where we started, *i.e.* if $f(x) = y$, then $f^{-1}(y) = x$. (Note that f^{-1} is a symbol and does not mean f to the power of -1 which is equal to $1/f$)

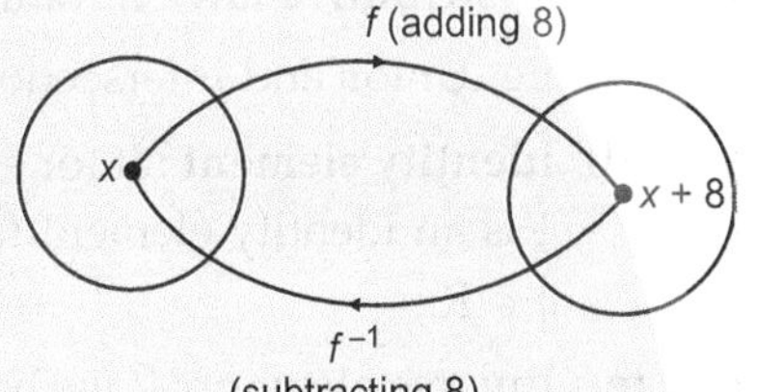

14. Domain and range of original function becomes the range and domain of inverse function respectively (look at eg. below) :

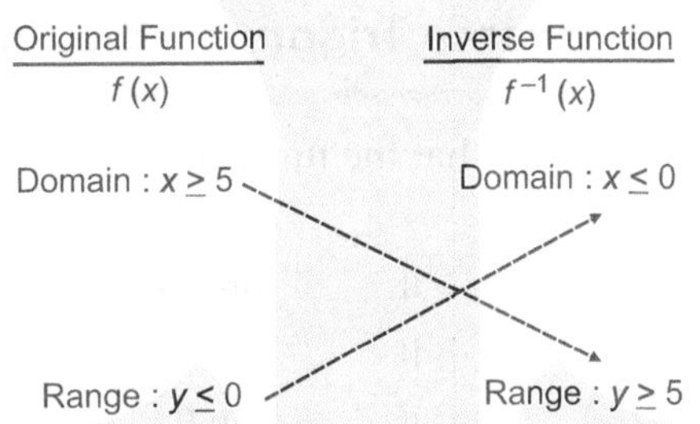

15. Not every function f has an inverse (or **invertible**), *i.e.* there are certain **conditions of invertibility** that f must satisfy for f^{-1} to exist which are as follows :
    - *If and only if it is an* ***one-to-one onto function*** *(bijective).*
    - *A function is one-to-one if $f(a) = f(b)$ only when $a = b$.*

16. *A composite function is generally a function that is written inside another function. Composition of a function is done by substituting one function into another function.*

    For example, $f[g(x)]$ is the composite function of $f(x)$ and $g(x)$. The composite function $f[g(x)]$ is read as "f of g of x". The function $g(x)$ is called an inner function and the function $f(x)$ is called an outer function. Hence, we can also read $f[g(x)]$ as "the function g is the inner function of the outer function f".

17. **Composition of inverse functions** : If the compositions of two functions is equal to the input value, the functions are inverses of each other, *i.e.* if $f(g(x)) = g(f(x)) = x$, then f and g are inverses of each other, *i.e.* $f = g^{-1}$, or $g = f^{-1}$.

18. *The composition of function is not commutative, i.e. $f(g(x)) \neq g(f(x))$ unless $f(x)$ and $g(x)$ are inverses of each other. However, converse is not true, i.e. if $f(g(x)) = g(f(x))$, it does not follow that f and g are inverses of each other.*

19. **Method for finding the inverse of a function** :

    Step 1 : Start with $y = f(x)$.

    Step 2 : Solve the equation $y = f(x)$ to obtain x in terms of y.

Step 3 : Replace x by $f^{-1}(y)$.

Step 4 : Put x for y.

Step 5 : Verify your work by checking that $f(f^{-1}(x)) = x$ and $f^{-1}(f(x)) = x$.

20. **Binary operations as function :** Binary operation can be defined as an operation * which is performed on a set A. The function is given by $* : A * A \rightarrow A$. So the operation * performed on operands a and b is denoted by $a * b$. There are four main types of binary operations which are :
   - **Binary Addition +** : $R \times R \rightarrow R$ is given by $(a, b) \rightarrow a + b$
   - **Binary Subtraction –** : $R \times R \rightarrow R$ is given by $(a, b) \rightarrow a - b$
   - **Binary Multiplication ×** : $R \times R \rightarrow R$ is given by $(a, b) \rightarrow a \times b$
   - **Binary Division /** : The division cannot be defined on real numbers. This is because $/ : R \times R \rightarrow R$ is given by $(a, b) \rightarrow a/b$. Now if we take b as 0 here, a/b is not defined.

21. **Properties of Binary Operations :** (* and ^ denotes operations on set A)

   (a) **Commutative law :** If for every $a, b \in A$, $a * b = b * a$,

   (b) **Associative law :** If for all $a, b, c \in A$, $(a * b) * c = a * (b * c)$

   (c) **Distributive law :** If for all $a, b, c \in A$, $a * (b \wedge c) = (a * b) \wedge (a * c)$

   Eg. Union and Intersection of sets are Commutative, Associative as well as Distributive.

   (d) **Identity element :** If for every $a \in A$, there exists an element $e \in A$ with the property $e * a = a * e = a$, eg. 1 is an identity element for multiplication operation of real numbers, since $1 \times a = a \times 1 = a$ for every $a \in R$.

   (e) **Inverse element :** If we have a set A having identity element e with respect to a binary operation*, then A is said to have an inverse element a^{-1} for each element a of A if $a * a^{-1} = a^{-1} * a = e$. eg. **additive inverse** and **multiplicative inverse** of real numbers.

Chapter 3. Inverse Trigonometric Functions

1. **Inverse trigonometric function :** We know that the function $x = \sin \theta$, means that θ is an angle whose sine is x or x is the sine of θ. Hence,

	$\theta = \sin^{-1} x$	if	$\sin \theta = x$
Similarly	$\theta = \cos^{-1} x$	if	$\cos \theta = x$
and	$\theta = \tan^{-1} x$	if	$\tan \theta = x$

   Then functions $\sin^{-1}x$, $\cos^{-1}x$, $\tan^{-1}x$, $\sec^{-1}x$, $\text{cosec}^{-1}x$ and $\cot^{-1}x$ are called *inverse trigonometric functions*.

   It is important to note that,

   (i) $\sin \theta$ is a number whereas $\sin^{-1} x$ is an angle.

   (ii) $\sin^{-1} x \neq (\sin x)^{-1}$ or $1/\sin x$

2. Inverses of trigonometric functions are all relations and not functions. eg. for $y = \sin x = 1/2$, the function $\{(x, y) : y = \sin x\}$ will be an infinite set of ordered pairs $\{(\pi/6 + 2n\pi, 1/2), (5\pi/6 + 2n\pi, 1/2), n \in I\}$, then to find inverse of $\sin x$, y is replaced by 1/2. Thus the inverse of this function is the set of ordered pairs $\{(1/2, \pi/6 + 2n\pi), (1/2, 5\pi/6 + 2n\pi, n \in I\}$, which is obviously not a function, because corresponding to a value of the independent variable (domain), there are more than one value of the dependent variable (range). ***Hence, we have to place certain restrictions on either the domain or range so that inverse of t-functions may be made a function.*** In the above example, since the domain is already restricted to $-1 \leq x \leq 1$, we consider restricting the range from $-\pi/2$ to $\pi/2$. Then defining the $\sin^{-1} x$ as $\{\theta : \pi - 2 \leq \theta \leq \pi/2\}$ denoted by $\sin^{-1} x = \theta$. Now we find that all values of x in domain are associated with one and only one value in the restricted range.

3. **Principal value :** The principal value of an inverse trigonometric function is the smallest numerical value, either positive or negative of the function.

   Limitations for principal values of inverse circular functions can be placed in a table given below :

Function	Domain (x)	Range (θ)
$\sin^{-1} x$	$[-1, 1]$	$[-\pi/2, \pi/2]$
$\cos^{-1} x$	$[-1, 1]$	$[0, \pi]$
$\tan^{-1} x$	$(-\infty, \infty)$	$(-\pi/2, \pi/2)$
$\cot^{-1} x$	$(-\infty, \infty)$	$(0, \pi)$
$\sec^{-1} x$	$(-\infty, -1) \cup (1, \infty)$	$[(0, \pi/2) \cup (\pi/2, \pi)]$
$\text{cosec}^{-1} x$	$(-\infty, -1) \cup (1, \infty)$	$[-\pi/2, 0) \cup (0, \pi/2]$

4. Graphs of inverse trigonometric functions :

(**Note :** '**arc sin***x*' is also used for $\sin^{-1}x$)

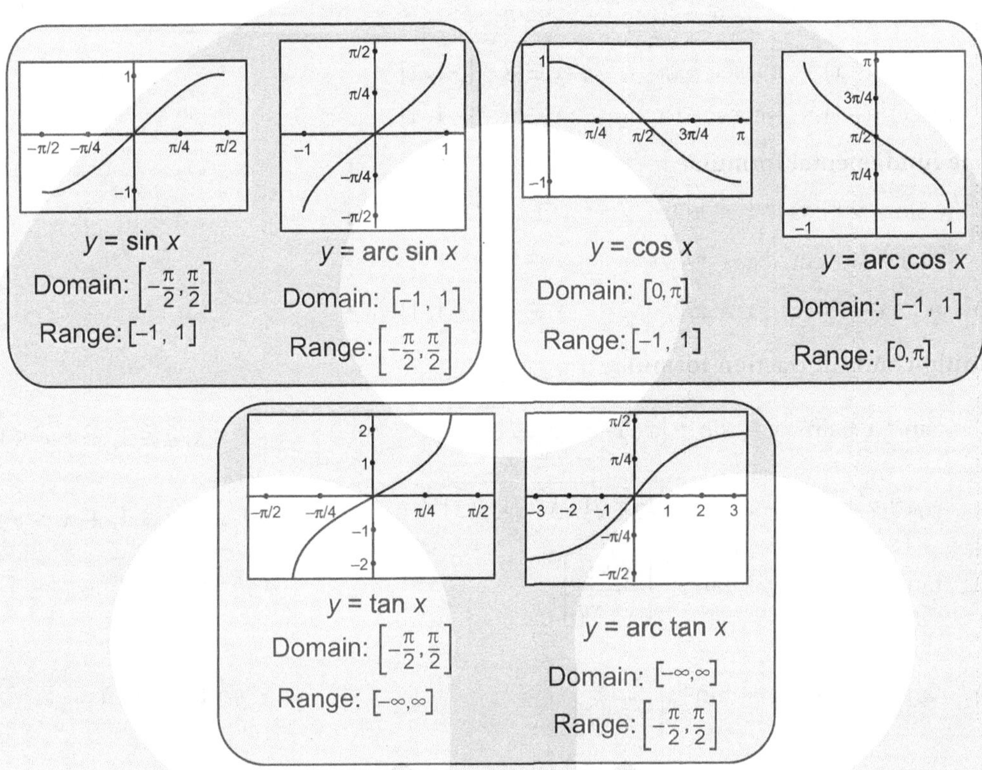

Inverse Trig Graphs

5. **Properties of inverse trigonometric functions :**

(I) Same angle can be expressed by different inverse trigonometric function.

$$30° = \sin^{-1}(1/2)$$

$$= \cos^{-1}(\sqrt{3}/2) = \tan^{-1}(1/\sqrt{3})$$

(II) Inverse property :

(A) (i) $\sin^{-1}(\sin\theta) = \theta,\quad -\pi/2 \le \theta \le \pi/2$

(ii) $\cos^{-1}(\cos\theta) = \theta,\quad 0 \le \theta \le \pi$

(iii) $\tan^{-1}(\tan\theta) = \theta,\quad -\pi/2 < \theta < \pi/2$

(B) (i) $\sin(\sin^{-1} x) = x,\quad -1 \le x \le 1$

(ii) $\cos(\cos^{-1} x) = x,\quad -1 \le x \le 1$

(iii) $\tan(\tan^{-1} x) = x,\quad x \in R$

(III) Principal of reciprocality : Following reciprocal relations exist between the inverse trigonometric functions.

(i) $\sin^{-1} x = \text{cosec}^{-1}(1/x), -1 \le x \le 1$ (ii) $\cos^{-1} x = \sec^{-1}(1/x), -1 \le x \le 1$

(iii) $\tan^{-1} x = \cot^{-1}(1/x), x > 0$

(iv) $\cot^{-1} x = \tan^{-1}(1/x), x > 0$

or

$\cot^{-1} x = \pi + \tan^{-1}\frac{1}{x}\ \ x < 0$

(v) $\sec^{-1} x = \cos^{-1}(1/x), x \le -1, x \ge 1$

(vi) $\text{cosec}^{-1} x = \sin^{-1}(1/x), x \le -1, x \ge 1.$

(IV) Inverse trigonometric functions are odd functions within the principal values :

(i) $\sin^{-1}(-x) = -\sin^{-1}(x),\quad -1 \le x \le 1$

(ii) $\cos^{-1}(-x) = \pi - \cos^{-1}(x),\quad -1 \le x \le 1$

(iii) $\tan^{-1}(-x) = -\tan^{-1} x,\quad x \in R$

(iv) $\cot^{-1}(-x) = \pi - \cot^{-1} x.\quad x \in R$

(v) $\sec^{-1}(-x) = \pi - \sec^{-1} x,\quad x \in R - [-1, 1]$

(vi) $\text{cosec}^{-1}(-x) = -\text{cosec}^{-1} x,\quad x \in R - [-1, 1]$

(V) Some fundamental formulae :

(i) $\sin^{-1} x + \cos^{-1} x = \pi/2,\quad -1 \le x \le 1$

(ii) $\tan^{-1} x + \cot^{-1} x = \pi/2,\quad x \in R$

(iii) $\sec^{-1} x + \text{cosec}^{-1} x = \pi/2,\quad x \in R - [-1, 1]$

(VI) Addition and subtraction formulae :

(i) $\sin^{-1} x \pm \sin^{-1} y = \sin^{-1}[x\sqrt{1-y^2} \pm y\sqrt{1-x^2}]\quad -1 \le x, y \le 1, x^2 + y^2 \le 1.$

(ii) $\cos^{-1} x \pm \cos^{-1} y = \cos^{-1}[xy \mp \sqrt{1-x^2}.\sqrt{1-y^2}]\quad -1 \le x, y \le 1, \mp x \le y$

(iii) $\tan^{-1} x + \tan^{-1} y = \tan^{-1}\left[\frac{x+y}{1-xy}\right]\quad x, y > 0, xy < 1$

(iv) $\tan^{-1} x - \tan^{-1} y = \tan^{-1}\left[\frac{x-y}{1+xy}\right]\quad x, y > 0, xy > -1$

(v) $\cot^{-1} x + \cot^{-1} y = \cot^{-1}\left[\frac{xy-1}{y+x}\right]$

(vi) $\cot^{-1} x - \cot^{-1} y = \cot^{-1}\left[\frac{xy+1}{y-x}\right]$

(vii) $\tan^{-1} x + \tan^{-1} y + \tan^{-1} z = \left[\frac{x+y+z-xyz}{1-yz-zx-xy}\right]$

(VII) Some important formulae :

(i) $2\tan^{-1} x = \sin^{-1}\left(\frac{2x}{1+x^2}\right) = \cos^{-1}\left(\frac{1-x^2}{1+x^2}\right) = \tan^{-1}\left(\frac{2x}{1-x^2}\right)$

(ii) $2\sin^{-1} x = \sin^{-1}(2x\sqrt{1-x^2})$

(iii) $2\cos^{-1} x = \cos^{-1}(2x^2 - 1)$

(iv) $3\sin^{-1} x = \sin^{-1}(3x - 4x^3)$

(v) $3\cos^{-1} x = \cos^{-1}(4x^3 - 3x)$

(vi) $3\tan^{-1} x = \tan^{-1}\left(\frac{3x-x^3}{1-3x^2}\right)$

(vii) $\sin^{-1} x = \cos^{-1} \sqrt{1-x^2} = \tan^{-1}\left(\frac{x}{\sqrt{1-x^2}}\right) = \cot^{-1}\left(\frac{\sqrt{1-x^2}}{x}\right) = \sec^{-1}\left(\frac{1}{\sqrt{1-x^2}}\right) = \text{cosec}^{-1}\left(\frac{1}{x}\right)$

(viii) $\cos^{-1} x = \sin^{-1} \sqrt{1-x^2} = \tan^{-1}\left(\frac{\sqrt{1-x^2}}{x}\right) = \sec^{-1}\left(\frac{1}{x}\right) = \text{cosec}^{-1}\left(\frac{1}{\sqrt{1-x^2}}\right) = \cot^{-1}\left(\frac{x}{\sqrt{1-x^2}}\right)$

(ix) $\tan^{-1} x = \sin^{-1}\left(\frac{x}{\sqrt{1+x^2}}\right) = \cos^{-1}\left(\frac{1}{\sqrt{1+x^2}}\right) = \cot^{-1}\left(\frac{1}{x}\right) = \sec^{-1} \sqrt{1+x^2} = \text{cosec}^{-1}\left(\frac{\sqrt{1+x^2}}{x}\right)$

Chapter 4. Determinants

1. The determinant of a matrix is a number that is *specially defined only for square matrices*. Determinants are very useful in the analysis and solution of systems of linear equations.
2. **Determinant of a square matrix of 2 × 2 order :**
   For a 2 × 2 matrix (2 rows and 2 columns) :

$$A = \begin{bmatrix} a & b \\ c & d \end{bmatrix}$$

   The determinant is : $|A| = ad - bc$ (Determinant may be denoted by D or Δ)
   (**Note :** The determinant $|-3|$ should not be confused with the absolute value of – 3)
3. **Determinant of a square matrix of 3 × 3 order :**
   For a 3 × 3 matrix (3 rows and 3 columns) :

$$A = \begin{bmatrix} a & b & c \\ d & e & f \\ g & h & i \end{bmatrix}$$

   The determinant is : $|A| = a\,(ei - fh) - b\,(di - fg) + c\,(dh - eg)$

$$\begin{bmatrix} a & & \\ & e & f \\ & h & i \end{bmatrix} - \begin{bmatrix} & b & \\ d & & f \\ g & & i \end{bmatrix} + \begin{bmatrix} & & c \\ d & e & \\ g & h & \end{bmatrix}$$

4. **Minors :** Let $A = [a_{ij}]$ be a square matrix of order n and let A_{ij} denote the square matrix of order $n - 1$, obtained from A by deleting the ith row and j^{th} column, then the number det A_{ij} is called *minor* of the entry a_{ij} and is denoted by M_{ij}.
   For example,

   Let $$A = \begin{bmatrix} 1 & -2 & 3 \\ -1 & 7 & 0 \\ 3 & 1 & 2 \end{bmatrix}$$

   then $$M_{11} = \begin{vmatrix} 7 & 0 \\ 1 & 2 \end{vmatrix} = 14 - 0 = 14.$$

$$M_{12} = \begin{vmatrix} -1 & 0 \\ 3 & 2 \end{vmatrix} = (-2 - 0) = -2 \text{ and so on}$$

5. **Co-factors :** Let $A = [a_{ij}]$ be a square matrix of order n, then the co-factor C_{ij} of element a_{ij} is defined as $C_{ij} = (-1)^{i+j} M_{ij}$

Chapter at a Glance

6. **The value of a determinant** is obtained by multiplying the elements with their respective co-factors and adding the resulting products, *i.e.* for a square matrix of order 3, $\Delta = a_{11}C_{11} + a_{21}C_{21} + a_{31}C_{31}$

7. **Sarrus rule :** The value of the determinant is obtained by finding the sum of the products of elements along the diagonal parallel to main diagonal and subtracting the sum of the products of elements along the diagonals which sum up from left to right.

For example, to evaluate $\begin{vmatrix} 1 & 3 & -4 \\ 2 & 5 & 0 \\ -2 & 4 & 7 \end{vmatrix}$, By Sarrus diagram,

$$\begin{matrix} 1 & 3 & -4 & 1 & 3 \\ 2 & 5 & 0 & 2 & 5 \\ -2 & 4 & 7 & -2 & 4 \end{matrix}$$

The value of the given determinant = [1.5.7 + 3.0 (– 2) + (– 4). 2.4] – [(– 2).5.(– 4) + 4.0.1 + 7.2.3]

= (35 + 0 – 32) – (40 + 0 + 42) = 3 – 82 = – 79.

8. **Properties of determinants :**

   (a) If each entry in any row, or any column, of a Δ is equal to 0, then $\Delta = 0$.

   (b) If rows be changed into columns and vice-versa, then **Δ remains *unaltered*.**

   (c) If any two rows (or columns) of a Δ are interchanged, **then resulting is the *negative* of the original Δ.**

   (d) If two rows (or columns) in a Δ are identical, then $\Delta = 0$.

   (e) If all the elements of any row (or column) be multiplied by a non-zero real number 'k', then the value of **new Δ is k times the value of original** Δ.

   (f) If each entry in a row (or column) of a Δ is written as the sum of two or more terms, then Δ **can be written as the sum of two or more** Δs.

   (g) If each entry of one row (or column) of a Δ is multiplied by a real number 'k' and the resulting product is added to the corresponding entry in another row (or column resp.) in the Δ, then the **resulting Δ = original Δ**.

   (h) If each element of a line (row or column) of a Δ be added to the equi-multiples of the corresponding elements of one or more parallel lines, then **Δ remains *unaltered*.**

   (i) **Factor Theorem :** If the elements of a Δ that involve x are polynomials in x, and if $\Delta = 0$ when 'a' is substituted for x, then **$(x - a)$ is a factor of the Δ.**

   (j) **Elementary Transformations :**

      1. Interchange of two rows or columns.
      2. Multiplication of a row or column of a Δ by a constant other than 0.
      3. Addition to the entries of a line of the Δ the constant multiples of a parallel line.

   (k) If two determinants are such that each can be obtained from the other by a finite number of elementary transformations, they are said to be **equivalent.**

9. **Product of two determinants of same order :** The two determinants of same order can be multiplied by any one of the following methods :

   (i) row by row, (ii) row by column,

   (iii) column by row, (iv) column by column.

10. **Area of a triangle :** Let the vertices of a triangle are (x_1, y_1), (x_2, y_2) and (x_3, y_3), then area of that triangle is given by

$$\Delta = \frac{1}{2}\begin{vmatrix} x_1 & y_1 & 1 \\ x_2 & y_2 & 1 \\ x_3 & y_3 & 1 \end{vmatrix}$$

$$= \frac{1}{2}[x_1(y_2 - y_3) + x_2(y_3 - y_1) + x_3(y_1 - y_2)]$$

Chapter 5. Matrices

1. **Order of Matrix :** The number of rows and columns that matrix has is called its order. By convention, rows are listed first and columns second.
   i.e., a matrix having m rows and n columns is called an $m \times n$ (read "m by n") matrix or a matrix of order $m \times n$.
2. **Comparable pair of matrices :** Two matrices are said to be comparable if each one of them contains as many rows and columns as the other *i.e.*, they are of same order.
3. **Types of matrices** :
   (a) Rectangular matrix : Any $m \times n$ matrix, where $m \neq n$
   (b) Row matrix : Any $m \times n$ matrix where $m = 1$
   (c) Column matrix : Any $m \times n$ matrix where $n = 1$
   (d) Square matrix : Any $m \times n$ matrix where $m = n$
   (e) Diagonal matrix : Any *square matrix* all of whose elements *except diagonal elements (i.e. elements along leading diagonal)*, are **zero.**

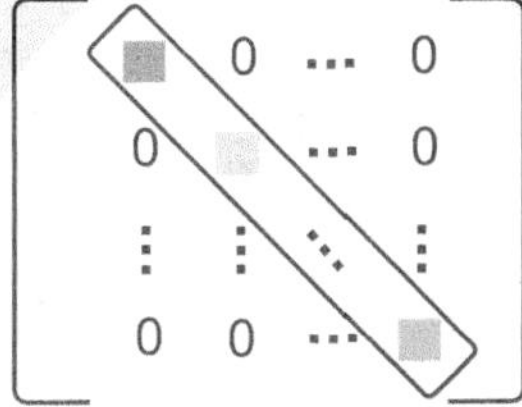

   (f) Scalar matrix : Any *diagonal matrix* in which *diagonal elements are all equal* is called a scalar matrix.

$$\begin{bmatrix} k & o & o \\ o & k & o \\ o & o & k \end{bmatrix}$$

   (g) Identity (or unit) matrix : Any *scalar matrix* in which *diagonal elements are all equal being unity* is called an identity or unit matrix (I).

$$\begin{bmatrix} 1 & 0 & 0 \\ 0 & 1 & 0 \\ 0 & 0 & 1 \end{bmatrix}$$

   (h) Equality of matrices : Two **matrices** are **equal** if they have the same order and the corresponding elements are identical.
   Equal Matrices :

$$A = \begin{bmatrix} 4 & 13 \\ -2 & 19 \end{bmatrix};\ B = \begin{bmatrix} 4 & 13 \\ -2 & 19 \end{bmatrix};\ C = \begin{bmatrix} 4 & 13 \\ 19 & -2 \end{bmatrix} \text{ and } M = \begin{bmatrix} 4 & 13 \\ -2 & 19 \\ 0 & -5 \end{bmatrix}$$

$$A = B;\ A \neq C;\ A \neq M.$$

   (i) Zero matrix (or null matrix) : Any matrix *all of whose elements are zero* and is denoted by $O_{m \times n}$, where $m \times n$ is the order of matrix.
4. **Operations on matrices :**
   (i) Addition of matrices : If A and B be two matrices of the same order $m \times n$, they are called conformable or compatible for the sum and their sum denoted by A + B, is a matrix of the same order $m \times n$ and is obtained by adding their corresponding elements. Thus, if $A = [a_{ij}]$ and $B = [b_{ij}]$ be two $m \times n$ matrices, then

$$A + B = [a_{ij}]_{m \times n} + [b_{ij}]_{m \times n}$$
$$= [a_{ij} + b_{ij}]_{m \times n}. \qquad 1 \leq i \leq m,\ 1 \leq j \leq n$$

   For example, if

$$A = \begin{bmatrix} 1 & 2 & 5 \\ 3 & 4 & -1 \end{bmatrix}$$

   and

$$B = \begin{bmatrix} -1 & 3 & 2 \\ 2 & 5 & 7 \end{bmatrix}$$

then
$$A+B=\begin{bmatrix}1 & 2 & 5\\ 3 & 4 & -1\end{bmatrix}+\begin{bmatrix}-1 & 3 & 2\\ 2 & 5 & 7\end{bmatrix}$$
$$=\begin{bmatrix}1-1 & 2+3 & 5+2\\ 3+2 & 4+5 & -1+7\end{bmatrix}=\begin{bmatrix}0 & 5 & 7\\ 5 & 9 & 6\end{bmatrix}$$

(ii) Difference of two matrices : If A and B are two matrices of the same order then we define A – B as the sum A + (– B) *i.e.*, if $A = [a_{ij}]_{m\times n}$ and $B = [b_{ij}]_{m\times n}$

Then
$$A-B=[a_{ij}-b_{ij}]_{m\times n}.$$

(iii) Negative of a matrix : Negative of a matrix $\mathbf{A}_{m\times n}$ is defined as a $\boldsymbol{m\times n}$ ***matrix*** *whose (i, j)th entry is* ***negative*** *(or* ***additive inverse****) of (i, j)th entry of* $\mathbf{A}_{m\times n}$, *i.e.*
$$-A=[-a_{ij}]_{m\times n}$$
$$A=\begin{bmatrix}3 & -5 & 7\\ -2 & 4 & 9\end{bmatrix}$$
$$-A=\begin{bmatrix}-3 & 5 & -7\\ 2 & -4 & -9\end{bmatrix}$$

(A) Properties of matrix addition :

(a) Addition of matrices is commutative : *i.e.*, if A and B are matrices of the same order, then
$$A+B=B+A.$$

(b) Addition of matrices is associative : *i.e.*, if A and B are matrices of the same order, then
$$(A+B)+C=A+(B+C).$$

(c) Existence of additive identity : For any matrix A, there exists a null matrix 0 of the same order such that
$$A+0=A=0+A.$$
0 is called the *additive identity.*

(d) Existence of additive inverse : For any matrix A, there exists a unique matrix –A of the same order such that
$$A+(-A)=0=(-A)+A,$$
– A is called additive inverse or negative of A.

(e) Cancellation laws under addition : If A, B and C are three matrices of the same order, then

(a) If $A+B=A+C$

$\Rightarrow$ $B=C.$ (left cancellation law)

(b) If $B+A=C+A$

$\Rightarrow$ $B=C.$ (right cancellation law)

(f) Scalar multiplication : The scalar multiple kA (or Ak) of a matrix A by a scalar k is the matrix obtained by multiplying each element of A by k.

Thus, if
$$A=\begin{bmatrix}2 & 3 & 1\\ 0 & 2 & -1\end{bmatrix}$$

then
$$2A=2\begin{bmatrix}2 & 3 & 1\\ 0 & 2 & -1\end{bmatrix}=\begin{bmatrix}2\times2 & 2\times3 & 2\times1\\ 2\times0 & 2\times2 & 2\times(-1)\end{bmatrix}=\begin{bmatrix}4 & 6 & 2\\ 0 & 4 & -2\end{bmatrix}$$

(B) Properties of scalar multiplication :

(a) If A and B are two matrices of the same type and k, m are any numbers, then

(i) $(k+m)A=kA+mA$

(ii) $k(A+B)=kA+kB$

Chapter at a Glance

(b) Let A be any matrix and k_1, k_2 are any scalars, then

(i) $(k_1k_2)\,A = k_1\,(k_2A) = k_2\,(k_1A)$

(ii) $1.A = A$

(iii) $(-1)\,A = -A.$

(C) Multiplication of matrices : Multiplication of two matrices is *defined only if* ***the number of columns of the 1st matrix must equal the number of rows of the 2nd matrix***, *i.e.* Let A and B be two matrices, then the product AB is defined only if the number of columns in A = number of rows in B. Such matrices are said to be **compatible** or **conformable** for multiplication.

Now, let us see how to multiply the two matrices if multiplication is defined :

This is a "Multiply row by column" process. We multiply the entries of a row by the corresponding entries of a column and then add the products.

Thus,

$$\begin{bmatrix} a_{11} & a_{12} & a_{13} \\ a_{21} & a_{22} & a_{23} \\ a_{31} & a_{32} & a_{33} \end{bmatrix}_{3\times 3} \times \begin{bmatrix} b_{11} & b_{12} \\ b_{21} & b_{22} \\ b_{31} & b_{32} \end{bmatrix}_{3\times 2}$$

$$\begin{bmatrix} a_{11}b_{11} + a_{12}b_{21} + a_{13}b_{31} & a_{11}b_{12} + a_{12}b_{22} + a_{13}b_{32} \\ a_{21}b_{11} + a_{22}b_{21} + a_{23}b_{31} & a_{21}b_{12} + a_{22}b_{22} + a_{23}b_{32} \\ a_{31}b_{11} + a_{32}b_{21} + a_{33}b_{31} & a_{31}b_{12} + a_{32}b_{22} + a_{33}b_{32} \end{bmatrix}_{3\times 2}$$

Note : From the above multiplication, we find that the product of an $m \times p$ and $p \times n$ matrix is an $m \times n$ matrix

$$A_{m\times p} \times B_{p\times n} = C_{m\times n}$$

inner dimensions equal — Product of the outer dimensions

(D) Properties of Matrix Multiplication :

(a) The product of matrices is not commutative, *i.e. whenever* AB *exists*, BA *is not always defined*. **Even if both AB and BA are defined, it is not necessary that AB = BA.**

(b) The product of two matrices can be zero without either factor being a zero matrix.

(c) Cancellation law for the multiplication of real numbers is not valid for the multiplication of matrices, *i.e.* **AB = AC does not imply that B = C** (However, if A, B, C are square matrices of same type and if A is non-singular, then AB = AC implies that B = C.

(d) Matrix multiplication is associative if conformability is assured, *i.e.* **A(BC) = (AB)C**

(e) Matrix multiplication is distributive with respect to matrix addition, *i.e.* **A (B + C) = AB + AC.**

(f) For any matrix $A_{p\times n}$, we have $\mathbf{O}_{m\times p}\,\mathbf{A}_{p\times n} = \mathbf{O}_{m\times n}$ (O denotes zero matrix)

(g) Multiplicative Identity for a square matrix : $\mathbf{A}_{n\times n}\,\mathbf{I}_{n\times n} = \mathbf{A}_{n\times n}$

(h) A^2 is defined only when A is a square matrix and is equal to AA.

5. Linear combination of matrices : Let A, B be two matrices of the same order and k_1, k_2 be two scalars, then the matrix $k_1A + k_2B$ is called a linear combination of the matrices A and B.

For example, if $A = \begin{bmatrix} 2 & 0 & 3 \\ 2 & 1 & 4 \end{bmatrix}$

and $B = \begin{bmatrix} 7 & -6 & 3 \\ 1 & 4 & 5 \end{bmatrix}$

Then $2A + 3B = 2\begin{bmatrix} 2 & 0 & 3 \\ 2 & 1 & 4 \end{bmatrix} + 3\begin{bmatrix} 7 & -6 & 3 \\ 1 & 4 & 5 \end{bmatrix}$

$$= \begin{bmatrix} 4 & 0 & 6 \\ 4 & 2 & 8 \end{bmatrix} + \begin{bmatrix} 21 & -18 & 9 \\ 3 & 12 & 15 \end{bmatrix}$$

Chapter at a Glance

$$= \begin{bmatrix} 4+21 & 0-18 & 6+9 \\ 4+3 & 2+12 & 8+15 \end{bmatrix}$$

$$= \begin{bmatrix} 25 & -18 & 15 \\ 7 & 14 & 23 \end{bmatrix}$$

is a linear combination of matrices A and B.

6. **Transpose of a matrix :** Let $A = [a_{ij}]$ be a $m \times n$ matrix, then the matrix obtained from A by changing rows into columns and columns into rows is called the *transpose* of A and is denoted by A^T or A'. Clearly A' is an $n \times m$ matrix whose $(j, i)^{th}$ entry = $(i, j)^{th}$ entry of A.

For example, if
$$A = \begin{bmatrix} 2 & -1 \\ 3 & 0 \\ 1 & 5 \end{bmatrix}$$

then
$$A_T = \begin{bmatrix} 2 & 3 & 1 \\ -1 & 0 & 5 \end{bmatrix}$$

(A) **Properties of transpose matrix** :

(a) $(A')' = A$, (b) $(-A)' = -A'$, (c) $(A + B)' = A' + B'$, (d) $(A - B)' = A' - B'$, (e) $(AB)' = B'A'$, (f) $(kA)' = kA'$.

7. **Symmetric matrix** : A square matrix $A = [a_{ij}]$ is symmetric if $a_{ij} = a_{ji}$.

Note : (a) Symmetric matrix is always a square matrix.

(b) A necessary and sufficient condition for a matrix to be symmetric is $A' = A$, where A' denotes transpose of matrix A.

(c) Diagonal matrices are always symmetric.

Skew-symmetric matrix : A square matrix $A = [a_{ij}]$ is skew-symmetric if $a_{ij} = -a_{ji}$. Note that *all diagonal elements of a skew-symmetric matrix is zero.* Also note that the necessary and sufficient condition for a matrix to be symmetric is $A' = -A$, where A' denotes transpose of matrix A. (A matrix which is both symmetric and skew-symmetric is called a square null matrix.)

8. **Singular and non-singular matrices :** A square matrix is singular if det [A] = 0 (or $\Delta = 0$), otherwise it is non-singular.

9. **Nilpotent matrix :** A square matrix A is called Nilpotent if there exists a positive integer m such that $A^m = 0$. If m is the least positive integer such that $A^m = 0$, then m is called the index of the nilpotent matrix A.

10. **Adjoint of a square matrix :** Let $A = [a_{ij}]$ be a square matrix of order n. Then the *adjoint of* A is the transpose of the matrix $[C_{ij}]_{m \times n}$ where C_{ij} is the co-factor of a_{ij} in $|A|$. It is denoted by adj. A.

For example, let
$$A = \begin{bmatrix} 2 & 3 \\ 1 & 5 \end{bmatrix}$$

then
$$|A| = \begin{vmatrix} 2 & 3 \\ 1 & 5 \end{vmatrix}$$

Now, $C_{11} = 5$, $C_{12} = -1$, $C_{21} = -3$, $C_{22} = 2$.

Then,
$$\text{adj } A = \begin{bmatrix} C_{11} & C_{12} \\ C_{21} & C_{22} \end{bmatrix}^T = \begin{bmatrix} C_{11} & C_{21} \\ C_{12} & C_{22} \end{bmatrix} = \begin{bmatrix} 5 & -3 \\ -1 & 2 \end{bmatrix}$$

Note : (i) If A is a square matrix of order n, then

$$A\,(\text{adj. } A) = |A|\, I_n = (\text{adj } A)\, A.$$

(ii) If A is a square matrix of order 3, then adj $(kA) = k^2$ (adj A).

1. **Relation between a square matrix and adj A :** If A be any n^{th} order square matrix, then (adj A) A = A (adj A) = $|A|I_n$, where I_n is n^{th} order unit matrix.
2. If A is a $n \times n$ *non-singular matrix,* then $|\text{adj } A| = |A|^{n-1}$
3. If A and B are two *non-singular matrices of same type,* then adj (AB) = (adj B) (adj A)
4. **Inverse of an n × n matrix : A^{-1} = adj A/ |A|,** but *inverse of a square matrix exists if and only if A is non-singular, i.e.* $|A| \neq 0$
5. **Properties of matrices and inverses :**
   (a) If A, B be two n^{th} order non-singular matrices, then AB is also non-singular and $(AB)^{-1} = B^{-1}A^{-1}$ and $(A')^{-1} = (A^{-1})'$
   (b) $AA^{-1} = A^{-1}A = I$
6. **Elementary operations of a matrix :**
   (a) Interchanging any two rows (or columns), denoted by $R_i \to R_j$ if i^{th} and j^{th} rows are interchanged.
   (b) Multiplication of the elements of any row (or column) by a non-zero scalar quantity, denoted by $R_i \to kR_i$
   (c) Addition of constant multiplication of any row to the corresponding element of any other row, *i.e.* $R_i \to R_i + kR_j$
7. **Method to find Inverse of a matrix by elementary operations :**
   Step 1 : Write the given matrix as A= IA for applying elementary row operations or AI = A for applying elementary column operations.
   Step 2 : Perform elementary row (or column) operations to convert A to I. We will obtain a new matrix equation of type I =PA. Then P is the inverse of A.
   Note : Only one type of elementary operations, *i.e.* either row or column operations are to be applied to solve a particular equation.
8. **Applications of matrices in solving linear equations (Martin's rule) :**
   Consider the three simultaneous equations in variables x, y and z.

$$a_1x + b_1y + c_1z = d_1$$
$$a_2x + b_2y + c_2z = d_2$$
$$a_3x + b_3y + c_3z = d_3$$

These can be written in matrix form as

$$\begin{bmatrix} a_1 & b_1 & c_1 \\ a_2 & b_2 & c_2 \\ a_3 & b_3 & c_3 \end{bmatrix} \begin{bmatrix} x \\ y \\ z \end{bmatrix} = \begin{bmatrix} d_1 \\ d_2 \\ d_3 \end{bmatrix}$$

or $$AX = B$$

where A is 3 × 3 matrix and X and B are 3 × 1 column matrices.

Now, if A is non-singular *i.e.,* $|A| \neq 0$ then we can left multiply both members of this equation by A^{-1} to obtain

$$A^{-1}AX = A^{-1}B$$
$$\Rightarrow \quad IX = A^{-1}B$$
$$\Rightarrow \quad X - A^{-1}B$$

The equations having one or more solutions are called consistent equations.

(i) If $|A| \neq 0$, the system of equations is consistent and has a unique solution.

(ii) If $|A| = 0$, the system of equations has either infinite number of solutions or no solution.

⇒ If $|A| = 0$ and (*adj* A)·B = 0 equations have infinitely many solutions *i.e.,* consistent but dependent.

⇒ If $|A| = 0$ but (*adj* A)·B ≠ 0, equations have no solution *i.e.,* inconsistent.

Chapter 6. Differential Calculus

1. **Continuity of a function at a point** : A function is continuous at $x = a$ if the graph around the poir $(a, f(a))$ is connected and unbroken. (otherwise it is discontinuous).

Now look at the graph below, we can see that $f(x)$ is continuous at $x = 0$ and $x = 3$, but it is discontinuous at $x = -2$.

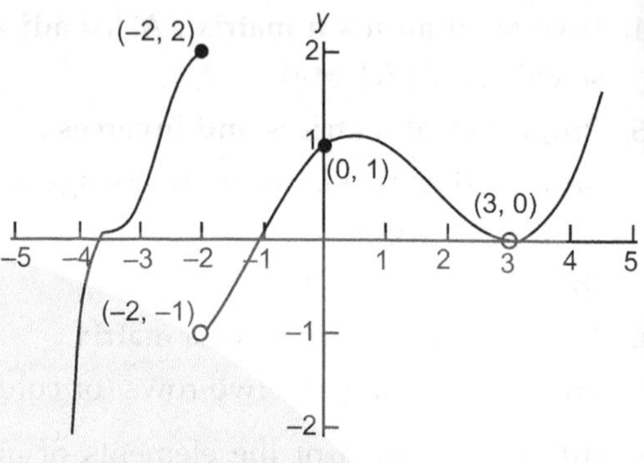

From the above graph, it is observed that :

A function $f(x)$ is said to be continuous at a point $x = a$, if and only if

$$\lim_{x \to a} f(x) = f(a)$$

$$\lim_{x \to a^-} f(x) = \lim_{x \to a^+} f(x) = f(a)$$

Note : It is obvious that all limits and $f(a)$ must be defined for continuity of $f(x)$

2. **Removable Discontinuity** : If $\lim_{x \to a^-} f(x) = \lim_{x \to a^+} f(x) \neq f(a)$, then the discontinuity is said to b removable discontinuity as it can be made continuous by redefining the function $f(x)$ at $x = a$.

(i) A function $f(x)$ is said to have a discontinuity of first kind at $x = a$ if

$$\lim_{x \to a^-} f(x) \neq \lim_{x \to a^+} f(x)$$

(ii) A function $f(x)$ is said to have a discontinuity of second kind at $x = a$ if

$\lim_{x \to a^-} f(x)$ or $\lim_{x \to a^+} f(x)$ does not exist or both do not exist.

3. **Continuity on an interval** :

Definition. For Open interval : A *function f is said to be continuous on an open interval (a, b), if it is continuou at every point in the interval.*

Definition. For Closed interval : A *function f (x) is said to be continuous on a closed interval [a, b], if it i continuous at every point of the open interval (a, b) and if it is continuous at the point 'a' from the right and continuou at 'b' from the left, i.e.,* $\lim_{x \to a^+} f(x) = f(a)$ *and* $\lim_{x \to b^-} f(x) = f(b)$.

Analogus definition may be given for continuity on intervals of the form $[a, b)$, $(a, b]$ (a, ∞), $(-\infty, b)$ $(-\infty, b]$ and $(-\infty, \infty)$.

4. **Algebra of continuous functions :**

(a) If f and g are two real valued continuous functions defined on point $x = a$, then : (a) $f + g$, (b) $f - g$ (c) $f.g$., (d) f/g (provided $g(a) \neq 0$), (e) $1/f$ (provided $f(a) \neq 0$) *are also continuous at* $x = a$.

(b) The composition of two continuous functions is continuous.

(c) The constant function is a continuous function.

(d) The identity function $f : R \to R$ such that $f(x) = x$ is continuous on R.

(e) Absolute value function $f(x) = |x|$ is continuous everywhere on R.

(f) The greatest integer function $f(x) = [x]$ is continuous at every real number *other than the integers.*

(g) A polynomial function is continuous everywhere on R.

(h) Any rational function is continuous *except at the point where the denominator becomes zero.*

(i) The function $\sin x$ and $\cos x$ are continuous everywhere on R.

(j) The function e^x is continuous everywhere on R.

(k) The function $\sin^{-1}x$, $\cos^{-1}x$ are continuous everywhere in their domain $[-1, 1]$.

(l) The function $\tan^{-1}x$, $\cot^{-1}x$ are continuous everywhere.

(m) The function $\sec^{-1}x$, $\text{cosec}^{-1}x$ are continuous in $(-\infty, -1] \cup [1, \infty)$.

Chapter at a Glance

5. **Properties of continuous functions :**
   - **(a)** The graph of a continuous function does not have any break in it.
   - **(b)** The values of a continuous function **cannot go from –ve to +ve without taking the value 0 in between,** *i.e.* it must cut x-axis at some **intermediate point.**
   - (c) A continuous function in an interval has a **least** and **greatest value**.

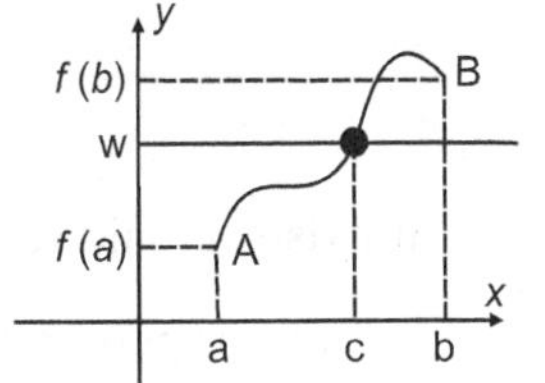

6. **Intermediate Value Theorem** : A function that is continuous on a closed interval $[a, b]$ takes on all values in between $f(a)$ and $f(b)$, *i.e.*, if w is any value between $f(a)$ and $f(b)$, then $w = f(c)$ for some c in $[a, b]$.

7. **Differentiability at a point :** A real valued function $f(x)$ is said to be differentiable at $x = a$, if

$$\lim_{x\to a}\frac{f(x)-f(a)}{x-a} \text{ exists finitely.}$$

8. **Differentiability on an interval :**
   - (i) **On an open interval :** A function $f(x)$ is said to be differentiable on an open interval (a, b) if it is differentiable at each point of (a, b).
   - (ii) **On a closed interval :** A function $f(x)$ is said to be differentiable on a closed interval $[a, b]$ if it is differentiable at each point of $[a, b]$ and it is differentiable at 'a' from right and at 'b' from left.

i.e., $$\lim_{x\to a^+}\frac{f(x)-f(a)}{x-a} \text{ and } \lim_{x\to b^-}\frac{f(x)-f(b)}{x-b} \text{ both exist.}$$

Note : A differentiable function is always continuous while a continuous function may or may not be differentiable.

9. **Some important limits :**

(i) $\lim_{x\to a}\frac{x^n-a^n}{x-a} = n\,a^{n-1}$ (ii) $\lim_{x\to 0}\sin x = 0$

(iii) $\lim_{x\to 0}\cos x = \lim_{x\to 0}\frac{\sin x}{x} = \lim_{x\to 0}\frac{\tan x}{x} = 1.$

(iv) $\lim_{x\to 0}(1+x)^{1/x} = e$ (v) $\lim_{x\to 0}(1+\lambda x)^{1/x} = e^{\lambda}$

(vi) $\lim_{x\to 0}\frac{\log_a(1+x)}{x} = \log_a e\ (a>0, a\neq 1)$ (vii) $\lim_{x\to 0}\frac{\log(1+x)}{x} = 1$

(viii) $\lim_{x\to 0}\frac{a^x-1}{x} = \log a\ (a>0)$ (ix) $\lim_{x\to 0}\frac{e^x-1}{x} = 1$

(x) $\lim_{x\to 0}\frac{(1+x)^m-1}{x} = m$ (xi) $\lim_{x\to 0}\frac{\log x}{x^m} = 0 \quad (m>0)$

(xii) $\lim_{x\to 0}\left(1+\frac{x}{\lambda}\right)^{1/x} = e^{1/\lambda}$ (xiii) $\lim_{x\to\infty}\left[1+\frac{1}{x}\right]^x = e$

(xiv) $\lim_{x\to\infty}\left[1+\frac{\lambda}{x}\right]^x = e^{\lambda}$ (xv) $\lim_{x\to\infty}\left(1+\frac{1}{\lambda x}\right)^2 = e^{1/\lambda}$

(xvi) $\lim_{x\to\infty} a^x = \begin{cases}\infty, & \text{if } a>1\\ 0, & \text{if } a<1\end{cases}$ (xvii) $\lim_{x\to 0}\frac{\sin^{-1}x}{x} = 1$

i.e., $a^{\infty} = \infty$, if $a > 1$ and $a^{\infty} = 0$, if $a < 1$.

(xviii) $\lim_{x\to 0}\frac{\tan^{-1}x}{x} = 1$ (xix) $\lim_{x\to 0}\frac{\sin x}{x} = \frac{\pi}{180}$

(xx) $\lim_{x \to a} \frac{\sin(x-a)}{x-a} = 1$ (xxi) $\lim_{x \to a} \frac{\tan(x-a)}{(x-a)} = 1$

10. Differential coefficient of a function : If $f(x)$ be a function of x, then

$$\lim_{h \to 0} \frac{f(x+h)-f(x)}{h}$$

if exists, is said to be differential coefficient of $f(x)$ and is denoted by $f'(x)$. The differential co-efficient of $y = f(x)$ is generally written as $\frac{d}{dx}[f(x)]$ or $\frac{df}{dx}$ $f'(x)$ or $Df(x)$ or $\frac{dy}{dx}$ or y_1 or $D(y)$ or y'.

11. Differentiation : The process of finding the differential co-efficient is called differentiation. The independent variable is indicated by saying differentiate $f(x)$ with respect to x.

12. Differentiation of Standard function :

(i) $\frac{d}{dx}[c] = 0$, where c is a constant.

(ii) $\frac{d}{dx}(x^n) = nx^{n-1}, n \neq 0.$

(iii) $\frac{d}{dx} u^n = nu^{n-1} \frac{du}{dx}$

(iv) $\frac{d}{dx}(\sqrt{x}) = \frac{1}{2\sqrt{x}}, x \neq 0.$

(v) $\frac{d}{dx}(c.\,u) = c.\frac{du}{dx}$

(vi) $\frac{d}{dx} e^x = e^x$

(vii) $\frac{d}{dx} a^x = a^x.\log_e a, a > 0$

(viii) $\frac{d}{dx} \log_a x = \frac{1}{x} \log_a e, x > 0$

(ix) $\frac{d}{dx} \log_e x = \frac{1}{x}, x > 0$

(x) $\frac{d}{dx} \sin x = \cos x.$

(xi) $\frac{d}{dx} \cos x = -\sin x.$

(xii) $\frac{d}{dx} \tan x = \sec^2 x.$

(xiii) $\frac{d}{dx} \cot x = -\operatorname{cosec}^2 x.$

(xiv) $\frac{d}{dx} \sec x = \sec x.\tan x$

(xv) $\frac{d}{dx} \operatorname{cosec} x = -\operatorname{cosec} x.\cot x$

(xvi) $\frac{d}{dx} \sin^{-1} x = \frac{1}{\sqrt{1-x^2}}, \quad -1 < x < 1$

(xvii) $\frac{d}{dx} \cos^{-1} x = \frac{-1}{\sqrt{1-x^2}}, \quad -1 < x < 1$

(xviii) $\frac{d}{dx} \tan^{-1} x = \frac{1}{\sqrt{1+x^2}}, \quad -\infty < x < \infty$

(xix) $\frac{d}{dx} \cot^{-1} x = \frac{-1}{1+x^2}, -\infty < x < \infty.$

(xx) $\frac{d}{dx} \sec^{-1} x = \frac{1}{|x|\sqrt{x^2-1}}, \quad |x| > 1$

(xxi) $\frac{d}{dx} \operatorname{cosec}^{-1} x = \frac{-1}{|x|\sqrt{x^2-1}}, \quad |x| > 1$

If u, v are the derivable functions of x, then

(xxii) Sum rule → $\frac{d}{dx}[u \pm v] = \frac{du}{dx} \pm \frac{dv}{dx}$

(xxiii) Product rule → $\frac{d}{dx}[u.\,v] = u.\frac{du}{dx} + v.\frac{dv}{dx}$

(xxiv) Quotient rule → $\frac{d}{dx}\left[\frac{u}{v}\right] = \frac{v.\frac{du}{dx} - u.\frac{dv}{dx}}{v^2}$

(xxv) $\frac{d}{dx}[u_1.u_2.u_3] = u_1.u_2.\frac{du_3}{dx} + u_1.u_3.\frac{du_2}{dx} + u_2.u_3.\frac{du_1}{dx}$

13. Differentiation by chain rule : If $y = f(t)$, $t = g(z)$, $z = h(x)$, then $= \frac{dy}{dx} = \frac{dy}{dt}.\frac{dt}{dz}.\frac{dz}{dx}$. This rule is called chain rule.

4. **Differentiation of parametric equations :** If $x = f(t)$, $y = g(t)$ then $= \dfrac{dy}{dx} = \dfrac{dy/dt}{dx/dt}$.

5. **Differentiation of implicit functions :** If the variables x and y are connected by a relation of the form $f(x, y) = 0$ and it is not possible or convenient to express y as a function of x in the form $y = \phi(x)$, then y is said to be implicit function of x. To find $\dfrac{dy}{dx}$ in such a case, we differentiate both sides of the given relation with respect to x, keeping in mind that the derivative of $\phi(y)$ w.r.t. x is $\dfrac{d\phi}{dy} \cdot \dfrac{dy}{dx}$. For example,

$$\frac{d}{dx}(\sin y) = \cos y \cdot \frac{dy}{dx}, \quad \frac{d}{dx}(y^2) = 2y\frac{dy}{dx}$$

It should be noted that $\dfrac{d}{dy}(\sin y) = \cos y$ but $\dfrac{d}{dx}(\sin y) = \cos y \cdot \dfrac{dy}{dx}$. Similarly $\dfrac{d}{dy}(y^3) = 3y^2$ whereas

$$\frac{d}{dx}(y^3) = 3y^2\frac{dy}{dx}$$

6. **Differentiation by using trigonometrical transformations :** Sometimes trigonometrical transformations before differentiation shortens the differentiation. For this we use the following formulae :

(i) $\sin 2x = 2 \sin x . \cos x = \dfrac{2\tan x}{1+\tan^2 x}$

(ii) $\cos 2x = 2\cos^2 x - 1 = 1 - 2\sin^2 x$
$= \cos^2 x - \sin^2 x$
$= \dfrac{1-\tan^2 x}{1+\tan^2 x} = \dfrac{\cot^2 x - 1}{\cot^2 x + 1}$

(iii) $\tan 2x = \dfrac{2\tan x}{1-\tan^2 x}$

(iv) $\sin 3x = 3\sin x - 4\sin^3 x$.

(v) $\cos 3x = 4\cos^3 x - 3\cos x$.

(vi) $\tan 3x = \dfrac{3\tan x - \tan^3 x}{1 - 3\tan^2 x}$

(vii) $\tan\left(\dfrac{\pi}{4}+\theta\right) = \dfrac{1+\tan\theta}{1-\tan\theta}$

(viii) $\tan\left(\dfrac{\pi}{4}-\theta\right) = \dfrac{1-\tan\theta}{1+\tan\theta}$

(ix) $\sin^{-1} x \pm \sin^{-1} y = \sin^{-1}\left[x\sqrt{1-y^2} \pm y\sqrt{1-x^2}\right]$

(x) $\cos^{-1} x \pm \cos^{-1} y = \cos^{-1}\left[xy \pm \sqrt{1-x^2}.\sqrt{1-y^2}\right]$

(xi) $\tan^{-1} x + \tan^{-1} y = \tan^{-1}\left[\dfrac{x+y}{1-xy}\right]$, $xy < 1$

(xii) $\tan^{-1} x - \tan^{-1} y = \tan^{-1}\left[\dfrac{x-y}{1+xy}\right]$

(xiii) $\sin^{-1} x + \cos^{-1} x = \tan^{-1} x + \cot^{-1} x$
$= \sec^{-1} x + \text{cosec}^{-1} x = \pi/2$.

7. **Method of substitution :** The functions may also be reduced to simpler forms by the substitutions as follows :

(i) If the function involve the term $\sqrt{a^2 - x^2}$, then put $x = a\sin\theta$ or $a\cos\theta$.

(ii) If the function involve the term $\sqrt{a^2+x^2}$, then put $x = a \tan\theta$ or $a \cot\theta$.

(iii) If the function involve the term $\sqrt{x^2-a^2}$, then put $x = a \sec\theta$ or $x = a \operatorname{cosec}\theta$.

(iv) If the function involve the term $\sqrt{\dfrac{a-x}{a+x}}$, then put $x = a\cos\theta$ or $x = a \operatorname{cosec} 2\theta$.

18. Logarithmic differentiation :

Let $y = [f(x)]^{g(x)}$. On taking logarithm of both sides,

we get, $\log y = \log [f(x)]^{g(x)} = g(x) \log [f(x)]$

Differentiating w.r.t. x we get

$$\frac{1}{y}\frac{dy}{dx} = \frac{g(x)}{f(x)}\frac{df(x)}{dx} + \log[f(x)].\frac{dg(x)}{dx}$$

$$\therefore \quad \frac{dy}{dx} = y\left[\frac{g(x)}{f(x)}\cdot\frac{df(x)}{dx} + \log[f(x)]\cdot\frac{dg(x)}{dx}\right]$$

Alternatively, we may write

$y = [f(x)]^{g(x)} = e^{g(x).\log[f(x)]}$ and then differentiating with respect to x, we may get

$$\frac{dy}{dx} = e^{g(x)\log\{f(x)\}}\left[g(x)\cdot\frac{1}{f(x)}\frac{df(x)}{dx} + \log\{f(x)\}\cdot\frac{dg(x)}{dx}\right]$$

$$= [f(x)]^{g(x)}\left[\frac{g(x)}{f(x)}\cdot\frac{df(x)}{dx} + \log\{f(x)\}\cdot\frac{dg(x)}{dx}\right]$$

19. Successive differentiation : If $y = f(x)$, then $\dfrac{dy}{dx}$, the differential co-efficient of y with respect to x, is itself, in general a function of x and can be differentiated again. Derivative of $\dfrac{dy}{dx}$ w.r.t., x is called the second order derivative of y and is denoted by $\dfrac{d^2y}{dx^2}$.

Similarly, the derivative of $\dfrac{d^2y}{dx^2}$ w.r.t. x is called the third order derivative of y and is denoted by $\dfrac{d^3y}{dx^3}$ and so on.

The successive differential coefficients of y are denoted by,

First derivative : $\dfrac{dy}{dx}$, y_1, y', Dy, $f'(x)$

Second derivative : $\dfrac{d^2y}{dx^2}$, y_2, y'', D^2y, $f''(x)$

---

n^{th} derivative : $\dfrac{d^ny}{dx^n}$, y_n, $y^{(n)}$, D^ny, $f^n(x)$

20. Indeterminate forms of limits and L' Hospital's Rule : If a function $f(x)$ takes the form $\dfrac{0}{0}$ at $x = a$, then we say that $f(x)$ is indeterminate at $x = a$. Other indeterminate forms are $\dfrac{\infty}{\infty}$, $\infty - \infty$, $0 \times \infty$, 1^∞, 0^0, ∞^0.

L' Hospital's Rule : If $\phi(x)$ and $\psi(x)$ are functions of x such that $\phi(a) = 0$, and $\psi(a) = 0$, then

$$\lim_{x\to a}\frac{\phi(x)}{\psi(x)} = \lim_{x\to a}\frac{\phi'(x)}{\psi'(x)}$$

Chapter at a Glance

Important Note : While applying L' Hospital Rule, we do not have to differentiate $\frac{\phi(x)}{\psi(x)}$ by the rule finding the differential co-efficient of the quotient of two functions. But we have to differentiate the numerator and denominator separately.

21. **Relation between** $\frac{dy}{dx}$ **and** $\frac{dx}{dy}$: $\frac{dy}{dx} = \frac{1}{dx/dy}$.

22. **Rolle's theorem :** Let $f(x)$ be a function of x such that
    (i) $f(x)$ is continuous in the closed interval $[a, b]$, *i.e.,* $a \le x \le b$.
    (ii) $f(x)$ is derivable in the open interval (a, b) *i.e.,* $a < x < b$.
    (iii) $f(a) = f(b)$.
    Then, there exists at least one real number $x = c$ in the open interval (a, b) (*i.e.,* $a < c < b$) such that $f'(c) = 0$.
    Remark : (i) Rolle's theorem fails for the function which does not even satisfies one condition.
    (ii) Every polynomial in x is a continuous function for each x, sin x, cos x, e^x are continuous for all values of x and log x is continuous for all $x > 0$.
    (iii) If f and g are both continuous in the closed interval $[a, b]$ then $f \pm g$ and fg are also continuous in $[a, b]$.
    (iv) If $f(x)$ is derivable for every point in a given interval, then it must be continuous in this interval.

23. **Lagrange's mean value theorem :** Let $f(x)$ be a function defined in the closed interval $[a, b]$ such that
    (i) $f(x)$ is continuous in the closed interval $[a, b]$.
    (ii) $f(x)$ is derivable in the open interval (a, b).
    Then, there exists at least one real number $c \in (a, b)$ such that

$$f'(c) = \frac{f(b) - f(a)}{b - a}$$

Chapter 7. Applications of Derivatives

1. **Tangents and Normals :**

Geometrical Interpretation of $\frac{dy}{dx}$: We know that $\frac{dy}{dx}$ at any point P(x, y) of the curve $y = f(x)$ represents the slope (or gradient) of the tangent at the point P, *i.e.,*

$\frac{dy}{dx} = \tan\theta =$ slope of tangent at (x, y)

Thus, if *tangent is parallel to x-axis,* then $\theta = 0°$, so $\frac{dy}{dx} = \tan 0° = 0$

And, if *tangent is perpendicular to x-axis,* then $\theta = 90°$, so $\frac{dy}{dx} = \tan 90° = \infty$ or $\frac{dy}{dx} = \tan 90° = \infty$.

2. **Equation of Tangent and Normal :** Let $y = f(x)$ be a continuous curve and P (x', y') be any point on it. Then, the equation of the tangent at P is

$$y - y' = m(x - x')$$

where, $m =$ slope of tangent at P $= \left(\frac{dy}{dx}\right)_{(x', y')}$

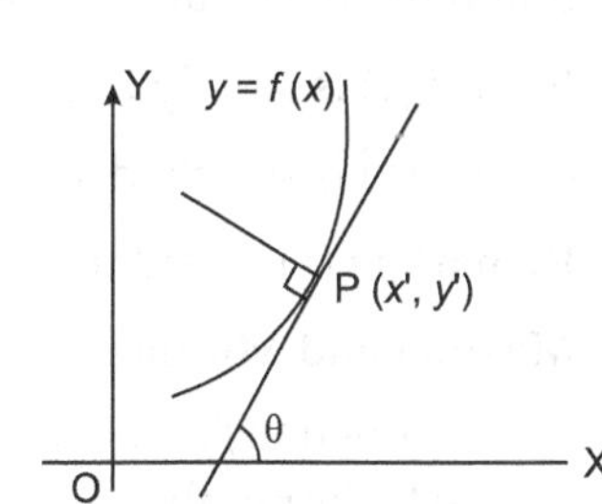

Since the normal at P (x', y') passes through P having slope

$$-\frac{1}{m} \text{ i.e., } \frac{-1}{\left(\frac{dy}{dx}\right)_{(x' y')}}$$

Then, the equation of the normal at P is $y - y' = \frac{-1}{m}(x - x')$

3. **Approximations :** Let $y = f(x)$ be a function of x and let δx and δy be a small change in x and y, respectively. Then,

$$\lim_{\delta x \to \infty} \frac{\delta y}{\delta x} = \frac{dy}{dx}$$

$$\frac{\delta y}{\delta x} = \frac{dy}{dx} + \in \text{ where, } \in \to 0 \text{ as } \delta x \to 0.$$

$$\delta y = \frac{dy}{dx} \delta x + \in \delta x$$

$$\delta y = \frac{dy}{dx} \delta x \text{ (approx.)}$$

4. **Derivative as a Rate Measurer :** Let $y = f(x)$ be a function of x and let δx represent a small change in x then, δy will be the corresponding change in y.

Hence, $\frac{\delta y}{\delta x}$ represents the average rate of change of y with respect to x in the interval $(x, x + \delta x)$.

Now, $\lim_{\delta x \to \infty} \frac{\delta y}{\delta x}$ = rate of change of y with respect to x.

i.e., $\frac{dy}{dx}$ represents the actual rate of change of y w.r.t. x for any particular value of x.

Remark : If $x = f(t)$ and $y = g(t)$ then,

$$\frac{dy}{dx} = \frac{\frac{dy}{dt}}{\frac{dx}{dt}} \text{ provided } \frac{dx}{dt} \neq 0.$$

5. **Monotonicity of functions :**

(i) A function $f(x)$ is said to be a monotonically increasing function on an interval $[a, b]$ if the values of $f(x)$ increases with the increase in x or decreases with the decrease in x.

(ii) A function $f(x)$ is said to be a monotonically decreasing function on an interval $[a, b]$ if the values of $f(x)$ decreases with the increase in x or vice versa.

6. **Strictly Increasing Function :** A function $f(x)$ is said to be a strictly increasing function on an interval (a, b), if

$$f(x_1) < f(x_2) \text{ whenever, } x_1 < x_2 \ \forall \ x_1, x_2 \in (a, b)$$

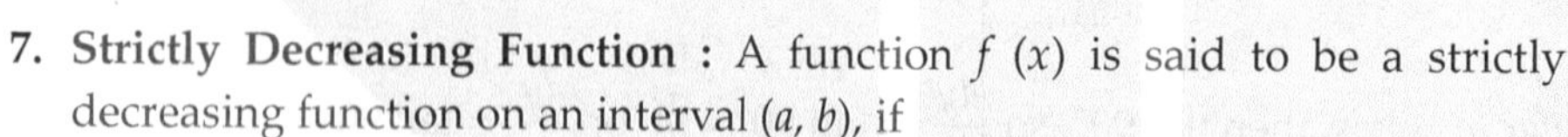

7. **Strictly Decreasing Function :** A function $f(x)$ is said to be a strictly decreasing function on an interval (a, b), if

$$f(x_1) > f(x_2), \text{ whenever } x_1 < x_2 \ \forall \ x_1, x_2 \in (a, b)$$

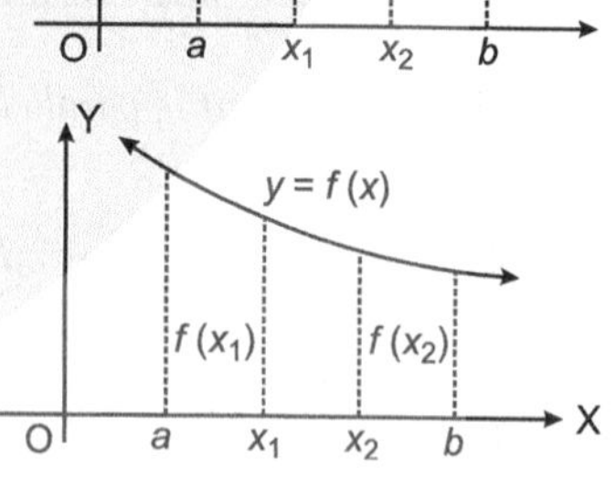

8. **Condition for Monotonic functions :** Let $f(x)$ be a continuous function on a closed interval $[a, b]$ and differentiable on an open interval (a, b).

(i) If $f(x)$ is strictly increasing on (a, b) then, $f'(x) > 0 \ \forall \ x \in (a, b)$.

(ii) If $f(x)$ is strictly decreasing on (a, b) then, $f'(x) < 0 \ \forall \ x \in (a, b)$.

Remark : $f(x)$ is said to be a constant on (a, b), if $f'(x) = 0 \ \forall \ x \in (a, b)$.

9. **Maxima and Minima :**

(i) A function $f(x)$ is said to have a maximum value in an interval I at x_0, if x_0 in I and if $f(x_0) \geq f(x)$ for all x in I. The number $f(x_0)$ is called the *maximum value* or the greatest value or the absolute maximum value of $f(x)$ in I and x_0 is called a point of maximum of $f(x)$ in I.

Consider the function,

$$f(x) = -(x-1)^2 + 10 \text{ for all } x \in R$$

$\because$ $$-(x-1)^2 \leq 0 \text{ for all } x \in R$$

$\therefore$ $$-(x-1)^2 + 10 \leq 10 \text{ for all } x \in R$$

$$f(x) \leq 10 \text{ for all } x \in R$$

Thus, 10 is maximum value of $f(x)$. Clearly $f(x)$ attains the value of $x = 1$ so $x = 1$ is the point of maximum or the point of absolute maximum.

(ii) A function $f(x)$ is said to have a minimum value in an interval I at x_0, if x_0 in I and if $f(x_0) \leq f(x)$ for all x in I. The number $f(x_0)$ is called the *minimum value* or the least value or the absolute minimum values of $f(x)$ in I and x_0 is called a point of minimum of $f(x)$ in I.

Consider the function $$f(x) = (3x-1)^2 + 5 \text{ for all } x \in R$$

$$(3x-1)^2 \geq 0 \text{ for all } x \in R$$

$$(3x-1)^2 + 5 \geq 5 \text{ for all } x \in R$$

Thus, 5 is the minimum value or the least value or the absolute minimum value of $f(x)$ in its domain. Clearly $f(x)$ attains this value at $x = 1/3$ is the point of minimum or the point of absolute minimum.

The general term used for *maxima* and *minima* is known as *extreme values.*

10. **Extreme points or turning points :** The points at which a function attains either the maximum values or the minimum values are known as the *extreme points or turning points* and both maximum and minimum values of $f(x)$ are called *extreme* or *extreme values.* Thus, a function attains an extreme value at $x = a$ if $f(a)$ is either a maximum or a minimum value.

11. **Points of inflexion** : A point on a curve where the slope $\frac{dy}{dx}$ is a maximum or minimum, *i.e.* has a turning value, is called a Point of Inflexion. Necessary conditions for point of inflexion :

1. $\frac{d^2y}{dx^2} = 0$, 2. $\frac{d^3y}{dx^3} \neq 0$

If in addition, $\frac{dy}{dx} = 0$ for values of x obtained from $\frac{d^2y}{dx^2} = 0$, the tangent at the inflexion point will be parallel to x-axis.

12. **General Conditions for turning points** : In general, *non-zero value of an odd differential coefficient decides in favour of a point of inflexion, and non-zero value of an even differential coefficient decides in favour of a maximum or minimum depending on its sign.*

13. **Relative maxima and minima :** (i) f is said to have a *relative* (or local) *maxima* at $c \in D_f$, if there exists a real number $\delta > 0$, such that

$$f(x) < f(c);\ \forall\ x \in (c-\delta, c+\delta), x \neq c.$$

Hence, c is called the *point of local maxima* and $f(c)$ is called the *local maximum value.*

(ii) f is said to have a *relative* (or local) *minima* at $d \in D_f$, if there exists a real number $\delta > 0$, such that

$$f(x) > f(d),\ \forall\ x \in (d-\delta, d+\delta), x \neq d.$$

Hence, d is called the *point of local minima* and $f(d)$ is called the *local minimum value.*

14. **Necessary and sufficient conditions for maximum and minimum values :** We know that $f(x)$ has an extreme value at

$$x = a \text{ if } f'(a) = 0.$$

Case I. When $f'(x)$ changes from +ve to –ve while passing through a.

Here, $f(x)$ increases in the left nighbourhood of a and decreases in the right neighbourhood of a. Hence, $f(x)$ has a maximum value at $x = a$.

Case II. When $f'(x)$ changes from –ve to +ve while passing through a.

Here, $f(x)$ decrease in the left neighbourhood of a and increases in the right neighbourhood of a. Hence, $f(x$ has a minimum value at $x = a$.

15. Working rule :

(i) For finding the maximum and minimum value of $y = f(x)$:

(A) Put $\frac{dy}{dx} = 0$. Solve it for x, given $x = a, b, c, \ldots$

(B) Select $x = a$, study the sign of $\frac{dy}{dx}$ when

(a) $x < a$ slightly (b) $x > a$ slightly.

(C) If the former is +ve and later is –ve then $f(x)$ is maximum at $x = a$.

(D) If the former is –ve and later is +ve, then $f(x)$ is minimum at $x = a$.

(E) Putting these values of x for which $f(x)$ is maximum or minimum and get the correspondin maximum or minimum values of $f(x)$.

(ii) Working rule to find the maximum or minimum values of a function in a closed interval :

(A) Put y = given function $f(x)$ and find $\frac{dy}{dx}$ *i.e.,* $f'(x)$.

(B) Put $\frac{dy}{dx} = 0$, *i.e.,* $f'(x) = 0$ and solve it for x, giving $x = a, b, c, \ldots$

(C) Select $x = a$, find $\frac{d^2y}{dx^2}$ *i.e,* $f''(x)$ at $x = a$.

(a) If $\left(\frac{d^2y}{dx^2}\right)_{x=a}$ *i.e.,* $f''(a)$ is –ve, $x = a$ gives the maximum value of the function.

(b) If $\left(\frac{d^2y}{dx^2}\right)_{x=a}$ *i.e.,* $f''(a)$ is +ve, $x = a$ gives the minimum value of the function.

(iii) Use of second derivative theorem :

(A) A function $f(x)$ is maximum at $x = a$ if

$$f'(a) = 0 \text{ and } f''(a) < 0 \text{ (–ve).}$$

(B) A function $f(x)$ is minimum at $x = a$ if

$$f'(a) = 0 \text{ and } f''(a) > 0 \text{ (+ve).}$$

16. Properties of maxima and minima :

(i) If $f(x)$ is a continuous function in its domain, then at least one maxima and one minima must lie betwee two equal values of $f(x)$.

(ii) Maxima and minima occur alternately, *i.e.,* between two maxima there is one minima and vice-versa.

(iii) If $f(x) \to \infty$ as $x \to a$ or b and $f'(x) = 0$, only for one value of x (say c) between a and b, then $f(c)$ i necessarily the minimum and the least value.

If $f(x) \to -\infty$ as $x \to a$ or b then $f(c)$ is necessarily the maximum and the greatest value.

17. Higher order derivative test : Let f be a differentiable function on an interval I and let c be an interio point of I such that

(i) $f'(c) = f''(c) = f'''(c) = \ldots\ldots = f^{n-1}(c) = 0$, and

(ii) $f^n(c)$ exists and is non-zero.

Then, if n is even and $f^n(c) < 0 \Rightarrow x = c$ is a point of local maximum

If n is even and $f^n(c) > 0 \Rightarrow x = c$ is a point of local minimum

If n is odd $\Rightarrow x = c$ is neither a point of local maximum nor a point of local minimum.

Algorithm for determining extreme values of a function : From the above test criteria we obtain th following rules for determining maxima and minima of $f(x)$:

Chapter at a Glance

Step I : Find $f'(x)$

Step II : Put $f'(x) = 0$ and solve this equation for x. Let $c_1, c_2, \ldots, c_n$ be the roots of this equation. $c_1, c_2, \ldots, c_n$ are stationary values of x and these are the possible points where the function can attain a local maximum or a local minimum. So we test the function at each one of these points.

Step III : Find $f''(x)$. Consider $x = c_1$.

If $f''(c_1) < 0$, then $x = c_1$ is a point of local maximum.

If $f''(c_1) > 0$, then $x = c_1$ is a point of local minimum.

If $f''(c_1) = 0$, we must find $f'''(x)$ and substitute in it c_1 for x.

If $f'''(c_1) \neq 0$, then $x = c_1$ is neither a point of local maximum nor a point of local minimum and is called the point of inflection.

If $f'''(c_1) = 0$, we must find $f''''(x)$ and substitute in it c_1 for x.

If $f''''(c_1) < 0$, then $x = c_1$ is a point of local maximum and $f''''(c_1) > 0$, then $x = c_1$ is a point of local minimum.

If $f''''(c_1) = 0$, we must find $f'''''(x)$, and so on. Similarly, the value of $c_2, c_3, \ldots$ may be tested.

Chapter 8. Integral Calculus

1. **Integration :** It is the inverse operation of differentiation.

   In differential calculus, we are given a function and we are required to differentiate it, but in *integral calculus* we are required to find a function whose differential coefficient (or derivative) is given.

   If the differential coefficient of a function $f(x)$ w.r.t. x is $F(x)$, then we can say that the integral (or primitive or antiderivative) of $F(x)$ w.r.t. x is $f(x)$.

2. **Constant of integration :** If $\frac{d}{dx}[f(x) + c] = \frac{d}{dx} f(x) + \frac{d}{dx}(c) = f(x) + 0 = F(x)$, then

$$\int F(x).\, dx = f(x) + c,$$

   where c is an *arbitrary constant.*

   The constant c is called the *constant of integration.* The constant of integration is generally omitted.

   Note : Different methods of integrating a function may give different answers apparently, but any two such answers differ only by a constant.

3. **Indefinite integral :** The integral of any function is not definite, therefore, it is called *indefinite integral* and denoted by $\int f(x)\, dx$, where $f(x)$ be the function.

   For example, x^3 is a primitive or indefinite integral of $3x^2$, there are infinitely many indefinite integrals of $3x^2$. *e.g.,* $x^3 + 2, x^3 - 1, x^3 + 5, x^3 - 5, \ldots$

4. **Standard forms :** All the standard forms given below are of great importance and the students are strongly advised to learn them.

   (i) $\int 0\, dx = c$ (constant).

   (ii) $\int 1\, dx = \int x^0\, dx = \frac{x^{0+1}}{0+1} = x + c.$

   (iii) $\int x^n\, dx = \frac{x^{n+1}}{n+1} + c$ if $n \neq -1$

   (iv) $\int \frac{1}{x}\, dx = \log_e + c.$

   (v) $\int e^x\, dx = e^x + c.$

   (vi) $\int a^x\, dx = \frac{a^x}{\log_e a} + c$

   (vii) $\int \sin x\, dx = -\cos x + c.$

   (viii) $\int \cos x\, dx = \sin x + c.$

   (ix) $\int \sec^2 x\, dx = \tan x + c.$

   (x) $\int \text{cosec}^2 x\, dx = -\cot x + c.$

   (xi) $\int \sec x . \tan x\, dx = \sec x + c.$

   (xii) $\int \text{cosec}\, x \cot x\, dx = -\text{cosec}\, x + c.$

(xiii) $\int \cot x\, dx = \log |\sin x| + c.$

(xiv) $\int \tan x\, dx = -\log |\cos x| + c.$

or $\int \tan x\, dx = \log |\sec x| + c.$

(xv) $\int \sec x\, dx = \log |\sec x + \tan x| + c = \log \left|\tan\left(\frac{\pi}{4}+\frac{x}{2}\right)\right| + c.$

(xvi) $\int \text{cosec}\, x\, dx = \log |\text{cosec}\, x - \cot x| + c = \log \left|\tan\frac{x}{2}\right| + c.$

(xvii) $\int \frac{1}{\sqrt{1-x^2}} dx = \sin^{-1} x + c = -\cos^{-1} x + c.$

(xviii) $-\int \frac{1}{\sqrt{1-x^2}} dx = \cos^{-1} x + c = -\sin^{-1} x + c.$

(xix) $\int \frac{dx}{x\sqrt{x^2-1}} = \sec^{-1} x + c = -\text{cosec}^{-1} x + c.$

(xx) $-\int \frac{1}{x\sqrt{x^2-1}} dx = \text{cosec}^{-1} x + c = -\sec^{-1} x + c.$

(xxi) $\int \frac{1}{1+x^2} dx = \tan^{-1} x + c = -\cot^{-1} x + c.$

(xxii) $-\int \frac{1}{1+x^2} dx = \cot^{-1} x + c = -\tan^{-1} x + c.$

5. If $f'(x)$ denotes the differentiation w.r.t., x, then

(i) $\int [f(x)]^n f'(x).dx = \frac{[f(x)]^{n+1}}{n+1} + c,\ n \neq 1.$

(ii) $\int \frac{f'(x)}{f(x)} dx = \log |f(x)| + c.$

(iii) $\int a^{f(x)} f'(x)\, dx = \frac{a^{f(x)}}{\log a} + c.$ $(a > 0)$

(iv) $\int e^{f(x)} f'(x)\, dx = e^{f(x)} + c.$

6. Integration by parts : If u and v be two functions of x, then

$$\int u.\, v.\, dx = u \int v.\, dx - \int \frac{du}{dx} \left(\int v.\, dx\right) dx.$$

In words, integral of the product of two functions = Ist function × integral of 2nd function – integral of [differentiation of Ist × integral of 2nd function].

Method to choose the first function : In finding integrals by this method, proper choice of 1st and 2nd functions is essential. Although there is no fixed rule for taking first and the second functions and their choice is possible by practice, yet the following rule is helpful in the choice of 1st and 2nd.

If the two functions are of different types, take that function as 1st which comes in the word, ILATE where

I : stands for inverse trigonometric function.
L : stands for logarithmic function.
A : stands for algebraic function.
T : stands for trigonometric function.
E : stands for exponential function.

7. Important relations :

(i) $\int e^x [f(x) + f'(x)]\, dx = e^x . f(x) + c.$

(ii) $\int e^{ax}.\sin(bx + c)\, dx = \frac{1}{\sqrt{a^2+b^2}} e^{ax}.\sin\left[bx + c - \tan^{-1}\left(\frac{b}{a}\right)\right] + c.$

$= \frac{e^{ax}}{a^2+b^2}[a \sin(bx + c) + b \cos(bx + c)] + C$, C being constant of integration.

(iii) $\int e^{ax}.\cos(bx + c).\, dx = \frac{1}{\sqrt{a^2+b^2}} e^{ax}.\cos\left[bx + c - \tan^{-1}\left(\frac{b}{a}\right)\right] + C.$

$= \frac{e^{ax}}{a^2+b^2}[a \cos(bx + c) + b \sin(bx + c)] + c$, C being constant of integration.

(iv) $\int e^{ax} \sin bx\, dx = \frac{e^{ax}}{a^2+b^2}(a \sin bx - b \cos bx) + c.$

(v) $\int e^{ax} \cos bx\, dx = \frac{e^{ax}}{a^2+b^2}(a \cos bx + b \sin bx) + c.$

8. Selection for proper substitution : In order to evaluate certain problems of integration, we make some substitutions which reduces the given integrand to some standard forms already known to us. There is no hard and fast rule for making substitutions. However, there are some useful substitutions.

(a) (i) If the integrand contains a *t*-ratio of $f(x)$ or logarithm of $f(x)$ or an exponential function in which the index is $f(x)$, put $f(x) = t$.

(ii) If the integrand is a rational function of e^x, put $e^x = t$.

(b) For evaluating integrals $\int \sin^n x.\, dx$ or $\int \cos^n x\, dx$, where n is a positive integer,

(i) Put $\sin x = t$, if the index of $\cos x$ is an odd positive integer.

(ii) Put $\cos x = t$, if the index of $\sin x$ is an odd positive integer.

(iii) Express $\sin^n x$ or $\cos^n x$ in terms of cosines of multiples of angles by using $2 \sin^2 x = 1 - \cos 2x$ or $2 \cos^2 x = 1 + \cos 2x$ if n is an even positive integer.

(c) For evaluating integrals $\int \sin^p x \cos^q x\, dx$, where p, q are positive integers :

(i) Put $\sin x = t$ if q is an odd positive integer.

(ii) Put $\cos x = t$ if p is an odd positive integer.

(iii) Express $\sin^p x$ or $\cos^q x$ in terms of cosines of multiples of angles by using the formula of $\cos 2x$ and $\sin 2x$ if both p and q are positive even integers.

(d) For evaluating integrals $\int \sec^n x\, dx$ or $\int \text{cosec}^n x\, dx$ where n is positive even integer :

(i) Put $\tan x = t$ if the index of $\sec x$ is a positive even integer.

(ii) Put $\cot x = t$ if the index of $\text{cosec}\, x$ is a positive even integer.

(e) (i) If $\sqrt{a^2 - x^2}$ occurs in the integrand put $x = a \sin \theta$.

(ii) If $\sqrt{a^2 + x^2}$ or $a^2 + x^2$ occurs in the integrand put $x = a \tan \theta$.

(iii) If $\sqrt{x^2 - a^2}$ occurs in the integrand, put $x = a \sec \theta$.

(iv) If $\sqrt{a + x}$ occurs in the intergrand, put $a + x = t^2$.

(v) If $\sqrt{\frac{a-x}{a+x}}$ occurs in the integrand, put $x = a \cos 2\theta$.

(f) If a power of x (*i.e.*, x^p) is a factor of the integrand and the remaining integrand is an algebraic functio of x, then put $x^{p+1} = t$.

9. List of important results which are frequently used :

(i) $\int (ax+b)^n\,dx = \dfrac{(ax+b)^{n+1}}{a(n+1)} + c,\ n \neq -1.$

(ii) $\int \dfrac{1}{(ax+b)}dx = \dfrac{1}{a}\log|ax+b| + c.$

(iii) $\int e^{ax}\,dx = \left(\dfrac{1}{a}\right).e^{ax} + c$

(iv) $\int e^{ax+b}\,dx = \dfrac{1}{a}e^{ax+b} + c.$

(v) $\int a^{bx+c}\,dx = \dfrac{1}{b}\dfrac{a^{bx+c}}{\log a} + c,\ a > 0$ and $a \neq 1$

(vi) $\int \sin(ax+b)\,dx = -\left(\dfrac{1}{a}\right)\cos(ax+b) + c.$

(vii) $\int \cos(ax+b)\,dx = \left(\dfrac{1}{a}\right)\sin(ax+b) + c.$

(viii) $\int \sec^2(ax+b)\,dx = \left(\dfrac{1}{a}\right)\tan(ax+b) + c.$

(ix) $\int \text{cosec}^2(ax+b)\,dx = -\left(\dfrac{1}{a}\right)\cot(ax+b) + c.$

(x) $\int \sec(ax+b).\tan(ax+b)\,dx = \left(\dfrac{1}{a}\right)\sec(ax+b) + c$

(xi) $\int \text{cosec}(ax+b).\cot(ax+b)\,dx = -\left(\dfrac{1}{a}\right)\text{cosec}(ax+b) + c.$

(xii) $\int \tan(ax+b)\,dx = -\dfrac{1}{a}\log|\cos(ax+b)| + c.$

(xiii) $\int \cot(ax+b)\,dx = \dfrac{1}{a}\log|\sin(ax+b)| + c.$

(xiv) $\int \sec(ax+b)\,dx = \dfrac{1}{a}\log|\sec(ax+b) + \tan(ax+b)| + c.$

(xv) $\int \text{cosec}(ax+b)\,dx = \log|\text{cosec}(ax+b) - \cot(ax+b)| + c.$

10. Some Special Integrals :

(i) $\int \dfrac{dx}{x^2+a^2} = \dfrac{1}{a}\tan^{-1}\dfrac{x}{a} + c.$

(ii) $\int \dfrac{dx}{x^2-a^2} = \dfrac{1}{2a}\log\left|\dfrac{x-a}{x+a}\right| + c.$

(iii) $\int \dfrac{dx}{a^2-x^2} = \dfrac{1}{2a}\log\left|\dfrac{a+x}{a-x}\right| + c.$

(iv) $\int \dfrac{dx}{\sqrt{x^2+a^2}} = \log|x+\sqrt{(x^2+a^2)}| + c$

(v) $\int \dfrac{dx}{\sqrt{x^2-a^2}} = \log|x+\sqrt{(a^2-x^2)}| + c.$

(vi) $\int \dfrac{dx}{\sqrt{a^2-x^2}} = \sin^{-1}\dfrac{x}{a} + c.$

(vii) $\int \sqrt{x^2+a^2}\,dx = \dfrac{1}{2}x\sqrt{(a^2+x^2)} + \dfrac{1}{2}a^2\log|x+\sqrt{(a^2+x^2)}| + c.$

(viii) $\int \sqrt{x^2-a^2}\,dx = \dfrac{1}{2}x\sqrt{(x^2-a^2)} - \dfrac{1}{2}a^2\log|x+\sqrt{(x^2-a^2)}| + c.$

(ix) $\int \sqrt{a^2-x^2}\,dx = \dfrac{1}{2}x\sqrt{(a^2-x^2)} + \dfrac{1}{2}a^2\sin^{-1}\left(\dfrac{x}{a}\right) + c.$

11. Partial fractions : Let $f(x) = \dfrac{g(x)}{h(x)}$ be a proper rational function. First of all, we split up the denominato $h(x)$ as the product of non-repeated linear or repeated linear factors or non-repeated or repeated quadratics (which cannot be split into real linear factors). Then $f(x)$ can be written as the sum of the fractions ir which the numerator is either a constant or real linear polynomials. For this the following points shoulc be kept in mind.

(i) Corresponding to each non-repeated linear factor $ax + b$, there is a partial fraction of the form $\frac{A}{ax+b}$

(ii) Corresponding to each repeated linear factor $(ax + b)^2$, there is sum of two terms of the form

$$\frac{A}{ax+b}+\frac{B}{(ax+b)^2}.$$

(iii) Corresponding to each repeated linear factor $(ax + b)^3$, there is a sum of three terms of the form

$$\frac{A}{ax+b}+\frac{B}{(ax+b)^2}+\frac{C}{(ax+b)^3}$$

(iv) Corresponding to each non-repeated quadratic factor $ax^2 + bx + c$ (which cannot be put as the product of the linear factors), there is a partial factor of the form

$$\frac{Ax+B}{ax^2+bx+c}.$$

(v) Corresponding to each repeated quadratic factor $(ax^2 + bx + c)^2$ (which cannot be put as the product of linear factors), there is a sum of two terms of the form

$$\frac{Ax+B}{ax^2+bx+c}+\frac{Cx+D}{(ax^2+bx+c)^2}.$$

(vi) When only even powers of x occur both in the numerator and denominator put $x^2 = y$ and resolve the resulting algebraic fraction in y into partial fractions and then put $y = x^2$ in the resulting partial fractions.

S. No.	Form of the Integrand	Method of reducting to known forms
1.	$\frac{1}{ax^2+bx+c}$	Make the coefft. of x^2 unity and express in standard form by completing the square.
2.	$\frac{px+q}{ax^2+bx+c}$	Express $px + q$ in the form A $(2ax + b)$ + B.
3.	$\frac{p(x)}{ax^2+bx+c}$ Where degree of $p(x)$ is ≥ 0	Carry out the division and express it as a polynomial + $\frac{px+q}{ax^2+bx+c}$.
4.	$\sqrt{ax^2+bx+c}$	Make the coefft. of x^2 unity and express in standard form by completing the square.
5.	$\frac{1}{\sqrt{ax^2+bx+c}}$	Make the coefft. of x^2 unity and express it in standard form by completing the square.
6.	$\frac{px+q}{\sqrt{ax^2+bx+c}}$	Express $px + q$ in the form A $(2ax + b)$ + B.
7.	$\int(px+q)\sqrt{ax^2+bx+c}$	Put $px + q = A\frac{d}{dx}(ax + bx + c) + B$ Find the values of constants A and B by comparing coefficients of like powers of x on both sides.
8.	$\int\frac{dx}{a+b\cos x}$	Put $\sin x = \frac{2\tan\frac{x}{2}}{1+\tan^2\frac{x}{2}}$, $\cos x = \frac{1-\tan^2\frac{x}{2}}{1+\tan^2\frac{x}{2}}$

Chapter at a Glance

9.	$\int \frac{dx}{a+b\sin x}$	Replace $1+\tan^2\frac{x}{2}$ by $\sec^2\frac{x}{2}$ in the numerator
10.	$\int \frac{dx}{a\cos x+b\sin x}$	Put $\tan\frac{x}{2}=t$
11.	$\int \frac{dx}{a\cos x+b\sin x+c}$	Put $t=\tan\frac{x}{2}$, so that $\cos x \frac{1-t^2}{4t^2}$, $\sin x=\frac{2t}{1+t^2}$
12.	$\int \frac{a\cos x+b\sin x}{c\cos x+d\sin x}dx$	1. Let numerator = A (denom) + B $\frac{d}{dx}$ (denom)
		2. Compare coefft. of $\cos x$ and $\sin x$ on both sides.
13.	$\int \frac{dx}{a+b\cos^2 x}$, $\int \frac{dx}{a+b\sin^2 x}$	1. Divide the numerator and denomenator by $\cos^2 x$
	$\int \frac{dx}{a\sin^2 x+b\cos^2 x}$, $\int \frac{dx}{(a\sin x+b\cos x)^2}$	2. Replace $\sec^2 x$, if present in the denom. by $1+\tan^2 x$.
	$\int \frac{dx}{a+b\sin^2 x+c\cos^2 x}$	3. Put $\tan x=t$ so that $\sec^2 x dx=dt$.
14.	$\int \frac{x^2+1}{x^4+kx^2+1}dx$ or $\int \frac{x^2-1}{x^4+kx^2+1}dx$	Divide the numerator and denominator by x^2 and
	Put,	
	Ex. $\int \frac{x^2+1}{x^4+1}dx$, $\int \frac{x^2+4}{x^4+16}dx$	$\left(x+\frac{1}{x}\right)=t$ or $\left(x-\frac{1}{x}\right)=t$, which ever on differentiation
	$\int \frac{x^2-1}{x^4+x^2+1}dx$, $\int \frac{dx}{x^4+1}$,	gives the numerator of the resulting integrand.
	$\int \frac{(x-1)^2}{x^4+x^2+1}dx$, $\int \frac{dx}{x^4+3x^2+1}$, etc.	**[Hint.** $\int \frac{(x-1)^2}{x^4+x^2+1}dx = \int \frac{x^2+1-2x}{x^4+x^2+1}dx$
		$= \int \frac{x^2+1}{x^4+x^2+1}dx - \int \frac{2x}{x^4+x^2+1}dx]$

13. **Definite integral :** If $f(x)$ is a continuous function on a closed interval $[a, b]$, then $\int_a^b f(x).dx$ is defined as a definite integral which is equal to $F(b)-F(a)$, where $F(x)$ is an integral (anti-derivative) of $f(x)$. Therefore,

$$\int_a^b f(x).\,dx = [F(x)+c]_a^b$$

$$= [F(b)+c-F(a)-c]$$

$$= F(b)-F(a).$$

The numbers a and b are called the limits of integration, 'a' is called the lower limit and 'b' is called the upper limit. The interval $[a, b]$ is called the interval of integration.

The value of $\int_a^b f(x).\,dx$ does not depend upon the value of arbitrary constant. Therefore, it is called definite integral.

4. Change of variable of integration in a definite integral : Method to evaluate $\int_a^b f(x)\,dx$ by substitution $x = g(t)$

(i) Let, $I = \int_a^b f(x)dx$ ... (i)

(ii) In the integrand put $x = g(t)$

then $dx = g'(t)\,dt$ ... (ii)

(iii) From (ii), when $x = a$, find the corresponding value of new variable 't', let it be α. Also find the corresponding value of 't' when $x = b$. Let it be β.

(iv) Substitute the values of x, a, b etc., in (i), we get $I = \int_\alpha^\beta f(g(t))\,g'(t)\,dt$ which can be calculated between two limits of integration.

5. Fundamental properties of definite integrals :

(i) $\int_a^b f(x)dx = \int_a^b f(t)\,dt$.

(ii) $\int_a^b f(x)dx = -\int_b^a f(x)dx$.

(iii) $\int_a^b f(x)\,dx = \int_a^c f(x)\,dx + \int_c^b f(x)\,dx$, where $a < c < b$.

(iv) $\int_0^a f(x)\,dx = \int_0^a f(a-x)\,dx$.

(v) $\int_{-a}^a f(x)\,dx = \begin{cases} 0, \text{ if } f(x) \text{ is an odd function} \\ 2\int_0^a f(x)\,dx, \text{ if } f(x) \text{ is an even function.} \end{cases}$

(vi) $\int_0^{2a} f(x)\,dx = \begin{cases} 0, \text{ if } f(2a-x) = -f(x) \\ 2\int_0^a f(x)\,dx, \text{ if } f(2a-x) = f(x). \end{cases}$

(vii) $\int_a^b f(x)\,dx = \int_a^b f(a+b-x)\,dx$.

(viii) $\left|\int_a^b f(x)\,dx\right| \le \int_a^b |f(x)|\,dx$

(ix) $\int_0^\infty f(x)\,dx = \lim_{t\to\infty} \int_0^t f(x)\,dx$.

(x) If m and M are the smallest and greatest values of a function $f(x)$ defined on an interval $[a, b]$, then

$$m(b-a) \le \int_a^b f(x)\,dx \le M(b-a).$$

(xi) If $f(t)$ is an odd function, then $\phi(x) = \int_a^x f(t)\,dt$ is an even function.

(xii) If $f(t)$ is an even function, then

$\phi(x) = \int_a^x f(t)\,dt$ is an odd function.

(xiii) If $f(x) \ge 0$, for all $x \in [a, b]$, then $\int_a^b f(x)\,dx \ge 0$.

(xiv) If $f(x) \le g(x)$ for all $x \in [a, b]$, then $\int_a^b f(x)\,dx \le \int_a^b g(x)\,dx$.

6. Integrals involving modulus functions :

Method. Use property 3, *i.e.;* $\int_a^b f(x)\,dx = \int_a^c f(x)\,dx + \int_c^b f(x)\,dx$, **where $a < c < b$.**

To integrate $\int_a^b |f(x)|\,dx$, the interval (a, b) is split into two parts such as (a, c) and (c, b) so that the value of the given function is positive in one part (a, c) and negative in the other part (c, b).

7. Integrals involving Greatest integer function :

Ex. Evaluate : $\int_{0.2}^{3.5} [x]\,dx$,

Sol. Since $[x]$ is the greatest integer function, we have,

$$f(x) = \begin{cases} 0; \text{ if } & 0.2 \le x < 1 \\ 1; & 1 \le x < 2 \\ 2; & 2 \le x < 3 \\ 3; & 3 \le x \le 3.5 \end{cases}$$

hence,
$$\int_{0.2}^{3.5} [x]\, dx = \int_{0.2}^{1} 0\, dx + \int_{1}^{2} 1\, dx + \int_{2}^{3} 2\, dx + \int_{3}^{3.5} 3\, dx$$
$$= 0 + [x]_1^2 + 2[x]_2^3 + 3[x]_3^{3.5}$$
$$= 1 + 2 + 1.5 = 4.5$$

18. Important integral

$$\int_0^{\frac{x}{2}} \log \sin x\, dx = \int_0^{\frac{x}{2}} \log \cos x\, dx$$
$$= -\frac{\pi}{2}\log 2 \longleftarrow \text{Remember}$$

19. Definite integral as a limit of a sum : If $f(x)$ be a single valued continuous function defined in the interva $[a, b]$ where $a < b$ and the interval (a, b) is divided into n equal parts of each length h so that $nh = b - a$, ther we define,

$$\int_a^b f(x)\, dx = \lim_{h \to 0} h\,[f(a) + f(a+h) + f(a+2h) + \ldots + f(a+\overline{n-1}\,h)] \text{ where } nh = b - a$$

or $\displaystyle\int_a^b f(x).\, dx = (b-a).\lim_{n\to\infty} \frac{1}{n}\,[f(a) + f(a+h) + f(a+2h) + \ldots + f(a+\overline{n-1}\,h)] \text{ where } h = \frac{b-a}{n}$

is called the definite integral of $f(x)$ between the limits $x = a$ and $x = b$.

Chapter 9. Differential Equations

1. A differential equation (D. E.) is an equation that involves *independent and dependent variables and derivative of the dependent variables.*
2. The **order of a differential equation** is the *order of the highest derivative* occurring in it.
3. The **degree of a differential equation** is the *degree of the derivative of the highest order occurring in it after th equation is freed from radical signs and fractions in the derivative (if any).*
   eg. $\frac{dy}{dx} + y \sin x = \cos x$ is a differential equation of first order and first degree., where y is the dependen variable and x is the independent variable.
4. A differential equation is said to be **linear** if the *dependent variable and all of its derivatives occurring in th equation occur only in the first degree and are not multiplied together.*
5. A differential equation is said to be **non-linear** if its degree > 1, dependent variable and any of its derivative have degree more than one and are multiplied together.
6. **Formation of differential equations** : To obtain the differential equation from this equation we follow th following steps :
   Step 1 : *Differentiate the given function w.r.t. the independent variable* present in the equation.
   Step 2 : *Keep differentiating number of times in such a way that (n + 1) equations are obtained if n arbitrary constant are present.*
   Step 3 : *Using the (n + 1) equations obtained, eliminate the constants* $(c_1, c_2 \ldots \ldots c_n)$.
   (**Remark** : The order of the differential equation is equal to the number of arbitrary constants in th given equation.)
7. **Solution of a differential equation (D. E.)** : A solution or an integral of a differential equation *is a functio of the form $y = f(x)$ which satisfies the given differential equation.* For eg : $y = e^x$ is a solution of the differentia equation $\frac{dy}{dx} = y$
   (**Note :** Solution of a D. E. does not contain the derivatives.)
   (**Remark :** A solution of nth order D. E. contains n arbitrary constants.)

8. **General and particular solutions** : A solution of a D. E. *with arbitrary constants equal in number to the order of the* D. E. is called **general** solution. Other solutions, *obtained by giving particular value to these constants* in the general solution are called particular solutions.

For eg. : $y = \text{A} \cos x + \text{B} \sin x$ is the general solution of the 2nd order D. E.

$$\frac{d^2y}{dx^2} + y = 0.$$

But, $y = \text{A}$

cos x is not the general solution despite satisfying the equation as it contains only one arbitrary constant.

And, $y = 3\cos x + 2\sin x$ is a particular solution of the D. E.

(**Note** : The general solution of a D. E. of the nth order must contain n and only n independent arbitrary constants.)

9. **Solving Different forms of differential equations :**

Type I : Differential equation $\frac{dy}{dx} = f(x)$.

We have $\frac{dy}{dx} = f(x)$, here we treat $\frac{dy}{dx}$ as $dy \div dx$

$$\therefore \quad dy = f(x).\, dx$$

Integrating both sides, we get

$$\int dy = \int f(x).\, dx + c$$

where c is a constant of integration.

$$\therefore \quad y = \int f(x).\, dx + c.$$

Type II : Differential equations of the type $\frac{dy}{dx} = f(y)$.

To solve this type of differential equations we integrate both sides to obtain the general solution as discussed under.

$$\frac{dy}{dx} = f(y) \Rightarrow \frac{dx}{dy} = \frac{1}{f(y)} \Rightarrow dx = \frac{1}{f(y)} dy$$

Integrating both sides, we obtain

$$\int dx = \int \frac{1}{f(y)} dy + c$$

or

$$x = \int \frac{1}{f(y)} dy + c.$$

Type III : Variables separable : If in any differential equation, it is possible to express all the functions of x and dx on one side and all the functions of y and dy on the other side, the variables are said to be *separable.*

Rule : To solve the equation $\frac{dy}{dx} = \text{XY}$, where X is a function of x only and Y is a function of y only.

(i) Given equation is $\frac{dy}{dx} = \text{XY}$.

(ii) Separating the variables, *i.e.,* $\frac{dy}{\text{Y}} = \text{X}.dx$

(iii) Integrating both sides and adding an arbitrary constant on one side.

i.e., $= \int \frac{dy}{\text{Y}}$ X. $dx + c$, which is the required solution.

Note : Addition of an arbitrary constant is must on one side of the differential equation.

Type IV : Homogeneous equations : A differential equation of the form $\frac{dy}{dx} = \frac{f_1(x,y)}{f_2(x,y)}$ where $f_1(x, y)$ and $f_2(x, y)$ are both homogeneous functions of the same degree in x and y, *i.e.,* an equation of the form $\frac{dy}{dx} = \text{F}\left(\frac{y}{x}\right)$ is called a *homogeneous differential equation.*

A function is said to be homogeneous of the nth degree in x and y if it can be put in the form $x^n \cdot f\left(\frac{y}{x}\right)$.

Rule : Put $y = vx$, then $\frac{dy}{dx} = v + x.\frac{dv}{dx}$ and on substituting these values of y and $\frac{dy}{dx}$ in the given differential equation it will be reducible to variable separable.

Type V : Equation reducible to the form in which variable can be separated :

Equation of the form $\frac{dy}{dx} = f(ax + by + c)$ can be reduced to the form in which the variables are separable.

Put $ax + by + c = z$, we have on differentation w.r.t. x,

$$a + b\frac{dy}{dx} = \frac{dz}{dx} \Rightarrow \frac{dy}{dx} = \frac{1}{b}\left[\frac{dz}{dx} - a\right]$$

$\therefore$ Given equation becomes, $\frac{1}{b}\left[\frac{dz}{dx} - a\right] = f(z)$.

$$\Rightarrow \quad \frac{dz}{dx} - a = bf(z) \Rightarrow \frac{dz}{dx} = a + b\,f(z).$$

Separating the variables $\frac{dz}{a + b\,f(z)} = dx$

which can now be integrated.

Type VI : Linear differential equation of the form $\frac{dy}{dx} + Rx = S$.

Sometimes a linear differential equation can be put in the form

$$\frac{dx}{dy} + Rx = S$$

where R and S are functions of y or constants.

Note that y is independent variable and x is a dependent variable.

The following algorithm is used to solve these types of equations.

10. First order linear equation with coefficients : A linear differential equation is that in which the dependent variables and its differential coefficients occur only in the first degree and are not multiplied together.

The standard form of linear differential equation of the first order is,

$$\frac{dy}{dx} + Py = Q$$

where P and Q are functions of x (and not of y) or constants.

Similarly, $$\frac{dx}{dy} + Px = Q$$

where P and Q are function of y only, is also called a linear differential equation of the first order.

Working rule to solve the linear equations :

(i) Write the given equation in the form $\frac{dy}{dx} + Py = Q$, P and Q are functions of x only.

(ii) Find the integrating factor (I. F.) $= e^{\int P.dx}$.

(iii) The solution of the differential equation is

$$y\,[\text{I.F.}] = \int Q.\,(\text{I.F.})\,dx + c.$$

Note : If the differential equation can be represented in the form $\frac{dy}{dx} + Px = Q$, where P and Q are functions of y only, I. F. $= e^{\int P.dy}$ and the solution is given by

$$x.\,(\text{I. F.}) = \int Q.(\text{I. F.})\,dy + c.$$

1. **Application of differential equations :** If $y = f(x)$ is a differential equation and $\frac{dy}{dx}$ is proportional to y, then we have $\frac{dy}{dx} = ky, k \neq 0$, a constant. This simple D. E. helps in solving problems related to growth and decay and rate of change. Differential equations also help in solving problems of coordinate geometry and velocity, acceleration, distance and time (acceleration $= \frac{dv}{dt}$)

Chapter 10. Probability

1. **Independent event** : An event is said to be independent of another event *when the actual happening on one does not influence in any degree the probability of the happening of other.* Let E and F be the events associated with a sample space S of a random experiment, E and F are called independent events if

$$P(E \cap F) = P(E).P(F)$$

2. Conditional event (or dependent event) : *If the probability of the happening of other is dependent or influenced by the previous happening,* then other event is said to be **dependent** or **conditional** on one. It is denoted by **A/B** (A occurs after occurrence of B). The concept of dependent events gives rise to the concept of conditional probability.
3. **Conditional Probability** : Let A and B be two events, then the probability of the occurrence of A under the condition that B has already occurred and $P(B) \neq 0$, is called the conditional probability of A given B and is written as $P(A/B)$.
4. Formula to evaluate $P(A/B)$: $P(A/B) = P(A \cap B)/P(B), P(B) \neq 0$

   Similarly, $P(B/A) = P(A \cap B)/P(A), P(A) \neq 0$

   Note : **1.** $P(A/B)$ is meaningful only when $P(B) \neq 0$, *i.e.,* when B is not an impossible event.

   2. If A and B are mutually exclusive events, then $P(A/B) = 0$ and $P(B/A) = 0$.
5. **Properties of Conditional Probability :**

   (a) The conditional probability of an event A given that B has occurred lies between 0 and 1.

   (b) If A and B are two events associated with a random experiment, then $P(A'/B) = 1 - P(A/B)$.
6. **Addition Theorem of Probability** :

   If A and B are any two events then

$$P(A \cup B) = P(A) + P(B) - P(A \cup B)$$

   If A, B and C are any three events then

$$P(A \cup B \cup C) = P(A) + P(B) + P(C) - P(A \cap B) - P(B \cap C) - P(A \cap C) + P(A \cap B \cap C)$$

7. **Multiplication Theorem(Rule) of Probability :** Let A and B be two events associated with a sample space S. Then,

$$P(A \cap B) = P(A) . P(B/A) = P(B) . P(A/B), \text{ provided } P(A) \neq 0 \text{ and } P(B) \neq 0$$

   If A, B and C are three events associated with a random experiment, then

$$P(A \cap B \cap C) = P(A) . P(B/A) . P(C/A \cap B)$$

   Similarly, if $A_1, A_2, \ldots, A_n$ are n events related to a random experiment, then

$$P(A_1 \cap A_2 \cap A_3 \cap \ldots\ldots A_n) = P(A_1) . P(A_2/A_1) . P(A_3/A_1 \cap A_2) \ldots\ldots P(A_n/ A_1 \cap A_2 \cap A_3 \cap \ldots\ldots A_{n-1})$$

8. **Probability of Independent Events** : If A and B are independent events, then $P(A/B)$ is precisely the same as $P(A)$ since A is not affected by B, *i.e.,* $P(A/B) = P(A)$. Similarly, $P(B/A) = P(B)$. Now by multiplication rule of probability, we have

$$P(A \cap B) = P(A) . P(B/A) = P(B) . P(A/B)$$

$$\Rightarrow \quad P(A \cap B) = P(A) . P(B)$$

   (Known as multiplication rule for independent events)

   Note : So any one of these three conditions may be used as a test for independence of events :

   (a) $P(A \quad B) = P(A) . P(B)$ (b) $P(A/B) = P(A)$ (c) $P(B/A) = P(B)$

9. Probability problems involving the use of infinite geometric progression (G. P.) :

In such problems, use the following formula :

If $S_\infty = a + ar + ar^2 + ar^3 + \ldots\ldots$(to infinity),

then, $S_\infty = a/(1-r)$ for $|r| < 1$,

where a = first term

and r = common ratio

10. Theorem of Total Probability :

Theorem : Let S be the sample space and E_1, E_2, ...E_n be n mutually exclusive and exhaustive event associated with a random experiment. Let A be any event associated with S, *i.e.*, which occurs with E_1 or E or ...... or E_n, then

$$P(A) = P(E_1).P(A/E_1) + P(E_2).P(A/E_2) + \ldots\ldots + P(E_n)\, P(A/E_n).$$

Proof. Let S be the sample space of a random experiment and E_1, E_2, ... E_n be n mutually exclusive and exhaustive events.

$$S = E_1 \cup E_2 \cup \ldots \cup E_n, \text{ where } E_1 \quad E_1 = \phi \text{ for}$$

$i \neq j$.

Since A is an event of S, *i.e.*, $A \subset S$, therefore,

$$A = S\ A = (E_1 \cup E_2 \cup E_3 \ldots \cup E_n)\ A$$

$$= (E_1\ A) \cup (E_2\ A) \cup (E_3\ A) \ldots. \cup (E_n\ A). \qquad \text{(Distributive law}$$

$$P(A) = P[(E_1\ A) \cup (E_2\ A) \cup \ldots \cup (E_n\ A)]$$

$$= P(E_1\ A) + P(E_2\ A) + \ldots + P(E_n\ A)$$

$$= P(A\ E_1) + P(A\ E_2) + \ldots + P(A\ E_n). \qquad \ldots(1$$

$[\because E_1\ A, E_2\ A, \ldots, E_n\ A$ are mutually exclusive events

$$P(A) = P(E_1)\,P(A/E_1) + P(E_2)\,P(A/E_2) + \ldots., + P(E_n)\,P(A/E_n)]$$

$$= \sum_{i=1}^{n} P(E_1)\,P(A/E_1), \qquad \ldots(2$$

Note : The rule of total probability can be depicted y the tree diagram shown alongside.

11. Bayes' Theorem :

$$P(E_1/A) = \frac{P(A \cap E_1)}{P(A \cap E_1) + P(A \cap E_2) + \ldots + P(A \cap E_n)} \qquad \text{[First Form}$$

We know that $P(A\ E_1) = P(E_1)\,P(A/E_1)$.

$$\therefore \quad P(E_1/A) = \frac{P(E_1)\,P(A/E_1)}{P(E_1)\,P(A/E_1) + P(E_2)\,P(A/E_2) + \ldots + P(E_n)\,P(A/E_n)}$$

$$= \frac{P(E_1)\,P(A/E_1)}{\sum P(E_1)\,P(A/E_1)} \qquad \text{[Second Form}$$

12. How to apply Bayes' theorem :

Procedure, Step 1. Identify mutually exclusive and exhaustive 'priori' events E_1, E_2, ... E_n,

Step 2. Find $P(E_1)$, $P(E_2)$, ... $P(E_n)$ and check that the sum of these probabilities is 1.

Chapter at a Glance

Step 3. Identify the event A and find $P(A/E_1)$, $P(A/E_2)$, ...$P(A/E_n)$.

Step 4. Identify the 'posteriori' event E_i/A whose probability is required.

Step 5. Calculate $P(E_i/A)$ by using 'Bayes' theorem.

13. Probability distribution :

Random variable : A random variable is a real valued function X defined over entire sample space of an experiment *i.e.,* it is a function which associates to each point of the sample space of an experiment, a unique real number.

14. Probability distribution table : Let a discrete random experiment assumes values $x_1, x_2, x_3, \ldots x_n$ with probabilities $p_1, p_2, p_3, \ldots, p_n$ respectively satisfying $p_1 + p_2 + p_3 + \ldots + p_n = 1$ and $0 \le p_i \le 1$, for $i = 2, 3, \ldots, n$ and in this case the following table describes the probability distribution of X .

X	x_1	x_2	x_3	...	x_n
P (X)	p_1	p_2	p_3	...	p_n

The different values of $x_1, x_2, \ldots, x_n$ of X together with corresponding probabilities form a probability distribution of the random variable X.

15. Mean and variance of a probability distribution : Let X be a random variable and let the probability distribution of X is as given in the above table, then

(i) *Mean* of probability distribution of a random variable X is defined as

$$\frac{p_1x_1 + p_2x_2 + p_3x_3 + \ldots\ldots\ldots + p_nx_n}{p_1 + p_2 + p_3 + \ldots + p_n} = \frac{\Sigma p_i\, x_i}{\Sigma p_i} \quad \text{where } p_i \ge 0.$$

$[\because p_1 + p_2 + p_3 + \ldots + p_n = \Sigma p_i = 1]$

It is denoted by μ.

(ii) *Variance of* probability distribution of a random variable X is defined as

Thus, $$\sigma^2 = \Sigma p_i\,(x_i - \mu)^2 = \Sigma p_i\, x_i^2 - \mu^2$$

It is denoted by σ^2.

Note : (i) The standard deviation of the probability distribution of the random variable is

$$\sigma = \sqrt{\Sigma p_i\, x_i^2 - \mu^2}\,.$$

(ii) Mean of a random variable is also called average or expected value and is denoted by E (X).

16. Repeated independent (Bernoulli's trials) : Bernoulli trial is also known as binomial trial where only two outcomes 'success or failure' of a random experiment is possible.

For example : Flipping of coin, rolling a die where appearance of a particular number is success etc.

17. Binomial distribution : If an experiment is repeated n times, the successive trials being independent of one another, then the probability of r successes (out of n trials) is ${}^nC_r p^r q^{n-r}$, where p is the probability of success in a single trial and $q = 1 - p$ is the probability of failure in a single trial.

In the above situation, probabilities of 0, 1, 2, 3,..., n successes out of n trials are given by the terms (in order) of the binomial expansion of $(q + p)^n$. Further,

(i) Mean of binomial distribution = np

(ii) Variance of binomial distribution = npq.

(iii) Standard deviation = $\sqrt{npq}$.

Note : (i) n and p (or q) occurring in the binomial distribution are called *parameters* of the distribution.

(ii) Since the random variable X can take only integral values 0, 1, 2, 3,............n, in the binomial distribution, therefore binomial distribution is *discrete.*

(iii) Probability of at least r successes in n trials.

$$P(x \ge r) = P(r) + P(r + 1) + \ldots + P(n)$$

$$= \sum_{k=r}^{n} P(k) = \sum_{k=r}^{n} {}^{n}C_k\, q^{n-k}\, p^k$$

(iv) Probability of at most r successes in n trials

$$P(x \le r) = P(0) + P(1) + \ldots + P(r)$$

$$= \sum_{k=0}^{n} P(k) = \sum_{k=0}^{n} {}^{n}C_k\, q^{n-k}\, p^k$$

Chapter 11. Vectors

1. **Scalars :** A quantity which has only magnitude but no direction is called a *scalar quantity or scalar.* Mass, length, speed, kinetic energy etc., are all scalar quantities.
2. **Vectors :** A quantity which has both magnitude as well as direction is called a *vector quantity* or *vector.* Weight, displacement, velocity, momentum etc., are all vector quantities.
3. **Representation of a vector :** A directed line segment represents a vector. A vector whose magnitude is proportional to the length AB and whose direction is from A to B and is denoted by length $\overrightarrow{AB}$ where A is called the initial point and B is the terminal point of the vector $\overrightarrow{AB}$. The magnitude of the vector $\overrightarrow{AB}$ is denoted by $|\overrightarrow{AB}|$ which is read as *modulus of* $\overrightarrow{AB}$ or *simply* $\overrightarrow{AB}$.

$$|\overrightarrow{AB}| = AB.$$

If the directed line segment AB is a part of line AB, then the line l is called the *support of vector* $\overrightarrow{AB}$.

4. **Types of vectors :**

(i) **Null vector or zero vector :** A vector whose magnitude is zero is called a *null* or *zero vector* and is represented by $\vec{0}$. Modulus of the null vector is zero.
The initial and terminal points of a zero vectors are coincident and its direction is arbitrary.
Thus, $\overrightarrow{AA} = \overrightarrow{BB} = \ldots\ldots\ldots\ldots = \vec{0}$ and $|\vec{0}| = 0$.
Note : A non-zero vector is called a *proper vector.*

(ii) **Like and unlike vector :** Two (or more than two) vectors are said to be *like vectors* if they have the same direction (no matter what their magnitudes are). *Unlike vectors* have opposite directions.

(iii) **Proper vector :** Any non-zero vector is called a *proper vector.* Thus, $\vec{a}$ is proper vector, if $|\vec{a}| \ne 0$.

(iv) **Unit Vector :** A vector whose length (magnitude or modulus) is unity is called a *unit vector.*
A unit vector is denoted by a single small letter with the sign over it, read as 'a cap'.

Thus, $\vec{a} = |\vec{a}|\,.\,\hat{a}$ $\qquad \therefore\ \hat{a} = \dfrac{\vec{a}}{|\vec{a}|}$

Note : Two unit vectors of same modulus are not equal if they have the different directions.

(v) **Equal Vectors :** Two vectors are said to be *equal vectors* if they have same magnitude and same direction. Thus

$$\overrightarrow{AB} = \overrightarrow{CD}$$

(vi) **Localised Vectors :** A vector having a fixed initial point is called a *localised vector.*

(vii) **Co-initial vectors :** Two or more vectors are said to be *co-initial vectors* if they have the same initial point. For example, $\overrightarrow{AB}, \overrightarrow{AC}, \overrightarrow{AD}$ are co-initial vectors with initial point A.

(viii) **Collinear vectors :** Two or more vectors are said to be *collinear vectors* if they are either parallel or coincident irrespective of their magnitude.

(ix) **Coplanar vectors :** Three or more vectors are said to be *coplanar* if they lie in the same plane or are parallel to the same plane.

(x) **Negative vector :** Two vectors are called *negative vector* of each other if they have same magnitude but opposite direction. Negative $\vec{a}$ of is denoted by $-\vec{a}$.

(xi) **Free vectors :** Vectors whose directions and magnitudes are known but the initial point and the support are not known are called *free vectors.*

(xii) **Reciprocal vectors :** Two vectors are said to be *reciprocal vectors* if they have the same direction but whose lengths are reciprocal to each other.

5. **Multiplication of vectors by scalars**: The product of a vector $\vec{a}$ by a scalar 'm' is denoted as $m\vec{a}$ such that:

(a) the support of $m\vec{a}$ is same or parallel to that of $\vec{a}$,

(b) $|m\vec{a}| = |m|\,|\vec{a}|$

(c) sense of $m\vec{a}$ is same or opposite to that of $\vec{a}$, according as m is +ve or –ve.

Note : $0\vec{a} = 0$, *i.e.* the product of a vector $\vec{a}$ by scalar 0 is the zero vector. Here 0 (left) stands for zero scalar and 0 (right) for zero vector.

6. **Addition of vectors :**

(a) **Triangle law** : $\vec{AC} = \vec{AB} + \vec{BC} \Rightarrow \vec{c} = \vec{a} + \vec{b}$

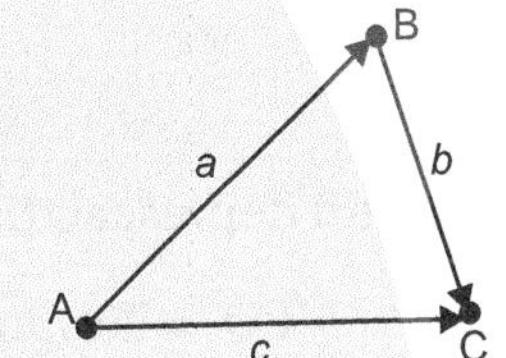

Note : 1. $\vec{c}$ is called the resultant of $\vec{a}$ and $\vec{b}$.

2. The addition of vectors is achieved by 'tail to nose' placing of the directed line segments in a triangle.

(a) **Parallelogram law**: *The result of adding two co-initial vectors is the vector represented by the diagonal of the parallelogram formed with the component vectors as adjacent sides.* eg : in the fig. below,

$$\vec{OC} = \vec{OB} + \vec{BC} = \vec{OB} + \vec{OA}$$

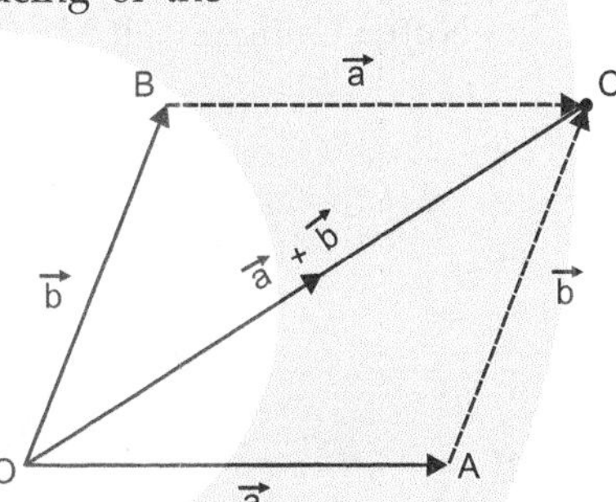

Note : Addition (or resultant) of more than two vectors can be found by repeated application of the triangle law. Eg. in fig. alongside vector AE is thus sum of given vectors and polygon ABCDE is a vector polygon.

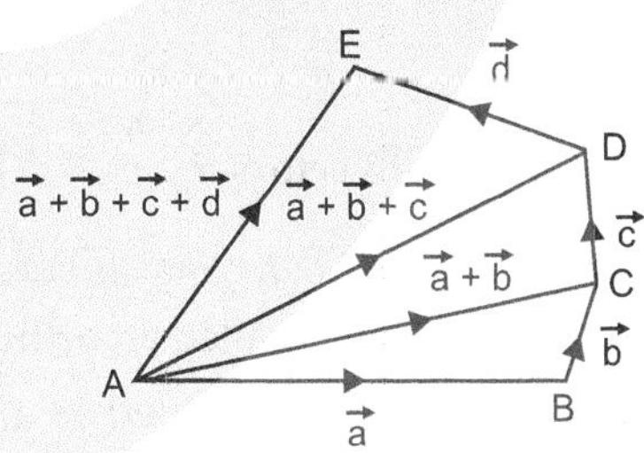

7. **Difference of two vectors :** $\vec{a} + (-\vec{b}) = \vec{a} - \vec{b}$

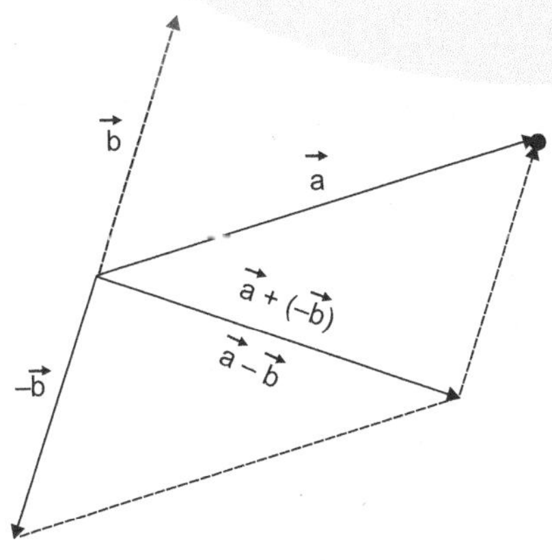

Note : To **subtract two vectors**, you put their feet (or tails, the non-pointy parts) together; then draw the resultant **vector**, which is the difference of the **two vectors**, from the head of the **vector** you're **subtracting** to the head of the **vector** you're **subtracting** it from.

8. **Position vector (p. v.) of a point** : The vector OP is said to be a position vector of P with respect to origin O, if P be any terminal point and O is the origin which is fixed.

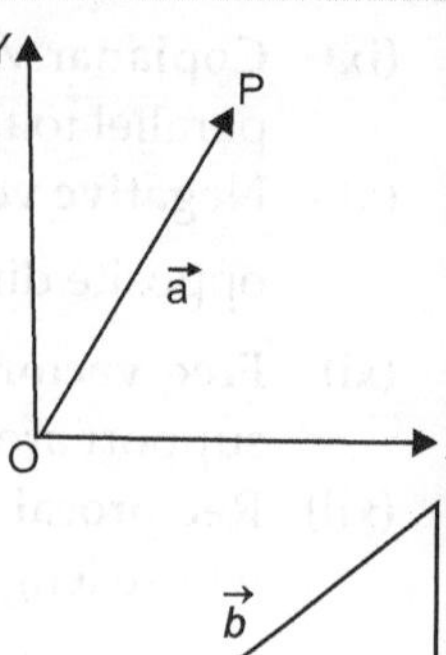

9. **Section Formula :**

(i) For internal division : Let $\vec{a}$ and $\vec{b}$ be the position vectors of two points A and B respectively and let C be a point with position vector $\vec{c}$ dividing AB internally in the ratio $m : n$. Then the position vector of C is

$$\vec{c} = \vec{OC} = \frac{m\vec{b} + n\vec{a}}{m+n}$$

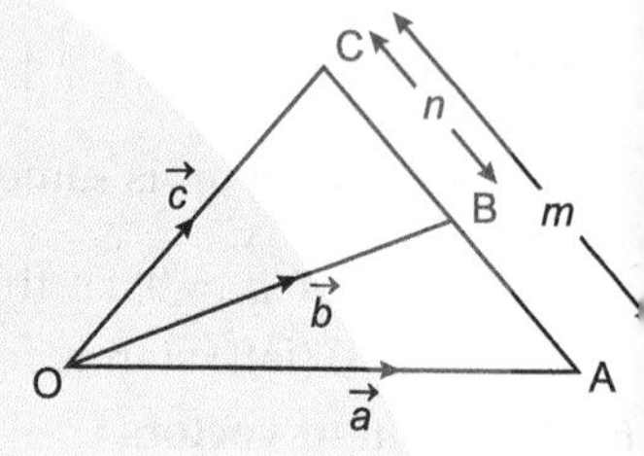

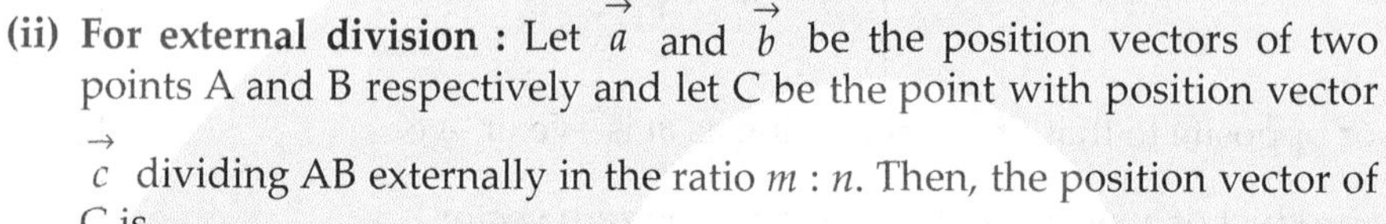

(ii) For external division : Let $\vec{a}$ and $\vec{b}$ be the position vectors of two points A and B respectively and let C be the point with position vector $\vec{c}$ dividing AB externally in the ratio $m : n$. Then, the position vector of C is

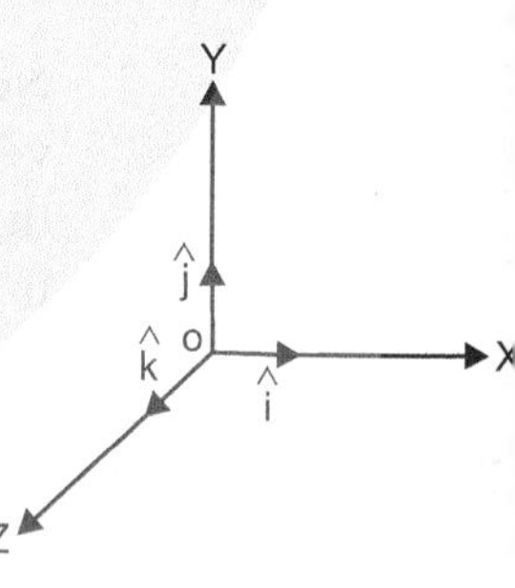

$$\vec{c} = \vec{OC} = \frac{m\vec{b} - n\vec{a}}{m+n}$$

(iii) Representation of a vector in terms of position vectors of its end points :

$\vec{PQ}$ = Position vector of Q – Position vector of P

Note : Using p.v., if M is the midpoint of PQ, then, $\vec{OM} = \frac{1}{2}(\vec{OP} + \vec{OQ})$

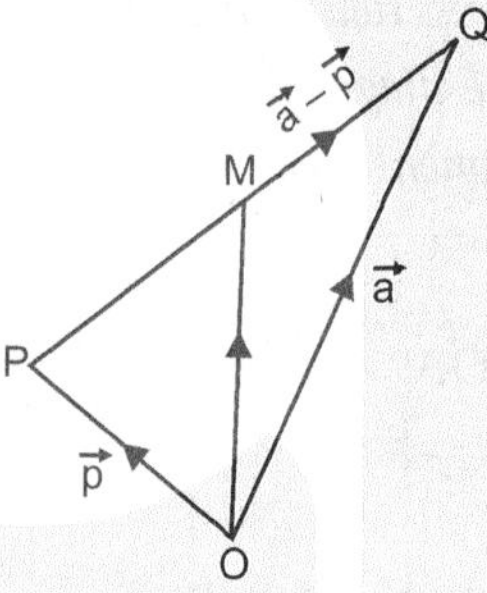

(iv) Unit vectors $\hat{i}, \hat{j}, \hat{k}$: Taking O as the origin, we denote unit vectors along x-axis(OX), y-axis(OY) and z-axis(OZ) as $\hat{i}$, $\hat{j}$ and $\hat{k}$ respectively. This system is called an **ortho-normal** system. Any vector along OX, OY, and OZ can be represented as $a\hat{i}$, $b\hat{j}$, $c\hat{k}$ respectively where a, b and c are scalars.

Note : If a point P in a plane XY has coordinates (x,y), then, $\vec{OP} = x\hat{i} + y\hat{j} + z$

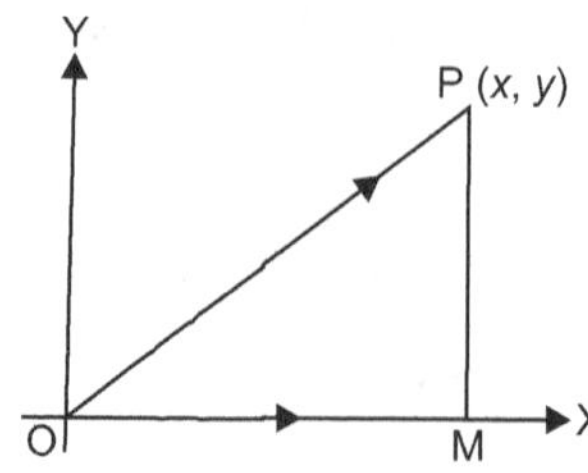

Similarly, in the 3-dimensional system, $\vec{OP} = \vec{r} = a\hat{j} + b\hat{j} + c\hat{k}$

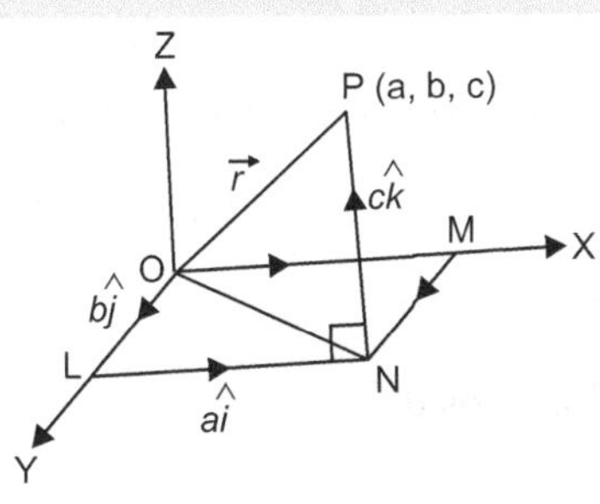

10. Components of a Vector in terms of the coordinates of its end points :

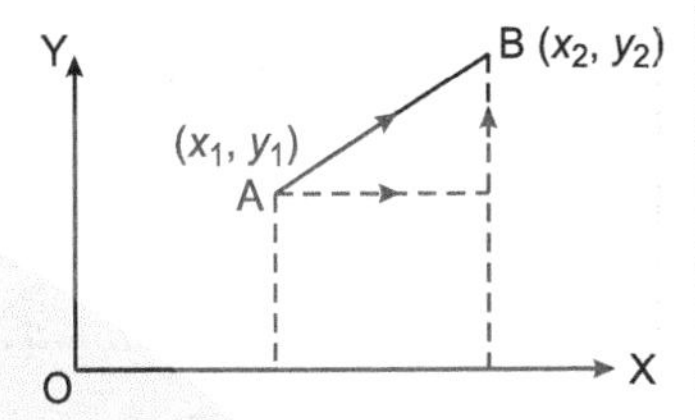

Consider A (x_1, y_1) and B (x_2, y_2) be any two points in XOY plane and let $\hat{i}$ and $\hat{j}$ be unit vectors along OX and OY respectively. Then,

$$|\vec{AB}| = \sqrt{(x_2 - x_1)^2 + (y_2 - y_1)^2}$$

(i) **Theorem** : Two vectors are equal if and only if their corresponding components along the axes are equal.

(ii) **Component (Projection) of a vector on a line** : The projection of AB on l is a vector of magnitude $|AB|\ |\cos\theta|$, in the direction of l or opposite to l according as cos θ is +ve or –ve.

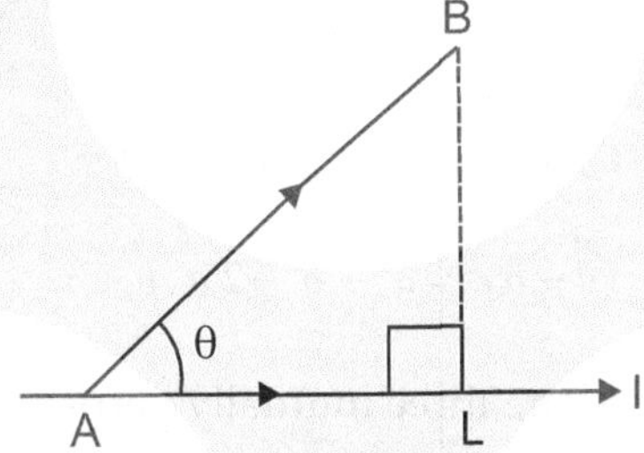

(iii) Modulus of the vector $a\hat{i} + b\hat{j} + c\hat{k} = \sqrt{(a^2 + b^2 + c^2)}$

(iv) If $a\hat{i} + b\hat{j} + c\hat{k} = 0$, then $a = b = c = 0$

(v) If two vectors are equal, then their resolved parts along the axes are also equal.

(vi) Sum of two vectors : $a\hat{i} + b\hat{j} + c\hat{k}$ and $p\hat{i} + q\hat{j} + r\hat{k} = (a + p)\hat{i} + (b + q)\hat{j} + (c + r)\hat{k}$

(vii) Scalar multiplication : $n(a\hat{i} + b\hat{j} + c\hat{k}) = an\hat{i} + bn\hat{j} + cn\hat{k}$

(viii) Additive inverse of vector $a\hat{i} + b\hat{i} + c\hat{k} = -(a\hat{i} + b\hat{j} + c\hat{k})$

Distance between two points : Let A $(x_1\hat{i} + y_1\hat{j} + z_1\hat{k})$ and B $(x_2\hat{i} + y_2\hat{j} + z_2\hat{k})$ be two points in space, then $AB^2 = (x_2 - x_1)^2 + (y_2 - y_1)^2 + (z_2 - z_1)^2$

11. Direction cosines of a vector $\vec{AB}$ *i.e.,* $a\hat{i} + b\hat{j} + c\hat{k}$ in terms of unit vectors $\hat{i}, \hat{j}, \hat{k}$ are

$$\frac{a}{\sqrt{a^2+b^2+c^2}}, \frac{b}{\sqrt{a^2+b^2+c^2}}, \frac{c}{\sqrt{a^2+b^2+c^2}}$$

and direction ratios of $\vec{AB}$ are proportional to a, b, c.

Condition for parallel vectors : For vectors A $(x_1\hat{i} + y_1\hat{j} + z_1\hat{k})$ and B $(x_2\hat{i} + y_2\hat{j} + z_2\hat{k})$ to be parallet, the ratio of the coefficients of $\hat{i}, \hat{j}, \hat{k}$ must be equal, *i.e.* $x_1/x_2 = y_1/y_2 = z_1/z_2$

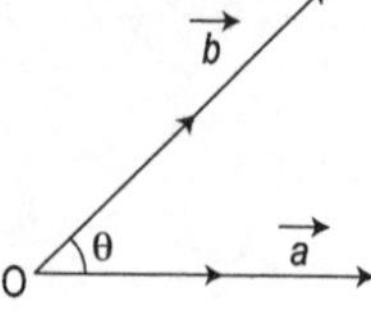

12. Scalar or dot product : Let $\vec{a}$ and $\vec{b}$ be two non-zero vectors inclined at an angle θ.

Then the scalar or dot product of $\vec{a}$ with $\vec{b}$ is denoted by $\vec{a}\cdot\vec{b}$ and is defined as the scalar $|\vec{a}|\,|\vec{b}|\cos\theta$ (where $0 \le \theta \le \pi$).

Thus, $$\vec{a}\cdot\vec{b} = |\vec{a}|\,|\vec{b}|\cos\theta$$

13. Properties of scalar product :

(i) $\vec{a}\cdot\vec{a} = |\vec{a}|^2 = \vec{a}^2$ (ii) $\vec{a}\cdot\vec{b} = \vec{b}\cdot\vec{a}$ (commutativity)

(iii) $\vec{a}\cdot 0 = 0$ (iv) $\vec{a}\cdot(\vec{b}+\vec{c}) = \vec{a}\cdot\vec{b} + \vec{a}\cdot\vec{c}$ (Distributivity)

(v) $\vec{a}\cdot(\vec{b}-\vec{c}) = \vec{a}\cdot\vec{b} - \vec{a}\cdot\vec{c}$

(vi) $\vec{a}\cdot\vec{b} = |\vec{a}|\,.\,|\vec{b}|\,.\cos\theta$

(vii) If x is any scalar, then

$$(x\vec{a})\cdot\vec{b} = x(\vec{a}\cdot\vec{b}) = \vec{a}\cdot(x\vec{b})$$

(viii) $(x\vec{a})\cdot(y\vec{b}) = xy(\vec{a}\cdot\vec{b})$ where x, y are scalars.

(ix)
$$(\vec{a}+\vec{b})\cdot(\vec{a}-\vec{b}) = |\vec{a}|^2 - |\vec{b}|^2$$
$$(\vec{a}+\vec{b})^2 = \vec{a}^2 + \vec{b}^2 + 2\,\vec{a}\cdot\vec{b}$$
$$(\vec{a}-\vec{b})^2 = \vec{a}^2 + \vec{b}^2 - 2\,\vec{a}\cdot\vec{b}$$

(x) If $\hat{i}, \hat{j}, \hat{k}$ are three unit vectors along three mutually perpendicular lines, then

(a) $\hat{i}\cdot\hat{i} = \hat{j}\cdot\hat{j} = \hat{k}\cdot\hat{k} = 1$

(b) $\hat{i}\cdot\hat{j} = \hat{j}\cdot\hat{k} = \hat{k}\cdot\hat{i} = \hat{j}\cdot\hat{i} = \hat{k}\cdot\hat{j} = \hat{i}\cdot\hat{k} = 0.$

$\hat{i}, \hat{j}, \hat{k}$ are said to form orthonormal basis.

(xi) If OX, OY, OZ are three mutually perpendicular axes and $\hat{i}, \hat{j}, \hat{k}$ are unit vectors along these axes. If co-ordinates of any point P are (x, y, z) then its position vector is $x\hat{i} + y\hat{j} + z\hat{k}$.

14. Scalar product in terms of components : Let $\vec{a} = a_1\hat{i} + a_2\hat{j} + a_3\hat{k}$ and $\vec{b} = b_1\hat{i} + b_2\hat{j} + b_3\hat{k}$. Then by using the distributivity of dot product over vector addition, we get

$$\vec{a}\cdot\vec{b} = a_1b_1 + a_2b_2 + a_3b_3.$$

Thus, the scalar product of two vectors is equal to the sum of the products of their corresponding components.

If θ be the angle between two vector, $\vec{a} = a_1\hat{i} + a_2\hat{j} + a_3\hat{k}$

and $\vec{b} = b_1\hat{i} + b_2\hat{j} + b_3\hat{k}$

then, $\vec{a}\cdot\vec{b} = a_1b_1 + a_2b_2 + a_3b_3$

and $|\vec{b}| = \sqrt{b_1^2 + b_2^2 + b_3^2}$

$$\therefore \quad \cos\theta = \frac{\vec{a}\cdot\vec{b}}{|\vec{a}|\,|\vec{b}|} = \frac{a_1b_1 + a_2b_2 + a_3b_3}{\sqrt{a_1^2+a_2^2+a_3^2}\,\sqrt{b_1^2+b_2^2+b_3^2}}$$

5. Components of a vector $\vec{b}$ along and perpendicular to vector $\vec{a}$:

The components of $\vec{b}$ along and perpendicular to $\vec{a}$ are $\left(\frac{\vec{a}.\vec{b}}{|\vec{a}|^2}\right)\vec{a}$ and $\vec{b}-\left(\frac{\vec{a}.\vec{b}}{|\vec{a}|^2}\right)\vec{a}$ respectively.

6. Vector or Cross Product : Let $\vec{a}$ and $\vec{b}$ be two non-zero non-parallel vectors. Then the vector product $\vec{a}\times\vec{b}$ in that order, is defined as a vector whose magnitude is $|\vec{a}||\vec{b}|\sin\theta$, where θ is the angle between $\vec{a}$ and $\vec{b}$ and whose direction is perpendicular to the plane of $\vec{a}$ and $\vec{b}$ in such a way that $\vec{a}$, $\vec{b}$ and this direction constitute a right handed system.

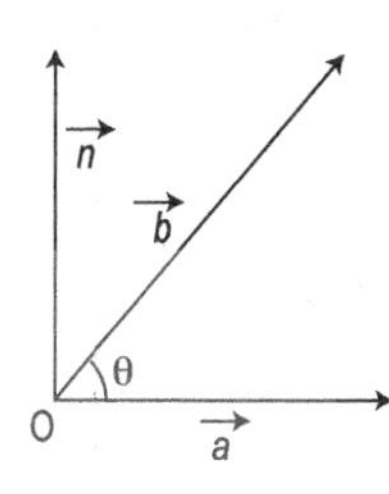

In other words $\vec{a}\times\vec{b} = |\vec{a}||\vec{b}|\sin\theta\,\hat{n}$, where θ is the angle between $\vec{a}$ and $\vec{b}$.

(where $0 \le \theta \le \pi$) and $\hat{n}$ is a unit vector perpendicular to the plane of $\vec{a}$ and $\vec{b}$ such that $\vec{a}$, $\vec{b}$, $\hat{n}$ form a right handed system.

When we say that $\vec{a}$, $\vec{b}$, $\hat{n}$ form a right handed system it means that if we rotate vector $\vec{a}$ into the vector $\vec{b}$, then $\hat{n}$ will point in the direction perpendicular to the plane of $\vec{a}$ and $\vec{b}$ in which a right handed screw will move if it turned in the same manner.

In particular, if either $\vec{a}=\vec{0}$ or $\vec{b}=\vec{0}$ we define $\vec{a}\times\vec{b}=\vec{0}$. From the definition, it follows that

(i) If $\vec{a}$ and $\vec{b}$ be unit vectors, then $|\vec{a}\times\vec{b}| = \sin\theta$. Thus, we find that the modulus of the vector product of two unit vectors is the sign of the angle between their directions.

(ii) If $\vec{a}$ and $\vec{b}$ are perpendiculars, the three vectors $\vec{a}$, $\vec{b}$ and $\vec{a}\times\vec{b}$ form a right handed triad of mutually perpendicular vectors.

(iii) *Condition of parallelism of two vectors :* The two non-zero (proper vectors) $\vec{a}$ and $\vec{b}$ are parallel if $\vec{a}\times\vec{b}=\vec{0}$.

(iv) If $\vec{a}=\vec{b}$ then $\theta = 0$, *i.e.,* $\sin\theta = 0$, and we get $\vec{a}\times\vec{a}=\vec{0}$.

7. Properties of vector product :

(i) $\vec{a}\times\vec{b} = -\vec{b}\times\vec{a}$ (non-commutative).

(ii) $\vec{a}\times\vec{a} = 0$.

(iii) $x(\vec{a}\times\vec{b}) = (x\vec{a})\times\vec{b} = \vec{a}\times(x\vec{b})$ where x is any scalar.

(iv) $(x\vec{a})\times(y\vec{b}) = xy(\vec{a}\times\vec{b})$.

(v) $(\vec{a}\times\vec{b})^2 = |\vec{a}|^2|\vec{b}|^2 - (\vec{a}.\vec{b})^2$.

(vi) $\hat{i}\times\hat{i} = \hat{j}\times\hat{j} = \hat{k}\times\hat{k} = 0$.

(vii) $\hat{i}\times\hat{j} = \hat{k}, \hat{j}\times\hat{k} = \hat{i}, \hat{k}\times\hat{i} = \hat{j}$.

(viii) $\hat{j}\times\hat{i} = -\hat{k},\ \hat{k}\times\hat{j} = -\hat{i},\ \hat{i}\times\hat{k} = -\hat{k}$.

(ix) Two non-zero vectors and are collinear if $x = 0$.

(x) If $\vec{a} = a_1\hat{i} + a_2\hat{j} + a_3\hat{k}$

and $\vec{b} = b_1\hat{i} + b_2\hat{j} + b_3\hat{k}$

then, $$\vec{a} \times \vec{b} = \begin{vmatrix} \hat{i} & \hat{j} & \hat{k} \\ a_1 & a_2 & a_3 \\ b_1 & b_2 & b_3 \end{vmatrix}$$

$$= (a_2b_3 - a_3b_2)\hat{i} - (a_1b_3 - a_3b_1)\hat{j} + (a_1b_2 - a_2b_1)\hat{k}$$

(xi) Vector area of a parallelogram whose adjacent sides are represented by vectors

$$\vec{OA} = \vec{a} \text{ and } \vec{OB} = \vec{b} \text{ is } \vec{a} \times \vec{b}.$$

$$\text{Area of parallelogram} = |\vec{a} \times \vec{b}|$$

$$\text{Area of triangle} = \frac{1}{2}|\vec{a} \times \vec{b}|.$$

(xii) A vector of magnitude λ, perpendicular to the plane of $\vec{a}$ and is $\lambda \dfrac{\vec{a} \times \vec{b}}{|\vec{a} \times \vec{b}|}$ and unit vecto

perpendicular to the plane of $\vec{a}$ and $\vec{b}$ is $\dfrac{\vec{a} \times \vec{b}}{|\vec{a} \times \vec{b}|}$.

(xiii) If $\vec{a}, \vec{b}, \vec{c}$ are three vectors, then

$$\vec{a} \times (\vec{b} + \vec{c}) = \vec{a} \times \vec{b} + \vec{a} \times \vec{c}$$ {Distributive law

and $$(\vec{b} + \vec{c}) \times \vec{a} = \vec{b} \times \vec{a} + \vec{c} \times \vec{a}.$$

18. Vector area of a triangle :

(i) The vector area of a triangle ABC is equal to $\frac{1}{2}\vec{AB} \times \vec{AC}$.

(ii) The vector area of a triangle ABC with vertices having position vectors $\vec{a}, \vec{b}, \vec{c}$ respectively is

$$\frac{1}{2}\cdot[\vec{b} \times \vec{c} + \vec{c} \times \vec{a} + \vec{a} \times \vec{b}]$$

19. Collinear Vector : If the points whose position vectors are $\vec{a}, \vec{b}, \vec{c}$ are collinear then the vector area o triangle formed by these points must be zero.

i.e. $$\vec{a} \times \vec{b} + \vec{b} \times \vec{c} + \vec{c} \times \vec{a} = 0$$

which is the required condition of collinearity.

20. Scalar triple product : Let $\vec{a}, \vec{b}, \vec{c}$ be three vectors, then the scalar $(\vec{a} \times \vec{b}).\vec{c}$ is called the scalar tripl product of $\vec{a}$, $\vec{b}$ and $\vec{c}$ and is denoted by $[\vec{a}\,\vec{b}\,\vec{c}]$.

Thus, $$[\vec{a}\,\vec{b}\,\vec{c}] = (\vec{a} \times \vec{b}).\vec{c}$$

Remark :

(i) The scalar triple product is a scalar quantity.

(ii) The scalar triple product $(\vec{a} \times \vec{b}).\vec{c}$ is also denoted by the symbol $[\vec{a}\,\vec{b}\,\vec{c}]$ or $[\vec{a}, \vec{b}, \vec{c}]$.

Chapter at a Glance

21. Geometrical interpretation of scalar triple product (volume of parallelopiped) : The scalar triple product is the volume of the parallelopiped whose adjacent sides are represented by vectors $\vec{a}$, $\vec{b}$ and $\vec{c}$ is $(\vec{a} \times \vec{b}).\vec{c}$ or $[\vec{a}, \vec{b}, \vec{c}]$.

22. Properties of scalar triple product :

(i) For an orthonormal right-handed vector triad $\hat{i}, \hat{j}, \hat{k}$

$$[\hat{i}, \hat{j}, \hat{k}] = [\hat{i} \times \hat{j}].\hat{k} = \hat{k}.\hat{k} = 1.$$

Note : The scalar triple product of the orthonormal right handed unit vectors is equal to unity.

(ii) In a scalar triple product the position of dot and cross can be interchanged provided the cyclic order of the factors is maintained.

$$(\vec{a} \times \vec{b}).\vec{c} = (\vec{b} \times \vec{c}).\vec{a} = (\vec{c} \times \vec{a}).\vec{b}.$$

Note : The change of cyclic order of factors brings about a change of sign in the value of the scalar triple product.

$$[\vec{a}\,\vec{b}\,\vec{c}] = [\vec{b}\,\vec{c}\,\vec{a}] = [\vec{c}\,\vec{a}\,\vec{b}] = V$$

and

$$[\vec{a}\,\vec{c}\,\vec{b}] = [\vec{b}\,\vec{a}\,\vec{c}] = [\vec{c}\,\vec{b}\,\vec{a}] = -V.$$

(iii) **Condition of coplanarity of three vectors :** If $\vec{a}, \vec{b}, \vec{c}$ are coplanar, then $\vec{b} \times \vec{c}$, being the vector perpendicular to the plane of $\vec{b}$ and $\vec{c}$ is also perpendicular to the vector $\vec{a}$.

∴ The scalar product of $\vec{a}$ and $(\vec{b} \times \vec{c})$ must be zero .

∴ $\vec{a}.(\vec{b} \times \vec{c}) = 0.$

or $[\vec{a}\,\vec{b}\,\vec{c}] = 0.$

Note : (a) The converse is also true.

(b) If three vectors are coplanar, their scalar triple product is zero.

(iv) **Scalar triple product when two of the vectors are equal :** If $\vec{a}, \vec{a}, \vec{b}$ be the three vectors.

∴ $[\vec{a}\,\vec{a}\,\vec{b}] = (\vec{a} \times \vec{a}).\vec{b} = 0$ $[\because \vec{a} \times \vec{a} = 0]$

Note : If two of the vectors are equal, the scalar triple product *vanishes*.

(v) If two of the vectors are parallel, the scalar triple product is zero.

(vi) **To express the scalar triple product $[\vec{a}\,\vec{a}\,\vec{b}]$ in terms of orthonormal triad of vectors $\hat{i}$, $\hat{j}$ and $\hat{k}$:**

If,

$$\vec{a} = a_1 \hat{i} + a_2 \hat{j} + a_3 \hat{k}$$

$$\vec{b} = b_1 \hat{i} + b_2 \hat{j} + b_3 \hat{k}$$

$$\vec{c} = c_1 \hat{i} \mid c_2 \hat{j} + c_3 \hat{k}$$

Then,

$$[\vec{a}\,\vec{b}\,\vec{c}] = \begin{vmatrix} a_1 & a_2 & a_3 \\ b_1 & b_2 & b_3 \\ c_1 & c_2 & c_3 \end{vmatrix}$$

(vii) Volume of tetrahedron ABCD is $\frac{1}{6} | \vec{AB}.(\vec{AC} \times \vec{AD}) |$.

(viii) If k is a scalar, then

$$[k\vec{a}, \vec{b}, \vec{c}] = k\,[\vec{a}, \vec{b}, \vec{c}]$$

(ix) $$[\vec{a}+\vec{b}, \vec{c}, \vec{d}] = [\vec{a}\,\vec{c}\,\vec{d}] + [\vec{b}\,\vec{c}\,\vec{d}].$$

(x) The vectors $\vec{a}, \vec{b}, \vec{c}$ form a right-handed screw system if $[\vec{a}, \vec{b}, \vec{c}] > 0$ and left handed screw system if $[\vec{a}, \vec{b}, \vec{c}] < 0$.

(xi) $$[\vec{a}, \vec{b}, \vec{c}]\,[\vec{a}', \vec{b}', \vec{c}'] = \begin{vmatrix} \vec{a}.\vec{a}' & \vec{b}.\vec{a}' & \vec{c}.\vec{a}' \\ \vec{a}.\vec{b}' & \vec{b}.\vec{b}' & \vec{c}.\vec{b}' \\ \vec{a}.\vec{c}' & \vec{b}.\vec{c}' & \vec{c}.\vec{c}' \end{vmatrix}$$

Chapter 12. Three Dimensional Geometry

1. **Origin :** If X′OX, Y′OY, Z′OZ the three mutually perpendicular fixed straight lines (whose positive directions are X′OX, Y′OY, Z′OZ as marked by arrows), intersect at O, then O is called *origin.*
2. **Co-ordinates axes :**
   (i) X′OX is called the x-axis and the equation of the x-axis are $y = 0, z = 0$.
   (ii) Y′OY is called the y-axis and the equation of the y-axis are $z = 0, x = 0$.
   (iii) Z′OZ is called the z-axis and the equation of the z-axis are $x = 0, y = 0$.
   These three axis taken together are called the co-ordinate axis.
   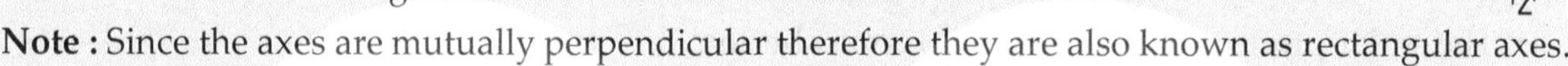
   Note : Since the axes are mutually perpendicular therefore they are also known as rectangular axes.
3. **Co-ordinate planes :**
   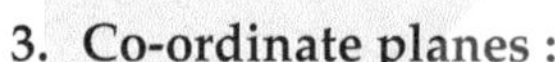
   (i) XOY is called the xy-plane and the equation of the xy-plane is $z = 0$.
   (ii) YOZ is called the yz-plane and the equation of the yz-plane is $x = 0$.
   (iii) ZOX is called the zx-plane and the equation of the zx-plane is $y = 0$.
   These three planes taken together are called co-ordinate planes.
4. **Co-ordinates :** The co-ordinates of any point are its perpendicular distances from the co-ordinate planes. The co-ordinates are positive if measured along OX, OY, OZ and negative if measured along OX′, OY′, OZ′ respectively. Co-ordinates of origin are naturally (0, 0, 0).
5. **Distance formula :** The distance 'd' between the two points A (x_1, y_1, z_1) and B (x_2, y_2, z_2) is

$$d = \sqrt{(x_2 - x_1)^2 + (y_2 - y_1)^2 + (z_2 - z_1)^2}$$

   Aid to memory : $d = \sqrt{(\text{diff. of } x\text{-co-ordinate})^2 + (\text{diff. of } y\text{-co-ordinate})^2 + (\text{diff. of } z\text{-co-ordinate})^2}$

   Corollary : *Distance of a point from the origin :* The distance from a point (x_1, y_1, z_1) from the origin (0, 0, 0)

$$= \sqrt{x_1^2 + y_1^2 + z_1^2}$$

6. **Section formula :** If A (x_1, y_1, z_1) and B (x_2, y_2, z_2) be two given points. A point P (x, y, z) which divides AB *internally* in the ratio of $m_1 : m_2$ is given by

$$x = \frac{m_1x_2 + m_2x_1}{m_1 + m_2}$$

$$y = \frac{m_1y_2 + m_2y_1}{m_1 + m_2}$$

$$z = \frac{m_1z_2 + m_2z_1}{m_1 + m_2}$$

If it divides *externally* in the ratio of $m_1 : m_2$, then

$$x = \frac{m_1x_2 - m_2x_1}{m_1 - m_2}$$

$$y = \frac{m_1y_2 - m_2y_1}{m_1 - m_2}$$

$$z = \frac{m_1z_2 - m_2z_1}{m_1 - m_2}.$$

Corollary : The co-ordinates of mid-point of the line joining (x_1, y_1, z_1) and (x_2, y_2, z_2) is

$$\left[\frac{x_1+x_2}{2}, \frac{y_1+y_2}{2}, \frac{z_1+z_2}{2}\right].$$

7. **Centroid formula of a triangle :** *Centroid of a triangle* is the point which divides all the medians in the ratio of 2 : 1.

The co-ordinates of the centroid of a triangle whose vertices are (x_1, y_1, z_1), (x_2, y_2, z_2), (x_3, y_3, z_3) is given as,

$$\left[\frac{x_1+x_2+x_3}{3}, \frac{y_1+y_2+y_3}{3}, \frac{z_1+z_2+z_3}{3}\right].$$

8. **Direction cosines :** If α, β, γ are the angles which a vector OP makes with the positive direction of the co-ordinate axes OX, OY, OZ respectively, then cos α, cos β, cos γ are known as the direction cosines of $\overrightarrow{OP}$ and are generally denoted by the letters l, m, n respectively *i.e.*, $l = \cos\alpha$, $m = \cos\beta$, $n = \cos\gamma$.

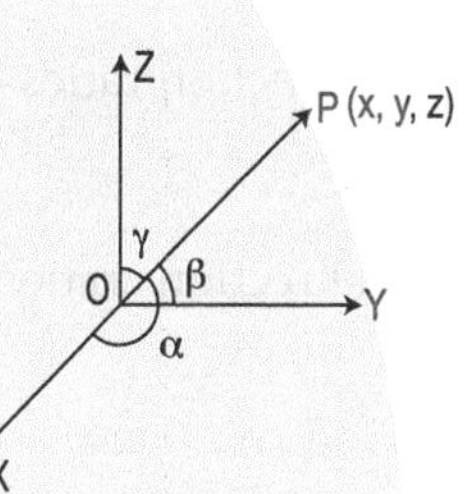

The angle α, β, γ are known as the direction angles and satisfy the condition that $0 \le \alpha, \beta, \gamma \le \pi$.

As the x-axis makes angle 0°, π/2, π/2 with OX, OY and OZ respectively. Therefore, direction cosines of x-axis are cos 0°, cos π/2, cos π/2 *i.e.*, 1, 0, 0. Similarly the direction cosines of y and z-axes are 0, 1, 0 and 0, 0, 1 respectively.

9. **Direction ratios :** A set of three numbers proportional to direction cosines are called *direction ratios or direction numbers.*

Let (l, m, n) be direction cosines of a line and direction ratios of the line be (a, b, c). Then

$$\frac{l}{a} = \frac{m}{b} = \frac{n}{c} = \frac{\sqrt{l^2+m^2+n^2}}{\sqrt{a^2+b^2+c^2}} = \pm\frac{1}{\sqrt{a^2+b^2+c^2}}$$

Thus,

$$l = \pm\frac{a}{\sqrt{a^2+b^2+c^2}}$$

$$m = \pm\frac{b}{\sqrt{a^2+b^2+c^2}}$$

$$n = \pm\frac{c}{\sqrt{a^2+b^2+c^2}}$$

10. **Direction ratios of the line joining two points :** P (x_1, y_1, z_1) and Q (x_2, y_2, z_2) are $x_2 - x_1$, $y_2 - y_1$, $z_2 - z_1$ and direction cosines are $\frac{x_2-x_1}{PQ}, \frac{y_2-y_1}{PQ}, \frac{z_2-z_1}{PQ}$ respectively.

11. **Angle between two lines :** If θ be the angle between two lines having direction cosines (l_1, m_1, n_1) and (l_2, m_2, n_2) respectively. Then

$$\cos\theta = l_1l_2 + m_1m_2 + n_1n_2.$$

Also

$$\sin^2\theta = (m_1n_2 - m_2n_1)^2 + (n_1l_2 - n_2l_1)^2 + (m_1l_2 - l_1m_2)^2$$

$$= \Sigma\,(l_1m_2 - l_2m_1)^2.z$$

12. Angle between two lines in vector form : Let the vector equations of the two lines be $\vec{r} = \vec{a_1} + \lambda \vec{b_1}$ and $\vec{r} = \vec{a_2} + \mu \vec{b_2}$. These two lines are parallel to the vectors $\vec{b_1}$ and $\vec{b_2}$ respectively. Therefore, angle between these two lines are equal to the angle between $\vec{b_1}$ and $\vec{b_2}$. Thus, if θ is the angle between the given lines then

$$\cos\theta = \frac{\vec{b_1} \cdot \vec{b_2}}{|\vec{b_1}||\vec{b_2}|}$$

(i) **Condition of perpendicularity :** If the lines are perpendicular, then $\vec{b_1} \cdot \vec{b_2} = 0$.

(ii) **Condition of parallelism :** If the lines are parallel, then $\vec{b_1}$ and $\vec{b_2}$ are parallel, therefore $\vec{b_1} = \lambda \vec{b_2}$ for some scalar λ.

13. Angle between two lines in cartesian form : Let the Cartesian equations of the two lines be,

$$\frac{x-x_1}{a_1} = \frac{y-y_1}{b_1} = \frac{z-z_1}{c_1} \quad \ldots\text{(i)}$$

and
$$\frac{x-x_2}{a_2} = \frac{y-y_2}{b_2} = \frac{z-z_2}{c_2} \quad \ldots\text{(ii)}$$

Direction ratios of (i) are a_1, b_1, c_1. Therefore,

$$\vec{m_1} = \text{vector parallel to (i)} = a_1\hat{i} + b_1\hat{j} + c_1\hat{k}$$

Directions ratios of (ii) are a_2, b_2, c_2. Therefore,

$$\vec{m_2} = \text{vector parallel to (ii)} = a_2\hat{i} + b_2\hat{j} + c_2\hat{k}$$

Let θ be the angle between (i) and (ii). Then θ is also the angle between $\vec{m_1}$ and $\vec{m_2}$. Therefore,

$$\cos\theta = \frac{\vec{m_1} \cdot \vec{m_2}}{|\vec{m_1}||\vec{m_2}|}$$

$$\Rightarrow \quad \cos\theta = \frac{a_1a_2 + b_1b_2 + c_1c_2}{\sqrt{a_1^2 + b_1^2 + c_1^2}\sqrt{a_2^2 + b_2^2 + c_2^2}}$$

(i) **Condition of perpendicularity :** If the lines are perpendicular, then

$$\vec{m_1} \cdot \vec{m_2} = 0 \Rightarrow = a_1a_2 + b_1b_2 + c_1c_2 = 0$$

(ii) **Condition of parallelism :** If the lines are parallel, then $\vec{m_1}$ and $\vec{m_2}$ are parallel. Therefore

$$\vec{m_1} = \lambda \vec{m_2} \quad \text{for same scalar } \lambda$$

$$\Rightarrow \quad \frac{a_1}{a_2} = \frac{b_1}{b_2} = \frac{c_1}{c_2}.$$

14. Lagrange's identity :

$$(l_1^2 + m_1^2 + n_1^2)(l_2^2 + m_2^2 + n_2^2) - (l_1l_2 + m_1m_2 + n_1n_2)^2$$
$$= (l_1m_2 - l_2m_1)^2 + (m_1n_2 - m_2n_1)^2 + (n_1l_2 - n_2l_1)^2$$
$$= \Sigma (l_1m_2 - l_2m_1)^2$$

15. Conditions for two lines to be (*a*) perpendicular (*b*) parallel :

(a) $l_1l_2 + m_1m_2 + n_1n_2 = 0$

or, $a_1a_2 + b_1b_2 + c_1c_2 = 0$

(b) $\frac{l_1}{l_2} = \frac{m_1}{m_2} = \frac{n_1}{n_2}$

Chapter at a Glance

or $$\frac{a_1}{a_2} = \frac{b_1}{b_2} = \frac{c_1}{c_2}$$

16. Perpendicular distance of a given point from a given line :

Cartesian form : To find the perpendicular distance of a given point (α, β, γ) from a given line

$$\frac{x - x_1}{a} = \frac{y - y_1}{b} = \frac{z - z_1}{c}.$$

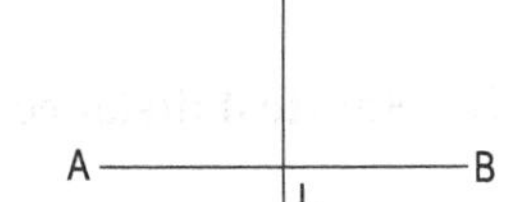

Let L be the foot of the perpendicular drawn from P (α, β, γ) on the line $\frac{x - x_1}{a} = \frac{y - y_1}{b} = \frac{z - z_1}{c}$.

Let the co-ordinates of L be $(x_1 + a\lambda, y_1 + b\lambda, z_1 + c\lambda)$. Then direction ratios of PL are

$$x_1 + a\lambda - \alpha, y_1 + b\lambda - \beta, z_1 + c\lambda - \gamma.$$

Direction ratio of AB are *a, b, c*. Since PL is perpendicular to AB, therefore

$(x_1 + a\lambda - \alpha)\, a + (y_1 + b\lambda - \beta)\, b + (z_1 + c\lambda - \gamma)\, c = 0$

$$\Rightarrow \quad \lambda = \frac{a(\alpha - x_1) + b(\beta - y_1) + c(\gamma - z_1)}{a^2 + b^2 + c^2}$$

Putting this value of λ in $(x_1 + a\lambda, y_1 + b\lambda, z_1 + c\lambda)$, we obtain co-ordinates of L. Now, using distance formula, we can obtain the length PL.

17. Reflection or image of a point in a straight line : If the perpendicular PL from P on the given line be produced to Q such that PL = QL, then Q is known as the image or reflection of P in the given line. Also, L is the foot of the perpendicular or the projection of P on the line.

Cartesian form : To find the reflection or image of a point in a straight line in Cartesian form.

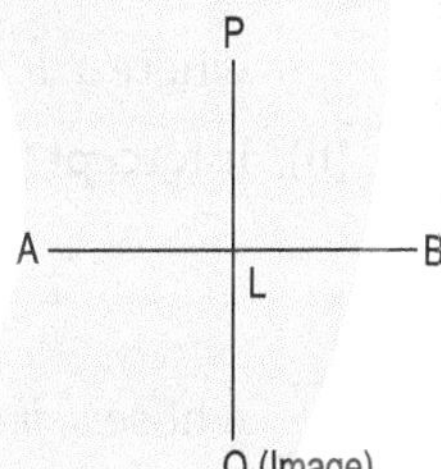

Let P (α, β, γ) be the given point and let = $\frac{x - x_1}{a} = \frac{y - y_1}{b} = \frac{z - z_1}{c}$ be the given line. Let L be the foot of the perpendicular from P on the given line, and let Q be the image of P in the given line. Let the co-ordinates of L be $(x_1 + a\lambda, y_1 + b\lambda, z_1 + c\lambda)$. Then direction ratios of PL are

$$x_1 + a\lambda - \alpha, y_1 + b\lambda - \beta, z_1 + c\lambda - \gamma.$$

Since, PL is perpendicular to the given line, whose direction ratios are *a, b, c*. Therefore

$$(x_1 + a\lambda - \alpha)\, a + (y_1 + b\lambda - \beta)\, b + (z + c\lambda - \gamma)\, c = 0$$

$$\Rightarrow \quad \lambda = -\frac{[a(x_1 - \alpha) + b(y_1 - \beta) + c(z_1 - \gamma)]}{a^2 + b^2 + c^2}$$

Substituting the value of λ in $(x_1 + a\lambda, y_1 + b\lambda, z_1 + c\lambda)$ we obtain co-ordinates of L. Let the co-ordinates of Q be (α', β', γ'). Since, L is the mid-point of PQ, therefore,

$$\frac{\alpha + \alpha'}{2} = x_1 + a\lambda, \quad \frac{\beta + \beta'}{2} = y_1 + b\lambda, \quad \frac{\gamma + \gamma'}{2} = z_1 + c\lambda$$

$$\Rightarrow \quad \alpha' = 2(x_1 + a\lambda) - \alpha, \beta' = 2(y_1 + b\lambda) - \beta, \gamma' = 2(z_1 + c\lambda) - \gamma.$$

18. Coplanar and Skew Lines : If two straight lines are either parallel or intersecting they are said to be coplanar.

Two straight lines are skew if they are neither parallel nor intersecting.

Note : Two skew lines are always non-coplanar.

19. Conditions for intersection of two lines : The two straight lines = $\frac{x - x_1}{a_1} = \frac{y - y_1}{b_1} = \frac{z - z_1}{c_1}$ and $\frac{x - x_2}{a_2} = \frac{y - y_2}{b_2} = \frac{z - z_2}{c_2}$ intersect (or coplanar)

Chapter at a Glance

if $a_1 : b_1 : c_1 \neq a_2 : b_2 : c_2$

and
$$\begin{vmatrix} x_2 - x_1 & y_2 - y_1 & z_2 - z_1 \\ a_1 & b_1 & c_1 \\ a_2 & b_2 & c_2 \end{vmatrix} = 0.$$

20. Shortest distance between two skew lines : The shortest distance between two skew lines

$$\frac{x - x_1}{l_1} = \frac{y - y_1}{m_1} = \frac{z - z_1}{n_1} \text{ and } \frac{x - x_2}{l_2} = \frac{y - y_2}{m_2} = \frac{z - z_2}{n_2}$$

is
$$\frac{\begin{vmatrix} x_2 - x_1 & y_2 - y_1 & z_2 - z_1 \\ l_1 & m_1 & n_1 \\ l_2 & m_2 & n_2 \end{vmatrix}}{\sqrt{(m_1 n_2 - m_2 n_1)^2 + (n_1 l_2 - l_1 n_2)^2 + (l_1 m_2 - l_2 m_1)^2}}$$

Remark : (i) The shortest distance between two intersecting lines is zero

i.e.,
$$\begin{vmatrix} x_2 - x_1 & y_2 - y_1 & z_2 - z_1 \\ l_1 & m_1 & n_1 \\ l_2 & m_2 & n_2 \end{vmatrix} = 0.$$

21. The Plane : A plane is a surface such that line joining any two points on the surface lies wholly on the surface.

Any first degree equation in x, y, z represents a plane.

Different forms of equation of a plane are :

(i) General form of equation of plane is

$$ax + by + cz + d = 0$$

where a, b, c, d are constants and not all are zero.

(ii) Intercept form is

$$\frac{x}{a} + \frac{y}{b} + \frac{z}{c} = 1$$

whose a, b, c are intercepts on co-ordinates axes.

where l, m and n are direction cosine.

(iii) The vector equation of a plane normal to unit vector $\hat{n}$ and at a distance d from the origin is $\vec{r} \cdot \hat{n} = d$.

(iv) Equation of any plane through a given point (x_1, y_1, z_1) is :

$a(x - x_1) + b(y - y_1) + c(z - z_1) = 0.$

(v) Equation of a plane through three given points (x_1, y_1, z_1), (x_2, y_2, z_2), (x_3, y_3, z_3) is

$$\begin{vmatrix} x - x_1 & y - y_1 & z - z_1 \\ x_2 - x_1 & y_2 - y_1 & z_2 - z_1 \\ x_3 - x_1 & y_3 - y_1 & z_3 - z_1 \end{vmatrix} = 0.$$

22. Intersection of a straight line and plane : The point of intersection of the line

$$\frac{x - x_1}{l} = \frac{y - y_1}{m} = \frac{z - z_1}{n} = 0 \quad \text{...(i)}$$

and the plane, $\quad ax + by + cz + d = 0 \quad$ ...(ii)

Any point on the line (i) is given by $(x_1 + lr, y_1 + mr, z_1 + nr)$ ...(iii)

It lies on the plane (ii), if

$$a(x_1 + lr) + b(y_1 + mr) + c(z_1 + nr) + d = 0$$

If, $\quad r[al + bm + cn] + [ax_1 + by_1 + cz_1 + d] = 0 \quad$ ...(iv)

If $r = -\left(\dfrac{ax_1 + by_1 + cz_1 + d}{al + bm + cn}\right)$, putting this value of r in (iii) we get the required point of intersection.

3. Distance of a point from a plane :

Vector form :

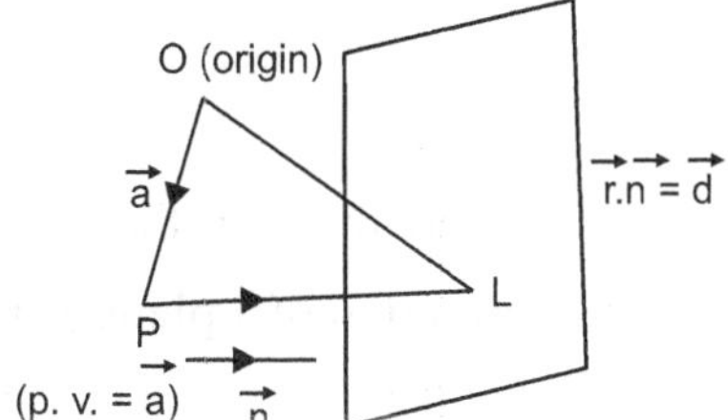

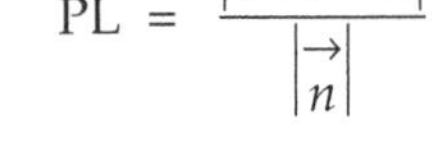

$$PL = \frac{|\vec{a}.\vec{n} - d|}{|\vec{n}|}$$

Cor : Perpendicular distance of origin :

$$= \frac{|0.\vec{n} - d|}{|\vec{n}|} = \frac{|d|}{|\vec{n}|}$$

Cartesian form : The length of the perpendicular from a point $P\ (x_1, y_1, z_1)$ to the plane $ax + by + cz + d = 0$ is given by

$$\frac{|ax_1 + by_1 + cz_1 + d|}{\sqrt{a^2 + b^2 + c^2}}.$$

4. Equation of a plane passing through a given point and parallel to two given vectors :

(i) Parametric form : The equation of the plane passing through a point having position vector $\vec{a}$ and parallel to vectors $\vec{b}$ and $\vec{c}$ is $\vec{r} = \vec{a} + \lambda\vec{b} + \mu\vec{c}$, where λ and μ are scalars.

(ii) Non-Parametric form : The equation of the plane passing through a point having position vector $\vec{a}$ and parallel to vectors $\vec{b}$ and $\vec{c}$ is $(\vec{r} - \vec{a}).(\vec{b} \times \vec{c}) = 0$ or, $\vec{r}.(\vec{b} \times \vec{c}) = \vec{a}.(\vec{b} \times \vec{c})$ or $[\vec{r}\ \vec{b}\ \vec{c}] = [\vec{a}\ \vec{b}\ \vec{c}]$

5. Angle between two planes in vector form : The angle between two planes is defined as the angle between their normals.

The angle θ between the planes $\vec{r}.\vec{n}_1 = d_1$ and $\vec{r}.\vec{n}_2 = d_2$ is given by

$$\cos\theta = \frac{\vec{n}_1 . \vec{n}_2}{|\vec{n}_1|.|\vec{n}_2|}$$

Thus, the planes $\vec{r}.\vec{n}_1 = d_1$ and $\vec{r}.\vec{n}_2 = d_2$ are perpendicular, if $\vec{n}_1 . \vec{n}_2 = 0$

Thus, the planes $\vec{r}.\vec{n}_1 = d_1$ and $\vec{r}.\vec{n}_2 = d_2$ are parallel if $= \vec{n}_1 = \lambda\vec{n}_2$.

6. Angle between two planes in Cartesian form : The angle θ between the planes $a_1x + b_1y + c_1z + d_1 = 0$ and $a_2x + b_2y + c_2z + d_2 = 0$ is given by

$$\Rightarrow \qquad \cos\theta = \frac{a_1a_2 + b_1b_2 + c_1c_2}{\sqrt{a_1^2 + b_1^2 + c_1^2}.\sqrt{a_2^2 + b_2^2 + c_2^2}}$$

Thus, the planes $a_1x + b_1y + c_1z + d_1 = 0$ and $a_2x + b_2y + c_2z + d_2 = 0$ are perpendicular, if

$$a_1a_2 + b_1b_2 + c_1c_2 = 0$$

Thus, the planes $a_1x + b_1y + c_1z + d_1 = 0$ and $a_2x + b_2y + c_2z + d_2 = 0$ are parallel, if

$$\frac{a_1}{a_2} = \frac{b_1}{b_2} = \frac{c_1}{c_2}.$$

(i) Equation of a plane passing through a given point and perpendicular to two given planes : Let the two planes be $\vec{r}.\vec{n}_1 = d_1$ and $\vec{r}.\vec{n}_2 = d_2$ and point be $\vec{a}$, then equation of the required plane is :

$$(\vec{r} - \vec{a}).(\vec{n}_1 \times \vec{n}_2) = 0$$

(ii) Equation of a plane passing through two given points and perpendicular to a given plane :

Let the two points be P $(\vec{a})$ and Q $(\vec{b})$ and the plane be $\vec{r}.\vec{n}_1 = d_1$, then equation of the required plane is :

$$(\vec{r} - \vec{a}).(\vec{n}_1 \times \vec{PQ}) = 0 \Rightarrow (\vec{r} - \vec{a}).\vec{n} = 0$$

where the $\vec{n}$ is the normal vector to the plane.

(iii) Equation of a plane parallel to a given plane :

Let the given plane be $ax + by + cz + d = 0$, then equation of the required plane is :

$$ax + by + cz + k = 0$$

where k is an arbitrary constant.

27. Angle between a line and a plane :

Cartesian form : Let the given line be $\frac{x-x_1}{l} = \frac{y-y_1}{m} = \frac{z-z_1}{n}$ and given plane be $ax + by + cz + d = 0$

The angle between a line and a plane is complement of the angle between the line and normal to the plane. If l, m, n be d.r's of line and a, b, c be d.r's of normal and θ be the angle between the line and plane, then

$$\cos(90° - \theta) = \frac{al + bm + cn}{\sqrt{a^2+b^2+c^2}\sqrt{l^2+m^2+n^2}}$$

i.e.,

$$\sin\theta = \frac{al + bm + cn}{\sqrt{a^2+b^2+c^2}\sqrt{l^2+m^2+n^2}}$$

(i) Line is parallel to the plane if $\theta = 0$, *i.e.,* if

$$al + bm + cn = 0.$$

(ii) Line is perpendicular to the plane if line is parallel to the normal of the plane *i.e.,* if

$$\frac{a}{l} = \frac{b}{m} = \frac{c}{n}.$$

Vector form :

$$\Rightarrow \quad \sin\theta = \frac{\vec{b}.\vec{n}}{|\vec{b}|.|\vec{n}|}$$

Condition of perpendicularity : If the line is perpendicular to the plane, then it is parallel to the normal to the plane. Therefore, $\vec{b}$ and $\vec{n}$ are parallel.

$$\Rightarrow \quad \vec{b} \times \vec{n} = 0$$

or $\vec{b} = \lambda\vec{n}$ for some scalar λ.

Condition of parallelism : If the line is parallel to the plane, then it is perpendicular to the normal to the plane. Therefore, $\vec{b}$ and $\vec{n}$ are perpendicular.

$$\Rightarrow \quad \vec{b}.\vec{n} = 0.$$

28. Equation of a plane passing through the intersection of two planes :

(i) **Cartesian form :** The equation of a plane passing through the intersection of the planes $a_1x + b_1y + c_1z + d_1 = 0$ and $a_2x + b_2y + c_2z + d_2 = 0$ is $(a_1x + b_1y + c_1z + d_1) + \lambda\,(a_2x + b_2y + c_2z + d_2) = 0$

where λ is arbitrary constant.

Chapter at a Glance

(ii) Vector form : The equation of a plane passing through the intersection of the planes

$$\vec{r}\cdot\vec{n}_1 = d_1 \text{ and } \vec{r}\cdot\vec{n}_2 = d_2 \text{ is } (\vec{r}\cdot\vec{n}_1 - d_1) + \lambda\,(\vec{r}\cdot\vec{n}_2 - d_2) = 0$$

or, $$\vec{r}\cdot(\vec{n}_1 + \lambda\vec{n}_2) = d_1 + \lambda d_2.$$

Chapter 13. Application of Integrals

1. **Area of bounded region :** Definite integrals find its application in computing the area bounded by curves, lines and coordinate axes. To find the area, rough sketches are need to be drawn of given curves which enclose the region.

2. The area bounded by the curve $y = f(x)$, the x-axis and the ordinates $x = a$ and $x = b$ is given by

$$\int_a^b y\,dx \text{ or, } \int_a^b f(x)\,dx$$

provided, $f(x)$ a continuous function defined on $[a, b]$

$\therefore$ Area of ACDB (A) $= \int_a^b y dx = \int_a^b f(x)\,dx$

3. If the curve $y = f(x)$ lies below the x-axis, then the area bounded by the curve $y = f(x)$, the x-axis and the lines $x = a$ and $x = b$ is given by

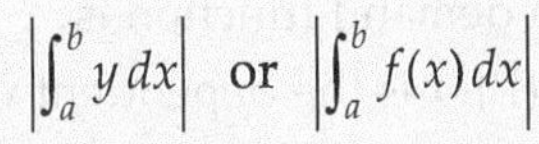

$$\left|\int_a^b y\,dx\right| \text{ or } \left|\int_a^b f(x)\,dx\right|$$

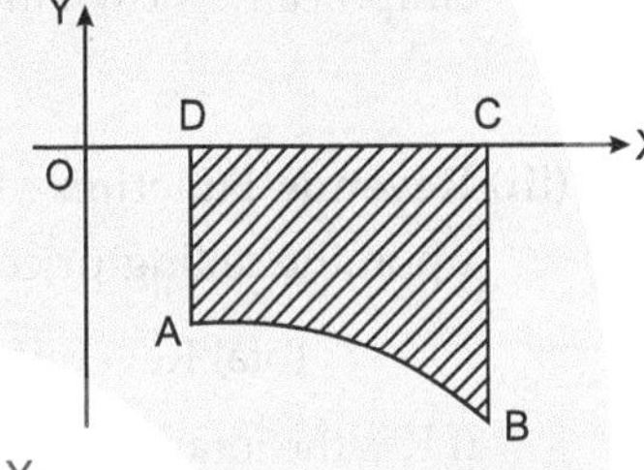

4. The area bounded by the curve $x = f(y)$, the y-axis and the abscissal $y = a$ and $y = b$ is given by

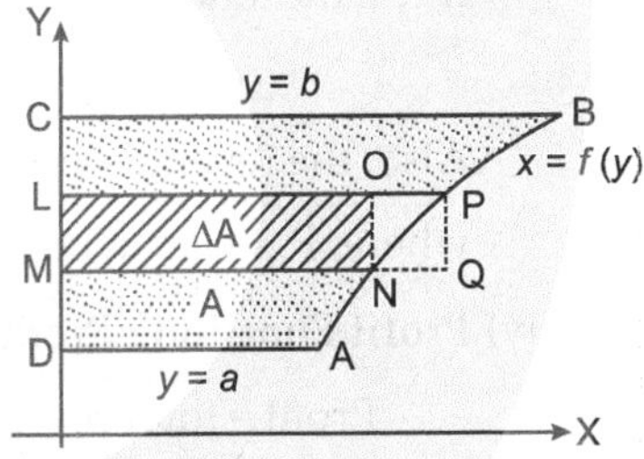

$$\int_a^b x\,dy \text{ or } \int_a^b f(y)\,dy$$

provided $f(y)$ a continuous function defined on $[a, b]$

$\therefore$ Area of DABC (A) $= \int_a^b xdy = \int_a^b f(y)\,dy$

5. **Area enclosed between two curves :** Provided $f(x)$ and $g(x)$ be two continuous function defined on $[a, b]$. Also, the area between the curve $y = f(x)$ and the x-axis from $x = a$ to $x = b$ is $\int_a^b f(x)\,dx$ and the area between the curve $y = g(x)$ and the x-axis from $x = a$ to $x = b$ is $\int_a^b g(x)\,dx$.

Then, the area enclosed between the two curves is $\int_a^b f(x)\,dx - \int_a^b g(x)\,dx$.

or $$\int_a^b [f(x) - g(x)]\,dx \text{ where, } f(x) > g(x),\ a \le b.$$

Working rule to find area defined by two curves :

Working rule : Calculate.

1. The x-coordinates of the points of intersection of the curves to give the limits of integration.
2. The area under each curve separately.
3. The difference between the areas. Steps 2 and 3 give

   Area between two curves :

   $$= \int_a^b y \text{ of } f(x)\,dx - \int_a^b y \text{ of } g(x)\,dx.$$

   where, $f(x) > g(x)$.

Chapter at a Glance

Chapter 14. Application of Calculus

1. Functions Related to Business and Economics

(i) Cost Function : Let x denote the quantity produced of a certain commodity at total cost C, then cost function is expressed as $C = C(x)$, explicitly and $f(C, x) = 0$ in the implicit form.

Total Cost = Fixed Cost + Variable Cost

i.e., $C = F + V$

and the cost function is

$$C = C(x) = F + V(x)$$

Graphically shown in the figure alongside,

(ii) Demand Function : Demand function represents the functional relationship between demand and price of a commodity.

If P denotes the price per unit and x is the number of units demanded by a consumer at that price.

Then demand function is

$x = f(P)$, x is the dependent variable and P is the independent variable

Graphically shown in the figure alongside,

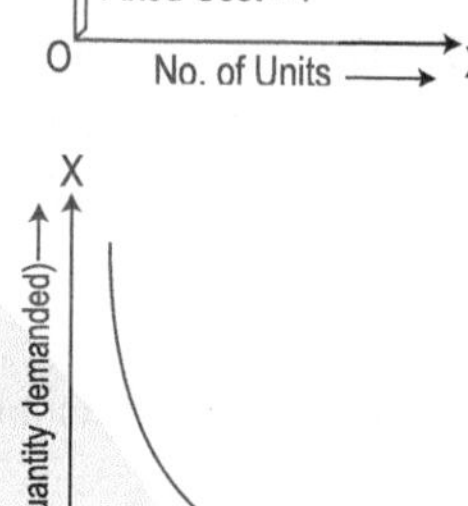

(iii) Revenue Function : Revenue is the amount of money received from the sales of goods. It depends upon the selling price of the product and actual quantity sold.

Total Revenue = Qty. Sold × selling price per unit of commodity

If R is the total revenue collected by the seller on selling x units on a commodity at price P per unit, then

$$R = Px$$

Also, the revenue function is $R(x) = Px = xg(x)$

where, $P = g(x)$ when expressed as a function of x (demand function)

(iv) Profit Function : Profit function is expressed as :

Profit = total revenue – total cost

$P(x) = R(x) - C(x)$

Where, $P(x)$ = Total profit

$R(x)$ = Total revenue collected by seller on selling x number of units of the commodity.

$C(x)$ = Total cost on producing x number of units of the commodity.

(v) The Break Even Point : It is that point where total revenue equals total cost incurred. Thus, it is the point at which a company begins to earn a profit. It is a point of no profit-no loss.

At Break Even Point, $P(x) = 0$

$\Rightarrow$ $R(x) - C(x) = 0$ for $x > 0$

$\Rightarrow$ $R(x) = C(x)$

Graphically shown in the figure alongside,

2. Average and Marginal Costs

(i) Average Cost

$$\text{Average Cost (AC)} = \frac{\text{Total Cost}}{\text{Total quantity of goods produced}}$$

$$= \frac{C}{x}$$

Graphically shown in the figure alongside,

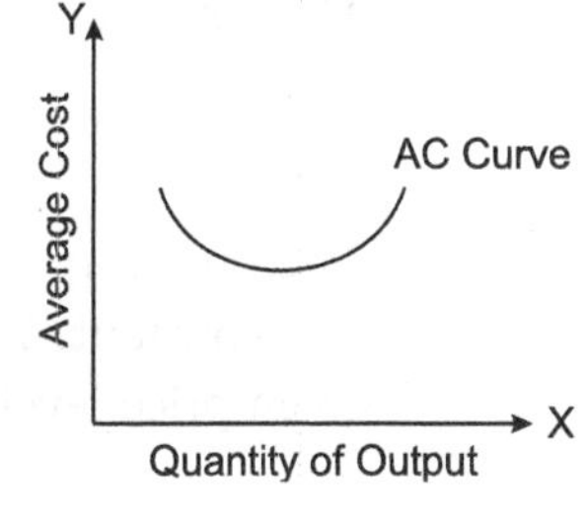

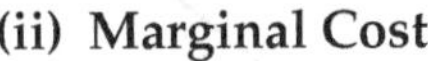

(ii) Marginal Cost

$$\text{Marginal Cost (MC)} = \frac{dC}{dx}$$

Graphically,

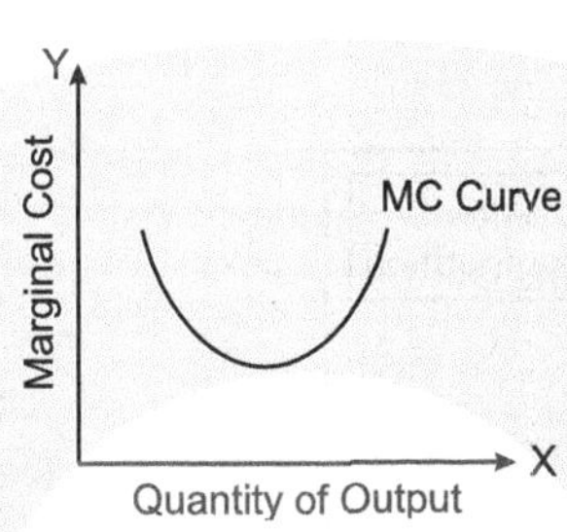

Marginal Average Cost : It is denoted by MAC

$$\text{MAC} = \frac{d}{dx}\left(\frac{C}{x}\right) = \frac{x\frac{dC}{dx} - C.1}{x^2}$$

$$= \frac{1}{x}\left(\frac{dC}{dx} - \frac{C}{x}\right)$$

$$\Rightarrow \qquad \text{MAC} = \frac{d}{dx}(\text{AC}) = \frac{1}{x}(\text{MC} - \text{AC}) \text{ for } x > 0.$$

(i) If MC > AC, $\frac{d}{dx}(\text{AC}) > 0 \Rightarrow$ AC increases with x.

(ii) If MC = AC, $\frac{d}{dx}(\text{AC}) = 0 \Rightarrow$ AC is constant.

(iii) If MC < AC, $\frac{d}{dx}(\text{AC}) < 0 \Rightarrow$ AC decreases with x.

3. Average and Marginal Revenue

(i) Average Revenue (AR)

$$\text{AR} = \frac{\text{Total Revenue}}{\text{Number of Units Sold}} = \frac{R}{x} \Rightarrow \text{AR} = \frac{Px}{x}$$

$$\Rightarrow \qquad \text{AR} = P.$$

(ii) Marginal Revenue (MR)

$$\text{MR} = \frac{dR}{dx} = \frac{d}{dx}(Px)$$

$$\Rightarrow \qquad \text{MR} - P + \frac{xdP}{dx} = P\left(1 + \frac{x}{P}\cdot\frac{dP}{dx}\right).$$

(iii) Sketching the total revenue (TR or R), average revenue (AR) and marginal revenue (MR) curves : If p is the price and x is the number of units sold, then

Case 1 : (Under perfect competition)

TR curve : TR is measured along y-axis and volume of sales is measured along x-axis. **TR curve is an upward sloping straight line from origin,** *i.e.* it increases proportionately with increase in sales. The proportionality is the result of constant price in the perfect competition market.

AR and MR curves : AR = TR/Qty. sold = $(p.x)/x = p$ (price)

Thus, *average income or AR is the average, market price of the output.*

Since price is constant in perfect competition market, AR is also constant. Thus, AR curve is a horizontal straight line at given price level.

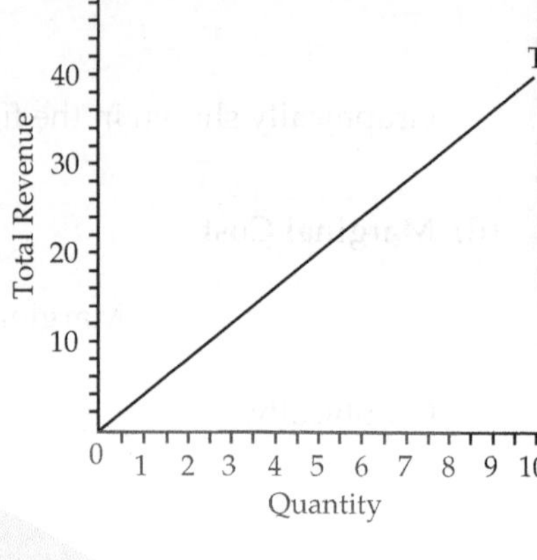

$$\text{MR} = \frac{d(px)}{dx} = p\frac{d(x)}{dx} = p \text{ (price)}$$

Hence, MR = AR = p *and thus* AR *and* MR *curves are identical and parallel to the output axis.*

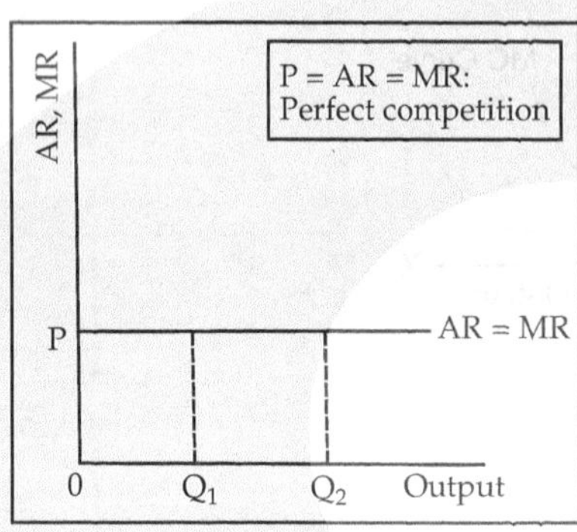

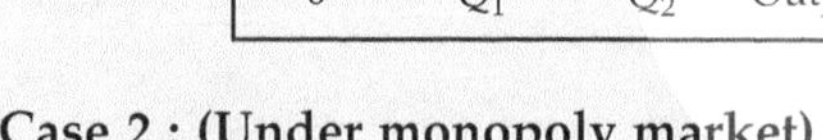

Case 2 : (Under monopoly market)

Here, we have price as a variable and the demand curve is downward sloping, *i.e.* $\frac{dp}{dx} < 0$

Thus, we have $\quad \text{AR} = (px)/p = p$

$$\text{MR} = p + x\frac{d(px)}{dx}$$

$\Rightarrow \quad \text{MR} < p$

$\Rightarrow \quad \text{MR} < \text{AR} \qquad (\because p = \text{AR})$

AR, MR Curves Under Monopoly.

Thus, for a monopoly, both AR and MR curves are negatively sloped and MR curve remains below AR curve at every +ve quantity.

TR = px

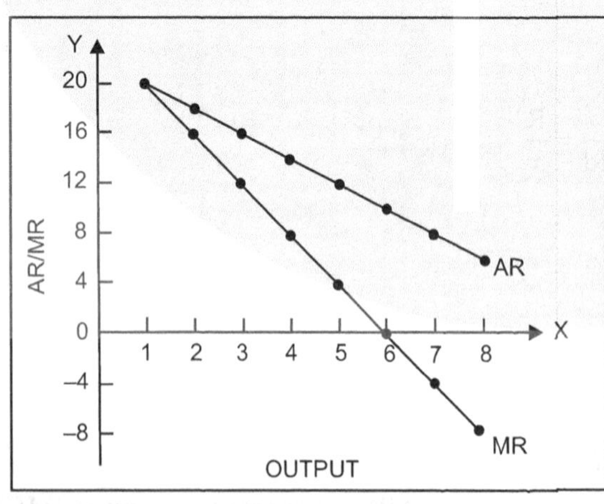

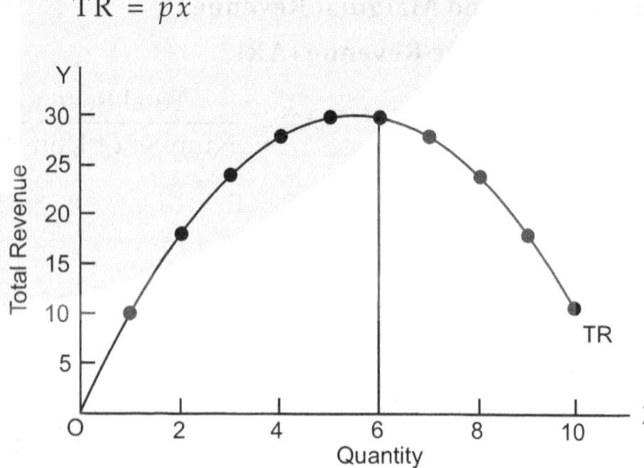

(iv) **Maximisation and Minimisation :**

- **Maximization of total revenue :**

If p is the price and x is the number of units sold, then total revenue of the firm,

$$\text{TR} = px = x\,f(x)$$

Procedure :

Step 1. Find the total revenue R.

Step 2. Differentiate R *w.r.t* x. So obtain $\frac{dR}{dx}$ and put $\frac{dR}{dx} = 0$.

Step 3. Find $\frac{d^2R}{dx^2}$.

Step 4. From the values of x obtained in Step 2, test which of these make $\frac{d^2R}{dx^2}$ negative, *i.e.*, < 0.

- **Maximisation of total profit :**

We know, $P(x) = R(x) - C(x)$, where P is profit function, R is total revenue function and C is total cost function of producing x units.

$$\Rightarrow \quad \frac{dP}{dx} = MR - MC$$

For P to be maximum,

$$\frac{dP}{dx} = 0$$

$$MR = MC$$

and

$$\frac{d^2P}{dx^2} < 0$$

$$\frac{d(MR)}{dx} < \frac{d(MC)}{dx}$$

- **Minimising average cost :**

If $C = C(x) = f(x)$ is the total cost function of producing x units

Then, Average Cost (A) = C/x

For minimum A, $\frac{dA}{dx} = 0$ and $\frac{d^2A}{dx^2} > 0$

(v) Determination of cost function and average cost function

Total production cost (C) for d units produced = $\int_0^d (\text{Marginal cost})\, dx$

Thus, average cost (AC) = C/x

(vi) Determination of Revenue function and demand function :

Total Revenue is given by :

$$\Rightarrow \quad \int dR = \int MR\, dx$$

$\Rightarrow \int MR\, dx + k$, where k is a constant of integration by using the fact that $R = 0$, where $x = 0$.

Subsequently, corresponding demand function can be found out by, $x = R/p$.

Chapter 15. Linear Regression

1. Scatter diagram : To study the relationship between two variables, a scatter diagram is quite helpful in understanding the form, direction and strength of their relationship. It is a graphical method, which is appealing to the eye. Our first step is to designate one of our variables as the *independent variable* and the other as the *dependent variable*. The problems of relationship ordinarily arise when someone try to estimate something they wish. For example, to estimate the price of wheat and in making their estimate they make use of data showing the number of bushels of wheat as harvested. One man may take the number of bushels of wheat as independent variable and the price of wheat as dependent variable. The other man may take the price as independent and the number of bushels as dependent. Both are correct.

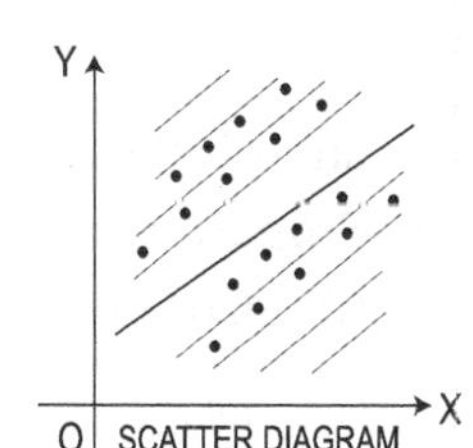

After deciding the designation of variables as independent and dependent we plot graph of one variable against the other, and we find that there is no simple functional relationship like a straight line or elegant looking curve. In other words we get points in our graph very much scattered at random, as shown in fig. This graph is called a **scattered diagram** or **scatter gram.** This scattering at random is always a bad sign. But the mathematicians most commonly think of trying to get a straight line law out of it. It is sufficient to say that the analysis of straight line trends is common most problem. This straight line which is obtained by the method of least squares is called the **Line of Best Fit.** It is a line from which the sum of the deviations of various points on either side is equal to zero. In other words, if the vertical distances of the various points from the line on one side of it are measured and totalled this figure will be equal to that which would be obtained if the vertical distances of the points on the other side of the line were measured and totalled. This being so, the sum of the squares of these deviations would be the least as compared to the sums of squares of the deviations obtained by using other lines. It is on account of this fact is known as the **Method of Least Squares.**

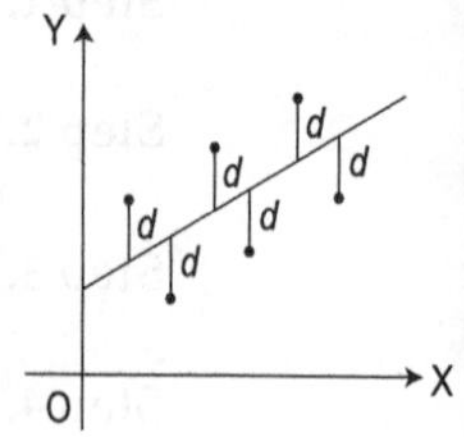

2. **Regression :** It is the estimate or prediction of unknown values of one variable from known values of another variable. It measures the nature and extent of correlation.

   If there are only two variables under consideration, then the regression is called simple *regression.* For example, the study of regression between 'income' and 'expenditure' for a group of families would be termed as *simple regression.* If there are more than two variables under study then regression is called *multiple regression.*

   The simple regression is called *linear regression* whereas multiple regression is called a *non-linear regression* or *curvilinear regression.*

3. **Lines of regression of x on y and regression of y on x :** A line of regression is the straight line which gives the best fit in the least squares sense to the given sets of data. The line of regression in algebraic form is $y-\bar{y} = b_{yx}\ (x-\bar{x})$ *i.e.,* $y = mx + c$ represents the line of regression of Y on X where the coefficient of x *i.e.,* b_{yx} (= m) denoting the regression coefficient of Y on X.
   Regression of X and Y — line of regression of algebraic form $x - \bar{x} = b_{xy}\ (y-\bar{y})$
   i.e., $x = my + c$ is said to be a regression line of X on Y. where, b_{xy} (= m) represents the regression coefficient of X on Y.

   Note : Both lines of regression passes through the point $(\bar{x}, \bar{y})$ where, x and y are the means of the series.

4. **Calculation of regression coefficient :**

$$bxy = \frac{\Sigma x_i y_i - \dfrac{\Sigma x_i + \Sigma y_i}{n}}{\Sigma y_i^2 - \dfrac{1}{n}(\Sigma y_i)^2} = \frac{x\sigma_x}{\sigma_y}$$

$$byx = \frac{\Sigma x_i\, y_i - \dfrac{\Sigma x_i\ \Sigma y_i}{n}}{\Sigma x_i^2 - \dfrac{1}{n}(\Sigma x_i)^2} = \frac{x\sigma_y}{\sigma_x}$$

   where,

$$\sigma_y^2 = \frac{1}{n}\ (\Sigma y_i^2 - (\bar{y})^2)$$

$$\sigma_x^2 = \frac{1}{n}\ (\Sigma x_i^2 - (\bar{x})^2)$$

   and r is the correlation coefficient.

5. **Properties of Regression Coefficients :**
   (i) Correlation coefficient is the geometric mean of two regression coefficients.

   i.e., $$r = \sqrt{b_{yx} \cdot b_{xy}}$$

   (ii) Arithmetic mean of two regression coefficients is always greater than the correlation coefficient

   i.e., $$\frac{b_{xy} + b_{xy}}{2} > r.$$

(iii) The two regression coefficients have the same sign as that of correlation coefficient.

(iv) The correlation coefficient lies between – 1 and 1.

i.e., $r^2 \le 1$ or $-1 \le r \le 1$.

6. Angle between two regression lines : If θ be the angle between two regression lines, then

$$\tan\theta = \frac{1-r^2}{r} \cdot \frac{\sigma_x \sigma_y}{\sigma_x^2 + \sigma_y^2}$$

Remark : (i) when $r = 0$, then $\theta = \frac{\pi}{2}$

∴ The two lines of regression are perpendicular to each other.

(ii) when $r = \pm 1$, then $\theta = 0$ or π

∴ there is a perfect correlation as the two lines of regression coincide.

	y on *x*	*x* on *y*
1. Regression Equations	$y = mx + c$ $y - \bar{y} = b_{yx}(x - \bar{x})$	$x = my + c$ $x - \bar{x} = b_{yx}(y - \bar{y})$
2. Normal Equations	$\Sigma y = nc + m\Sigma x$ $\Sigma xy = c\Sigma x + m\Sigma x^2$	$\Sigma x = nc + m\Sigma y$ $\Sigma xy = c\Sigma y + m\Sigma y^2$
3. Regression Coefficients	$b_{yx} = r\frac{\sigma_y}{\sigma_x}$	$b_{xy} = r\frac{\sigma_x}{\sigma_y}$
(i) When deviations are taken from the mean :	$b_{yx} = \frac{\Sigma d_x d_y}{\Sigma d_x^2}$	$b_{xy} = \frac{\Sigma d_x d_y}{\Sigma d_y^2}$
$d_x = x - \bar{x}, d_y = y - \bar{y}$	*or* $b_{yx} = \frac{\Sigma xy - n\bar{x}\bar{y}}{\Sigma x^2 - n.\bar{x}^2}$	*or* $b_{xy} = \frac{\Sigma xy - n\bar{x}\bar{y}}{\Sigma y^2 - n.\bar{y}^2}$
(ii) When deviations are taken from the assumed mean : $u = x - a, v - y - b$ *a* and *b* being the assumed means for *x* and *y* series respectively.	$b_{yx} = \frac{\Sigma uv - \frac{\Sigma u - \Sigma v}{n}}{\Sigma u^2 - \frac{(\Sigma u)^2}{n}}$	$b_{xy} = \frac{\Sigma uv - \frac{\Sigma u - \Sigma v}{n}}{\Sigma v^2 - \frac{(\Sigma v)^2}{n}}$
(iii) When the original values are used	$b_{yx} = \frac{n\Sigma xy - (\Sigma x)(\Sigma y)}{n\Sigma x^2 - (\Sigma x)^2}$ $= \frac{\Sigma xy = \frac{(\Sigma x)(\Sigma y)}{n}}{\Sigma x^2 - \frac{(\Sigma x)^2}{n}}$	$b_{xy} = \frac{n\Sigma xy - (\Sigma x)(\Sigma y)}{n\Sigma y^2 - (\Sigma y)^2}$ $= \frac{\Sigma xy = \frac{(\Sigma x)(\Sigma y)}{n}}{\Sigma y^2 - \frac{(\Sigma y)^2}{n}}$
4. The two lines of regression (*y* on x and x on *y*) intersect at the point $(\bar{x}, \bar{y})$; coefficient of correlation r or $\rho(x, y) = \sqrt{b_{yx} \times b_{xy}}$; $-1 \le r 1$; r, b_{yx}, b_{xy} are of the same sign. $\text{Cov}(x, y) = \frac{\Sigma d_x d_y}{n}$.		

Chapter 16. Linear Programming

1. Solution of Simultaneous Linear Inequation in two variables : Any statement of the form $ax + by > c$, $ax + by < c$, $ax + by \ge c$, $ax + by \le c$ and $ax + by + c \ge 0$ are known as inequalities.

The set of all ordered pairs of real numbers which satisfy a given inequation is called the solution se of inequation e.g. If the given inequation is $2x + y \leq 5$ then, (1, 2), (2, 0) are the members of solution se because,

$$2 \times 1 + 2 = 4 < 5$$

and

$$2 \times 2 + 0 = 4 < 5$$

For the complete solution set we draw the graph of inequation then we find the portion of plane above th line or below the line according to given inequation.

For the linear equation, we take

$$2x + y = 5$$
$$x = 1, \; y = 3$$
$$x = 2, \; y = 1$$

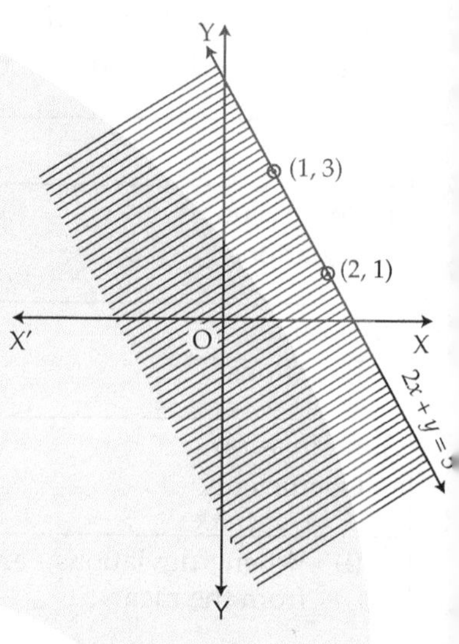

Now, we plot the points (1, 3) and (2, 1) and join them to get the graph of the equation $2x + y = 5$ on the co-ordinate plane.

If we put the coordinates of the origin (0, 0) in $2x + y \leq 5$, we get $0 + 0 \leq 5$, which is correct.

$\therefore$ All the points lying in the shaded region (below the line) satisfies inequation.

If two or more inequations are given, then common portion satisfied with all inequations is called solution of inequations.

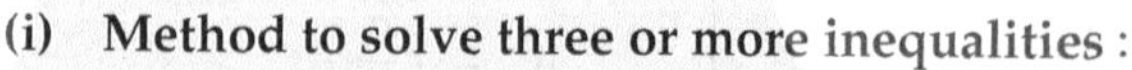

(i) Method to solve three or more inequalities :

Step 1. Graph each of the inequality on the same coordinate plane.

Step 2. The solution set is given by the common region of intersection of graphs of all the inequalities (which are closed half planes).

(ii) Theorem :

If a and b are any real numbers, and if the linear expression $ax + by$ has a maximum (greatest) valu over a feasibility region that is the intersection of finite number of closed half-planes and that ha corner points, then the maximum occurs for the coordinates of some corner point.

Similarly, if $ax + by$ has a minimum (least) value over the region, then a minimum occures for th coordinates of some corner point.

2. **Linear programming :** This is a method of determining extreme values (maximum or minimum value) of a linear function subject to constraints expressed as a system of linear equations or inequations.

3. **Terminologies related to the solution of LPP :**

(a) **Objective function :** It is the linear function $ax + by + c$ whose maximum or minimum value is determined

(b) **Optimize** means to maximize or minimize.

(c) **Constraints** : The system of inequations or equations involving variables of a LPP which describe th conditions under which optimisation is to be attained are called constraints. These involve the sign $\leq, =, \geq$.

(d) **Non-negativity restriction** : These are the constraints which imply that variables involved in LPP ar non-negative.

(e) **Convex region** : A region is said to be **convex** if whenever you choose any two points in the region and draw the segment joining them, the segment is contained in the region.

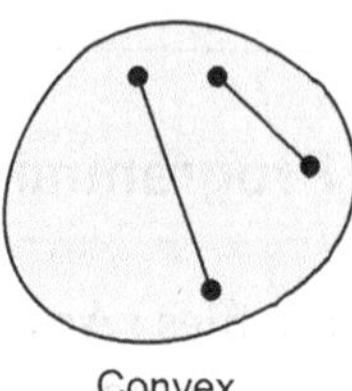
Convex

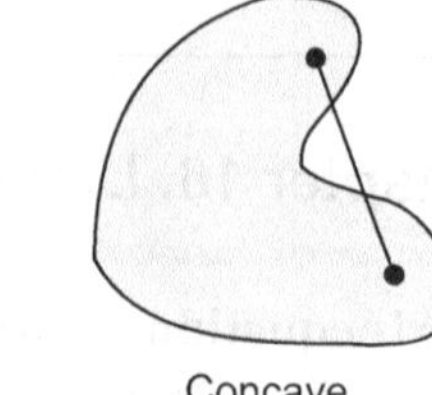
Concave

(f) **Feasible region** : The solution region, *i.e.*, the common region determined by the graphical representation of all the constraints is called the *feasible region* of the LPP. It is always a convex set.

(g) **Feasible Solution** : A solution which satisfies the given system of inequalities is called a *feasible solution* to LPP.

(h) **Infeasible region** : The region other than the feasible region is called *infeasible region* of the LPP.

(i) **Infeasible Solution** : If the system of constraints has no point which satisfies all the constraints and non-negativity restriction, the solution is called an infeasible solution to LPP.

(j) **Optimal feasible solution** : A feasible solution which optimizes (*i.e.*, maximizes or minimizes) the objective function is called an *optimal feasible solution.*

4. **Formulation of a linear programming problem in two variables x_1 and x_2 :**

Step 1. Firstly, identify the decision variables by assigning a particular symbol (*i.e.*, x_1, x_2 etc.) to those quantities whose values are to be calculated.

Step 2. All the constraints are need to be identified as a linear equations/inequations involving the decision variables.

Step 3. Identify the objective function as a linear combination of decision variables, which is to be maximized or minimized.

Step 4. At the end, the non-negative restrictions are required to be imposed on decision variables. These restrictions imply the variables involved in a L.P.P. are non-negative.

❑❑

Formulae Based Questions

| Set 2 |

Chapter 4. Determinants

1. *Using determinants, find the values of k, if the area of triangle with vertices (– 2,0), (0, 4) and (0, k) is 4 square units.**

Sol. Given, Area of $\Delta ABC = 4$

$$\Rightarrow \quad \frac{1}{2}\begin{vmatrix} x_1 & y_1 & 1 \\ x_2 & y_2 & 1 \\ x_3 & y_3 & 1 \end{vmatrix} = 4$$

$$\Rightarrow \quad \frac{1}{2}\begin{vmatrix} -2 & 0 & 1 \\ 0 & 4 & 1 \\ 0 & k & 1 \end{vmatrix} = 4$$

$$\Rightarrow \quad \frac{1}{2}\,|[-2(4-k)]| = 4$$

[$\because$ Area cannot be negative]

$$\Rightarrow \quad |-8+2k| = 8$$

$$\Rightarrow \quad -8+2k = \pm 8$$

$$\Rightarrow \quad -8+2k = 8$$

$$\text{or} \quad -8+2k = -8$$

$$\Rightarrow \quad 2k = 16$$

$$\text{or} \quad 2k = 0$$

$$\Rightarrow \quad k = 8 \text{ or } k = 0$$

Ans.

Chapter 5. Matrices

1. *Construct a 3 × 4 matrix whose elements are $a_{ij} = i + j$.*

Sol. $a_{11} = 1+1 = 2$, $\quad a_{12} = 1+2 = 3$

$a_{13} = 1+3 = 4$, $\quad a_{14} = 1+4 = 5$

$a_{21} = 2+1 = 3$, $\quad a_{22} = 2+2 = 4$

$a_{23} = 2+3 = 5$, $\quad a_{24} = 2+4 = 6$

$a_{31} = 3+1 = 4$, $\quad a_{32} = 3+2 = 5$

$a_{33} = 3+3 = 6$, $\quad a_{34} = 3+4 = 7$

Hence, the required matrix is $\begin{bmatrix} 2 & 3 & 4 & 5 \\ 3 & 4 & 5 & 6 \\ 4 & 5 & 6 & 7 \end{bmatrix}$

Ans.

2. *Construct a matrix of order 3 × 2 whose element in ith row and jth column is given by $a_{ij} = \dfrac{3i+j}{2}$.*

Sol. Here, i varies from 1 to 3 and j varies from 1 to 2.

$\therefore$ Elements of matrix will be

$$a_{11} = \frac{3\times1+1}{2} = \frac{4}{2} = 2$$

$$a_{12} = \frac{3\times1+2}{2} = \frac{5}{2} = 2{\cdot}5$$

$$a_{21} = \frac{3\times2+1}{2} = 3{\cdot}5$$

$$a_{22} = \frac{3\times2+2}{2} = 4$$

$$a_{31} = \frac{3\times3+1}{2} = 5$$

$$a_{32} = \frac{3\times3+2}{2} = 5{\cdot}5$$

Hence, the required matrix is

$$\begin{bmatrix} 2 & 2{\cdot}5 \\ 3{\cdot}5 & 4 \\ 5 & 5{\cdot}5 \end{bmatrix}$$

Ans.

3. *Using matrix method, solve the following equations :*

$$5x + 3y + z = 16$$
$$2x + y + 3z = 19$$
$$x + 2y + 4z = 25$$

Sol. The given system of equation is

$$5x + 3y + z = x - \text{It}$$
$$2x + y + 3z = x - \text{It}$$
$$x + 2y + 4z = 25$$

This system can be written as AX = B

where $A = \begin{bmatrix} 5 & 3 & 1 \\ 2 & 1 & 3 \\ 1 & 2 & 4 \end{bmatrix}$, $X = \begin{bmatrix} x \\ y \\ z \end{bmatrix}$ and $B = \begin{bmatrix} 16 \\ 19 \\ 25 \end{bmatrix}$

* Frequently asked previous years Board Exam Questions.

Given, $A = \begin{bmatrix} 5 & 3 & 1 \\ 2 & 1 & 3 \\ 1 & 2 & 4 \end{bmatrix}$

$\therefore$ $|A| = 5(4-6) - 3(8-3) + 1(4-1)$

$= -10 - 15 + 3 = -22 \neq 0$

As $|A| \neq 0$, the given system has a unique solution.

$\therefore$ A^{-1} exists

Now cofactors of A,

$A_{11} = \begin{vmatrix} 1 & 3 \\ 2 & 4 \end{vmatrix} = -2,$

$A_{12} = -\begin{vmatrix} 2 & 3 \\ 1 & 4 \end{vmatrix} = -5,$

$A_{13} = \begin{vmatrix} 2 & 1 \\ 1 & 2 \end{vmatrix} = 3,$

$A_{21} = -\begin{vmatrix} 3 & 1 \\ 2 & 4 \end{vmatrix} = -10,$

$A_{22} = \begin{vmatrix} 5 & 1 \\ 1 & 4 \end{vmatrix} = 19,$

$A_{23} = -\begin{vmatrix} 5 & 3 \\ 1 & 2 \end{vmatrix} = -7,$

$A_{31} = \begin{vmatrix} 3 & 1 \\ 1 & 3 \end{vmatrix} = 8,$

$A_{32} = -\begin{vmatrix} 5 & 1 \\ 2 & 3 \end{vmatrix} = 13,$

$A_{33} = \begin{vmatrix} 5 & 3 \\ 2 & 1 \end{vmatrix} = -1$

$$\text{adj}(A) = \begin{bmatrix} -2 & -5 & 3 \\ -10 & 19 & -7 \\ 8 & -13 & -1 \end{bmatrix}^T$$

$$\therefore \quad \text{adj}(A) = \begin{bmatrix} -2 & -10 & 8 \\ -5 & 19 & -13 \\ 3 & -7 & -1 \end{bmatrix}$$

$$\therefore \quad A^{-1} = \frac{1}{|A|}\text{adj}(A)$$

$$= \frac{1}{-22}\begin{bmatrix} -2 & -10 & 8 \\ -5 & 19 & -13 \\ 3 & -7 & -1 \end{bmatrix}$$

$X = A^{-1}B$

$$= -\frac{1}{22}\begin{bmatrix} -2 & -10 & 8 \\ -5 & 19 & -13 \\ 3 & -7 & -1 \end{bmatrix}\begin{bmatrix} 16 \\ 19 \\ 25 \end{bmatrix}$$

$$\Rightarrow \begin{bmatrix} x \\ y \\ z \end{bmatrix} = -\frac{1}{22}\begin{bmatrix} -32-190+200 \\ -80+361-325 \\ 48-133-25 \end{bmatrix}$$

$$= -\frac{1}{22}\begin{bmatrix} -22 \\ -44 \\ -110 \end{bmatrix}$$

$$\Rightarrow \begin{bmatrix} x \\ y \\ z \end{bmatrix} = \begin{bmatrix} 1 \\ 2 \\ 5 \end{bmatrix}$$

$\Rightarrow$ $x = 1, y = 2, z = 5.$ **Ans.**

4. *Using matrix method solve the following system of equations.*

$x - 2y + 3z = 6$

$x + 4y + z = 12$

$x - 3y + 2z = 1$

Sol. The given equations can be written as $AX = B$

$$\begin{bmatrix} 1 & -2 & 3 \\ 1 & 4 & 1 \\ 1 & -3 & 2 \end{bmatrix}\begin{bmatrix} x \\ y \\ z \end{bmatrix} = \begin{bmatrix} 6 \\ 12 \\ 1 \end{bmatrix}$$

where, $A = \begin{bmatrix} 1 & -2 & 3 \\ 1 & 4 & 1 \\ 1 & -3 & 2 \end{bmatrix}$, $X = \begin{bmatrix} x \\ y \\ z \end{bmatrix}$

and $B = \begin{bmatrix} 6 \\ 12 \\ 1 \end{bmatrix}$

$\Rightarrow$ $AX = B$

Now, $A = \begin{bmatrix} 1 & -2 & 3 \\ 1 & 4 & 1 \\ 1 & -3 & 2 \end{bmatrix}$

$\Rightarrow$ $|A| = 1(4 \times 2 + 3 \times 1)$

$- (-2)(1 \times 2 - 1) + 3[(1 \times (-3) - 4]$

$= 1 \times 11 + 2(1) + 3(-7)$

$= 11 + 2 - 21 = -8$

$\Rightarrow$ $|A| = -8 \neq 0$, the given system has unique solution.

$\therefore$ A^{-1} exists

Now cofactors of A,

$A_{11} = \begin{vmatrix} 4 & 1 \\ -3 & 2 \end{vmatrix} = 4 \times 2 + 3 \times 1 = 11$

$A_{12} = -\begin{vmatrix} 1 & 1 \\ 1 & 2 \end{vmatrix} = -(1 \times 2 - 1 \times 1) = -1$

$$A_{13} = \begin{vmatrix} 1 & 4 \\ 1 & -3 \end{vmatrix} = (1 \times (-3) - 4 \times 1) = -7$$

$$A_{21} = -\begin{vmatrix} -2 & 3 \\ -3 & 2 \end{vmatrix} = -(-2 \times 2 + 3 \times 3) = -5$$

$$A_{22} = \begin{vmatrix} 1 & 3 \\ 1 & 2 \end{vmatrix} = (1 \times 2 - 1 \times 3) = 2 - 3 = -1$$

$$A_{23} = -\begin{vmatrix} 1 & -2 \\ 1 & -3 \end{vmatrix} = -(1 \times (-3) - 1 \times (-2)) = 1$$

$$A_{31} = \begin{vmatrix} -2 & 3 \\ 4 & 1 \end{vmatrix} = (-2 \times 1 - 3 \times 4) = -14$$

$$A_{32} = -\begin{vmatrix} 1 & 3 \\ 1 & 1 \end{vmatrix} = -(1 \times 1 - 3 \times 1) = 2$$

$$A_{33} = \begin{vmatrix} 1 & -2 \\ 1 & 4 \end{vmatrix} = 1 \times 4 + 1 \times 2 = 6$$

$$\text{adj}(A) = \begin{bmatrix} 11 & -1 & -7 \\ -5 & -1 & 1 \\ -14 & 2 & 6 \end{bmatrix}^T$$

$$\therefore \quad \text{adj}(A) = \begin{bmatrix} 11 & -5 & -14 \\ -1 & -1 & 2 \\ -7 & 1 & 6 \end{bmatrix}$$

$$\therefore \quad A^{-1} = \frac{1}{|A|}\text{adj}(A)$$

$$\Rightarrow \quad A^{-1} = -\frac{1}{8}\begin{bmatrix} 11 & -5 & -14 \\ -1 & -1 & 2 \\ -7 & 1 & 6 \end{bmatrix}$$

$$\therefore \quad X = A^{-1}B$$

$$\Rightarrow \quad X = -\frac{1}{8}\begin{bmatrix} 11 & -5 & -14 \\ -1 & -1 & 2 \\ -7 & 1 & 6 \end{bmatrix}\begin{bmatrix} 6 \\ 12 \\ 1 \end{bmatrix}$$

$$= -\frac{1}{8}\begin{bmatrix} 11\times6 - 5\times12 - 14\times1 \\ -1\times6 - 1\times12 + 2\times1 \\ -7\times6 + 1\times12 + 6\times1 \end{bmatrix}$$

$$= -\frac{1}{8}\begin{bmatrix} -8 \\ -16 \\ -24 \end{bmatrix} = \begin{bmatrix} 1 \\ 2 \\ 3 \end{bmatrix}$$

$$\Rightarrow \quad \begin{bmatrix} x \\ y \\ z \end{bmatrix} = \begin{bmatrix} 1 \\ 2 \\ 3 \end{bmatrix}$$

Equating corresponding element of equa matrices, we get

$\Rightarrow$ $x = 1, y = 2$ and $z = 3$. **Ans**

5. *Using matrices, solve the following system o equations :*

$$x_1 - 2x_2 - 3x_3 = 4$$
$$2x_1 + x_2 - 3x_3 = 5$$
$$-x_1 + x_2 + 2x_3 = 3.$$

Sol. The given system of equations is

$$x_1 - 2x_2 - 3x_3 = 4$$
$$2x_1 + x_2 - 3x_3 = 5$$
$$-x_1 + x_2 + 2x_3 = 3$$

This system can be written as

$$AX = B$$

where $A = \begin{bmatrix} 1 & -2 & -3 \\ 2 & 1 & -3 \\ -1 & 1 & 2 \end{bmatrix}, X = \begin{bmatrix} x_1 \\ x_2 \\ x_3 \end{bmatrix}$

and $B = \begin{bmatrix} 4 \\ 5 \\ 3 \end{bmatrix}$

Now, $|A| = 1(2+3) + 2(4-3) - 3(2+1$
$= 5 + 2 - 9 = -2 \neq 0$, the giver system has unique solution.

$\therefore$ A^{-1} exists.

Now, cofactors of A,

$$A_{11} = \begin{vmatrix} 1 & -3 \\ 1 & 2 \end{vmatrix} = 5$$

$$A_{12} = -\begin{vmatrix} 2 & -3 \\ -1 & 2 \end{vmatrix} = -1$$

$$A_{13} = \begin{vmatrix} 2 & 1 \\ -1 & 1 \end{vmatrix} = 3$$

$$A_{21} = -\begin{vmatrix} -2 & -3 \\ 1 & 2 \end{vmatrix} = 1$$

$$A_{22} = \begin{vmatrix} 1 & -3 \\ -1 & 2 \end{vmatrix} = -1$$

$$A_{23} = -\begin{vmatrix} 1 & -2 \\ -1 & 1 \end{vmatrix} = 1$$

$$A_{31} = \begin{vmatrix} -2 & -3 \\ 1 & -3 \end{vmatrix} = 9$$

$$A_{32} = -\begin{vmatrix} 1 & -3 \\ 2 & -3 \end{vmatrix} = -3$$

$$A_{33} = \begin{vmatrix} 1 & -2 \\ 2 & 1 \end{vmatrix} = 5$$

$$\text{adj A} = \begin{bmatrix} 5 & -1 & 3 \\ 1 & -1 & 1 \\ 9 & -3 & 5 \end{bmatrix}^T$$

$\therefore$ $$\text{adj A} = \begin{bmatrix} 5 & 1 & 9 \\ -1 & -1 & -3 \\ 3 & 1 & 5 \end{bmatrix}$$

And, $$A^{-1} = \frac{1}{|A|}\text{adj}(A)$$

$$A^{-1} = \frac{-1}{2}\begin{bmatrix} 5 & 1 & 9 \\ -1 & -1 & -3 \\ 3 & 1 & 5 \end{bmatrix}$$

$$X = A^{-1}B = -\frac{1}{2}\begin{bmatrix} 5 & 1 & 9 \\ -1 & -1 & -3 \\ 3 & 1 & 5 \end{bmatrix}\begin{bmatrix} 4 \\ 5 \\ 3 \end{bmatrix}$$

$$\begin{bmatrix} x_1 \\ x_2 \\ x_3 \end{bmatrix} = -\frac{1}{2}\begin{bmatrix} 52 \\ -18 \\ 32 \end{bmatrix} = \begin{bmatrix} -26 \\ 9 \\ -16 \end{bmatrix}$$

Equating corresponding element of equal matrices, we get

$x_1 = -26, x_2 = 9, x_3 = -16.$ **Ans.**

6. *Using matrix method, solve the following system of equations :*

$x - 2y = 10,\ 2x + y + 3z = 8$

and $-2y + z = 7$

Sol. The given system of equations is

$$x - 2y = 10$$
$$2x + y + 3z = 8$$
$$-2y + z = 7$$

where, $A = \begin{bmatrix} 1 & -2 & 0 \\ 2 & 1 & 3 \\ 0 & -2 & 1 \end{bmatrix}, X = \begin{bmatrix} x \\ y \\ z \end{bmatrix}, B = \begin{bmatrix} 10 \\ 8 \\ 7 \end{bmatrix}$

Now, $|A| = 1(1+6) + 2(2-0) + 0$

$= 7 + 4 = 11 \neq 0,$

$\therefore$ A^{-1} exists, the given system has unique solution.

Now cofactors of A,

$$A_{11} = \begin{vmatrix} 1 & 3 \\ -2 & 1 \end{vmatrix} = (1+6) = 7,$$

$$A_{12} = -\begin{vmatrix} 2 & 3 \\ 0 & 1 \end{vmatrix} = -2,$$

$$A_{13} = \begin{vmatrix} 2 & 1 \\ 0 & -2 \end{vmatrix} = -4$$

$$A_{21} = -\begin{vmatrix} -2 & 0 \\ -2 & 1 \end{vmatrix} = -(-2) = 2,$$

$$A_{22} = \begin{vmatrix} 1 & 0 \\ 0 & 1 \end{vmatrix} = 1,$$

$$A_{23} = -\begin{vmatrix} 1 & -2 \\ 0 & -2 \end{vmatrix} = -(-2) = 2$$

$$A_{31} = \begin{vmatrix} -2 & 0 \\ 1 & 3 \end{vmatrix} = -6,$$

$$A_{32} = -\begin{vmatrix} 1 & 0 \\ 2 & 3 \end{vmatrix} = -(3) = -3,$$

$$A_{33} = \begin{vmatrix} 1 & -2 \\ 2 & 1 \end{vmatrix} = 1 + 4 = 5$$

$\therefore$ $$\text{adj A} = \begin{bmatrix} 7 & -2 & -4 \\ 2 & 1 & 2 \\ -6 & -3 & 5 \end{bmatrix}^T$$

$$= \begin{bmatrix} 7 & 2 & -6 \\ -2 & 1 & -3 \\ -4 & 2 & 5 \end{bmatrix}$$

$\therefore$ $$A^{-1} = \frac{1}{|A|}\text{adj}(A)$$

$$= \frac{1}{11}\begin{bmatrix} 7 & 2 & -6 \\ -2 & 1 & -3 \\ -4 & 2 & 5 \end{bmatrix}$$

$\therefore$ $AX = B \Rightarrow X = A^{-1}B$

$\Rightarrow$ $$\begin{bmatrix} x \\ y \\ z \end{bmatrix} = \frac{1}{11}\begin{bmatrix} 7 & 2 & -6 \\ -2 & 1 & -3 \\ -4 & 2 & 5 \end{bmatrix}\begin{bmatrix} 10 \\ 8 \\ 7 \end{bmatrix}$$

$\Rightarrow$ $$\begin{bmatrix} x \\ y \\ z \end{bmatrix} = \frac{1}{11}\begin{bmatrix} 70+16-42 \\ -20+8-21 \\ -40+16+35 \end{bmatrix}$$

$$= \frac{1}{11}\begin{bmatrix} 44 \\ -33 \\ 11 \end{bmatrix} = \begin{bmatrix} 4 \\ -3 \\ 1 \end{bmatrix}$$

Equating corresponding element of equal matrices, we get

$\therefore$ $x = 4, y = -3, z = 1.$ **Ans.**

7. *Using matrices, solve the following system of equations:*

$$x + 2y = 5$$
$$y + 2z = 8$$
$$2x + z = 5$$

Sol. The given system of equations is

$$x + 2y = 5$$
$$y + 2z = 8$$
$$2x + z = 5$$

This system can be written as

$$AX = B$$

where $A = \begin{bmatrix} 1 & 2 & 0 \\ 0 & 1 & 2 \\ 2 & 0 & 1 \end{bmatrix}, X = \begin{bmatrix} x \\ y \\ z \end{bmatrix}$

and $B = \begin{bmatrix} 5 \\ 8 \\ 5 \end{bmatrix}$

Now, $|A| = 1(1-0) - 2(0-4)$
$= 1 + 8 = 9 \neq 0$

$\therefore$ A^{-1} exists, thus the given system has unique solution.

Now cofactors of A,

$$A_{11} = \begin{vmatrix} 1 & 2 \\ 0 & 1 \end{vmatrix} = 1 - 0 = 1$$

$$A_{12} = -\begin{vmatrix} 0 & 2 \\ 2 & 1 \end{vmatrix} = -(0-4) = 4$$

$$A_{13} = \begin{vmatrix} 0 & 1 \\ 2 & 0 \end{vmatrix} = (0-2) = -2$$

$$A_{21} = -\begin{vmatrix} 2 & 0 \\ 0 & 1 \end{vmatrix} = -(2-0) = -2$$

$$A_{22} = \begin{vmatrix} 1 & 0 \\ 2 & 1 \end{vmatrix} = 1 - 0 = 1$$

$$A_{23} = -\begin{vmatrix} 1 & 2 \\ 2 & 0 \end{vmatrix} = -(0-4) = 4$$

$$A_{31} = \begin{vmatrix} 2 & 0 \\ 1 & 2 \end{vmatrix} = 4 - 0 = 4$$

$$A_{32} = -\begin{vmatrix} 1 & 0 \\ 0 & 2 \end{vmatrix} = -(2-0) = -2$$

$$A_{33} = \begin{vmatrix} 1 & 2 \\ 0 & 1 \end{vmatrix} = 1 - 0 = 1$$

$$\therefore \quad \text{adj}(A) = \begin{bmatrix} 1 & 4 & -2 \\ -2 & 1 & 4 \\ 4 & -2 & 1 \end{bmatrix}^T$$

$$\text{adj}(A) = \begin{bmatrix} 1 & -2 & 4 \\ 4 & 1 & -2 \\ -2 & 4 & 1 \end{bmatrix}$$

$$\therefore \quad A^{-1} = \frac{\text{adj}(A)}{|A|}$$

$$= \frac{1}{9}\begin{bmatrix} 1 & -2 & 4 \\ 4 & 1 & -2 \\ -2 & 4 & 1 \end{bmatrix}$$

The given system of equation can be written as,

$$\begin{bmatrix} 1 & 2 & 0 \\ 0 & 1 & 2 \\ 2 & 0 & 1 \end{bmatrix}\begin{bmatrix} x \\ y \\ z \end{bmatrix} = \begin{bmatrix} 5 \\ 8 \\ 5 \end{bmatrix}$$

As $|A| \neq 0$, So, A^{-1} Exists thus the system has a unique solution.

$$\therefore \quad X = A^{-1}B$$

$$X = \frac{1}{9}\begin{bmatrix} 1 & -2 & 4 \\ 4 & 1 & -2 \\ -2 & 4 & 1 \end{bmatrix}\begin{bmatrix} 5 \\ 8 \\ 5 \end{bmatrix}$$

$$= \frac{1}{9}\begin{bmatrix} 5-16+20 \\ 20+8-10 \\ -10+32+5 \end{bmatrix}$$

$$= \frac{1}{9}\begin{bmatrix} 9 \\ 18 \\ 27 \end{bmatrix} = \begin{bmatrix} 1 \\ 2 \\ 3 \end{bmatrix}$$

Equating corresponding element of equal matrices, we get

$x = 1, y = 2$ and $z = 3$. **Ans.**

8. *Solve the following system of linear equations using matrix method :** 2019 - Q11

$$\frac{1}{x} + \frac{1}{y} + \frac{1}{z} = 9$$

$$\frac{2}{x} + \frac{5}{y} + \frac{7}{z} = 52$$

$$\frac{2}{x} + \frac{1}{y} - \frac{1}{z} = 0$$

Sol. Let $\frac{1}{x} = a, \frac{1}{y} = b$ and $\frac{1}{z} = c$.

$\therefore$ The given equations become,

$$a + b + c = 9$$
$$2a + 5b + 7c = 52$$
$$2a + b - c = 0$$
$$AX = B$$

* Frequently asked previous years Board Exam Questions.

Let $A = \begin{bmatrix} 1 & 1 & 1 \\ 2 & 5 & 7 \\ 2 & 1 & -1 \end{bmatrix}$ $B = \begin{bmatrix} 9 \\ 52 \\ 0 \end{bmatrix}$ and $X = \begin{bmatrix} a \\ b \\ c \end{bmatrix}$

$|A| = 1(-5-7) - 1(-2-14) + 1(2-10)$

$= -12 + 16 - 8$

$= -4$

Cofactors of A are

$A_{11} = -12 \quad A_{12} = 16 \quad A_{13} = -8$

$A_{21} = 2 \quad A_{22} = -3 \quad A_{23} = 1$

$A_{31} = 2 \quad A_{32} = -5 \quad A_{33} = 3$

$$\text{adj } A = \begin{bmatrix} -12 & 16 & -8 \\ 2 & -3 & 1 \\ 2 & -5 & 3 \end{bmatrix}^T$$

$$= \begin{bmatrix} -12 & 2 & 2 \\ 16 & -3 & -5 \\ -8 & 1 & 3 \end{bmatrix}$$

$$A^{-1} = \frac{1}{|A|}$$

$$\text{adj } A = \frac{1}{-4}\begin{bmatrix} -12 & 2 & 2 \\ 16 & -3 & -5 \\ -8 & 1 & 3 \end{bmatrix}$$

$X = A^{-1}B$

$$\begin{bmatrix} a \\ b \\ c \end{bmatrix} = \frac{1}{-4}\begin{bmatrix} -12 & 2 & 2 \\ 16 & -3 & -5 \\ -8 & 1 & 3 \end{bmatrix}\begin{bmatrix} 9 \\ 52 \\ 0 \end{bmatrix}$$

$$\begin{bmatrix} a \\ b \\ c \end{bmatrix} = \frac{1}{-4}\begin{bmatrix} -108 + 104 + 0 \\ 144 - 156 - 0 \\ -72 + 52 + 0 \end{bmatrix}$$

$$\begin{bmatrix} a \\ b \\ c \end{bmatrix} = \frac{1}{-4}\begin{bmatrix} -4 \\ -12 \\ -20 \end{bmatrix}$$

$\therefore \quad a = \frac{-1}{4} \times (-4) = 1,\ b = \frac{-12}{-4} = 3$

and $\quad c = \frac{-20}{-4} = 5$

$\therefore \quad x = \frac{1}{a} = 1,\ y = \frac{1}{b} = \frac{1}{3} \quad z = \frac{1}{c} = \frac{1}{5}$

$\Rightarrow \quad x = 1,\ y = \frac{1}{3}$ and $z = \frac{1}{5}$ **Ans.**

Chapter 6. Differential Calculus

1. *Verify Rolle's theorem for the function* $f(x) = e^{2x}(\sin 2x - \cos 2x)$ *defined in the interval* $\left[\frac{\pi}{8}, \frac{5\pi}{8}\right]$.

Sol. Given, $\quad f(x) = e^{2x}(\sin 2x - \cos 2x)$,

$$x \in \left[\frac{\pi}{8}, \frac{5\pi}{8}\right]$$

(i) Sine and cosine functions and exponential functions are always continuous.

$\therefore$ Given function is continuous in $\left[\frac{\pi}{8}, \frac{5\pi}{8}\right]$

(ii) Differentiable, $f'(x) = e^{2x} \times 2(\sin 2x - \cos 2x)$

$+ e^{2x}(2\cos 2x + 2\sin 2x)$

$= 2e^{2x}[\sin 2x - \cos 2x + \cos 2x + \sin 2x]$

$= 2e^{2x}(2\sin 2x) = 4e^{2x}\sin 2x$

exists in the given limits.

(iii) $f\left(\frac{\pi}{8}\right) = e^{\pi/4}\left(\sin\frac{\pi}{4} - \cos\frac{\pi}{4}\right)$

$= e^{\pi/4} \times 0 = 0$

$f\left(\frac{5\pi}{8}\right) = e^{5\pi/4}\left(\sin\frac{5\pi}{4} - \cos\frac{5\pi}{4}\right)$

$= e^{5\pi/4} \times 0 = 0$

$\therefore \quad f\left(\frac{\pi}{8}\right) - f\left(\frac{5\pi}{8}\right)$

Now, $\quad f'(c) = 0$

$\Rightarrow \quad 4e^{2c}\sin 2c = 0$

since, $\quad e^{2c} \neq 0$

$\therefore \quad \sin 2c = 0$

$\Rightarrow \quad 2c = 0, \pi, 2\pi, 3\pi, \ldots$

$\Rightarrow \quad c = 0, \frac{\pi}{2}, \pi, \frac{3\pi}{2}, \ldots$

$\because \quad \frac{\pi}{2} \in \left[\frac{\pi}{8}, \frac{5\pi}{8}\right]$

Hence, Rolle's theorem is verified. **Ans.**

2. *Examine the validity and conclusion of Rolle's theorem for the function.*

$f(x) = e^x \sin x,\ \forall\ x \in [0, \pi]$

Sol. Given, $f(x) = e^x \sin x, \forall x \in [0, \pi]$

(i) $f(x)$ is a product of an exponential function and a trignometrical function in $[0, \pi]$

Hence, $f(x)$ is continuous in the given interval.

(ii) $f(x) = e^x \sin x$

$\Rightarrow \quad f'(x) = e^x \sin x + e^x \cos x$

Clearly, $f(x)$ exist in open interval $(0, \pi)$.

(iii) $f(0) = e^0 \sin 0 = 0$

$f(\pi) = e^\pi \sin \pi = 0$

$\Rightarrow \quad f(0) = f(\pi) = 0$

Hence, all conditions of Rolle's theorem are satisfied.

Hence, there exist at least one value of x, say c, in $0 < x < \pi$.

such that $f'(c) = 0$

$\therefore \quad e^c (\sin c + \cos c) = 0$

$\therefore \quad \sin c + \cos c = 0 \quad [\because e^c \neq 0]$

$\Rightarrow \quad \sin c = -\cos c$

$\Rightarrow \quad \tan c = -1$

$\Rightarrow \quad c = \frac{3\pi}{4} \quad \left[\tan \frac{3\pi}{4} = -1\right]$

$\therefore \quad \frac{3\pi}{4}$ lies between 0 and π

Hence, Rolle's theorem is verified. **Ans.**

3. *Verify Rolle's Theorem for the function*

$f(x) = e^x(\sin x - \cos x)$ *on* $\left[\frac{\pi}{4}, \frac{5\pi}{4}\right]$

Sol. Given, $f(x) = e^x (\sin x - \cos x), x \in \left[\frac{\pi}{4}, \frac{5\pi}{4}\right]$

(i) Sine, cosine and exponential functions are always continuous.

$\therefore$ Given function is continuous in $\left[\frac{\pi}{4}, \frac{5\pi}{4}\right]$

(ii) Differentiating w.r. t. x, we get

$$f'(x) = e^x (\cos x + \sin x) + (\sin x - \cos x) e^x$$
$$= e^x [\cos x + \sin x + \sin x - \cos x]$$
$$= 2 e^x \sin x$$

which exists for all x.

(iii) $f(\pi/4) = e^{\pi/4}\left(\frac{1}{\sqrt{2}} - \frac{1}{\sqrt{2}}\right) = 0$

and $f(5\pi/4) = e^{5\pi/4}\left(-\frac{1}{\sqrt{2}} + \frac{1}{\sqrt{2}}\right) = 0$

$\therefore \quad f(\pi/4) = f\left(\frac{5\pi}{4}\right) = 0$

$\therefore$ The given function satisfies all three conditions of Rolle's theorem.

For maxima or minima,

$f'(x) = 0$

$\Rightarrow \quad 2e^x \sin x = 0$

$\Rightarrow \quad \sin x = 0$

$\Rightarrow \quad x = n\pi$

$\Rightarrow \quad x = \pi$

$\because \pi$ lies between $\left[\frac{\pi}{4}, \frac{5\pi}{4}\right]$ so Rolle's theorem is verified. **Hence Proved**

4. *Verify Rolle's theorem for the function*

$f(x) = \log\left\{\frac{x^2 + ab}{(a+b)x}\right\}$ in the interval $[a, b]$ where $0 < a < b$.

Sol. Given, $f(x) = \log\left(\frac{x^2 + ab}{(a+b)\, x}\right), x \in [a, b]$

$= \log (x^2 + ab) - \log (a + b) - \log x \quad ...(i)$

(i) Since $a > 0$ and $\log x$ is continuous for all $x > 0$ therefore, $f(x)$ is continuous in $[a, b]$.

(ii) $f(x)$ is derivable in (a, b), and

(iii) $f(a) = \log\left(\frac{a^2 + ab}{(a+b)a}\right) = \log 1 = 0$

$f(b) = \log\left(\frac{b^2 + ab}{(a+b)\, b}\right) = \log 1 = 0$

$\Rightarrow \quad f(a) = f(b)$.

Thus, all the three conditions of Rolle's theorem are satisfied. Therefore, there exists at least one real number c in (a, b) such that $f'(c) = 0$.

Differentiating (i) w.r.t. x, we get

$$f'(x) = \frac{1}{x^2 + ab} \cdot 2x - 0 - \frac{1}{x}$$
$$= \frac{2x}{x^2 + ab} - \frac{1}{x}$$
$$= \frac{x^2 - ab}{x\,(x^2 + ab)}$$

Now, $f'(c) = 0 \Rightarrow \frac{c^2 - ab}{c\,(c^2 + ab)} = 0$

$\Rightarrow \quad c^2 - ab = 0$

$\Rightarrow \quad c^2 = ab$

$\Rightarrow \quad c = \pm\sqrt{ab}$

$\because \quad 0 \notin [a, b] \Rightarrow -ab \notin [a, b]$

$\therefore \quad c = \sqrt{ab}$

So, there exists a real number

$c = \sqrt{ab}$ in (a, b) such that

$f'(c) = 0$

Hence, Rolle's theorem is verified and $c = \sqrt{ab}$.

Ans.

5. *Using Rolle's theorem, find a point on the curve* $y = \sin x + \cos x - 1, x \in \left[0, \frac{\pi}{2}\right]$ *where the tangent is parallel to the x-axis.*

Sol. Given, $f(x) = \sin x + \cos x - 1, x \in \left[0, \frac{\pi}{2}\right]$.

(i) $f(x)$ is continuous in $\left[0, \frac{\pi}{2}\right]$.

(ii) $f(x)$ is derivable in $\left(0, \frac{\pi}{2}\right)$.

(iii) $f(0) = \sin 0 + \cos 0 - 1$

$= 0 + 1 - 1 = 0$

And, $f\left(\frac{\pi}{2}\right) = \sin\frac{\pi}{2} + \cos\frac{\pi}{2} - 1$

$= 1 + 0 - 1 = 0$

$\therefore \quad f(0) = f\left(\frac{\pi}{2}\right)$

Thus, all the three conditions of Rolle's theorem are satisfied, therefore, there exist at least one real number c in $\left(0, \frac{\pi}{2}\right)$ such that $f'(c) = 0$.

Differentiating $f(x) = \sin x + \cos x - 1$ with respect to 'x', we get

$f'(x) = \cos x - \sin x$

Now $f'(c) = 0 \Rightarrow \cos c - \sin c = 0 \Rightarrow \sin c = \cos c$

$\Rightarrow \tan c = 1$

$\Rightarrow \quad c = \frac{\pi}{4}, \frac{5\pi}{4}, \frac{9\pi}{4}, \frac{-3\pi}{4}$ .... but

$c \in \left(\frac{\pi}{4}\right)$

$\therefore \quad c = \frac{\pi}{4}$

$\therefore$ There exists, $\left(\frac{\pi}{4}\right)$ in $\left(0, \frac{\pi}{2}\right)$ such that

$f'\left(\frac{\pi}{4}\right) = 0$

Thus, the Rolle's theorem is verified and $c = \frac{\pi}{4}$.

Ans.

6. *Verify the conditions of Rolle's theorem for the following function :*

$$f(x) = \log(x^2 + 2) - \log 3 \text{ on } [-1, 1]$$

Find a point in the interval, where the tangent to the curve is parallel to x-axis.

Sol. (i) Since $x^2 + 2 > 0, \forall x \in [-1, 1]$, therefore, the function $f(x) = \log(x^2 + 2) - \log 3$ is continuous in $[-1, 1]$

(ii) $f'(x) = \frac{1}{x^2+2} \cdot 2x = \frac{2x}{x^2+2}$ which exists for all x. Thus, the function $f(x)$ is differentiable in the open interval $(-1, 1)$.

(iii) $f(-1) = \log(1 + 2) - \log 3 = 0,$

$f(1) = \log(1 + 2) - \log 3 = 0$

$\therefore \quad f(-1) = f(1)$

Thus, $f(x)$ satisfies all the conditions of Rolle's theorem.

$\therefore$ There must exist at least one value of x, say c, in the open interval $(-1, 1)$ such that $f'(c) = 0$

Now, $\quad f'(c) = \frac{2c}{c^2+2} = 0$

$\Rightarrow \quad c = 0 \in (-1, 1)$

Hence, Rolle's theorem is verified.

When $c = 0, f(0) = \log 2 - \log 3$

$= \log\left(\frac{2}{3}\right)$

So, there exist a point $\left(0, \log\left(\frac{2}{3}\right)\right)$ on the given curve $f(x) = \log(x^2 + 2) - \log 3$ where the tangent is parallel to the x-axis. **Ans.**

7. *Use Lagrange's mean value theorem to determine a point P on the curve* $y = \sqrt{x^2 - 4}$ *defined in the interval [2, 4] where the tangent is parallel to the chord joining the end-points on the curve.*

Sol. Given,

$f(x) = y = \sqrt{x^2 - 4}$ in the interval $[2, 4]$.

Since, $f(x)$ is continuous in the interval $[2, 4]$.

Differentiating $f(x)$ w.r.t. x, we get

$$\frac{dy}{dx} = \frac{1}{2}\frac{2x}{\sqrt{x^2-4}} = \frac{x}{\sqrt{x^2-4}}$$

which exists in the open interval (2, 4).

Let P be the point with abscissa c.

Where the tangent is parallel to the chord joining the end-points on the curve.

By Lagrange's mean value theorem,

$$f'(c) = \frac{f(b)-f(a)}{b-a}$$

where $a = 2, b = 4$.

i.e., $$\frac{c}{\sqrt{c^2-4}} = \frac{\sqrt{12}-0}{4-2} = \frac{2\sqrt{3}}{2} = \sqrt{3}.$$

i.e., $$\frac{c}{\sqrt{c^2-4}} = \sqrt{3}$$

$\Rightarrow \quad 2c^2 = 12$

$\Rightarrow \quad c^2 = 6$

$\therefore \quad c = \pm\sqrt{6}$

Since $-\sqrt{6} \notin [2, 4]$, $\therefore c = \sqrt{6}$

when $x = \sqrt{6}, y = \sqrt{6-4} = \sqrt{2}$

Hence, the required point is $(\sqrt{6}, \sqrt{2})$. **Ans.**

8. *Verify Lagrange's Mean Value Theorem for the function $f(x) = \sqrt{x^2 - x}$ in the interval [1, 4].*

Sol. Given, $f(x) = \sqrt{x^2-x}, x \in [1, 4]$

(i) Since $x^2 - x$ is continuous on R

$\Rightarrow \sqrt{x^2-x}$ is continuous $\forall\ x \in R$

$\Rightarrow f(x)$ is continuous in [1, 4].

(ii) Differentiating the given function w.r.t. x, we get

$$f'(x) = \frac{1}{2}(x^2-x)^{-1/2}\cdot(2x-1)$$

$$= \frac{2x-1}{2\sqrt{x^2-x}}$$

which exists for $x \in R$.

$\therefore f(x)$ is derivable in (1, 4).

Thus, both the conditions of Lagrange's mean value theorem is satisfied therefore, there exists c in (1, 4), such that

$$f'(c) = \frac{f(4)-f(1)}{4-1}$$

$$\Rightarrow \frac{2c-1}{2\sqrt{c^2-c}} = \frac{\sqrt{12}}{3}$$

$\Rightarrow \quad 3(2c-1) = 2\sqrt{c^2-c}.\ \sqrt{12}$

$\Rightarrow 9(4c^2 - 4c + 1) = 48(c^2 - c)$

$\Rightarrow 3(4c^2 - 4c + 1) = 16(c^2 - c)$

$\Rightarrow \quad 12c^2 - 12c + 3 = 16c^2 - 16c$

$\Rightarrow \quad 4c^2 - 4c - 3 = 0$

$\Rightarrow (2c - 3)(2c + 1) = 0$

$\Rightarrow \quad c = \frac{3}{2}, \frac{-1}{2}$

Thus, there exist $c = \frac{3}{2} \in (1, 4)$

Such that $f'\left(\frac{3}{2}\right) = \frac{f(4)-f(1)}{4-1}$

Hence, Lagrange's mean value theorem is verified and $c = \frac{3}{2}$. **Ans.**

9. *Verify Lagrange's mean value theorem for the function $f(x) = \sin x - \sin 2x$ in the interval $[0, \pi]$.*

Sol. Given, $f(x) = \sin x - \sin 2x, x \in [0, \pi]$

The function $f(x) = \sin x - \sin 2x$ is derivable for all values of x, and hence derivable in $[0, \pi]$

Also $\quad f'(x) = \cos x - 2\cos 2x$

Thus, both conditions of Lagrange's mean value theorem is satisfied therefore, there exists c in $(0, \pi)$

Such that, $$f'(c) = \frac{f(\pi)-f(0)}{\pi-0}$$

$$\Rightarrow \quad \cos c - 2\cos 2c = \frac{0-0}{\pi} = 0$$

$\Rightarrow \quad \cos c - 2(2\cos^2 c - 1) = 0$

$[\because \cos 2x = 2\cos^2 x - 1]$

$\Rightarrow \quad 4\cos^2 c - \cos c - 2 = 0$

$$\Rightarrow \quad \cos c = \frac{1 \pm \sqrt{1+32}}{8}$$

$$= \frac{1 \pm \sqrt{33}}{8}$$

$$\Rightarrow \quad \cos c = \frac{1 \pm 5{\cdot}7}{8}$$

$$\Rightarrow \quad \cos c = \frac{1 + 5{\cdot}7}{8}$$

$$\text{or} \quad \cos c = \frac{1 - 5{\cdot}7}{8}$$

$\Rightarrow \quad \cos c = 0{\cdot}84$

or $\quad \cos c = -0{\cdot}59$

$\Rightarrow \quad c = 33°$

or $\quad c = 126{\cdot}2°$ (approx.)

Since, both values of c lie in the interval $[0, \pi]$.

Hence, Lagrange's Mean Value theorem is verified. **Ans.**

10. *Examine the validity and conclusion of Lagrange's mean value theorem for the function*

$$f(x) = x(x-1)(x-2) \text{ for every } x \in \left(0, \frac{1}{2}\right).$$

Sol. Given, $f(x) = x(x-1)(x-2),\ x \in \left(0, \frac{1}{2}\right)$

(i) The given function is continuous in the interval $\left(0, \frac{1}{2}\right)$

(ii) Now, $f(x) = (x^2 - x)(x-2)$

$$= x^3 - 2x^2 - x^2 + 2x$$

$$= x^3 - 3x^2 + 2x$$

$$\therefore \quad f'(x) = 3x^2 - 6x + 2$$

$$\Rightarrow \quad f'(c) = 3c^2 - 6c + 2$$

differentiable in $\left(0, \frac{1}{2}\right)$

$$f\left(\frac{1}{2}\right) = \frac{1}{2}\left(\frac{1}{2}-1\right)\left(\frac{1}{2}-2\right) = \frac{3}{8}$$

By Lagrange's mean value theorem

$$\therefore \quad \frac{f(b)-f(a)}{b-a} = f'(c)$$

$$\Rightarrow \quad \frac{\frac{3}{8}-0}{\frac{1}{2}-0} = 3c^2 - 6c + 2$$

$$\Rightarrow \quad \frac{3}{4} = 3c^2 - 6c + 2$$

$$\Rightarrow \quad 3 = 12c^2 - 24c + 8$$

$$\Rightarrow 12c^2 - 24c + 5 = 0$$

$$\Rightarrow \quad c = \frac{24 \pm \sqrt{576-240}}{24}$$

$$= \frac{24 \pm \sqrt{336}}{24}$$

$$\Rightarrow \quad c = \frac{24 + 4\sqrt{21}}{24}$$

$$\text{or} \quad c = \frac{24 - 4\sqrt{21}}{24}$$

$$\therefore \quad c = \frac{4(6+\sqrt{21})}{24}$$

$$\text{or} \quad c = \frac{4(6-\sqrt{21})}{24}$$

$$\Rightarrow \quad c = \frac{6+\sqrt{21}}{6} = 1 + \frac{\sqrt{21}}{6}$$

$$\text{or} \quad c = \frac{6-\sqrt{21}}{6} = 1 - \frac{\sqrt{21}}{6}$$

$$= 0{\cdot}236 \text{ (approx).}$$

Since, value of 'c' must lie within 0 and $\frac{1}{2}$, but $c = 1 + \frac{\sqrt{21}}{6}$ is beyond this value, hence it is not acceptable. **Ans.**

11. *Verify Lagrange's Mean Value Theorem for the following function :*

$f(x) = 2\sin x + \sin 2x$ on $[0, \pi]$.

Sol. Given, $f(x) = 2\sin x + \sin 2x,\ x \in [0, \pi]$

$f(x)$ is continuous in $[0, \pi]$

$f(x)$ is differentiable in $(0, \pi)$

Thus, both the conditions of Lagrange's mean value theorem are satisfied by the function $f(x)$ in $[0, \pi]$, therefore, there exists at least one real number c in $(0, \pi)$ such that

$$f'(c) = \frac{f(\pi) - f(0)}{\pi - 0}$$

$$f(\pi) = 2\sin\pi + \sin 2\pi = 0$$

$$f(0) = 2\sin 0 + \sin 0 = 0$$

Differentiating $f(x)$ w.r.t. x, we get

$$f'(x) = 2\cos x + 2\cos 2x$$

Now, $f'(c) = \frac{0}{\pi}$

$$\Rightarrow \quad 2\cos c + 2\cos 2c = 0$$

$$\cos c + \cos 2c = 0$$

$$\Rightarrow \quad 2\cos^2 c + \cos c - 1 = 0$$

$$(\because \cos 2x = 2\cos^2 x - 1)$$

$$\Rightarrow \quad 2\cos^2 c + 2\cos c - \cos c - 1 = 0$$

$$\Rightarrow 2\cos c(\cos c + 1) - 1(\cos c + 1) = 0$$

$$\Rightarrow \quad (2\cos c - 1)(\cos c + 1) = 0$$

$$2\cos c - 1 = 0$$

$$\text{or} \quad \cos c + 1 = 0$$

$$2\cos c = 1 \quad \text{or} \quad \cos c = -1$$

$$\cos c = \frac{1}{2} \quad \text{or} \quad \cos c = -1$$

$$\Rightarrow \quad c = \frac{\pi}{3} \text{ or } \pi$$

$\therefore \quad c = \frac{\pi}{3} \qquad \left[\because \frac{\pi}{3} \in (0, \pi)\right]$

Thus, Lagrange's mean value theorem is verified. **Ans.**

12. *Verify Lagrange's mean value theorem for the function :*

f(x) = x (1 – log x) and find the value of 'c' in the interval [1, 2]

Sol. Given, $f(x) = x(1 - \log x)$ in [1, 2]

The function is continuous in the closed interval [1, 2] as x and $\log x$ are continous in [1, 2].

Now, $f'(x) = 1 - \log x + x\left(-\frac{1}{x}\right)$

$= 1 - \log x - 1$

$= -\log x$ which exists for all values in (1, 2).

So, $f(x)$ is differentiable in (1, 2)

By Lagrange's mean value theorem, we have

$$f'(c) = \frac{f(2) - f(1)}{2-1}$$

$$\Rightarrow -\log c = \frac{2(1-\log 2) - 1(1 - \log 1)}{2-1}$$

$$\Rightarrow -\log c = \frac{2 - 2\log 2 - 1 + \log 1}{1}$$

$$\Rightarrow -\log c = 1 - \log 2^2 \qquad (\because \log 1 = 0)$$

$$\Rightarrow \log c = \log 4 - \log e$$

$$\Rightarrow \log c = \log\left(\frac{4}{e}\right)$$

$c = \frac{4}{e}$ which lies in the interval (1, 2

Hence, the theorem is verified. **Hence verified**

Chapter 7. Applications of Derivatives

1. (a) *The volume of a closed rectangular metal box with a square base is 4096 cm³. The cost of polishing the outer surface of the box is ₹ 4 per cm². Find the dimensions of the box for the minimum cost of polishing it.*

(b) *Find the point on the straight line 2x + 3y = 6, which is closest to the origin.**

Sol. (a) Let Length of the box = x cm

Height of the box = h cm

$\therefore$ Breadth of the box = x cm [$\because$ square base]

$\because$ Volume of the box = 4096 cm³

$$x^2h = 4096$$

$$h = \frac{4096}{x^2} \qquad \ldots(i)$$

Surface area of outer surface $= 2[lb + bh + lh]$

$= 2[x^2 + xh + hx]$

$= 2x^2 + 4hx$

Cost of polishing the surface = ₹ 4 per cm²

So, Total cost, $C = 4(2x^2 + 4hx)$

$$\Rightarrow C = 8x^2 + 16xh$$

$$\Rightarrow C = 8x^2 + 16x\left[\frac{4096}{x^2}\right] \qquad \text{[From (i)]}$$

$$\Rightarrow C = 8x^2 + \frac{65536}{x}$$

Differentiating w.r.t. x, we get

$$\frac{dC}{dx} = 16x - \frac{65536}{x^2}$$

$$\frac{d^2C}{dx^2} = 16 + \frac{65536 \times 2}{x^3}$$

$$\frac{d^2C}{dx^2} > 0$$

$\because$ Cost of polishing the surface is minimum,

$$\frac{dC}{dx} = 0$$

$$\Rightarrow 16x - \frac{65536}{x^2} = 0$$

$$\Rightarrow 16x^3 - 65536 = 0$$

$$\Rightarrow x^3 = \frac{65536}{16}$$

$$\Rightarrow x^3 = \frac{16 \times 4096}{16}$$

$$\Rightarrow x^3 = 4096$$

$$\Rightarrow x = 16 \text{ cm}$$

$$\therefore h = \frac{4096}{16 \times 16}$$

$= 16$ cm

$\therefore$ Length = 16 cm, breadth = 16 cm and height = 16 cm. **Ans.**

(b) The given equation of the line is $2x+3y=6$. Let the point on the straight line be P (x, y) and O (0, 0) be the origin.

$$\therefore \quad OP = \sqrt{(x-0)^2+(y-0)^2}$$

$$OP = \sqrt{x^2+y^2}$$

$$\therefore \quad OP^2 = x^2+y^2$$

Now, OP is minimum when OP^2 is minimum

Let $\quad OP^2 = f(x)$

$$\therefore \quad f(x) = x^2+y^2$$

$$\because \quad 2x+3y = 6$$

$$\therefore \quad y = \frac{6-2x}{3}$$

$$\Rightarrow \quad f(x) = x^2 + \left(\frac{6-2x}{3}\right)^2$$

Differentiating w.r.t. x,

$$\Rightarrow \quad f'(x) = 2x + \frac{2}{9}(6-2x)(-2)$$

For $f(x)$ to be minimum,

$$f'(x) = 0$$

$$\Rightarrow 2x - \frac{4(6-2x)}{9} = 0$$

$$\Rightarrow \quad 2x = \frac{4(6-2x)}{9}$$

$$\Rightarrow \quad 9x = 12-4x$$

$$\Rightarrow \quad 13x = 12$$

$$\Rightarrow \quad x = \frac{12}{13}$$

$$\therefore \quad y = \frac{6-2x}{3} = \frac{6-2\times\frac{12}{13}}{3}$$

$$y = \frac{54}{39} = \frac{18}{13}$$

$\therefore$ The point on the straight line P(x, y) is $\left(\frac{12}{13}, \frac{18}{13}\right)$

Now, $$f''(x) = 2 - \frac{4}{9}\times(-2)$$

$$= 2+\frac{8}{9} = \frac{26}{9}$$

$$\because \quad f''(x) > 0$$

$\therefore$ P$\left(\frac{12}{13}, \frac{18}{13}\right)$ is closest to the origin. **Ans.**

2. *A box is to be constructed from a square metal sheet of side 60 cm, by cutting out identical squares from the four corners and turning up the sides. Find the length of side of square to be cut out so that box has maximum volume.*

Sol. Let the length of side of square to be cut out be x.

$\therefore$ Volume of the box (V) = $l \times b \times h$

$$= x(60-2x)^2$$

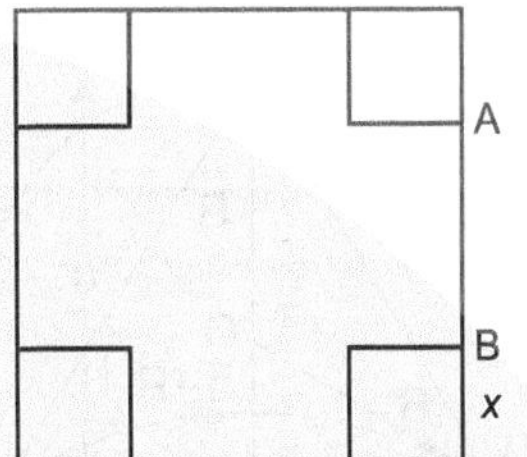

Differentiating w.r.t. x, we get

$$\frac{dV}{dx} = (60-2x)^2 \times 1 + x \, . \, 2(60-2x)(-2)$$

$$= (60-2x)^2 - 4x(60-2x)$$

$$= (60-2x)(60-2x-4x)$$

$$= (60-2x)(60-6x)$$

For maxima or minima, $\frac{dV}{dx} = 0$

$$\Rightarrow \quad (60-2x)(60-6x) = 0$$

$$\Rightarrow \quad x = 30, \; x = 10$$

Since 30 is impossible because then volume of the box V = 0 (not possible), therefore

$$x = 10$$

Differentiating $\frac{dV}{dx}$ w.r.t x, we get

$$\frac{d^2V}{dx^2} = (60-2x)(-6)+(60-6x)(-2)$$

$$= -360+12x-120+12x$$

$$= 24x-480$$

At $x = 10$,

$$\frac{d^2V}{dx^2} = 240-480 < 0$$

$\therefore$ Maxima exists at $x = 10$

$\therefore$ At $x = 10$, V = $x(60-2x)^2$

$$= 10(60-20)^2$$

$$= 10 \times 1600$$

$$= 16000 \text{ cm}^3$$

Thus, the length of the side of square is 10 cm.

Ans.

3. *Find the volume of the largest cone that can be inscribed in a sphere of radius R.*

Sol. Let cone of base radius r and height h is inscribed in a sphere of radius R.

In right Δ OAB, by Pythagoras theorem

$$R^2 = (h - R)^2 + r^2$$

$$\Rightarrow \quad r^2 = R^2 - h^2 + 2hR - R^2$$

$$\Rightarrow \quad r^2 = 2hR - h^2 \quad \text{...(i)}$$

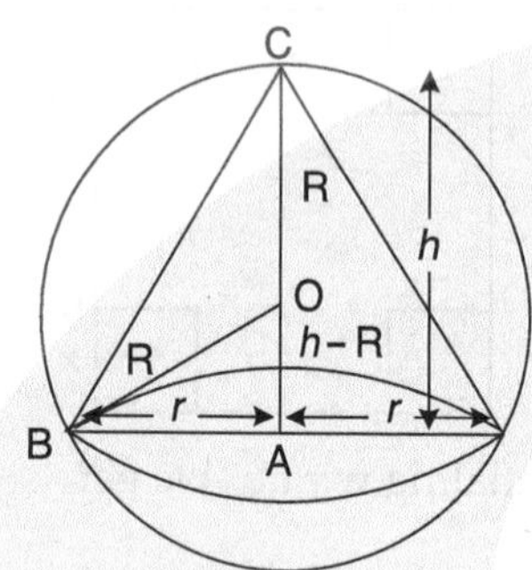

Now, Volume of cone,

$$V = \frac{1}{3}\pi r^2 h$$

$$\Rightarrow \quad V = \frac{1}{3}\pi(2hR - h^2)h \quad \text{[from (i)]}$$

$$\Rightarrow \quad V = \frac{1}{3}\pi(2h^2R - h^3)$$

On differentiating w.r.t. h, we get

$$\frac{dV}{dh} = \frac{1}{3}\pi(4hR - 3h^2)$$

For maxima or minima, $\frac{dV}{dh} = 0$

$$\therefore \quad 4hR - 3h^2 = 0$$

$$\Rightarrow \quad h(4R - 3h) = 0$$

$$\Rightarrow \quad h = 0,\ h = \frac{4R}{3}$$

Differentiating $\frac{dV}{dh}$ w.r.t. h, we get

$$\frac{d^2V}{dh^2} = \frac{1}{3}\pi(4R - 6h)$$

At $h = \frac{4R}{3}$

$$\Rightarrow \quad \frac{d^2V}{dh^2} = \frac{1}{3}\pi\left(4R - 6 \times \frac{4R}{3}\right)$$

$$= \frac{1}{3}\pi(-4R) < 0$$

∴ For $h = \frac{4R}{3}$, volume is maximum.

$$\therefore \quad r^2 = 2hR - h^2$$

$$= 2R \times \frac{4R}{3} - \frac{16R^2}{9}$$

$$= \frac{8R^2}{3} - \frac{16}{9}R^2$$

$$= \frac{24R^2 - 16R^2}{9} = \frac{8R^2}{9}$$

Volume of cone $= \frac{1}{3}\pi\frac{8R^2}{9} \times \frac{4R}{3}$

Hence, Maximum volume of cone

$$= \frac{32}{81}\pi R^3 \text{ cu. units.}$$ **Ans.**

4. *A wire of length 50 m is cut into two pieces. One piece of the wire is bent in the shape of a square and the other in the shape of a circle. What should be the length of each piece so that the combined area of the two is minimum ?*

Sol. Length of wire = 50 m (Given)

Let length of one piece to make a square = x m

∴ Length of other piece to make a circle

$$= (50 - x) \text{ m}$$

Now, perimeter of square = $4a = xm$

⇒ Side of a square, $a = \frac{x}{4} m$

And, circumference of circle = $2\pi r = (50 - x)m$

⇒radius of circle, $r = \frac{50 - x}{2\pi}$

∴Combined Area, $A = a^2 + \pi r^2$

$$= \frac{x^2}{16} + \pi . \left(\frac{50 - x}{2\pi}\right)^2$$

$$= \frac{x^2}{16} + \pi . \frac{(50 - x)^2}{4\pi^2}$$

$$\Rightarrow \quad A = \frac{x^2}{16} + \frac{(50 - x)^2}{4\pi}$$

Differentiating w.r. t. x, we get

$$\frac{dA}{dx} = \frac{2x}{16} + \frac{2(50 - x)(-1)}{4\pi}$$

$$\Rightarrow \quad \frac{dA}{dx} = \frac{x}{8} + \frac{(x - 50)}{2\pi}$$

$$= \frac{\pi x + 4x - 200}{8\pi}$$

$$= \frac{x(4 + \pi) - 200}{8\pi}$$

Formulae Based Questions

For maxima or minima,

$$\frac{dA}{dx} = 0$$

$\therefore x(4+\pi) - 200 = 0$

$$\Rightarrow \quad x = \frac{200}{4+\pi}$$

$$\frac{d^2A}{dx^2} = \frac{(4+\pi)}{8\pi}$$

$$\because \quad \frac{d^2A}{dx^2} > 0$$

Thus, area is minimum when

Length of square wire, $x = \frac{200}{4+\pi}$ m

and length of circle wire $= 50 - \frac{200}{4+\pi}$

$$= \frac{50(4+\pi)-200}{4+\pi}$$

$$= \frac{50\pi}{4+\pi} \text{ m}$$ **Ans.**

5. *Find the maximum volume of the cylinder which can be inscribed in a sphere of radius $3\sqrt{3}$ cm. (Leave the answer in terms of π).*

Sol. Let h be the height of the cylinder and R be the radius of the cylinder. Given that $3\sqrt{3}$ cm is the radius of the sphere.

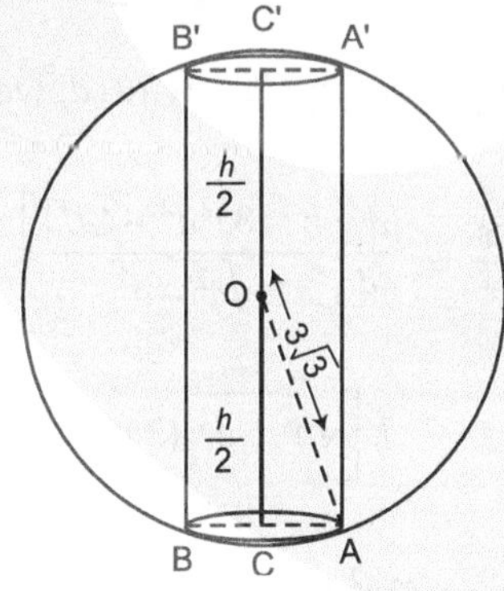

If V is the volume of the cylinder, then

$$V = \pi R^2 h \quad \ldots(i)$$

Let O be the centre of the sphere and $C'OC \perp B'A'$ as well as BA.

In right angled Δ OCA, by Pythagoras theorem

$$OA^2 = OC^2 + CA^2$$

$$(3\sqrt{3})^2 = \left(\frac{h}{2}\right)^2 + R^2$$

$$\Rightarrow \quad R^2 = 27 - \frac{h^2}{4} \quad \ldots(ii)$$

$$\therefore \quad V = \pi\left(27 - \frac{h^2}{4}\right)\cdot h$$

$$\Rightarrow \quad V = 27\pi h - \pi\frac{h^3}{4}$$

Differentiating w.r.t. h, we get

$$\frac{dV}{dh} = 27\pi - \frac{3\pi h^2}{4}$$

Differentiating again w.r.t. h, we get

$$\frac{d^2V}{dh^2} = -\frac{6\pi h}{4}$$

For maxima or minima, = 0

$$\Rightarrow \quad \frac{dV}{dh} = 0$$

$$\therefore \quad 27\pi = \frac{3\pi h^2}{4}$$

$$\Rightarrow \quad h^2 = \frac{4\times 27}{3}$$

$$\Rightarrow \quad h^2 = 36$$

$$\Rightarrow \quad h = 6, -6$$

Ignore $h = -6$, as height cannot be negative

and $\left(\frac{d^2V}{dh^2}\right)_{h=6} = \frac{-6\pi\times 6}{4} < 0$

$\therefore$ V is maximum when $h = 6$,

Putting, $h = 6$ in equation (ii), we get

$$\Rightarrow \quad R^2 = 27 - \frac{36}{4}$$

$$\Rightarrow \quad R^2 = 18$$

$\therefore$ From (i), $V = \pi R^2 h$

$$= \pi \times 18 \times 6$$

$$= 108\pi \text{ cm}^3$$ **Ans.**

6. *A closed right circular cylinder has volume $\frac{539}{2}$ cubic units. Find the radius and the height of the cylinder so that the total surface area is minimum.*

Sol. Let r be the radius and h be the height of the cylinder.

Given, $V = \pi r^2 h = \frac{539}{2}$

$$\Rightarrow \quad h = \frac{539}{2\pi r^2}$$

Let S be the total surface area, then

$$S = 2\pi rh + 2\pi r^2$$

$$= 2\pi r\left(\frac{539}{2\pi r^2}\right)+2\pi r^2$$

$$= \frac{539}{r}+2\pi r^2$$

Differentiating w.r.t. r, we get

$$\therefore \quad \frac{dS}{dr} = -\frac{539}{r^2}+4\pi r \quad \ldots(i)$$

For maxima or minima $\frac{dS}{dr} = 0$

$$\Rightarrow \quad -\frac{539}{r^2}+4\pi r = 0$$

$$\Rightarrow \quad r = \left(\frac{539}{4\pi}\right)^{1/3} = \frac{7}{2} \text{ units}$$

Differentiating equation (i) w.r.t. r, we get

$$\therefore \quad \frac{d^2S}{dr^2} = \frac{1078}{r^3}+4\pi\text{, which is positive}$$

$\therefore$ S is minimum, when $r = \frac{7}{2}$ and

$$h = \frac{539}{2\pi\left(\frac{7}{2}\right)^2} = \frac{539}{7\times\frac{7}{2}\times\frac{22}{7}} = 7 \text{ units.}$$ **Ans.**

7. *Assuming that the stiffness of a beam of a rectangular cross-section varies as the breadth and as the cube of depth, what must be the breadth of stiffest beam that can be cut from a log of diameter a.*

Sol. Let s be the stiffness of beam of rectangular cross-section.

Given, $\quad s \propto b \times d^3$

$\Rightarrow \quad s = k\,.\,b\,.\,d^3,$

where k is a constant

Given that beam is cut of a cylinder of diameter 'a' as shown in the figure.

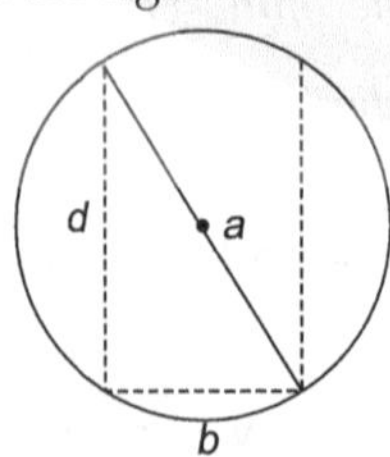

At any time, by Pythagoras theorem

$$b^2 = a^2 - d^2$$

$$\Rightarrow \quad b = \sqrt{a^2-d^2}$$

$$\Rightarrow \quad s = k\,.\,b\,.\,d^3 = k\,.\,\sqrt{a^2-d^2}\,.\,d^3$$

On differentiating w.r.t. d, we get

$$\Rightarrow \frac{ds}{dd} = k\left[\frac{1}{2\sqrt{a^2-d^2}}\cdot(-2d)\,.\,d^3+\sqrt{a^2-d^2}\,.\,3d^2\right]$$

$$= k\left[\frac{-d^4+(a^2-d^2)\,.\,3d^2}{\sqrt{a^2-d^2}}\right]$$

$$= k\left[\frac{-d^4+3a^2d^2-3d^4}{\sqrt{a^2-d^2}}\right]$$

$$= k\left[\frac{-4d^4+3a^2d^2}{\sqrt{a^2-d^2}}\right]$$

For maxima or minima, $\frac{ds}{dd} = 0$

$$\Rightarrow -4d^4+3a^2d^2 = 0$$

$$\Rightarrow \quad 3a^2d^2 = 4d^4$$

$$\Rightarrow \quad 3a^2 = 4d^2$$

$$\Rightarrow \quad a = \frac{2}{\sqrt{3}}d$$

$$\Rightarrow \quad d = \frac{\sqrt{3}}{2}a$$

$$\Rightarrow \quad b = \sqrt{a^2-d^2} = \sqrt{a^2-\frac{3}{4}a^2}$$

$$= \sqrt{\frac{a^2}{4}} = \frac{a}{2}$$

Differentiating $\frac{ds}{dd}$ w.r.t. d, we get

$$\frac{d^2s}{dd^2} = \frac{d}{dd}\left[\frac{k\,.\,\{3a^2d^2-4d^4\}}{\sqrt{a^2-d^2}}\right]$$

$$= k\left[\frac{\sqrt{a^2-d^2}\,.(3a^2\times 2d-16d^3) - (3a^2d^2-4d^4)\left(-\frac{1}{2}\right)(a^2-d^2)^{-3/2}.(-2d)}{(a^2-d^2)}\right.$$

$$= k\left[\frac{\sqrt{a^2-d^2}\,.(6a^2d-16d^3)-(3a^2d^3-4d^5)}{}\right.$$

$$\frac{(a^2-d^2)^{-3/2}}{(a^2-d^2)}$$

$$\because \quad d = \frac{\sqrt{3}}{2}a$$

$$= k\left[\sqrt{a^2-\frac{3}{4}a^2}\left(6a^2\times\frac{\sqrt{3}}{2}a-\frac{16\times3\sqrt{3}\,a^3}{8}\right)\right.$$

$$\left.-\frac{\left(3a^2\times\frac{3\sqrt{3}}{8}a^3-\frac{4\times9\sqrt{3}}{32}a^5\right)\left(a^2-\frac{3}{4}a^2\right)^{-3/2}}{\left(a^2-\frac{3}{4}a^2\right)}\right]$$

$$= k\left[\frac{a}{2}\left(3\sqrt{3}\,a^3-6\sqrt{3}\,a^3\right)\right.$$

$$\left.-\frac{\left(\frac{9\sqrt{3}}{8}a^5-\frac{9\sqrt{3}}{8}a^5\right)\left(\frac{a^2}{4}\right)^{-3/2}}{\frac{a^2}{4}}\right]$$

$$= \frac{k\left[\frac{a}{2}\times(-3\sqrt{3}\,a^3)-0\right]}{\frac{a^2}{4}}$$

$$= \frac{-k.\,6\sqrt{3}\,a^4}{a^2}$$

$$= -6\sqrt{3}\ ka^2 < 0$$

Hence, it is point of maxima.

$\therefore$ For maximum stiffness $d = \frac{\sqrt{3}}{2}a,\ b = \frac{a}{2}$ **Ans.**

8. *How should a wire 20 cm long be divided into two parts, if one part is to be bent into a circle, the other part is to be bent into a square and the two plane figures are to have areas the sum of which is minimum.*

Sol. Let one piece be x cm.

Then, the other piece will be $(20 - x)$ cm.

Let x cm be bent into a circle and $(20 - x)$ be bent into a square.

$\therefore \quad 2\pi r = x \quad$ or $\quad r = \frac{x}{2\pi}$

$\Rightarrow$ Area of circle $= \pi\left(\frac{x}{2\pi}\right)^2 = A_1$ (say)

Perimeter of the square,

$$4a = 20 - x$$

$$\Rightarrow \quad a = \frac{20-x}{4}$$

$\Rightarrow$ Area of square $= \left(\frac{20-x}{4}\right)^2 = A_2$ (say)

Let A be the combined area $= A_1 + A_2$

$$\therefore \quad A = \pi\left(\frac{x}{2\pi}\right)^2+\left(\frac{20-x}{4}\right)^2$$

$$= \frac{x^2}{4\pi}+\frac{(20-x)^2}{16}$$

Differentiating w.r.t. x, we get

$$\frac{dA}{dx} = \frac{1}{4\pi}\times2x+\frac{1}{16}[2\times(20-x)\times(-1)]$$

$$\frac{dA}{dx} = \frac{x}{2\pi}-\frac{(20-x)}{8}$$

For area to be minimum, $\frac{dA}{dx} = 0$ and $\frac{d^2A}{dx^2}$ is positive.

i.e., $\quad \frac{x}{2\pi} - \frac{(20-x)}{8} = 0$

$$\Rightarrow \quad 8x - 40\pi + 2x\pi = 0$$

$$\Rightarrow \quad x(8 + 2\pi) = 40\pi$$

On solving, we get

$$x = \frac{20\pi}{\pi+4}$$

Differentiating $\frac{dA}{dx}$ w.r.t. x, we get

$$\frac{d^2A}{dx^2} = \frac{1}{2\pi}+\frac{1}{8} > 0$$

Hence, the wire should be bent at a distance of $\frac{20\pi}{\pi+4}$ from one end. **Ans.**

9. *A printed page is to have a total area of 80 sq. cm with a margin of 1 cm at the top and on each side and a margin of 1·5 cm at the bottom. What should be the dimensions of the page so that the printed area will be maximum ?*

Sol. Let x cm and y cm be the dimensions of the printed page.

$$\Rightarrow \quad xy = 80 \quad \ldots\text{(i)}$$

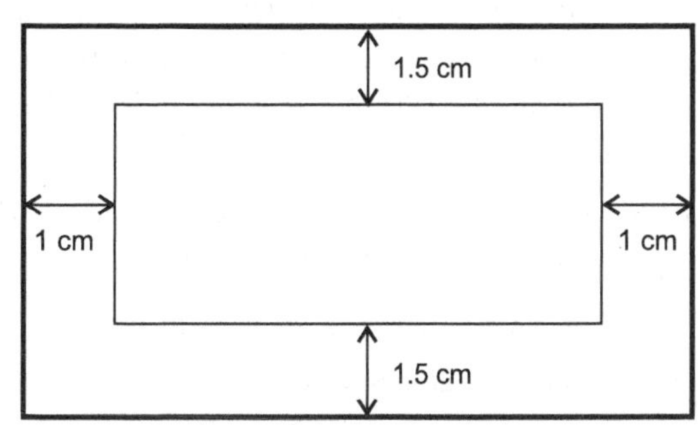

Let A (x sq. cm) be the printed area then,

$$A = (x-2)(y-5/2)$$

$$\Rightarrow \quad A = xy - \frac{5x}{2} - 2y + 5$$

$$\Rightarrow \quad A = 80 - \frac{5x}{2} - \frac{2\times 80}{x} + 5$$

[Using (i)]

$$\Rightarrow \quad A = 85 - \frac{5x}{2} - \frac{160}{x}$$

Differentiating w.r.t. x, we get

$$\frac{dA}{dx} = \frac{-5}{2} + \frac{160}{x^2}$$

Differentiating once again w.r.t. x, we get

$$\frac{d^2A}{dx^2} = \frac{-320}{x^3}$$

For Maximum and Minimum Area, $\frac{dA}{dx} = 0$

$$\Rightarrow \quad \frac{-5}{2} + \frac{160}{x^2} = 0$$

$$\Rightarrow \quad \frac{160}{x^2} = \frac{5}{2}$$

$$\Rightarrow \quad x^2 = 64$$

$$\Rightarrow \quad x = \pm\sqrt{64}$$

$$\Rightarrow \quad x = 8, -8,$$

But x cannot be negative

Also, $\left(\frac{d^2A}{dx^2}\right)_{x=8} = -\frac{320}{512} < 0$

Therefore, A is maximum at $x = 8$

From (i), $xy = 80$

$$\Rightarrow \quad 8y = 80$$

$$\Rightarrow \quad y = 10$$

Therefore, the dimensions of the printed page are 10 cm and 8 cm. **Ans.**

10. *The length of the perimeter of a section of a circle is 20 cm. Give an expression for the area of the sector in terms of r (the radius of the circle) and hence find the maximum area of the sector.*

Sol. Let O be the centre of the circle and OAB be the sector whose perimeter is 20 cm.

Let $\widehat{AB} = S$ and $\angle AOB = \theta$

Then, $2r + S = 20$

$S = r\theta$ [FromTrignometory]

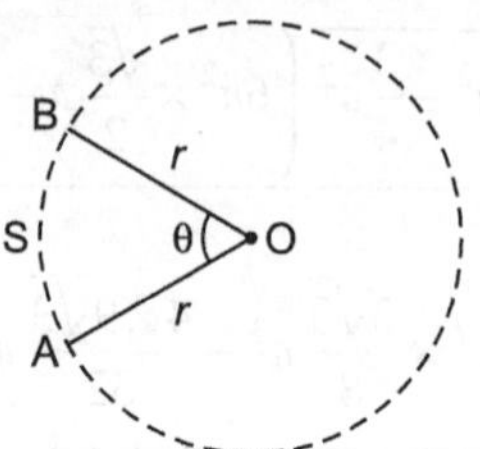

Let A be the area of the sector $= \frac{1}{2}r^2\theta$

$$= \frac{1}{2}r^2 \times \frac{S}{r} = \frac{1}{2}Sr$$

$$= \frac{1}{2}r\,(20-2r) = 10r - r^2$$

For the area A to be maximum, we must have

$$\frac{dA}{dr} = 0$$

Differentiating A w.r.t. r, we get

$$\frac{dA}{dr} = 10 - 2r = 0$$

$$\Rightarrow \quad r = 5$$

Differentiating $\frac{dA}{dr}$ w.r.t. r, we get

$$\frac{d^2A}{dr^2} = -2 < 0$$

Hence, the area is maximum when $r = 5$.

Area $= 10 \times 5 - 5 \times 5$

$= 25$ sq. cm. **Ans.**

11. *ABC is a right angled triangle of given area S. Find the sides of the triangle for which the area of circumscribed circle is east.*

Sol. Let ABC be a right–angled triangle whose area 'S' is constant.

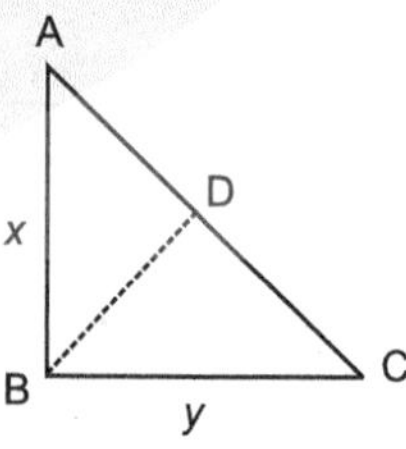

$$\therefore \quad S = \frac{1}{2}\times \text{Base} \times \text{Height}$$

$$= \frac{1}{2}x \times y = \frac{1}{2}xy$$

Circumscribed circle of ΔABC is one which passes through A, B and C.

Formulae Based Questions

Let 'D' be the centre of circumscribing circle.

$\therefore$ DA = DB = DC = radius

$\Rightarrow$ DA = DC

$\therefore$ D is the mid–point of AC.

$\therefore$ $AD = DC = \frac{1}{2} \times AC$

Area of circumscribed circle

$$= \pi \times (\text{Radius of Circle})^2$$

$$= \pi \times \left\{\frac{1}{2}\sqrt{x^2+y^2}\right\}^2$$

$$\Rightarrow \quad A = \frac{\pi}{4}(x^2+y^2)$$

Since, S is constant and $S = \frac{1}{2}xy$

$$\therefore \quad y = \frac{2S}{x} \quad \text{...(i)}$$

$\therefore$ A can be changed as a function of x only

$$\Rightarrow \quad A = \frac{\pi}{4}\left(x^2 + \left(\frac{2S}{x}\right)^2\right) \quad \text{[Using (i)]}$$

$$\Rightarrow \quad A = \frac{\pi}{4}\left\{x^2 + \frac{4S^2}{x^2}\right\}$$

Differentiating A w.r.t. x, we get

$$\frac{dA}{dx} = \frac{\pi}{4}\{2x + 4S^2 \times (-2)\,x^{-3}]$$

$$= \frac{\pi}{4}\left\{2x - \frac{8S^2}{x^3}\right\}$$

For A being least, $\frac{dA}{dx} = 0$

$$\therefore \quad 0 = \frac{\pi}{4}\left\{2x - \frac{8S^2}{x^3}\right\}$$

$$\Rightarrow \quad 0 = 2x^4 - 8S^2$$

$$\Rightarrow \quad x^4 = 4S^2$$

$$\Rightarrow \quad S^2 = \frac{x^4}{4} \quad \text{or} \quad S = \frac{1}{2}x^2$$

But, $S = \frac{1}{2}xy$

$$\therefore \quad \frac{1}{2}xy = \frac{1}{2}x^2$$

$$\Rightarrow \quad y = x$$

$\therefore$ Area of circumscribed circle is least, when

$$c = y.$$

Thus, right angled–triangle is isosceles triangle whose two sides forming right angle are equal.

Ans.

Chapter 8. Integral Calculus

1. *Evaluate :* $\int_0^{\pi} \frac{x \tan x}{\sec x + \tan x} dx$ *

Sol. Let

$$I = \int_0^{\pi} \frac{x \tan x}{\sec x + \tan x} dx \quad \text{...(i)}$$

Using $\int_0^a f(x)dx = \int_0^a f(a-x)dx$

$$I = \int_0^{\pi} \frac{(\pi - x)\tan(\pi - x)}{\sec(\pi - x) + \tan(\pi - x)} dx$$

$$I = \int_0^{\pi} \frac{(\pi - x)(-\tan x)}{(-\sec x) + (-\tan x)} dx$$

$$\Rightarrow \quad I = \int_0^{\pi} \frac{(\pi - x)\tan x}{\sec x + \tan x} dx \quad \text{...(ii)}$$

Adding equations (i) and (ii), we get

$$2I = \int_0^{\pi} \frac{\pi \tan x}{\sec x + \tan x} dx$$

$$\Rightarrow \quad 2I = \pi \int_0^{\pi} \frac{\tan x}{\sec x + \tan x} dx$$

$$\Rightarrow \quad 2I = \pi \int_0^{\pi} \frac{\tan x}{\sec x + \tan x} \times \frac{(\sec x - \tan x)}{(\sec x - \tan x)} dx$$

$$\Rightarrow \quad 2I = \pi \int_0^{\pi} \frac{\tan x(\sec x - \tan x)}{\sec^2 x - \tan^2 x} dx$$

$$\Rightarrow \quad 2I = \pi \int_0^{\pi} \frac{\sec x \tan x - \tan^2 x}{1} dx$$

$$[\because \sec^2 x - \tan^2 x = 1]$$

$$\Rightarrow \quad 2I = \pi \int_0^{\pi} \sec x \tan x - \sec^2 x + 1\, dx$$

$$[\because \tan^2 x = \sec^2 x - 1]$$

$$\Rightarrow \quad 2I = \pi\left[\sec x - \tan x + x\right]_0^{\pi}$$

$$= \pi[-1 - 0 + \pi - (1 - 0 + 0)]$$

$$\Rightarrow \quad = \pi[\pi - 2] \Rightarrow I = \frac{\pi}{2}[\pi - 2]$$

Ans.

* Frequently asked previous years Board Exam Questions.

Chapter 9. Differential Equations

1. *Solve the differential equation**

$$(1+x^2)\frac{dy}{dx} = 4x^2 - 2xy$$

Sol. Given differential equation is,

$$(1+x^2)\frac{dy}{dx} = 4x^2 - 2xy$$

$$\Rightarrow \quad \frac{dy}{dx} = \frac{4x^2}{1+x^2} - \frac{2xy}{1+x^2}$$

$$\text{or} \quad \frac{dy}{dx} + \frac{2xy}{1+x^2} = \frac{4x^2}{1+x^2}$$

This is of the form $\frac{dy}{dx} + \text{P}y = \text{Q}$

where, $\text{P} = \frac{2x}{1+x^2}$ and $\text{Q} = \frac{4x^2}{1+x^2}$

Thus, the solution of given differential equation is,

$$y \times \text{I.F.} = \int (\text{Q} \times \text{I.F.})\, dx + c$$

where, $\text{I.F.} = e^{\int p dx}$

$$= e^{\int \frac{2x}{1+x^2} dx}$$

$$= e^{\log(1+x2)}$$

$$\left[\because \int \frac{f'(x)}{f(x)} dx = \log[f(x)] + c\right]$$

$$= 1 + x^2 \qquad [\because e^{\log(a)} = a$$

$$y \times (1+x^2) = \int \left[\frac{4x^2}{1+x^2} \times (1+x^2)\right] dx + c$$

$$\Rightarrow \quad (1+x^2)y = \int 4x^2\, dx + c$$

$$\Rightarrow \quad (1+x^2)y = \frac{4x^3}{3} + c$$ **Ans.**

Chapter 10. Probability

1. *If events A and B are independent, such that $P(A) = \frac{3}{5}$, $P(B) = \frac{2}{3}$, find $P(A \cup B)$.**

Sol. Given, $\text{P(A)} = \frac{3}{5}$ and $\text{P(B)} = \frac{2}{3}$

$\because$ A and B are independent

$$\therefore \quad \text{P}(\text{A} \cap \text{B}) = \text{P(A)}.\,\text{P(B)} = \frac{3}{5} \times \frac{2}{3} = \frac{2}{5}$$

$$\because \quad \text{P}(\text{A} \cup \text{B}) = \text{P(A)} + \text{P(B)} - \text{P}(\text{A} \cap \text{B})$$

$$= \frac{3}{5} + \frac{2}{3} - \frac{2}{5}$$

$$= \frac{9+10-6}{15}$$

$$= \frac{13}{15} = 0{\cdot}8666$$ **Ans.**

2. *Let A and B be two events such that**

$$P(A) = \frac{1}{2}, \; P(B) = p \text{ and } P(A \cup B) = \frac{3}{5}$$

*find 'p' if A and B are independent events.**

Sol. Given : $\text{P(A)} = \frac{1}{2}$, $\text{P(B)} = p$ and $\text{P}(\text{A} \cup \text{B}) = \frac{3}{5}$.

We know,

$\text{P}(\text{A} \cup \text{B}) = \text{P(A)} + \text{P(B)} - \text{P}(\text{A} \cap \text{B})$

$$\Rightarrow \quad \frac{3}{5} = \frac{1}{2} + p - \text{P}(\text{A} \cap \text{B})$$

$$\Rightarrow \quad \text{P}(\text{A} \cap \text{B}) = \frac{1}{2} - \frac{3}{5} + p = p + \frac{1}{2} - \frac{3}{5}$$

$$= p + \frac{5-6}{10}$$

$$= p - \frac{1}{10}$$

Now, A and B are independent events.

$$\therefore \quad \text{P}(\text{A} \cap \text{B}) = \text{P(A)} \cdot \text{P(B)}$$

$$\Rightarrow \quad p - \frac{1}{10} = \frac{1}{2} \times p$$

$$\Rightarrow \quad p - \frac{p}{2} = \frac{1}{10}$$

$$\Rightarrow \quad \frac{p}{2} = \frac{1}{10}$$

$$\therefore \quad p = \frac{2}{10} = \frac{1}{5}$$ **Ans.**

3. *If two balls are drawn from a bag containing three red balls and four blue balls, find the probability that :*

(i) They are of the same colour.

(ii) They are of different colours.

Sol. Total number of ways of drawing 2 balls = ${}^7C_2$

(i) P (Balls are of same colour)

$$= \frac{{}^3C_2}{{}^7C_2} + \frac{{}^4C_2}{{}^7C_2} = \frac{{}^3C_2 + {}^4C_2}{{}^7C_2}$$

---

* Frequently asked previous years Board Exam Questions.

$$= \frac{\frac{3!}{2!\cdot 1!} + \frac{4!}{2!\cdot 2!}}{\frac{7!}{2!\cdot 5!}}$$

$$= \frac{3+6}{7\times 3} = \frac{9}{21} = \frac{3}{7}$$ **Ans.**

(ii) P (Balls are of different colour)

$$= \frac{{}^3C_1 \times {}^4C_1}{{}^7C_2} = \frac{\frac{3!}{1!\cdot 2!} \times \frac{4!}{1!\cdot 3!}}{\frac{7!}{2!\cdot 5!}}$$

$$= \frac{3\times 4}{7\times 3} = \frac{4}{7}$$ **Ans.**

4. *Three persons A, B and C shoot to hit a target. If A hits the target four times in five trials, B hits it three times in four trials and C hits it two times in three trials, find the probability that :*

(i) Exactly two persons hit the target.

(ii) At least two persons hit the target.

(iii) None hit the target.

Sol. Given, $P(A) = \frac{4}{5}$, $P(B) = \frac{3}{4}$, $P(C) = \frac{2}{3}$

$\therefore P(\bar{A}) = \frac{1}{5}$, $P(\bar{B}) = \frac{1}{4}$, $P(\bar{C}) = \frac{1}{3}$

(i) P (exactly two persons hit the target)

$$= P(A \cap B \cap \bar{C}) + P(A \cap \bar{B} \cap C) + P(\bar{A} \cap B \cap C)$$

$$= P(A)\cdot P(B)\cdot P(\bar{C}) + P(A)\cdot P(\bar{B})\cdot P(C) + P(\bar{A})\cdot P(B)\cdot P(C)$$

$$= \frac{4}{5}\cdot\frac{3}{4}\left(1-\frac{2}{3}\right) + \frac{4}{5}\left(1-\frac{3}{4}\right)\cdot\frac{2}{3} + \left(1-\frac{4}{5}\right)\cdot\frac{3}{4}\cdot\frac{2}{3}$$

$$= \frac{4}{5}\cdot\frac{3}{4}\cdot\frac{1}{3} + \frac{4}{5}\cdot\frac{1}{4}\cdot\frac{2}{3} + \frac{1}{5}\cdot\frac{3}{4}\cdot\frac{2}{3}$$

$$= \frac{12+8+6}{60} = \frac{26}{60} = \frac{13}{30}.$$ **Ans.**

(ii) P (at least two persons hit the target)

= P (all three persons hit the target) + P (exactly two persons hit the target)

$$= P(A \cap B \cap C) + \frac{13}{30}$$ [Using (i)]

$$= P(A)\,P(B)\,P(C) + \frac{13}{30}$$

$$= \frac{4}{5}\cdot\frac{3}{4}\cdot\frac{2}{3} + \frac{13}{30}$$

$$= \frac{2}{5} + \frac{13}{30} = \frac{12+13}{30} = \frac{25}{30} = \frac{5}{6}.$$ **Ans.**

(iii) P (None hit the target)

$$= P(\bar{A} \cap \bar{B} \cap \bar{C})$$

$$= P(\bar{A})\cdot P(\bar{B})\cdot P(\bar{C})$$

$$= \left(1-\frac{4}{5}\right)\cdot\left(1-\frac{3}{4}\right)\cdot\left(1-\frac{2}{3}\right)$$

$$= \frac{1}{5}\cdot\frac{1}{4}\cdot\frac{1}{3} = \frac{1}{60}.$$ **Ans.**

5. *There are 10 persons who are to be seated around a circular table. Find the probability that two particular person will always sit together.*

Sol. If 'n' objects are arranged in a circle, the total number of ways in which they can be arranged are $(n-1)!$.

Here, $n = 10$.

$\therefore$ Total number of ways in which 10 persons can be seated around a circular table = (10 – 1) ! = 9!.

If two particular persons always sit together, then we have to arrange only 9 persons around a circular table.

$\therefore$ No. of ways in which 9 persons can be seated around a circular table = (9 – 1)! = 8!

But these two particular persons can interchange their seats.

$\therefore$ Favourable cases = 2 × 8!

$\therefore$ Probability that two particular persons always sit together out of 10 persons around a circular table

$$= \frac{\text{Favourable cases}}{\text{Total cases}}$$

$$= \frac{2\times 8!}{9!} = \frac{2\times 8!}{9\times 8!} = \frac{2}{9}$$ **Ans.**

6. *A bag contains 20 balls marked from 1 to 20. One ball is drawn at random from the bag. What is the probabilily that the ball drawn is marked with a number which is multiple of 5 or 7 ?*

Sol. Multiples of 5 are 5, 10, 15 and 20

Multiples of 7 are 7, 14

Total favourable events = 4 + 2 = 6

Total number of possible outcomes = 20

∴ Probability that the ball drawn is marked with a number multiple of 5 or 7

$$= \frac{6}{20} = \frac{3}{10}. \quad \textbf{Ans.}$$

7. *What is the probability that a leap year has 53 Sundays.*

Sol. There are 366 days in a leap year having 52 full weeks and 2 days. Last two days could be Sunday and Monday, Monday and Tuesday, Tuesday and Wednesday, Wednesday and Thursday, Thursday and Friday, Friday and Saturday, Saturday and Sunday.

So, there are seven possible outcomes. First and last are in favour of the event.

Thus, the required probability is

$$= \frac{2}{7} \quad \textbf{Ans.}$$

8. *The probability that a boy will not pass MBA exam is $\frac{3}{5}$ and that a girl will not pass is $\frac{4}{5}$. Calculate the probabilily that at least one of them passes the exam.*

Sol. Let p_1 be the probability that a boy will pass MBA exam and q_1 be the probability that the boy will not pass MBA exam.

Let p_2 be the probability that a girl will pass MBA exam and q_2 be the probability that the girl will not pass MBA exam.

$$\therefore p_1 = \frac{2}{5},\ q_1 = \frac{3}{5},\ p_2 = \frac{1}{5},\ q_2 = \frac{4}{5}$$

Then, the probability that at least one of them will pass the exam is given as,

$$= p_1p_2 + p_1q_2 + p_2q_1$$

$$= \frac{2}{5}\times\frac{1}{5}+\frac{2}{5}\times\frac{4}{5}+\frac{1}{5}\times\frac{3}{5}$$

$$= \frac{2}{25}+\frac{8}{25}+\frac{3}{25} = \frac{13}{25} \quad \textbf{Ans.}$$

9. *The probability of A, B and C solving a problem are $\frac{1}{3}, \frac{2}{7}$ and $\frac{3}{8}$, respectively. If all the three try and solve the problem simultaneously, find the probability that only one of them will solve it.*

Sol. Given, $P(A) = \frac{1}{3} \quad \therefore P(\bar{A}) = \frac{2}{3}$

$$P(B) = \frac{2}{7} \quad \therefore P(\bar{B}) = \frac{5}{7}$$

$$P(C) = \frac{3}{8} \quad \therefore P(\bar{C}) = \frac{5}{8}$$

Probability that only one will solve the problem,

$$P = P(A)\times P(\bar{B})\times P(\bar{C}) + P(\bar{A})\times P(B) \times P(\bar{C}) + P(\bar{A})\times P(\bar{B})\times P(C)$$

$$= \frac{1}{3}\times\frac{5}{7}\times\frac{5}{8}+\frac{2}{3}\times\frac{2}{7}\times\frac{5}{8}+\frac{2}{3}\times\frac{5}{7}\times\frac{3}{8}$$

$$= \frac{25}{168}+\frac{20}{168}+\frac{30}{168}$$

$$= \frac{75}{168} = \frac{25}{56} \quad \textbf{Ans.}$$

10. *A bag contains 5 white and 4 black balls and another bag contains 7 white and 9 black balls. A ball is drawn from the first bag and two balls drawn from the second bag. What is the probability of drawing one white and two black balls ?*

Sol. Given,

Bag I	Bag II
5W	7W
4B	9B
Total balls in Bag I = 9	Total balls in Bag II = 16

Total number of ways of drawing two balls $= {}^{16}C_2$

Required probability

$$= \left(\frac{5}{9}\right)\left(\frac{{}^9C_2}{{}^{16}C_2}\right)+\left(\frac{4}{9}\right)\left(\frac{{}^7C_1\times{}^9C_1}{{}^{16}C_2}\right)$$

$$= \frac{5}{9}\times\frac{9\times8}{16\times15}+\frac{4}{9}\times\frac{7\times9}{8\times15}$$

$$= \frac{1}{6}+\frac{7}{30} = \frac{5+7}{30} = \frac{12}{30} = \frac{2}{5} \quad \textbf{Ans.}$$

11. *A bag contains 5 black and 6 red balls. Another bag contains 8 black and 5 red balls. A ball is then drawn from the first bag and put in the second. A ball is then drawn from the second. Find the probability that the ball drawn is black.*

Sol. The first bag contains 5 black and 6 red balls. The second bag contains 8 black and 5 red balls. The ball drawn from the first bag can be black or red.

Formulae Based Questions

(i) Let the ball drawn be black.

$\therefore$ P_1 (the probability of drawing black)

$$= \frac{5}{11}$$

If this is put in the second bag, the second bag now contains 9 black and 5 red balls.

$\therefore$ P_2 (the probability of drawing black from second bag)

$$= \frac{9}{14}$$

Hence, the probability in the first case

$$= \frac{5}{11} \times \frac{9}{14} = \frac{45}{154}$$

(ii) Let the ball drawn be red.

P_3 (the probability of drawing red) $= \frac{6}{11}$

Now, the second bag has 6 red and 8 black balls.

P_4 (the probability of drawing black from second bag)

$$= \frac{8}{14}$$

Hence, the probability in the second case

$$= \frac{6}{11} \times \frac{8}{14} = \frac{48}{154}$$

The required probability is

$$= \frac{45}{154} + \frac{48}{154} = \frac{93}{154}.$$ **Ans.**

12. *A word consists of 9 different alphabets, in which there are 4 consonants and 5 vowels. Three alphabets are chosen at random. What is the probability that more than one vowel will be selected ?*

Sol. Number of alphabets in the word = 9

Number of consonants (C) = 4

Number of vowels (V) = 5

Total number of ways in which three alphabets can be chosen at random = ${}^9C_3$

$\therefore$ Probability that more than one vowel will be selected

$$= P(2V \& 1C) + P(3V)$$

$$= \frac{{}^5C_2 \times {}^4C_1}{{}^9C_3} + \frac{{}^5C_3}{{}^9C_3}$$

$$= \frac{40}{84} + \frac{10}{84} = \frac{50}{84} = \frac{25}{42}$$ **Ans.**

13. *A and B throw two dices each. If A gets a sum of 9 on his two dice, then find the probability of B getting a higher sum.*

Sol. Sum of 9 on two dice

$$= \{(3, 6), (6, 3), (4, 5), (5, 4)\}$$
$$= 4$$

$\therefore$ Probability P (A) $= \frac{4}{36} = \frac{1}{9}$

Sum of more than 9 $= \{(4, 6), (6, 4), (5, 5), (5, 6), (6, 5), (6, 6)\}$

$$= 6$$

$\therefore$ Probability P (B) $= \frac{6}{36} = \frac{1}{6}$

$\therefore$ Required probability

$$= P(A) \cdot P(B)$$

[$\because$ 2 events are independent]

$$= \frac{1}{9} \times \frac{1}{6} = \frac{1}{54}.$$ **Ans.**

14. *A speaks truth in 55 percent cases and B speaks truth in 75 percent cases. Determine the probability of cases in which they are likely to contradict each other in stating the same fact.*

Sol. Let P_A is the event that A speaks the truth and P_B is the event that B speaks the truth. Q_A is the event that A speaks lies and Q_B is the event that B speak lies.

$\therefore$ $P_A = \frac{55}{100}$, $Q_A = 1 - \frac{55}{100} = \frac{45}{100}$

$P_B = \frac{75}{100}$, $Q_B = 1 - \frac{75}{100} = \frac{25}{100}$

They will contradict each other in following two cases :

(i) A speaks truth and B lies.

(ii) B speaks truth and A lies.

In case (i) $P = P_A \cdot Q_B$

$$= \frac{55}{100} \times \frac{25}{100} = \frac{11}{80}$$

$\therefore$ (ii) $P = Q_A \cdot P_B$

$$= \frac{45}{100} \times \frac{75}{100} = \frac{27}{80}$$

So, total probability that A and B will contradict each other

$$= \frac{11}{80} + \frac{27}{80} = \frac{38}{80} = \frac{19}{40}$$

$$= 47.5\%$$ **Ans.**

15. *Tickets numbered from 1 to 20 are mixed up together and then a ticket is drawn at random. What is the* *probability that the ticket has a number which is a multiple of 3 or 7 ?*

Sol. Total number of cases, $n(S) = 20$

Multiple of 3, $n(E_1) = \{3, 6, 9, 12, 15, 18\} = 6$

Multiple of 7, $n(E_2) = \{7, 14\} = 2$

$$\therefore \text{Required probability} = \frac{n(E_1)}{n(S)} + \frac{n(E_2)}{n(S)}$$

$$= \frac{6}{20} + \frac{2}{20} = \frac{3}{10} + \frac{1}{10}$$

$$= \frac{4}{10} = \frac{2}{5}.$$ **Ans.**

16. *A purse contains 4 silver and 5 copper coins. Second purse contains 3 silver and 7 copper coins. If a coin is taken out at random from one of the purses, what is the probability that it is a copper coin ?*

Sol. Probability of choosing first purse = Probability of choosing second purse

$$= \frac{1}{2}$$

Now, probability of taking out copper coin from first purse

$$= \frac{1}{2} \times \frac{5}{9} = \frac{5}{18}$$

and probability of taking out copper coin from second purse

$$= \frac{1}{2} \times \frac{7}{10} = \frac{7}{20}$$

Hence, the probability of taking out copper coin from either of the two purses

$$= \frac{5}{18} + \frac{7}{20} = \frac{50+63}{180} = \frac{113}{180}$$ **Ans.**

17. *A pair of dice is thrown. If the two numbers appearing on them are different, find the probability that the sum of number is 6.*

Sol. Numbers on a dice are 1, 2, 3, 4, 5, 6.

If a pair of dice is thrown, there are 36 pair of numbers possible as under :

(1, 1), (1, 2), (1, 3), (1, 4), (1, 5), (1, 6),
(2, 1), (2, 2), (2, 3), (2, 4), (2, 5), (2, 6),
(3, 1), (3, 2), (3, 3), (3, 4), (3, 5), (3, 6),
(4, 1), (4, 2), (4, 3), (4, 4), (4, 5), (4, 6),
(5, 1), (5, 2), (5, 3), (5, 4), (5, 5), (5, 6),
(6, 1), (6, 2), (6, 3), (6, 4), (6, 5), (6, 6)

We have to choose those pair of marks out of 36 pairs whose sum of numbers is 6. These pairs are our favourable pairs for the event.

i.e., (1, 5), (2, 4), (3, 3), (4, 2), (5, 1)

$\therefore$ Number of favourable pairs = 5

Number of total possible pairs = 36

$\therefore$ Probability that the sum of number appearing on a pair of dice is 6

$$= \left\{\frac{\text{No. of favourable outcomes}}{\text{No. of total outcomes}}\right\}$$

$$= \frac{5}{36}$$ **Ans**

18. *A candidate is selected for interview of management trainees for 3 companies. For the first company there are 12 candidates, for the second there are 15 candidates and for the third, there are 10 candidates. Find the probability that he is selected in at least one of the companies.*

Sol. Probability of getting selected in Ist company,

$$P(A) = \frac{1}{12}$$

Probability of getting selected in 2nd company,

$$P(B) = \frac{1}{15}$$

Probability of getting selected in 3rd company

$$P(C) = \frac{1}{10}$$

$\therefore$ Probability of not getting selected in any company

$$= P(\overline{A}) \cdot P(\overline{B}) \cdot P(\overline{C})$$

$$= \left(1-\frac{1}{12}\right)\left(1-\frac{1}{15}\right)\left(1-\frac{1}{10}\right)$$

$$= \frac{11}{12} \cdot \frac{14}{15} \cdot \frac{9}{10} = \frac{77}{100}$$

$\therefore$ Probability of selection in at least one company

$$= 1 - P(\overline{A})\, P(\overline{B})\, P(\overline{C})$$

$$= 1 - \frac{77}{100} = \frac{23}{100}$$ **Ans**

19. *One number is choosen at random from the numbers 1 to 21, find the probability that it may be a prime number.*

Sol. 2, 3, 5, 7, 11, 13, 17, 19 are the prime numbers in between 1 to 21.

$$\therefore \quad \text{Probability} = \frac{8}{21}$$ **Ans**

20. *There are 3 urns A, B and C. Urn A contains 4 red balls and 3 black balls. Urn B contains 5 red balls and 4 black balls. Urn C contains 4 red balls and*

Formulae Based Questions

4 black balls. One ball is drawn from each of these urns. What is the probability that the 3 balls drawn consist of 2 red balls and 1 black ball ?

Sol. P (3 balls drawn consist of 2 red balls and 1 black ball)

$= RRB + BRR + RBR$

$$=\left(\frac{4}{7}\times\frac{5}{9}\times\frac{4}{8}\right)+\left(\frac{3}{7}\times\frac{5}{9}\times\frac{4}{8}\right)+\left(\frac{4}{7}\times\frac{4}{9}\times\frac{4}{8}\right)$$

$$=\frac{80+60+64}{504}=\frac{204}{504}=\frac{102}{252}=\frac{51}{126}.$$ **Ans.**

21. *Two horses are considered for a race. The proability of selection of the first horse is $\frac{1}{4}$ and that of the second is $\frac{1}{3}$. What is the porbability that :*

(a) both of them will be selected.

(b) only one of them will be slected.

(c) none of them will be selected.

Sol. Let the probability of selecting first horse $= P(A) = \frac{1}{4}$

and the probability of selecting second horse $= P(B) = \frac{1}{3}$

then, $P(\bar{A}) = 1 - P(A) = 1 - \frac{1}{4} = \frac{3}{4}$

and $P(\bar{B}) = 1 - P(B) = 1 - \frac{1}{3} = \frac{2}{3}$

(a) P(Both of them selected) $= P(A)\cdot P(B)$

$= \frac{1}{4}\times\frac{1}{3} = \frac{1}{12}$ **Ans.**

(b) P(Only one selected)

$= P(A)\cdot P(\bar{B}) + P(\bar{A})\cdot P(B)$

$= \frac{1}{4}\times\frac{2}{3}+\frac{1}{3}\times\frac{3}{4} = \frac{5}{12}$

(c) P(None of them selected)

$= P(\bar{A})\cdot P(\bar{B})$

$= \frac{3}{4}\times\frac{2}{3}$

$= \frac{1}{2}$ **Ans.**

22. *An urn contains 10 white and 3 black balls while another urn contains 3 white and 5 black balls. Two balls are drawn from the first urn and put into the second urn and then a ball is drawn from the second urn. Find the probability that the ball drawn from the second urn is a white ball.*

Sol. There are three mutually exclusive and exhaustive ways in which 2 balls can be transferred from first bag to second bag.

First way : Two white balls are transferred from first bag to second bag, probability of which is $\frac{^{10}C_2}{^{13}C_2}$.

In the second bag we now have 5 white and 5 black balls and probability of getting a white ball is $\frac{5}{10}$.

$\therefore$ Required probability

$$= \frac{^{10}C_2}{^{13}C_2}\times\frac{5}{10} = \frac{45}{78}\times\frac{5}{10} = \frac{225}{780}$$

Second way : Two black balls are transferred from first bag to the second bag probability of which is $\frac{^{3}C_2}{^{13}C_2}$.

Now we have 3 white and 7 black balls in second bag and probability of getting a white ball is $\frac{3}{10}$.

$\therefore$ Required probability

$$= \frac{^{3}C_2}{^{13}C_2}\times\frac{3}{10} = \frac{3}{78}\times\frac{3}{10}$$

$$= \frac{9}{780}.$$

Third way : One black and one white ball are transferred from first bag to second bag, the probability of which is $\frac{^{10}C_1\times{}^{3}C_1}{^{13}C_2}$.

In the second bag, there are now 4 white and 6 black balls and the probability of drawing a white ball is $\frac{4}{10}$.

$\therefore$ Required probability

$$= \frac{^{10}C_1\times{}^{3}C_1}{^{13}C_2}\times\frac{4}{10}$$

$$= \frac{30}{78}\times\frac{4}{10} = \frac{120}{780}$$

Since, these three cases are mutually exclusive, the required probability of drawing a white ball is

$$= \frac{225}{780}+\frac{9}{780}+\frac{120}{780}$$

$$= \frac{354}{780} = \frac{177}{390} = \frac{59}{130}$$ **Ans.**

23. *A bag has 4 red and 5 black balls, a second bag has 3 red and 7 black balls. One ball is drawn from the first bag and two from the second. Find the probability that two balls are black and one is red.*

Sol. Given,

B_1	B_2
4 Red	3 Red
5 Black	7 Black

(i) Either we can draw 1 black ball from B_1 and 1 red and 1 black from B_2.

$$\therefore \quad P_1 = \frac{^5C_1}{^9C_1} \times \frac{^3C_1 \times {^7C_1}}{^{10}C_2}$$

$$= \frac{5}{9} \times \frac{3 \times 7}{45} = \frac{7}{27}$$

(ii) Or we can draw 1 red ball from B_1 and 2 black balls from B_2.

$$\therefore \quad P_2 = \frac{^4C_1}{^9C_1} \times \frac{^7C_2}{^{10}C_2}$$

$$= \frac{4}{9} \times \frac{21}{45} = \frac{28}{135}$$

Event (i) and event (ii) are mutually exclusive events *i.e.,* happening of both events simultaneously is impossible. One of the two events will occur at one time.

∴ Probability of event of drawing 2 black balls and one red is given by,

Ball from the bags $= P_1 + P_2$

$$= \frac{7}{27} + \frac{28}{135} = \frac{63}{135}$$

$$= \frac{7}{15}$$ **Ans.**

24. **(a)** *Akhil and Vijay appear for an interview for two vacancies. The probability of Akhil's selection is $\frac{1}{4}$ and Vijay's selection is $\frac{2}{3}$. Find the probability that only one of them will be selected.*

(b) *There are two bags. One bag contains six green and three red balls. The second bag contains five green and four red balls. One ball is transferred from the first bag to the second bag. Then one ball is drawn from the second bag. Find the probability that it is a red ball.*

Sol. Let A be the event of Akhil's selection and B be the event of Vijay's selction.

(a) $$P(A) = \frac{1}{4}$$

$$\therefore \quad P(\bar{A}) = 1 - \frac{1}{4} = \frac{3}{4}$$

$$P(B) = \frac{2}{3}$$

$$\therefore \quad P(\bar{B}) = 1 - \frac{2}{3} = \frac{1}{3}$$

∴ Probability that only one of them will be selected

$$= P(A) \cdot P(\bar{B}) + P(\bar{A}) \cdot P(B)$$

$$= \frac{1}{4} \times \frac{1}{3} + \frac{3}{4} \times \frac{2}{3}$$

$$= \frac{1+6}{12} = \frac{7}{12}.$$ **Ans.**

(b) Case I : When a green ball is transferred, then

$$P_1 = \frac{6}{9} = \frac{2}{3},$$

$$P_2 = \frac{4}{10} = \frac{2}{5}$$

∴ Probability of both the events

$$= P_1 P_2 = \frac{2}{3} \times \frac{2}{5} = \frac{4}{15}$$

Case II : When a red ball is transferred, then

$$P_3 = \frac{3}{9} = \frac{1}{3}, \quad P_4 = \frac{5}{10} = \frac{1}{2}$$

∴ Probability of both events

$$= P_3 P_4 = \frac{1}{3} \times \frac{1}{2} = \frac{1}{6}$$

∴ The required probability

$$= P_1P_2 + P_3P_4$$

$$= \frac{4}{15} + \frac{1}{6} = \frac{8+5}{30}$$

$$= \frac{13}{30}.$$ **Ans.**

25. *The probability that a teacher will give an unannounced test during any class meeting is $\frac{1}{5}$. If a student is absent twice, find the probability that the student will miss at least one test.*

Sol. Given, $p = \frac{1}{5}, q = \frac{4}{5}, n = 2$

Required probability

$$= {^2C_1}\left(\frac{1}{5}\right)^1\left(\frac{4}{5}\right) + {^2C_2}\left(\frac{1}{5}\right)^2$$

$$= 2 \times \frac{4}{25} + \frac{1}{25}.$$

$$= \frac{8}{25} + \frac{1}{25} = \frac{9}{25}.$$ **Ans.**

26. *In a college, 70% students pass in Physics, 75% pass in Mathematics and 10% students fail in both. One student is chosen at random. What is the probability that :*

(i) *He passes in Physics and Mathematics.*

(ii) *He passes in Mathematics given that he passes in Physics.*

(iii) *He passes in Physics given that he passes in Mathematics.*

Sol. Let x% students pass in both the subjects

Students who pass in Physics = 70%

$$\Rightarrow \quad P(P) = \frac{70}{100}$$

Students who pass in Maths = 75%

$$\Rightarrow \quad P(M) = \frac{75}{100}$$

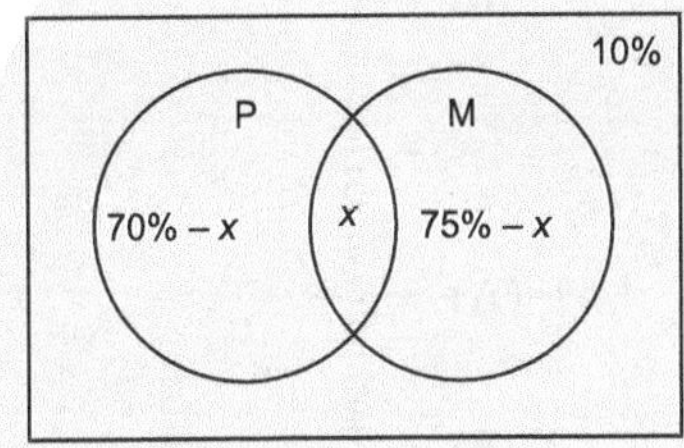

Students who fail in both = 10%

Students who pass in both at least in one

$$= 100 - 10\%$$

$$= 90\%$$

$$\Rightarrow \quad P(M \cup P) = \frac{90}{100} = 0{\cdot}90\%$$

$$\therefore \quad P(M) + P(P) - P(M \cap P) = P(M \cup P)$$

$$0{\cdot}75 + 0{\cdot}70 - P(M \cap P) = 0{\cdot}90$$

$$P(M \cap P) = 0{\cdot}55 \text{ or } \frac{55}{100}$$

$$\Rightarrow \quad P(M \cap P) = \frac{55}{100}$$

(i) P (passes Physics and Mathematics)

$$= \frac{55}{100} = \frac{11}{20}$$

$$= 0{\cdot}55$$ **Ans.**

(ii) $$P(M/P) = \frac{P(M \cap P)}{P(P)}$$

$$= \frac{55/100}{70/100}$$

$$= \frac{55}{70} = \frac{11}{14}$$ **Ans.**

(iii) $$P(P/M) = \frac{P(M \cap P)}{P(M)}$$

$$= \frac{55/100}{75/100}$$

$$= \frac{55}{75} = \frac{11}{14}$$ **Ans.**

27. (a) *A bag contains 8 red and 5 white balls. Two successive draws of 3 balls are made at random from the bag without replacement. Find the probability that the first draw yields 3 white balls and the second draw 3 red balls.*

(b) *A box contains 30 bolts and 40 nuts. Half of the bolts and half of the nuts are rusted. If two items are drawn at random from the box, what is the probability that either both are rusted or both are bolts ?*

Sol. (a) Let A : Event that 3 balls in the first draw are all white.

B : Event that 3 balls in the second draw are all red.

3 balls can be drawn out of 13 in ${}^{13}C_3$ and 3 white ball can be drawn out of 5 in ${}^5C_3$ ways.

$$\therefore \quad P(A) = \frac{{}^5C_3}{{}^{13}C_3} = \frac{5}{143}$$

Since 3 balls are not replaced before the second draw, we are left with 8 red and 2 white balls.

Now, 3 balls can be drawn in ${}^{10}C_3$ ways and 3 red balls can be drawn in ${}^8C_3$ ways.

$$\therefore \quad P(B/A) = \frac{{}^8C_3}{{}^{10}C_3} = \frac{7}{15}$$

$$\therefore \quad P(A \cap B) = P(A) \cdot P(B/A)$$

$$= \frac{5}{143} \cdot \frac{7}{15} = \frac{7}{429}$$ **Ans.**

(b) The number of bolts and nuts are 30 and 40 respectively. The number of rusted bolts and rusted nuts are 15 $\left(= \frac{30}{2}\right)$ and 20 $\left(= \frac{40}{2}\right)$, respectively.

Total number of items = 30 + 40 = 70

Total number of rusted items = 15 + 20 = 35

Total number of ways of drawing 2 items

$$= {}^{70}C_2.$$

Let R and B be the events that the both items drawn are rusted items and both are bolts respectively.

R and B are not mutually exclusive events because there are 15 rusted bolts.

∴ P (Items are both rusted or both bolts)

$$= P(R \cup B)$$

$$= P(R) + P(B) - P(R \cap B)$$

$$= \frac{^{35}C_2}{^{70}C_2} + \frac{^{30}C_2}{^{70}C_2} - \frac{^{15}C_2}{^{70}C_2}$$

$$= \frac{1 \times 2}{70 \times 69}\left(\frac{35 \times 34}{1 \times 2} + \frac{30 \times 29}{1 \times 2} - \frac{15 \times 14}{1 \times 2}\right)$$

$$= \frac{1850}{70 \times 69} = \frac{185}{483}.$$ **Ans.**

28. *A bag contains 20 balls numbered from 1 to 20. One ball is drawn at random from the bag. What is the probability that the ball drawn is marked with a number which is multiple of 3 or 4 ?*

Sol. Number of balls = 20

∴ $n(S) = 20$

Let A : A ball marked with multiple of 3 is drawn

i.e., 3, 6, 9, 12, 15, 18

$n(A) = 6$

∴ $P(A) = \frac{n(A)}{n(S)} = \frac{6}{20}$

and B : A ball marked with multiple of 4 is drawn

i.e., 4, 8, 12, 16, 20

$n(B) = 5$

∴ $P(B) = \frac{n(B)}{n(S)} = \frac{5}{20}$

Now, $n(A \cap B) = 1$

$P(A \cap B) = \frac{n(A \cap B)}{n(S)} = \frac{1}{20}$

∴ Required probability,

$$P(A \cup B) = P(A) + P(B) - P(A \cap B)$$

$$= \frac{6}{20} + \frac{5}{20} - \frac{1}{20}$$

$$= \frac{11}{20} - \frac{1}{20} = \frac{10}{20}$$

$$= \frac{1}{2}$$ **Ans.**

29. *In a certain city, the probability of not reading the morning newspaper by the residents is $\frac{1}{2}$ and that of not reading the evening newspaper is $\frac{2}{5}$. The probability of reading both the newspaper is $\frac{1}{5}$. Find the probability that a resident read either the morning or evening or both the papers.*

Sol. Probability of not reading morning newspaper

$(A') = \frac{1}{2}$

∴ Probability of reading morning newspaper

$(A) = 1 - \frac{1}{2}$

$= \frac{1}{2}$

Probability of not reading evening newspaper

$(B') = \frac{2}{5}$

Probability of reading evening newspaper

$(B) = 1 - \frac{2}{5}$

$= \frac{3}{5}$

∴ $P(A) = \frac{1}{2}$

$P(B) = \frac{1}{5}$

Probability of reading both,

$P(A \cap B) = \frac{1}{5}$

Probability of reading either of newspapers is given as,

$$P(A \cup B) = P(A) + P(B) - P(A \cap B)$$

$$= \frac{1}{2} + \frac{3}{5} - \frac{1}{5} = \frac{1}{2} + \frac{2}{5}$$

$$= \frac{9}{10}$$ **Ans.**

30. *A fair die is thrown once. What is the probability that either an even number or a number greater than three will turn up ?*

Sol. Sample space for one throw = 6

Probability to get even number, P (A)

$= \frac{3}{6} = \frac{1}{2}$

Probability to get greater than 3, P (B)

$= \frac{3}{6} = \frac{1}{2}$

Common numbers are 4 and 6.

∴ $P(A \cap B) = \frac{2}{6} = \frac{1}{3}$

Probability that either an even number or a number greater than 3 will turn up,

$$P(A \cup B) = P(A) + P(B) - P(A \cap B)$$

$$= \frac{1}{2} + \frac{1}{2} - \frac{1}{3}$$

$$= 1 - \frac{1}{3} = \frac{2}{3}$$ **Ans.**

31. *In an examination 30% of the students failed in mathematics, 20% failed in chemistry and 10% failed in both. A student is selected at random find the probability that :*

(a) the student has failed either in mathematics or in chemistry.

(b) the student has failed in mathematics, it is known that he has failed in chemistry.

Sol. Let A and B denote the events that the student has failed in mathematics and chemistry, respectively. Then, $P(A) = 30/100 = 3/10$, $P(B) = 20/100 = 2/10$ and, $P(A \cap B) = 10/100 = 1/10$

(a) The probability that the student has failed either in chemistry or in mathematics is

$$P(A \cup B) = P(A) + P(B) - P(A \cap B)$$

$$= \frac{3}{10} + \frac{2}{10} - \frac{1}{10} = \frac{4}{10} = \frac{2}{5}$$ **Ans.**

(b) The probability that the student has failed in mathematics, knowing that he has failed in chemistry is

$$P(A/B) = \frac{P(A \cap B)}{P(B)}$$

$$= \frac{1/10}{2/10} = \frac{1}{2}$$ **Ans.**

32. *The probability that a contractor will get a plumbing contract is $\frac{2}{3}$ and an electric contract is $\frac{4}{9}$. If the probability of getting at least one contract is $\frac{4}{5}$, find the probability that he will get both the contracts.*

Sol. $P(A) = \frac{2}{3}$, $P(B) = \frac{4}{9}$, $P(A \cup B) = \frac{4}{5}$

We have to find $P(A \cap B)$.

$\therefore$ $P(A \cup B) = P(A) + P(B) - P(A \cap B)$

$\Rightarrow$ $\frac{4}{5} = \frac{2}{3} + \frac{4}{9} - P(A \cap B)$

$\Rightarrow$ $P(A \cap B) = \frac{14}{45}$ **Ans.**

33. *A card is drawn at random from a pack of 52 playing cards. What is the probability that the card drawn is neither a spade nor a queen.*

Sol. This is a case of non-mutually exclusive event.

The probability that a card drawn at random from a pack of 52 playing cards is a spade or a queen is given as,

$$P(A \cup B) = P(A) + P(B) - P(A \cap B)$$

Probability of card is spade, $P(A) = \frac{13}{52}$

Probability of card is queen, $P(B) = \frac{4}{52}$

Probability of card is spade and queen

$$P(A \cap B) = \frac{1}{52}$$

$\therefore$ $P(A \cup B) = \frac{13}{52} + \frac{4}{52} - \frac{1}{52} = \frac{16}{52}$

$\Rightarrow$ The probability that the card drawn is neither a spade nor a queen

$$= 1 - \frac{16}{52} = \frac{36}{52} = \frac{9}{13}$$ **Ans.**

34. *A card is drawn from a well shuffled pack of playing cards. What is the probability that it is either a spade or an ace or both ?*

Sol. Let,

A = getting a spade

B = getting an ace

$$P(A) = \frac{13}{52} = \frac{1}{4}, \quad P(B) = \frac{4}{52} = \frac{1}{13}$$

$$P(\overline{A}) = 1 - \frac{1}{4} = \frac{3}{4}$$

$$P(\overline{B}) = 1 - \frac{1}{13} = \frac{12}{13}$$ **Ans.**

Required probability.

$$= 1 - P(\overline{A})\,P(\overline{B})$$

$$= 1 - \frac{3}{14} \times \frac{12}{13}$$

$$= 1 - \frac{9}{13} = \frac{4}{13}$$

Alternate Solution :

$$P(A \cap B) = \frac{1}{52}$$

$$P(A \cup B) = P(A) + P(B) - P(A \cap B)$$

$$= \frac{1}{4} + \frac{1}{13} - \frac{1}{52}$$

$$= \frac{13 + 4 - 1}{52}$$

$$= \frac{16}{52} = \frac{4}{13}$$ **Ans.**

Formulae Based Questions

35. *A pair of dice is thrown. What is the probability of getting an even number on the first die or a total of 8 ?*

Sol. The total number of possible outcomes is 36.

i.e., $n(S) = 36$

Let A : getting a total of 8.

The favourable outcomes are :

(2, 6), (4, 4), (6, 2), (3, 5), (5, 3)

$\therefore \quad n(A) = 5$

$$\therefore P(A) = \frac{n(A)}{n(S)} = \frac{5}{36}$$ **Ans.**

B : getting an even number on the first die

The favourable outcomes are {(2, 1), (2, 2), (2, 3), (2, 4), (2, 5), (2, 6), (4, 1), (4, 2), (4, 3), (4, 4), (4, 5), (4, 6), (6, 1), (6, 2), (6, 3), (6, 4), (6, 5), (6, 6)}.

$$P(B) = \frac{n(B)}{n(S)} = \frac{18}{36} = \frac{1}{2}$$

Probability that both events will occur simultaneously,

$$P(A \cap B) = \frac{n(A \cap B)}{n(S)} = \frac{3}{36} = \frac{1}{12}$$

$$P(A \cup B) = P(A) + P(B) - P(A \cap B)$$

$$= \frac{5}{36} + \frac{1}{2} - \frac{1}{12}$$

$$= \frac{5 + 18 - 3}{36}$$

$$= \frac{20}{36} = \frac{5}{9}$$ **Ans.**

36. *IA committee of 4 persons has to be chosen from 8 boys and 6 girls, consisting of at least one girl. Find the probability that the committee consists of more girls than boys.*

Sol. Let A : Number of girls is more than number of boys.

B : The 4 member committee consists of one girl.

Then, Required probility = P(A/B)

$$= \frac{P(A \cap B)}{P(B)} = \frac{n(A \cap B)/n(S)}{n(B)/n(S)} = \frac{n(A \cap B)}{n(B)}$$

Now, $n(A \cap B) = {}^6C_4 \cdot {}^8C_0 + {}^6C_3 \cdot {}^8C_1$

$= 15 + 160 = 175$

$n(B) = {}^6C_1 \cdot {}^8C_3 + {}^6C_2 \cdot {}^8C_2 + {}^6C_3 \cdot {}^8C_1 + {}^6C_4 \cdot {}^8C_0$

$= 1$

$$\therefore P(A/B) = \frac{n(A \cap B)}{n(B)} = \frac{175}{931} = 0.19$$ **Ans.**

37. (a) *Bag A contains three red and four white balls; bag B contains two red and three white balls. If one ball is drawn from bag A and two Balls from bag B, find the probability that :*

(i) One ball is red and two balls are white.

(ii) All the three balls are of the same colour.

(b) *Three persons, Aman, Bipin and Mohan attempt a Mathematics problem independently. The odds in favour of Aman and Mohan solving the problem are 3 : 2 and 4 : 1 respectively and the odds against Bipin solving the problem are 2 : 1. Find :*

(i) The probability that all the three will solve the problem.

(ii) The probability that problem will be solved.

Sol. (a) Possible selection are as follows :

(i) 1 red from bag A, 2 white from bag B, or 1 white from bag A, 1 white from bag B, 1 red from B

$\therefore$ P (one ball is red and two balls are white)

$$= \frac{{}^3C_1}{{}^7C_1} \times \frac{{}^3C_2}{{}^5C_2} + \frac{{}^4C_1}{{}^7C_1} \times \frac{{}^4C_1 \times {}^2C_1}{{}^5C_2}$$

$$= \frac{3}{7} \times \frac{3}{10} + \frac{4}{7} \times \frac{3 \times 2}{10}$$

$$= \frac{9}{70} + \frac{24}{70} = \frac{33}{70}.$$

(ii) Possible selection are as follows :

1 red from bag A, 2 red from bag B, or

1 white from bag A, 2 white from bag B

Formulae Based Questions

∴ P (All the three balls are of the same colour)

$$= \frac{^3C_1}{^7C_1} \times \frac{^2C_2}{^5C_2} + \frac{^4C_1}{^7C_1} \times \frac{^3C_2}{^5C_2}$$

$$= \frac{3}{7} \times \frac{1}{10} + \frac{4}{7} \times \frac{3}{10}$$

$$= \frac{3}{70} + \frac{12}{70} = \frac{15}{70} = \frac{3}{14}.$$

(b) Odd against $= \frac{1-p}{p}$

Odd in favour $= \frac{p}{1-p}$

A : Event Aman solves problem

B : Event Bipin solves problem

C : Event Mohan solves problem

Aman	Mohan	Bipin
$\frac{3}{2} = \frac{p_1}{1-p_1}$	$\frac{4}{1} = \frac{p_2}{1-p_2}$	$\frac{2}{1} = \frac{1-p_3}{p_3}$
$\Rightarrow p_1 = \frac{3}{5}$	$p_2 = \frac{4}{5}$	$\Rightarrow p_3 = \frac{1}{3}$
$\Rightarrow p(A) = \frac{3}{5}$	$\Rightarrow p(B) = \frac{4}{5}$	$\Rightarrow p(C) = \frac{1}{3}$

(i) Probability that all the there will solve the problem, $P(A \cap B \cap C)$

$$= P(A).\ P(B).\ P(C)$$

$$= \frac{3}{5} \times \frac{4}{5} \times \frac{1}{3} = \frac{4}{25}$$

(ii) Probability that the problem is not solved is equal to the probability that all the three fail to solve it.

$$\therefore\ P(\overline{A} \cap \overline{B} \cap \overline{C}) = P(\overline{A}).\ P(\overline{B}).\ P(\overline{C})$$

$$= \left(1-\frac{3}{5}\right) \cdot \left(1-\frac{4}{5}\right) \cdot \left(1-\frac{1}{3}\right)$$

$$= \frac{2}{5} \cdot \frac{1}{5} \cdot \frac{2}{3} = \frac{4}{75}$$

∴ The probability that problem will be solved :

$$= 1 - \frac{4}{75} = \frac{71}{75}.$$ **Ans.**

38. *A problem in mathematics is given to four students A, B, C, and D. Their chances of solving the problem respectively are $\frac{1}{2}, \frac{1}{3}, \frac{1}{4}$ and $\frac{1}{5}$. What is the probability that the problem is solved ?*

Sol. P (A can solve the problem) $= \frac{1}{2}$

P (B can solve the problem) $= \frac{1}{3}$

P (C can solve the problem) $= \frac{1}{4}$

P (D can solve the problem) $= \frac{1}{5}$

P (A cannot solve the problem) $= 1 - \frac{1}{2} = \frac{1}{2}$

P (B cannot solve the problem) $= 1 - \frac{1}{3} = \frac{2}{3}$

P (C cannot solve the problem) $= 1 - \frac{1}{4} = \frac{3}{4}$

P (D cannot solve the problem) $= 1 - \frac{1}{5} = \frac{4}{5}$

P (A, B, C and D cannot solve the problem)

$$= \frac{1}{2} \times \frac{2}{3} \times \frac{3}{4} \times \frac{4}{5} = \frac{1}{5}$$

∴ P (Problem will be solved) = 1 – P (Problem is not solved by any of them)

$$= 1 - \frac{1}{5} = \frac{4}{5}$$ **Ans.**

39. *Aman and Bhuvan throw a pair of dice alternately. In order to win, they have to get a sum of 8. Find their respective probabilities of winning if Aman starts the game.*

Sol. Success of Aman and Bhuvan is getting a total of 8 in a pair of dice.

∴ Favourable cases

= {(2, 6), (3, 5), (4, 4), (5, 3), (6, 2)}

= 5

Let p = Probability of getting a total of 8

and q = Probability of not getting a total of 8

$$\therefore\ p = \frac{5}{36} \text{ and } q = \frac{31}{36}$$

If Aman starts the game, he can win it in the I[st] throw, 3rd throw, 5th throw, 7th throw and so on

Now, Probability of Aman's winning in the I[st] throw $= p = \frac{5}{36}$

Probability of Aman winning in the 3rd throw

$$= qqp = \frac{31}{36}\times\frac{31}{36}\times\frac{5}{36} = \left(\frac{31}{36}\right)^2\times\frac{5}{36}$$

Probability of Aman's winning in 5th throw

$$= qqqqp = \left(\frac{31}{36}\right)^4\times\frac{5}{36} \text{ and so on}$$

Since all cases are mutually exclusive therefore the chances of Aman's winning the game first

$$= p + qqp + qqqqp\ldots\ldots\ldots \infty$$

$$= p\,[1 + q^2 + q^4 + \ldots\ldots\ldots \infty\,]$$

$$= p\left[\frac{1}{1-q^2}\right]$$

$$\left[\text{In a G.P. } S_\infty = \frac{a}{1-r}\right]$$

$$= \frac{5}{36}\left[\frac{1}{1-\left(\frac{31}{36}\right)^2}\right]$$

$$= \frac{5}{36}\left[\frac{36\times36}{36\times36-31\times31}\right]$$

$$= \frac{5}{36}\times\frac{36\times36}{(5\times67)} = \frac{5\times36}{5\times67} = \frac{36}{67}$$

Since, Aman and Bhuvan are throwing the pair of dice alternately, hence the events are mutually exclusive and therefore the sum of their probabilities of winning is 1.

∴ Probability of Bhuvan's winning

$$= 1-\frac{36}{67} = \frac{31}{67}$$ **Ans.**

40. *In a bolt factory, three machines A, B and C manufacture 25%, 35% and 40% of the total production respectively. Of their respective outputs, 5%, 4% and 2% are defective. A bolt is drawn at random from the total production and it is found to be defective. Find the probability that it was manufactured by machine C.*

Sol. Let E_1 : the event that bolt is produced by machine A

E_2 : the event that bolt is produced by machine B

E_3 : the event that bolt is produced by machine C

Here, E_1, E_2 and E_3 are mutually exclusive and exhaustive events.

∴ We have,

$$P(E_1) = 25\% = \frac{25}{100}$$

$$P(E_2) = 35\% = \frac{35}{100}$$

$$P(E_3) = 40\% = \frac{40}{100}$$

Let E : the event that bolt chosen is found to b defective.

$$\therefore P(E/E_1) = 5\% = \frac{5}{100},\ P(E/E_2) = 4\% = \frac{4}{100}$$

and $P(E/E_3) = 2\% = \frac{2}{100}$

Using Bayes' theorem, we have

P (defective bolt is produced by machine C)

$$P(E_3/E) = \frac{P(E_3)\cdot P(E/E_3)}{P(E_1)\cdot P(E/E_1)+P(E_2)\cdot P(E/E_2)+P(E_3)\cdot P(E/E_3)}$$

$$= \frac{\frac{40}{100}\times\frac{2}{100}}{\left(\frac{25}{100}\right)\times\frac{5}{100}+\frac{35}{100}\times\frac{4}{100}+\frac{40}{100}\times\frac{2}{100}}$$

$$= \frac{80}{125+140+80}$$

$$= \frac{80}{345} = 0{\cdot}23$$ **Ans**

41. *A company has two plants which manufactur scooters. Plant I manufactures 80% of the scooter while Plant II manufactures 20% of the scooters. A plant I, 85 out of 100 scooters are rated as being o standard quality, while at Plant II only 65 out of 10 scooters are rated as being of standard quality. If scooter is of standard quality, what is the probabilit that it came from Plant I ?*

Sol. Let A and B denote scooters manufacture by Plant I and Plant II respectively, the probability is

$$P(A) = \frac{80}{100} = 0{\cdot}8$$

$$P(B) = \frac{20}{100} = 0{\cdot}2$$

If X represent the event that scoote manufactured is of standard quality. Then,

$$P(X/A) = \frac{85}{100} = 0{\cdot}85$$

$$P(X/B) = \frac{65}{100} = 0{\cdot}65$$

Now, selected scooter is of standard quality produced by Plant I

Using Bayes' theorem, we have

$$P(A/X) = \frac{P(A)\ P(X/A)}{P(A)\ P(X/A)+P(B)\ P(X/B)}$$

$$= \frac{0{\cdot}8\times0{\cdot}85}{0{\cdot}8\times0{\cdot}85+0{\cdot}2\times0{\cdot}65}$$

$$= \frac{0{\cdot}68}{0{\cdot}68+0{\cdot}13}$$

$$= \frac{0{\cdot}68}{0{\cdot}81}$$

$$= \frac{68}{81} = 0{\cdot}8390 \approx 0{\cdot}84$$ **Ans.**

42. *A firm produces steel pipes in three plants A, B and C with daily production of 500, 1000 and 2000 units, respectively. It is known that fractions of defective output produced by the three plants are respectively 0·005, 0·008 and 0·010. A pipe is selected at random from a days total production and found to be defective. What is probability that it came from first plant.*

Sol. Let A, B and C denotes the selection of pipe from plant A, plant B and plant C respectively; then probability is

$$P(A) = \frac{500}{3500} = \frac{1}{7}$$

$$P(B) = \frac{1000}{3500} = \frac{2}{7}$$

and $$P(C) = \frac{2000}{3500} = \frac{4}{7}$$

Let D represent the event of defective steel pipe then,

$$P(D/A) = 0{\cdot}005 = \frac{5}{1000}$$

$$P(D/B) = 0{\cdot}008 = \frac{8}{1000}$$

and $$P(D/C) = 0{\cdot}010 = \frac{10}{1000}$$

Now, selected pipe is defective and it came from first plant.

Using Bayes' theorem, we have

P (A/D)

$$= \frac{P(A)\,P(D/A)}{P(A)\cdot P(D/A)+P(B)\cdot P(D/B)+P(C)\cdot P(D/C)}$$

$$= \frac{\frac{1}{7}\times\frac{5}{1000}}{\frac{1}{7}\times\frac{5}{1000}+\frac{2}{7}\times\frac{8}{1000}+\frac{4}{7}\times\frac{10}{1000}}$$

$$= \frac{5}{5+16+40} = \frac{5}{61}.$$ **Ans.**

43. (a) *A factory has three machines A, B and C producing 1,500, 2,500 and 3,000 bulbs per day, respectively. Machine A produces 1·5% defective bulbs, machine B produces 2% defective bulbs and machine C produces 2·5% defective bulbs. At the end of the day, a bulb is drawn at random and is found to be defective. What is the probability that this defective bulb has been produced by machine B ?*

(b) *Five bad eggs are mixed with 10 good ones. If three eggs are drawn one by one with replacement, find the probability distribution of the number of goods eggs drawn.*

Sol. (a) Total daily production of bulbs

= 1500 + 2500 + 3000

= 7000 bulbs.

$$\therefore \quad P(A) = \frac{1500}{7000} = \frac{3}{14},$$

$$P(B) = \frac{2500}{7000} = \frac{5}{14},$$

$$P(C) = \frac{3000}{7000} = \frac{3}{7}$$

Let F be the event of producing defective bulbs.

$$P(F/A) = 1{\cdot}5\% = \frac{1{\cdot}5}{100} = \frac{3}{200}$$

$$P(F/B) = 2\% = \frac{2}{100} = \frac{1}{50}$$

$$P(F/C) = 2{\cdot}5\% = \frac{2{\cdot}5}{100} = \frac{1}{40}$$

Using Bayes' theorem, we have

P (B/F)

$$= \frac{P(B)\cdot P(F/B)}{P(B)\cdot P(F/B)+P(A)\cdot P(F/A)+P(C)\cdot P(F/C)}$$

$$= \frac{\frac{5}{14}\times\frac{1}{50}}{\frac{5}{14}\times\frac{1}{50}+\frac{3}{14}\times\frac{3}{200}+\frac{3}{7}\times\frac{1}{40}}$$

$$= \frac{\frac{1}{140}}{\frac{1}{140}+\frac{9}{2800}+\frac{3}{280}} = \frac{20}{59}$$ **Ans.**

(b) Here, total eggs = 15

$$p = \frac{10}{15} = \frac{2}{3},$$

$$q = \frac{1}{3}, n = 3.$$

Let X denote the number of good eggs in 3 draws. Then X can take values 0, 1, 2 and 3.

$$\therefore\ P(X=0) = {}^3C_0\left(\frac{2}{3}\right)^0\left(\frac{1}{3}\right)^3 = \frac{1}{27}$$

$$P(X=1) = {}^3C_1\left(\frac{2}{3}\right)^1\left(\frac{1}{3}\right)^2 = 3\times\frac{2}{3}\times\frac{1}{9} = \frac{2}{9}$$

$$P(X=2) = {}^3C_2\left(\frac{2}{3}\right)^2\left(\frac{1}{3}\right)^1 = 3\times\frac{4}{9}\times\frac{1}{3} = \frac{4}{9}$$

$$P(X=3) = {}^3C_3\left(\frac{2}{3}\right)^3\left(\frac{1}{3}\right)^0 = \frac{8}{27}$$

Hence, probability distribution of X is :

X	0	1	2	3
P(X)	1/27	2/9	4/9	8/27

Ans.

44. *Bag A contains 2 white, 1 black and 3 red balls, Bag B contains 3 white, 2 black and 4 red balls and Bag C contains 4 white, 3 black and 2 red balls. One Bag is chosen at random and 2 balls are drawn at random from that Bag. If the randomly drawn balls happen to be red and black, what is the probability that both balls come from Bag B ?*

Sol. Given,

Bag A	Bag B	Bag C
2 W	3 W	4 W
1 B	2 B	3 B
3 R	4 R	2 R

Let E_1, E_2 and E_3 be the events of choosing bag A, bag B and bag C respectively.

$$\therefore\quad P(E_1) = P(E_2) = P(E_3) = \frac{1}{3}$$

Let E be the event of drawing 1 red and 1 black balls from the bags, then

$P(E/E_1)$ = Probability that 1 red and 1 black balls are drawn from bag A

$$= \frac{{}^3C_1\times{}^1C_1}{{}^6C_2} = \frac{3}{15} = \frac{1}{5}$$

$$\text{Similarly, } P(E/E_2) = \frac{{}^4C_1\times{}^2C_1}{{}^9C_2} = \frac{4\times2}{36} = \frac{2}{9}$$

$$\text{and}\quad P(E/E_3) = \frac{{}^2C_1\times{}^3C_1}{{}^9C_2} = \frac{2\times3}{36} = \frac{1}{6}$$

∴ Using Bayes' theorem,

Required probability = Probability that 2 balls (1 red, 1 black) are drawn from bag B is given by

$$P(E_2/E) = \frac{P(E_2)\cdot P(E/E_2)}{P(E_1)\cdot P(E/E_1) + P(E_2)\cdot P(E/E_2) + P(E_3)\cdot P(E/E_3)}$$

$$= \frac{\frac{1}{3}\times\frac{2}{9}}{\frac{1}{3}\times\frac{1}{5}+\frac{1}{3}\times\frac{2}{9}+\frac{1}{3}\times\frac{1}{6}}$$

$$= \frac{\frac{1}{3}\times\frac{2}{9}}{\frac{1}{3}\left[\frac{1}{5}+\frac{2}{9}+\frac{1}{6}\right]}$$

$$= \frac{\frac{2}{9}}{\frac{18+20+15}{90}}$$

$$= \frac{2\times90}{9\times53} = \frac{20}{53}.$$

Ans.

45. *If a bulb factory machine A, B and C manufactures 60%, 30% and 10% respecitvely. 1%, 2% and 3% of the bulbs produced by A, B and C are defective. A bulb is drawn at random from the total production and found to be defective. Find the probalility that bulb has been produced by machine A.*

Sol. Let E_1, E_2 and E_3 be the events of drawing a bulb produced by *A*, *B*, and C respectively. Then,

$$P(E_1) = \frac{60}{100} = 0{\cdot}6$$

$$P(E_2) = \frac{30}{100} = 0{\cdot}3$$

$$P(E_3) = \frac{10}{100} = 0{\cdot}1$$

Let *F* be the event of drawing a defective bulb then,

$$P(F/E_1) = \frac{1}{100} = 0{\cdot}01,$$

$$P(F/E_2) = \frac{2}{100} = 0{\cdot}02$$

$$\text{and}\quad P(F/E_3) = \frac{3}{100} = 0{\cdot}03$$

Here, we have to find the probability that the bulb has been produced by *A* given that it is defective *i.e.*, we have to find $P(E_1/F)$.

Using Bayes' theorem, we have

$P(E_1/F)$

$$= \frac{P(F/E_1).P(E_1)}{P(F/E_1)P(E_1)+P(F/E_2).P(E_2)+P(F/E_3).P(E_3)}$$

$$= \frac{0{\cdot}01\times0{\cdot}6}{(0{\cdot}01)\times(0{\cdot}6)+(0{\cdot}02)\times(0{\cdot}3)+(0{\cdot}03)\times(0{\cdot}1)}$$

$$= \frac{2}{5}$$ **Ans.**

46. *Bag 1 contains 5 green and 3 red balls, another bag 2 contains 4 green 6 red balls. A red ball has been drawn from one of the bags. Find the probability that it was drawn from bag 1.*

Sol. Let

A = A red ball is drawn

B_1 = Bag 1 is chosen

B_2 = Bag 2 is chosen

$$P(B_1) = \frac{1}{2}, \quad P(B_2) = \frac{1}{2}$$

$$P(A/B_2) = \frac{3}{8}, P(A/B_2) = \frac{6}{10}$$

∴ Using Bayes' theorem,

P (Bag 1 is choosen / ball drawn is red)

$$P(B_1/A) = \frac{P(B_1)\cdot P(A/B_1)}{P(B_1/A)P(A/B_1)+P(B_2)P(A/B_2)}$$

$$= \frac{\frac{1}{2}\cdot\frac{3}{8}}{\frac{1}{2}\cdot\frac{3}{8}+\frac{1}{2}\cdot\frac{6}{10}}$$

$$= \frac{\frac{3}{16}}{\frac{3}{16}+\frac{3}{10}}$$

$$= \frac{5}{13}$$ **Ans.**

47. *Box I contains two white and three black balls. Box II contains four white and one black balls and box III contains three white and four black balls. A dice having three red, two yellow and one green face, is thrown to select the box. If red face turns up, we pick up box I, if a yellow face turns up we pick up box II, otherwise, we pick up box III. Then, we draw a ball from the selected box. If the ball drawn is white, what is the probability that the dice had turned up with a red face ?*

Sol. Let E_1, E_2, E_3 and A be the events defined as follows :

E_1 : red face turns up

E_2 : yellow face turns up

E_3 : green face turns up

A : white ball is drawn

We wish to calculate the probability that dice had turn up with red face when A has occured (white ball is selected)

i.e., $P(E_1/A)$

By Baye's Theorem,

$$P(E_1/A) = \frac{P(E_1)\times P(A/E_1)}{P(E_1)\times P(A/E_1)+P(E_2)\times P(A/E_2)+P(E_3)\times P(A/E_3)}$$

$$= \frac{\frac{1}{2}\times\frac{2}{5}}{\frac{1}{2}\times\frac{2}{5}+\frac{1}{3}\times\frac{4}{5}+\frac{1}{6}\times\frac{3}{7}}$$

$$= \frac{\frac{1}{5}}{\frac{1}{5}+\frac{4}{15}+\frac{1}{14}}$$

$$= \frac{\frac{1}{5}}{\frac{113}{210}} = \frac{1}{5}\times\frac{210}{113} = \frac{42}{113}.$$ **Ans.**

48. *In an automobile factory, certain parts are to be fixed into the chassis in a section before it moves into another section. On a given day, one of the three persons A, B and C carries out this task. A has 45% chance, B has 35% chance and C has 20% chance of doing the task. The probability that A, B and C will take more than the allotted time is $\frac{1}{6}$, $\frac{1}{10}$ and $\frac{1}{20}$ respectively. If it is found that the time taken is more than the allotted time, what is the probability that A has done the task ?*

Sol. Let E_1, E_2, E_3 denote the events of carrying out the task by A, B and C respectively.

Let H denote the event of taking more time.

Then, $P(E_1) = 0{\cdot}45$, $P(E_2) = 0{\cdot}35$, $P(E_3) = 0{\cdot}20$

$$P(H/E_1) = \frac{1}{6},$$

$$P(H/E_2) = \frac{1}{10},$$

$$P(H/E_3) = \frac{1}{20}$$

Then, using Bayes' theorem

$$P(E_1/H) = \frac{P(E_1)\cdot P(H/E_1)}{P(E_1)\,P(H/E_1)+P(E_2)\,P(H/E_2)+P(E_3)\cdot P(H/E_3)}$$

$$= \frac{\frac{45}{600}}{\frac{45}{600}+\frac{35}{1000}+\frac{20}{2000}} = \frac{\frac{3}{40}}{\frac{3}{40}+\frac{7}{200}+\frac{1}{100}}$$

$$= \frac{3}{40}\times\frac{200}{24} = \frac{5}{8}$$ **Ans.**

49. *An unbiased die is thrown 3 times. If getting 3 or 5 is considered a success. Find the probability of at least 2 successes.*

Sol. An unbiased die is thrown. Sample space for one throw = 6

Getting 3 or 5 is considered a success

Let p = probability of success $= \frac{2}{6} = \frac{1}{3}$

Let q = probability of not getting 3 or 5

= probability of failure $= 1-\frac{1}{3} = \frac{2}{3}$

Probability of at least 2 successes in 3 times = probability of 2 success + probability of 3 success.

∴ By binomial distribution theorem,

P (r success in n independent trials)

$$= {}^nC_r\,(q)^{n-r}\,p^r$$

Hence, probability of 2 successes in 3 trials

$$= {}^3C_2\left(\frac{2}{3}\right)^{3-2}\left(\frac{1}{3}\right)^2 = \frac{3\times 2}{27} = \frac{6}{27}$$

∴ Probability of 3 successes in 3 trials

$$= {}^3C_3\left(\frac{1}{3}\right)^3 = \frac{1}{27}$$

Adding probability of at least 2 success, we get

$$= \frac{6}{27}+\frac{1}{27} = \frac{7}{27}.$$ **Ans.**

50. *Four dices are thrown simultaneously. If the occurrence of an odd number in a single dice is considered a success, find the probability of at most 2 successes.*

Sol. Number of dice = 4

∴ $n = 4$

Probability of getting odd number in a dice,

$$p = \frac{3}{6} = \frac{1}{2}$$

Probability of not getting odd number, $q = \frac{1}{2}$

Applying binomial distribution, we have

$$\sum_{r=0}^{n} {}^nC_r\, q^{n-r}\, p^r$$

For at most 2 success $r \le 2$.

∴ Required probability

$$= {}^4C_0\, q^4 + {}^4C_1\, q^3\, p + {}^4C_2\, q^2\, p^2$$

$$= {}^4C_0\left(\frac{1}{2}\right)^4 + {}^4C_1\left(\frac{1}{2}\right)^3\left(\frac{1}{2}\right) + {}^4C_2\left(\frac{1}{2}\right)^2\left(\frac{1}{2}\right)^2$$

$$= \left(\frac{1}{2}\right)^4 \{{}^4C_0 + {}^4C_1 + {}^4C_2\}$$

$$= \frac{1}{16}\{1+4+6\} = \frac{11}{16}$$ **Ans.**

51. *The probability that a bulb produced by a factory will fuse in 100 days of use is 0·05. Find the probability that out of 5 such bulbs, after 100 days of use :*

(i) *None fuse.*

(ii) *Not more than one fuses.*

(iii) *More than one fuses.*

(iv) *At least one fuses.*

Sol. Here, $p = 0{\cdot}05 = \frac{1}{20}$

$$q = \frac{19}{20}$$

(i) P (None fuse) $= {}^5C_0\, p^0 q^{5-0}$

$$= \left(\frac{1}{20}\right)^0\left(\frac{19}{20}\right)^5$$

$$= \left(\frac{19}{20}\right)^5$$ **Ans.**

(ii) P (Not more than one fuse)

$$= \sum_{r=0}^{1} {}^nC_r\, p^r\, q^{n-r}$$

$$= {}^5C_0\left(\frac{1}{20}\right)^0\left(\frac{19}{20}\right)^5 + {}^5C_1\left(\frac{1}{20}\right)\left(\frac{19}{20}\right)^4$$

$$= \left(\frac{19}{20}\right)^4\left(\frac{19}{20}+\frac{5}{20}\right) = \left(\frac{19}{20}\right)^4\left(\frac{24}{20}\right)$$

$$= \frac{6}{5}\left(\frac{19}{20}\right)^4$$ **Ans.**

Formulae Based Questions

(iii) P (more than one fuses)

$$= \sum_{r=2}^{5} {}^nC_r\, p^r q^{n-r}$$

$$= {}^5C_2\, p^2 q^3 + {}^5C_3\, p^3 q^2 + {}^5C_4\, p^4 q^1 + {}^5C_5\, p^5 q^0$$

$$= \frac{5!}{2!\cdot 3!}\left(\frac{1}{20}\right)^2\left(\frac{19}{20}\right)^3 + \frac{5!}{3!\cdot 2!}\left(\frac{1}{20}\right)^3\left(\frac{19}{20}\right)^2$$

$$+ \frac{5!}{4!\cdot 1!}\left(\frac{1}{20}\right)^4\left(\frac{19}{20}\right)^1 + \frac{5!}{5!\cdot 0!}\left(\frac{1}{20}\right)^5\left(\frac{19}{20}\right)^0$$

$$= 10\cdot\left(\frac{1}{20}\right)^2\left(\frac{19}{20}\right)^3 + 10\cdot\left(\frac{1}{20}\right)^3\left(\frac{19}{20}\right)^2$$

$$+ 5\cdot\left(\frac{1}{20}\right)^4\left(\frac{19}{20}\right) + \left(\frac{1}{20}\right)^5$$

$$= \left(\frac{1}{20}\right)^5 + [10(19)^3 + 10(19)^2 + 5 \times 19 + 1]$$

$$= 72296\cdot\left(\frac{1}{20}\right)^5$$ **Ans.**

(iv) P (at least one fuses) $= \sum_{r=1}^{5} {}^nC_r\, p^r q^{n-r}$

$$= {}^5C_1\, p^1 q^4 + {}^5C_2\, p^2 q^3 + {}^5C_3\, p^3 q^2 + {}^5C_4\, p^4 q^1 + {}^5C_5\, p^5 q^0$$

$$= {}^5C_1\, p^1 q^4 + 72296\cdot\left(\frac{1}{20}\right)^5$$ [from (ii)]

$$= 5\cdot\left(\frac{1}{20}\right)\cdot\left(\frac{19}{20}\right)^4 + 72296\cdot\left(\frac{1}{20}\right)^5$$

$$= \left(\frac{1}{20}\right)^5 [5 \times (19)^4 + 72296]$$

$$= 72296\cdot\left(\frac{1}{20}\right)^5$$ **Ans.**

52. *A and B play a game in which A's chances of winning the game are $\frac{3}{5}$. In a series of 6 games, find the probability that A will win at least 4 games.*

Sol. $p = \frac{3}{5}$, $q = 1 - \frac{3}{5} = \frac{2}{5}$. There are 6 games and probability of winning at least 4 games is given as,

$${}^6C_0\, p^6 + {}^6C_1 p^5 q + {}^6C_2 p^4 q^2$$

$$= 1\left(\frac{3}{5}\right)^6 + 6\left(\frac{3}{5}\right)^5\left(\frac{2}{5}\right) + 15\left(\frac{3}{5}\right)^4\left(\frac{2}{5}\right)^2$$

$$= \frac{729}{5^6} + \frac{2916}{5^6} + \frac{4860}{5^6} = \frac{8505}{5^6}$$

$$= \frac{8505}{15625} = \frac{1701}{3125}$$ **Ans.**

53. *On dialling certain telephone numbers, assume that on an average, one telephone number out of five is busy, Ten telephone numbers are randomly selected and dialled. Find the probability that at least three of them will be busy.*

Sol. Here, $p = \frac{1}{5}$, $q = 1 - \frac{1}{5} = \frac{4}{5}$, $n = 10$

Required probability (at least three phones busy)

= 1 – (Probability maximum two phones are busy)

$$= 1 - \left[{}^{10}C_0\left(\frac{1}{5}\right)^0\left(\frac{4}{5}\right)^{10} + {}^{10}C_1\left(\frac{1}{5}\right)^1\left(\frac{4}{5}\right)^9 + {}^{10}C_2\left(\frac{1}{5}\right)^2\left(\frac{4}{8}\right)^8\right]$$

$$= 1 - \left[1 \times 1 \times \left(\frac{4}{5}\right)^{10} + 10 \times \left(\frac{1}{5}\right) \times \left(\frac{4}{5}\right)^9 + 45 \times \left(\frac{1}{2}\right)^2 \times \left(\frac{4}{5}\right)^8\right]$$

$$= 1 - \left(\frac{4}{5}\right)^8 \left[\frac{16}{25} + \frac{8}{5} + \frac{9}{5}\right]$$

$$= 1 - \left[\left(\frac{4}{5}\right)^8 \left(\frac{16 + 40 + 45}{25}\right)\right]$$

$$= 1 - 0{\cdot}1678 \left(\frac{101}{25}\right)$$

$$= 1 - 0{\cdot}1678 \times 4{\cdot}04$$

$$= 1 - 0{\cdot}68 = 0{\cdot}32$$ **Ans.**

54. **(a)** *If the sum and the product of the mean and variance of a Binomial Distribution are 1·8 and 0·8 respectively, find the probability distribution and the probability of at least one success.*

(b) *For A, B and C, the chances of being selected as the manager of a firm are 4 : 1 : 2, respectively. The probabilities for them to introduce a radical change in the marketing strategy are 0·3, 0·8 and 0·5 respectively. If a change takes place; find the probability that it is due to the appointment of B.*

Sol. **(a)** According to question, we have

$$np + npq = 1{\cdot}8$$

$$\Rightarrow \quad np(1 + q) = 1{\cdot}8 \quad \text{...(i)}$$

And $\quad np.npq = 0{\cdot}8$

$\Rightarrow \quad n^2p^2q = 0{\cdot}8 \quad \ldots\text{(ii)}$

Dividing the square of (i) by (ii) we get

$$\frac{n^2p^2(1+q)^2}{n^2p^2q} = \frac{1{\cdot}8\times1{\cdot}8}{0{\cdot}8}$$

$$\Rightarrow \frac{(1+q)^2}{q} = \frac{3{\cdot}24}{0{\cdot}8}$$

$$\Rightarrow \frac{(1+q)^2}{q} = \frac{324}{80}$$

$$\Rightarrow \frac{(1+q)^2}{q} = \frac{81}{20}$$

$$\Rightarrow 20 + 2\times 20q + 20q^2 = 81q$$

$$\Rightarrow 20q^2 - 41q + 20 = 0$$

$$\Rightarrow 20q^2 - 25q - 16q + 20 = 0$$

$$\Rightarrow 5q(4q-5) - 4(4q-5) = 0$$

$$\Rightarrow (4q-5)(5q-4) = 0$$

$$\Rightarrow q = \frac{5}{4} \text{ or } q = \frac{4}{5} \text{ but } q \neq \frac{5}{4}$$

$$\therefore q = \frac{4}{5}$$

$$\Rightarrow p = \frac{1}{5}$$

From (i),

$$n\times\frac{1}{5}\left(1+\frac{4}{5}\right) = 1{\cdot}8$$

$$\Rightarrow n\times\frac{1}{5}\times\frac{9}{5} = 1{\cdot}8$$

$$\Rightarrow 9n = 25\times1{\cdot}8$$

$$\Rightarrow n = \frac{25\times1{\cdot}8}{9}$$

$$\Rightarrow n = \frac{45}{9}$$

$$\therefore n = 5$$

Hence, the binomial distribution is

$$(p+q)^n = \left(\frac{1}{5}+\frac{4}{5}\right)^5$$

Probability of getting at least 1 success

$= P(r \geq 1)$

$= P(1) + P(2) + P(3) + P(4) + P(5)$

$$= {}^5C_1\cdot\frac{1}{5}\left(\frac{4}{5}\right)^4 + {}^5C_2\left(\frac{1}{5}\right)^2\left(\frac{4}{5}\right)^3 + {}^5C_3\left(\frac{1}{5}\right)^3\left(\frac{4}{5}\right)^2 + {}^5C_4\left(\frac{1}{5}\right)^4\left(\frac{4}{5}\right) + {}^5C_5\left(\frac{1}{5}\right)^5$$

$$= 5\cdot\frac{1}{5}\times\frac{256}{625} + 10\cdot\frac{1}{25}\cdot\frac{64}{125} + 10\cdot\frac{1}{125}\cdot\frac{16}{25} + 5\cdot\frac{1}{625}\cdot\frac{4}{5} + 1\cdot\frac{1}{3125}$$

$$= \frac{256}{625} + \frac{128}{625} + \frac{32}{625} + \frac{4}{625} + \frac{1}{3125}$$

$$= \frac{1280+640+160+20+1}{3125}$$

$$= \frac{2101}{3125}$$ **Ans.**

(b) Let E_1, E_2, E_3 and E be the events as defined below :

E_1 = A is selected as Manager

E_2 = B is selected as Manager

E_3 = C is selected as Manager

and E = radical change occurs in marketing strategy

$$\therefore P(E_1) = \frac{4}{4+1+2} = \frac{4}{7}$$

$$P(E_2) = \frac{1}{4+1+2} = \frac{1}{7}$$

$$P(E_3) = \frac{2}{4+1+2} = \frac{2}{7}$$

Given, $P(E/E_1) = 0{\cdot}3$

$P(E/E_2) = 0{\cdot}8$

$P(E/E_3) = 0{\cdot}5$

We want to find the probability that the radical change in marketing strategy occured due to the appointment of B *i.e.*, we want to find $P(E_2/E)$.

Using Bayes' theorem, we have

$$P(E_2/E) = \frac{P(E_2).P(E/E_2)}{P(E_1)P(E/E_1)+P(E_2).P(E/E_2)+P(E_3)P(E/E_3)}$$

$$= \frac{\frac{1}{7}\times0{\cdot}8}{\frac{4}{7}\times0{\cdot}3+\frac{1}{7}\times0{\cdot}8+\frac{2}{7}\times0{\cdot}5}$$

$$= \frac{0{\cdot}8}{1{\cdot}2+0{\cdot}8+1}$$

$$= \frac{0{\cdot}8}{3} = \frac{8}{30}$$

$$= \frac{4}{15}$$ **Ans.**

55. *A box contains 4 red and 5 black marbles. Find the probability distribution of the red marbles in a random draw of three marbles. Also find the mean and standard deviation of the distribution.*

Sol. Let X denotes the number of red marbles drawn from the box. Since there are 4 red marbles therefore X can take the values 0, 1, 2 and 3.

Now, P(X = 0)

= Probability of getting no red marbles

= Probability of 3 marbles drawn are black

$$= \frac{^5C_3}{^9C_3} = \frac{10}{84} = \frac{5}{42}$$

P(X = 1)

= Probability of getting 1 red marbles

$$= \frac{^4C_1 \times {}^5C_2}{^9C_3} = \frac{4\times10}{84} = \frac{10}{21}$$

P(X = 2)

= Probability of getting 2 red marbles

$$= \frac{^4C_2 \times {}^5C_1}{^9C_3} = \frac{6\times5}{84} = \frac{5}{14}$$

P(X = 3)

= Probability of getting 3 red marbles

$$= \frac{^4C_3}{^9C_3} = \frac{4}{84} = \frac{1}{21}$$

∴ The probability distribution of X is given as :

X	0	1	2	3
P(X)	5/42	10/21	5/14	1/21

Again,

X_i	$P_i(x)$	$P_i \times X_i$	X_i^2	$P_i \times X_i^2$
0	5/42	0	0	0
1	10/21	10/21	1	10/21
2	5/14	5/7	4	10/7
3	1/21	1/7	9	3/7
		$\Sigma P_iX_i = 4/3$		$\Sigma P_i \times X_i^2 = 7/3$

Now, Mean $= \mu = \Sigma P_iX_i = \frac{4}{3}$

Standard Deviation, $= \sqrt{\Sigma P_iX_i^2 - \mu^2}$

$$= \sqrt{\frac{7}{3} - \frac{16}{9}}$$

$$= \sqrt{\frac{21-16}{9}}$$

$$= \frac{\sqrt{5}}{3} = 0{\cdot}745$$ **Ans.**

56. *Five dice are thrown simultaneously. If the occurrence of an odd number in a single dice is considered a success, find the probability of maximum three successes.*

Sol. Let A be the event occuring an odd number and it is success.

$$p = P(A) = \frac{1}{2}, q = 1 - p = 1 - \frac{1}{2} = \frac{1}{2}$$

∴ P (at most 3 success) or P (maximum of 3 success)

$$= \sum_{r=0}^{3} {}^nC_r\, p^r q^{n-r} \text{ where } n \text{ is 5.}$$

$$= {}^5C_0\left(\frac{1}{2}\right)^5 + {}^5C_1\left(\frac{1}{2}\right)\left(\frac{1}{2}\right)^4 + {}^5C_2\left(\frac{1}{2}\right)^2\left(\frac{1}{2}\right)^3 + {}^5C_3\left(\frac{1}{2}\right)^3\left(\frac{1}{2}\right)^2$$

$$= \frac{1}{32} + \frac{5}{32} + \frac{10}{32} + \frac{10}{32} = \frac{26}{32} = \frac{13}{16}.$$ **Ans.**

57. *The difference between mean and variance of a binomial distribution is 1 and the difference of their squares is 11. Find the distribution.*

Sol. Le the binomial distribution be $(q + p)^n$

Here, mean = np and variance = npq

Now, $np - npq = 1$

$np(1 - q) = 1$ ...(i)

and $(np)^2 - (npq)^2 = 11$

$(np)^2 (1 - q^2) = 11$ ...(ii)

Dividing equation (ii) by the square of equation (i), we get

$$\frac{(np)^2 (1-q^2)}{(np)^2 (1-q)^2} = \frac{11}{1}$$

$$\Rightarrow \frac{(1+q)(1-q)}{(1-q)(1-q)} = 11$$

$$\Rightarrow 1 + q = 11 - 11q$$

$$\Rightarrow q = \frac{10}{12} = \frac{5}{6}$$

$$\therefore p = 1 - q = 1 - \frac{5}{6} = \frac{1}{6}$$

Putting the values of p and q in (i), we get

$$n\left(\frac{1}{6}\right)\left(1 - \frac{5}{6}\right) = 1$$

$$n = 36$$

Hence, the binomial distribution is $\left(\frac{5}{6} + \frac{1}{6}\right)^{36}$.

Ans.

Chapter 11. Vectors

1. *If $\vec{a}, \vec{b}, \vec{c}, \vec{d}$ are the position vectors of four points A B C and D respectively and $\vec{b}-\vec{a}=\vec{c}-\vec{d}$, show that ABCD is a parallelogram.*

Sol. Let, ABCD be a quadrilateral

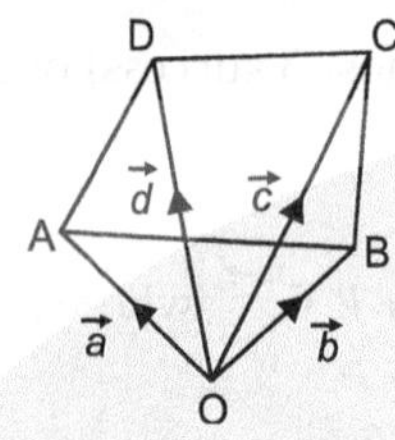

Let $\vec{OA}=\vec{a},\ \vec{OB}=\vec{b},$

$\vec{OC}=\vec{c},\ \vec{OD}=\vec{d}.$

Then, $\vec{AB}=\vec{OB}-\vec{OA}=\vec{b}-\vec{a}$

$\vec{DC}=\vec{OC}-\vec{OD}=\vec{c}-\vec{d}$

Since, $\vec{b}-\vec{c}=\vec{c}-\vec{d}$

$\Rightarrow \vec{AB}=\vec{DC}$

$\Rightarrow |AB|=|DC|$ and $AB \,||\, DC$.

Thus, ABCD is a parallelogram. **Hence Proved.**

2. *The vectors $2\hat{i}-\hat{j}+\hat{k}$, $\hat{i}-3\hat{j}-5\hat{k}$ and $3\hat{i}-4\hat{j}-4\hat{k}$ are the position vectors of vertices A, B and C respectively of a triangle ABC. Prove that ABC is a right angled triangle.*

Sol. Let

$$2\hat{i}-\hat{j}+\hat{k}=\vec{OA}$$

$$\hat{i}-3\hat{j}-5\hat{k}=\vec{OB}$$

$$3\hat{i}-4\hat{i}-4\hat{k}=\vec{OC}$$

$\therefore$ $\vec{AB}=\vec{OB}-\vec{OA}$

$=\hat{i}-3\hat{j}-5\hat{k}-2\hat{i}+\hat{j}-\hat{k}$

$=-\hat{i}-2\hat{j}-6\hat{k}$

$\Rightarrow |\vec{AB}|=\sqrt{1+4+36}=\sqrt{41}$

$\Rightarrow |\vec{AB}|^2=(\sqrt{41})^2=41$

And, $\vec{BC}=\vec{OC}-\vec{OB}$

$=3\hat{i}-4\hat{j}-4\hat{k}-\hat{i}+3\hat{j}+5\hat{k}$

$=2\hat{i}-\hat{j}+\hat{k}$

$\Rightarrow |\text{BC}|=\sqrt{4+1+1}=\sqrt{6}$

$\Rightarrow |\vec{BC}|^2=(\sqrt{6})^2=6$

And, $\vec{CA}=\vec{OA}-\vec{OC}$

$=2\hat{i}-\hat{j}+\hat{k}-3\hat{i}+4\hat{j}+4\hat{k}$

$=-\hat{i}+3\hat{j}+5\hat{k}$

$\Rightarrow |\vec{CA}|=\sqrt{1+9+25}=\sqrt{35}$

$\Rightarrow |\vec{CA}|^2=(\sqrt{35})^2=35$

$\therefore |\vec{AB}|^2=|\vec{BC}|^2+|\vec{CA}|^2$

$=(\sqrt{6})^2+(\sqrt{35})^2$

$=6+35=41$

Hence, it is right–angled triangle.

Hence Proved

3. *Write a unit vector perpendicular to the plane of two vectors $\vec{x}$ and $\vec{y}$ where $\vec{x}=5\hat{i}+2\hat{j}-3\hat{k}$ and $\vec{y}=-\hat{i}-3\hat{j}+\hat{k}$.*

Sol. Given,

$\vec{x}=5\hat{i}+2\hat{j}-3\hat{k}$

and $\vec{y}=-\hat{i}-3\hat{j}+\hat{k}$

A vector perpendicular to vector $\vec{x}$ and $\vec{y}$ is $\vec{x}\times\vec{y}$

$$\therefore \text{Unit vector} = \frac{\vec{x}\times\vec{y}}{|\vec{x}\times\vec{y}|}$$

$$\text{Now } \vec{x}\times\vec{y}=\begin{vmatrix}\hat{i} & \hat{j} & \hat{k}\\ 5 & 2 & -3\\ -1 & -3 & 1\end{vmatrix}$$

$=\hat{i}(2-9)-\hat{j}(5-3)+\hat{k}(-15+2)$

$=-7\hat{i}-2\hat{j}-13\hat{k}$

and $|\vec{x}\times\vec{y}|=\sqrt{(-7)^2+(-2)^2+(-13)^2}$

$=\sqrt{49+4+169}=\sqrt{222}$

$$\therefore \text{Unit vector} = \frac{-7\hat{i}-2\hat{j}-13\hat{k}}{\sqrt{222}}$$ **Ans.**

Formulae Based Questions

4. *Prove that the perpendicular bisectors of the sides of triangle are concurrent.*

Sol. Let D, E, F be the mid-points of the sides BC, CA, AB of the triangle ABC and P, the point of intersection of the perpendicular bisectors of BC and CA. FP is joined. (fig.)

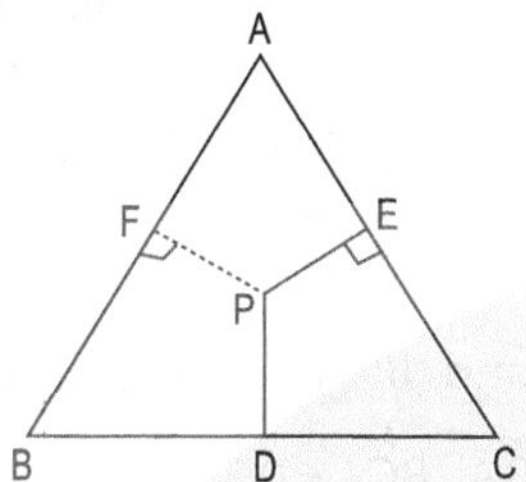

If a, b, c, p be the position vectors of A, B, C, P respectively, relative to some origin, then those of D, E, F are

$\frac{1}{2}(b+c), \frac{1}{2}(c+a), \frac{1}{2}(a+b)$, so that

$$\vec{DP} = p - \frac{1}{2}(b+c),$$

$$\vec{EP} = p - \frac{1}{2}(c+a),$$

$$\vec{FP} = p - \frac{1}{2}(a+b)$$

The condition of perpendicularity gives

$\left(p - \frac{b+c}{2}\right)\cdot(c-b) = 0$ and $\left(p - \frac{c+a}{2}\right).(a-c)$

$= 0$, when we derive, on adding,

$$\left(p - \frac{a+b}{2}\right)\cdot(a-b) = 0.$$

This shows that FP is perpendicular bisector of AB and hence, the perpendicular bisector of the sides of triangle are concurrent. **Hence Proved.**

5. *In a right-angled triangle prove that the square of the hypotenuse is equal to the sum of squares on sides containing the right angle.*

State and prove the converse.

Sol. Let ABC be a right-angled triangle at B.

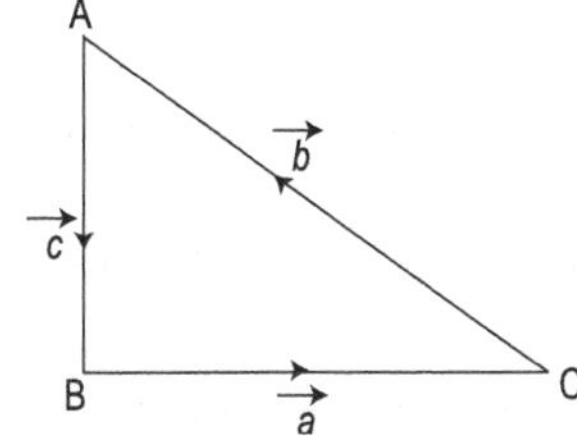

To prove,

$$AC^2 = AB^2 + BC^2$$

i.e., $|\vec{b}|^2 = |\vec{a}|^2 + |\vec{c}|^2$

Since, $\vec{a} + \vec{b} + \vec{c} = 0$

$\Rightarrow \quad \vec{b} = -(\vec{a} + \vec{c})$

$\Rightarrow \quad \vec{b}\cdot\vec{b} = [-(\vec{a} \times \vec{c})]\cdot[-(\vec{a} + \vec{c})]$

$\Rightarrow \quad |\vec{b}|^2 = |\vec{a}|^2 + |\vec{c}|^2 + 2\vec{a}.\vec{c}$

Since, AB is perpendicular to BC,

$\therefore \quad \vec{a}.\vec{c} = 0.$

Hence, $|\vec{b}|^2 = |\vec{a}|^2 + |\vec{c}|^2$

Converse is given by the statement.

In a triangle, the square of one side is equal to the sum of squares of the other two sides, then it is a right-angled triangle.

i.e., Given $|\vec{b}|^2 = |\vec{a}|^2 + |\vec{c}|^2$

We have to prove that $\vec{a}$ is $\perp$ to $\vec{c}$

Since, $\vec{a} + \vec{b} + \vec{c} = 0$

$\Rightarrow \quad \vec{b} = -(\vec{a} + \vec{c})$

$\Rightarrow \quad |\vec{b}|^2 = \vec{b}.\vec{b}$

$= |\vec{a}|^2 + |\vec{c}|^2 + 2\vec{a}.\vec{c}$

Since, $|\vec{b}|^2 = |\vec{a}|^2 + |\vec{c}|^2$

$\vec{a}.\vec{c} = 0$

i.e., BC is perpendicular to AB. **Hence Proved.**

6. *Using vectors , prove that angle in a semi–circle is a right angle.*

Sol. Let O be the centre of the circle and AB be its diameter

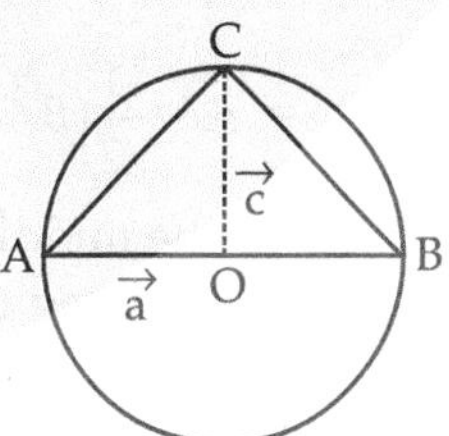

Let c be any point on semi-circle.

Position vector of $\vec{OA} = \vec{a}$ and $\vec{OC} = \vec{c}$

Then, $\vec{OB} = -\vec{OA} = -\vec{a}$

$$\vec{AC} = \vec{OC} - \vec{OA}$$

$$= \vec{c} - \vec{a}$$

And $\quad \vec{BC} = \vec{OC} - \vec{OB}$

$$= \vec{c} - (-\vec{a}) = \vec{c} + \vec{a}$$

Now, $\quad \vec{AC} \cdot \vec{BC} = (\vec{c} - \vec{a}) \cdot (\vec{c} + \vec{a})$

$$= \vec{c}.\vec{c} + \vec{c}.\vec{a} - \vec{a}.\vec{c} - \vec{a}.\vec{a}$$

$$= \vec{c}^2 - \vec{a}^2 = |\vec{c}|^2 - |\vec{a}|^2$$

Since, OA and OC are radii of circle

$\Rightarrow \quad |\vec{a}| = |\vec{c}|$

$\Rightarrow \quad \vec{AC}.\vec{BC} = 0$

$\Rightarrow \quad AC \perp BC$

Thus, angle in semi-circle is a right angle.

Hence Proved.

7. *Prove by vector method that the middle point of the hypotenuse of a right angled triangle is equidistant from the vertices of the triangle.*

Sol. Let ABC be a right–angled triangle. Taking B as origin. Let the position vectors of A and C are $\vec{BA} = \vec{a}$ and $\vec{BC} = \vec{b}$ respectively

Let D be the mid–point of hypotenuse AC.

position vector of D

$$\vec{BD} = \vec{BA} - \vec{DA} = \vec{BA} - \frac{1}{2} \times \vec{CA}$$

$$= \vec{BA} - \frac{1}{2}(\vec{BA} - \vec{BC}) = \frac{1}{2}(\vec{BA} + \vec{BC})$$

Point D exists in the middle of CA and also the mid–point, hence it is equidistant from A and C.

i.e., $\quad DA = DC$

$$\vec{BA} = \vec{BD} + \vec{DA}$$

$$\vec{BC} = \vec{BD} + \vec{DC}$$

$\therefore \quad \vec{BA}.\vec{BC} = (\vec{BD} + \vec{DA})\ (\vec{BD} + \vec{DC})$

$$= \vec{BD}.\vec{BD} + \vec{BD}.\vec{DC} + \vec{DA}.\vec{BD} + \vec{DA}.\vec{DC}$$

$\vec{DA}$ and $\vec{DC}$ are opposite vectors of equal magnitute.

$\therefore \quad \vec{DA} = -\vec{DC} = 0$

$\therefore \quad \vec{BA}.\vec{BC} = |\vec{BD}|^2 + \vec{BD}.\vec{DC} - \vec{BD}.\vec{DC} + \vec{DA}(-\vec{DA})$

$$= |\vec{BD}|^2 - |\vec{DA}|^2$$

$$= \vec{BD}^2 - \vec{DA}^2$$

But $\vec{BA}$ and $\vec{BC}$ are at right angle.

$\therefore \quad \vec{BA}.\vec{BC} = 0$

$\therefore \quad \vec{BD}^2 - \vec{DA}^2 = 0$

$\therefore \quad BD = DA$

But, $\quad DA = DC$

$\therefore \quad BD = DA = DC$

$\therefore$ D is equidistant from A, B and C.

Hence Proved.

8. *Prove that sum of squares of the diagonals of a parallelogram is equal to the sum of squares of the sides. (use vector method)*

Sol. Let $ABCD$ be a parallelogram of whose AC and BD are diagonals

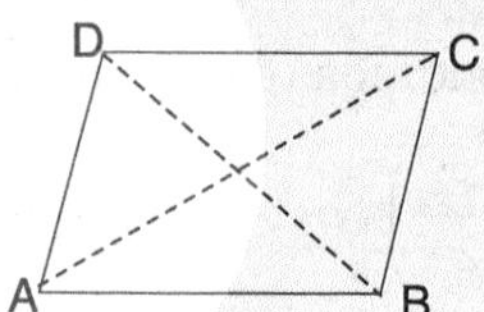

$\therefore \quad \vec{AC} = \vec{AB} + \vec{BC}$

$$\vec{BD} = \vec{DA} + \vec{AB}$$

$$\vec{DA} = -\vec{BC}$$

$\therefore \quad \vec{BD} = -\vec{BC} + \vec{AB} = \vec{AB} - \vec{BC}$

$$\vec{AC} \cdot \vec{AC} = (\vec{AB} + \vec{BC}) \cdot (\vec{AB} + \vec{BC})$$

$$= \vec{AB} \cdot \vec{AB} + \vec{AB} \cdot \vec{BC} + \vec{BC} \cdot \vec{AB} + \vec{BC} \cdot \vec{BC}$$

$\Rightarrow \quad AC^2 = AB^2 + BC^2 + 2\,\vec{AB} \cdot \vec{BC} \qquad$ ...(i)

$[\because \vec{AB} \cdot \vec{BC} = \vec{BC} \cdot \vec{AB}]$

$\Rightarrow \quad \vec{BD} \cdot \vec{BD} = (\vec{AB} - \vec{BC}) \cdot (\vec{AB} - \vec{BC})$

$$= \vec{AB} \cdot \vec{AB} - \vec{AB} \cdot \vec{BC} - \vec{BC} \cdot \vec{AB} + \vec{BC} \cdot \vec{BC}$$

$\Rightarrow \quad BD^2 = AB^2 + BC^2 - 2\,\vec{AB} \cdot \vec{BC} \qquad$ ...(ii)

By adding (i) and (ii), we get

$$AC^2 + BD^2 = AB^2 + BC^2 + 2\,\vec{AB}\cdot\vec{BC} + AB^2 + BC^2 - 2\,\vec{AB}\cdot\vec{BC}$$

$$= 2\,AB^2 + 2\,BC^2$$

$$= AB^2 + DC^2 + BC^2 + AD^2$$

Since, $\vec{AB} = \vec{DC}$ and $\vec{BC} = \vec{AD}$

$\therefore$ $|AB| = |DC|$ and $|BC| = |AD|$

Hence Proved.

9. *Two forces $\vec{F_1} = \hat{i}+2\hat{j}+\hat{k}$ and $\vec{F_2} = 2\hat{i}+3\hat{j}$ act on a particle to displace it from the point $3\hat{i}+2\hat{j}+\hat{k}$ to $12\hat{i}+3\hat{k}$, find the total work done by these forces.*

Sol. Let F be the resultant of F_1 and F_2

$$F = F_1 + F_2 = (\hat{i}+2\hat{j}+\hat{k})+(2\hat{i}+3\hat{j})$$

$$= 3\hat{i}+5\hat{j}+\hat{k}$$

Let d be the displacement from one point to the other.

$$d = (12\hat{i}+3\hat{k})-(3\hat{i}+2\hat{j}+\hat{k})$$

$$= 9\hat{i}-2\hat{j}+2\hat{k}$$

Therefore the work done is

$$F.d = (3\hat{i}+5\hat{j}+\hat{k}).(9\hat{i}-2\hat{j}+2\hat{k})$$

$$= 27 - 10 + 2$$

$$= 19 \text{ units.}$$ **Ans.**

10. *If $\vec{a}$ and $\vec{b}$ are two vectors such that $|\vec{a}+\vec{b}| = |\vec{a}|$ then show that vector $(2\vec{a}+\vec{b})$ is perpendicular to $\vec{b}$.*

Sol. Given $|\vec{a}+\vec{b}| = |\vec{a}|$

$\Rightarrow$ $|\vec{a}+\vec{b}|^2 = |\vec{a}|^2$

$\Rightarrow$ $|\vec{a}|^2+|\vec{b}|^2+2\vec{a}\cdot\vec{b} = |\vec{a}|^2$

$\Rightarrow$ $|\vec{b}|^2+2\vec{a}\cdot\vec{b} = 0$

$\Rightarrow$ $\vec{b}\cdot\vec{b}+2\vec{a}\cdot\vec{b} = 0$

$\Rightarrow$ $\vec{b}\,(\vec{b}+2\vec{a}) = 0$

$\Rightarrow$ $\vec{b}\,(2\vec{a}+\vec{b}) = 0$

$\Rightarrow$ $(2\vec{a}+\vec{b})$ is perpendicular to $\vec{b}$.

Hence Proved.

11. *$\vec{a},\vec{b},\vec{c},\vec{d}$ are vector quantities such that $\vec{a}\times\vec{b} = \vec{c}\times\vec{d}$ and $\vec{a}\times\vec{c} = \vec{b}\times\vec{d}$ then prove that $(\vec{a}-\vec{d})$ and $(\vec{b}-\vec{c})$ are parallel.*

Sol. Consider, $(\vec{a}-\vec{d})\times(\vec{b}-\vec{c})$

$$= \vec{a}\times\vec{b}-\vec{a}\times\vec{c}-\vec{d}\times\vec{b}+\vec{d}\times\vec{c}$$

$$= \vec{a}\times\vec{b}-\vec{a}\times\vec{c}+\vec{b}\times\vec{d}-\vec{c}\times\vec{d}$$

$$= \vec{a}\times\vec{b}-\vec{c}\times\vec{d} \quad \left[\text{given } \vec{a}\times\vec{b}=\vec{c}\times\vec{d},\ \vec{a}\times\vec{c}=\vec{b}\times\vec{d}\right]$$

$$= 0$$

Hence, $(\vec{a}-\vec{d})$ and $(\vec{b}-\vec{c})$ are parallel.

Hence Proved.

12. *Three vectors $\vec{a}$, $\vec{b}$ and $\vec{c}$ are given to be mutually perpendicular and $|\vec{a}| = |\vec{b}| = |\vec{c}|$. Show that vector $(\vec{a}+\vec{b}+\vec{c})^2$ is equally inclined to $(\vec{a}+\vec{b}+\vec{c}).(\vec{a}+\vec{b}+\vec{c})$.*

Sol. Given, $|\vec{a}| = |\vec{b}| = |\vec{c}| = k$ (let)

Now, $(\vec{a}+\vec{b}+\vec{c})^2$

$$= (\vec{a}+\vec{b}+\vec{c}).(\vec{a}+\vec{b}+\vec{c})$$

$$= |\vec{a}|^2+|\vec{b}|^2+|\vec{c}|^2+2\vec{a}.\vec{b}+2\vec{b}.\vec{c}+2\vec{c}.\vec{a}$$

$$= k^2+k^2+k^2+0+0+0 = 3k^2.$$

$\Rightarrow$ $|\vec{a}+\vec{b}+\vec{c}| = \sqrt{3}\,k$

Let $|\vec{a}+\vec{b}+\vec{c}|$ makes θ_1, θ_2, θ_3 angle with $\vec{a}$, $\vec{b}$ and $\vec{c}$, respectively.

$$\cos\theta_1 = \frac{(\vec{a}+\vec{b}+\vec{c}).\vec{a}}{|\vec{a}+\vec{b}+\vec{c}|.|\vec{a}|} = \frac{k^2}{\sqrt{3}k.k} = \frac{1}{\sqrt{3}}$$

$\therefore$
$$\cos\theta_2 = \frac{(\vec{a}+\vec{b}+\vec{c}).\vec{b}}{|\vec{a}+\vec{b}+\vec{c}|.|\vec{b}|} = \frac{k^2}{\sqrt{3}k.k} = \frac{1}{\sqrt{3}}$$

$$\therefore \quad \cos\theta_3 = \frac{(\vec{a}+\vec{b}+\vec{c}).\vec{c}}{|\vec{a}+\vec{b}+\vec{c}|.|\vec{c}|}$$

$$= \frac{k^2}{\sqrt{3}k.k} = \frac{1}{\sqrt{3}}$$

Hence, $(\vec{a}+\vec{b}+\vec{c})$ is equally inclined to $\vec{a}$, $\vec{b}$ and $\vec{c}$. **Hence Proved.**

13. *Using vectors, find the area of the triangle whose vertices are :**

A (3, –1, 2), B (1, –1, –3) and C (4, –3, 1)

Sol. Given : Vertices of ΔABC are A (3, –1, 2), B (1, –1, –3) and C (4, –3, 1).

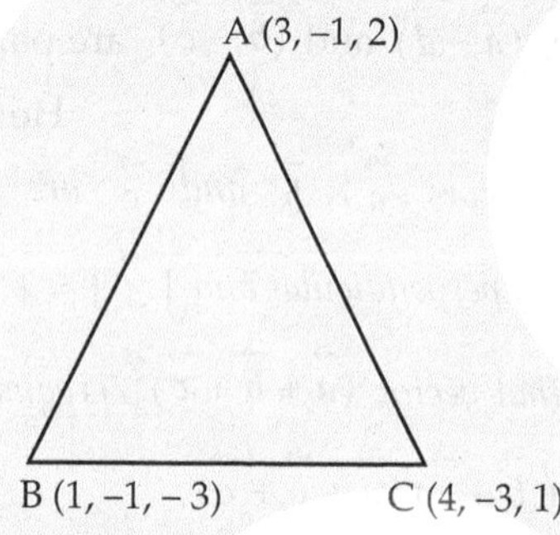

So, Position vector of A = $(3\hat{i}-\hat{j}+2\hat{k})$

Position vector of B = $(\hat{i}-\hat{j}-3\hat{k})$

Position vector of C = $(4\hat{i}-3\hat{j}+\hat{k})$

$\therefore \quad \vec{AB}$ = (p. v. of B) – (p. v. of A)

$= (\hat{i}-\hat{j}-3\hat{k})-(3\hat{i}-\hat{j}+2\hat{k})$

$= -2\hat{i}+0-5\hat{k}$

and $\vec{AC}$ = (p. v. of C) – (p. v. of A)

$= (4\hat{i}-3\hat{j}+\hat{k})-(3\hat{i}-\hat{j}+2\hat{k})$

$= \hat{i}-2\hat{j}-\hat{k}$

We know, area of ΔABC = $\frac{1}{2}|\vec{AB}\times\vec{AC}|$

Now, $\vec{AB}\times\vec{AC} = \begin{vmatrix} \hat{i} & \hat{j} & \hat{k} \\ -2 & 0 & -5 \\ 1 & -2 & -1 \end{vmatrix}$

$= (0-10)\hat{i} - (2+5)\hat{j} + (4-0)\hat{k}$

$= -10\hat{i} - 7\hat{j} + 4\hat{k}$

$\therefore \quad |\vec{AB}\times\vec{AC}| = \sqrt{(-10)^2+(-7)^2+(4)^2}$

$= \sqrt{100+49+16}$

$= \sqrt{165}$

So, Area of ΔABC = $\frac{1}{2}|\vec{AB}\times\vec{AC}|$

$= \frac{1}{2}\times\sqrt{165}$

$= \frac{\sqrt{165}}{2}$ sq. units **Ans**

14. *The vectors $-2\hat{i}+4\hat{j}+4\hat{k}$ and $-4\hat{i}-2\hat{k}$ represent the diagonals BD and AC of a parallelogram ABCD. Find the area of the parallelogram.*

Sol. Area of parallelogram

$= \frac{1}{2}|\vec{BD}\times\vec{AC}|$

$= \frac{1}{2}|(-2\hat{i}+4\hat{j}+4\hat{k})\times(-4\hat{i}-2\hat{k})|$

$= \frac{1}{2}\begin{vmatrix} \hat{i} & \hat{j} & \hat{k} \\ -2 & 4 & 4 \\ -4 & 0 & -2 \end{vmatrix}$

$= \frac{1}{2}|\hat{i}(-8)-\hat{j}(4+16)+\hat{k}(16)|$

$= \frac{1}{2}|-8\hat{i}-20\hat{j}+16\hat{k}|$

$= \frac{1}{2}\sqrt{(-8)^2+(-20)^2+16^2}$

$= \frac{1}{2}\times\sqrt{720}$

$= \frac{1}{2}\times 12\sqrt{5}$

$= 6\sqrt{5}$ sq. units. **Ans**

15. *Using vectors, show that medians of a triangle are concurrent.*

Sol. Let *ABC* be a triangle and *DEF* are the mid points of sides *BC, CA* and *AB* of Δ *ABC*.

Let *O* be the instersection point of medians *AD* and *BE*.

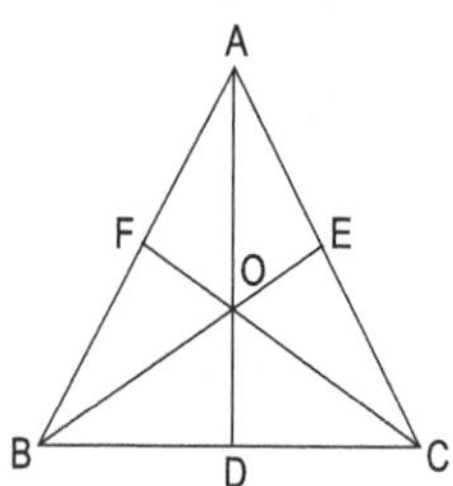

Formulae Based Questions

We have to prove that third median CF also passes through O.

Let position vectors of A, B, C with respect to O are $\vec{a}$, $\vec{b}$ and $\vec{c}$

In triangle OAB we see,

$$\vec{OF} = \frac{\vec{OA}+\vec{OB}}{2} = \frac{\vec{a}+\vec{b}}{2}$$

[because F is mid-point of AB]

Similarly, $\vec{OD} = \dfrac{\vec{b}+\vec{c}}{2}$

and $\vec{OE} = \dfrac{\vec{a}+\vec{c}}{2}$

$\vec{OA}$ and $\vec{OD}$ are opposite in direction hence, their cross product must be zero.

i.e., $\vec{OA}\times\vec{OD} = 0$

$$\Rightarrow \vec{a}\times\frac{1}{2}(\vec{b}+\vec{c}) = 0$$

$$\Rightarrow \frac{1}{2}[\vec{a}\times(\vec{b}+\vec{c})] = 0$$

$$\Rightarrow \vec{a}\times[\vec{b}+\vec{c}] = 0 \quad \ldots \text{(i)}$$

In the same way, $\vec{b}\times[\vec{c}+\vec{a}] = 0$ ... (ii)

Adding equation (i) and equation (ii), we get

$$\vec{a}\times(\vec{b}+\vec{c})+\vec{b}\times(\vec{c}+\vec{a}) = 0$$

$$\Rightarrow \vec{a}\times\vec{b}+\vec{a}\times\vec{c}+\vec{b}\times\vec{c}+\vec{b}\times\vec{a} = 0$$

$$\Rightarrow \vec{a}\times\vec{c}+\vec{b}\times\vec{c} = 0$$

$$\Rightarrow (\vec{a}+\vec{b})\times\vec{c} = 0$$

$$\Rightarrow \left(\frac{\vec{a}+\vec{b}}{2}\right)\times\vec{c} = 0$$

$$\Rightarrow \vec{OF}\times\vec{OC} = 0$$

$\therefore$ $\vec{OC}$ and $\vec{OF}$ are in opposite directions *i.e.,* median CF also passes through the point O which is point of intersection of medians AD and BE.

Hence Proved.

16. (a) *Find the value of λ for which the four points with position vectors $2\hat{i}+5\hat{j}+\hat{k}, -\hat{j}-4\hat{k}$, $3\hat{i}+\lambda\hat{j}+8\hat{k}$ and $-4\hat{i}+3\hat{j}+4\hat{k}$ are coplanar.*

(b) *In any ΔABC, prove by vector method that*

$$\cos B = \frac{c^2+a^2-b^2}{2ca}.$$

Sol. (a) Let position vectors of the point A, B, C, D are $(2\hat{i}+5\hat{j}+\hat{k}), (-\hat{j}-4\hat{k}), (3\hat{i}+\lambda\hat{j}+8\hat{k})$ and $(-4\hat{i}+3\hat{j}+4\hat{k})$ respectively.

Now, $\vec{AB} = \vec{B}-\vec{A}$

$$\Rightarrow \vec{a} = -\hat{j}-4\hat{k}-(2\hat{i}+5\hat{j}+\hat{k})$$

$$= -2\hat{i}-6\hat{j}-5\hat{k}$$

And, $\vec{AC} = \vec{C}-\vec{A}$

$$\Rightarrow \vec{b} = (3\hat{i}+\lambda\hat{j}+8\hat{k})-(2\hat{i}+5\hat{j}+\hat{k})$$

$$\Rightarrow \vec{b} = \hat{i}+(\lambda-5)\hat{j}+7\hat{k}$$

Position vector of

$$\vec{AD} = \vec{D}-\vec{A}$$

$$\Rightarrow \vec{c} = (-4\hat{i}+3\hat{j}+4\hat{k})-(2\hat{i}+5\hat{j}+\hat{k})$$

$$= -6\hat{i}-2\hat{j}+3\hat{k}$$

$\because$ $\vec{AB}$, $\vec{AC}$ and $\vec{AD}$ are coplanar,

$$\therefore [\vec{AB}\ \vec{AC}\ \vec{AD}] = 0$$

$$\Rightarrow \vec{AB}\cdot(\vec{AC}\times\vec{AD}) = 0$$

$$\begin{vmatrix} -2 & -6 & -5 \\ 1 & \lambda-5 & 7 \\ -6 & -2 & 3 \end{vmatrix} = 0$$

$$\Rightarrow -2(3\lambda-15+14)+6(3+42) -5(-2+6\lambda-30) = 0$$

$$\Rightarrow -6\lambda+2+270+160-30\lambda = 0$$

$$\Rightarrow 36\lambda = 432$$

$$\Rightarrow \lambda = 12.$$ **Ans.**

(b) In Δ ABC, $\vec{b} = \vec{c}+\vec{a}$

Squaring both sides, we get

$$(\vec{b}\cdot\vec{b}) = (\vec{c}+\vec{a})\cdot(\vec{c}+\vec{a})$$

$$\Rightarrow |\vec{b}|^2 = |\vec{c}|^2+|\vec{a}|^2+2\vec{a}\cdot\vec{c}$$

$$\Rightarrow |\vec{b}|^2 = |\vec{c}|^2+|\vec{a}|^2+2|\vec{a}||\vec{c}|\cos(\pi-B)$$

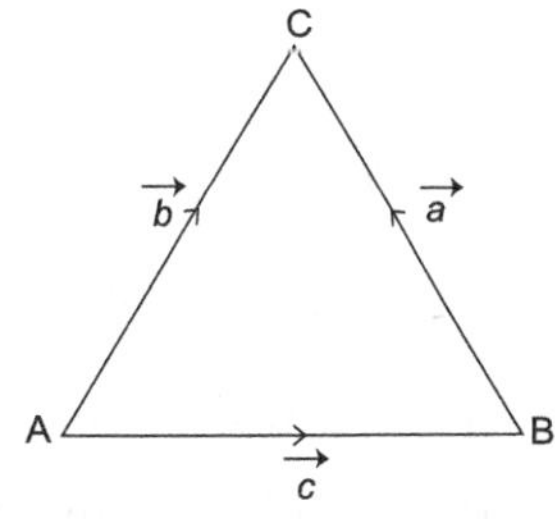

$\Rightarrow \quad |\vec{b}|^2 = |\vec{c}|^2 + |\vec{a}|^2 + 2|\vec{a}||\vec{c}|(-\cos B)$

$\because \quad |\vec{a}| = a, |\vec{b}| = b, |\vec{c}| = c$

$\Rightarrow 2ac \cos B = c^2 + a^2 - b^2$

$\Rightarrow \quad \cos B = \dfrac{c^2 + a^2 - b^2}{2ca}$ **Hence Proved.**

17. (a) *If $\vec{a}$ and $\vec{b}$ are unit vectors and θ is the angle between them, then show that $|\vec{a} - \vec{b}| = 2 \sin \frac{\theta}{2}$.*

(b) *Find the value of λ for which the four points A, B, C, D with position vectors $-\hat{j} - \hat{k}$; $4\hat{i} + 5\hat{j} + \lambda\hat{k}$; $3\hat{i} + 9\hat{j} + 4\hat{k}$ and $-4\hat{i} + 4\hat{j} + 4\hat{k}$ are coplanar.*

Sol. (a) We know that,

$$\vec{a} \cdot \vec{b} = |\vec{a}||\vec{b}|\cos\theta$$

$$= 1.\ 1.\ \cos\theta$$

$$= \cos\theta$$

Now, $|\vec{a} - \vec{b}|^2 = (\vec{a} - \vec{b})^2$

$$= |\vec{a}|^2 + |\vec{b}|^2 - 2\vec{a}.\vec{b}$$

$$= 1 + 1 - 2\cos\theta$$

$$= 2(1 - \cos\theta)$$

$\Rightarrow \quad |\vec{a} - \vec{b}| = 2{\cdot}2 \sin^2 \dfrac{\theta}{2}$

$\Rightarrow \quad 2 \sin \dfrac{\theta}{2} = |\vec{a} - \vec{b}|$ **Hence Proved.**

(b) Let A, B, C and D be the given points whose position vectors are $-\hat{j} - \hat{k}$, $4\hat{i} + 5\hat{j} + \lambda\hat{k}$, $3\hat{i} + 9\hat{j} + 4\hat{k}$ and $-4\hat{i} + 4\hat{j} + 4\hat{k}$ respectively.

$$\vec{AB} = 4\hat{i} + 5\hat{j} + \lambda\hat{k} - (-\hat{j} - \hat{k})$$

$$= 4\hat{i} + 6\hat{j} + (\lambda + 1)\hat{k}$$

$$\vec{AC} = 3\hat{i} + 9\hat{j} + 4\hat{k} - (-\hat{j} - \hat{k})$$

$$= 3\hat{i} + 10\hat{j} + 5\hat{k}$$

$$\vec{AD} = -4\hat{i} + 4\hat{j} + 4\hat{k} - (-\hat{j} - \hat{k})$$

$$= -4\hat{i} + 5\hat{j} + 5\hat{k}$$

Since, the points A, B, C and D ar0e coplanar, vectors $\vec{AB}, \vec{AC}$ and $\vec{AD}$ are coplanar.

$\Rightarrow [\vec{AB}\ \vec{AC}\ \vec{AD}] = 0$

$$\Rightarrow \begin{bmatrix} 4 & 6 & \lambda+1 \\ 3 & 10 & 5 \\ -4 & 5 & 5 \end{bmatrix} = 0$$

$\Rightarrow \quad 4(50 - 25) - 3(30 - 5 - 5\lambda)$
$\quad -4(30 - 10 - 10\lambda) = 0$

$\Rightarrow 100 - 75 + 15\lambda - 80 + 40\lambda = 0$

$\Rightarrow \quad -55 + 55\lambda = 0$

$\Rightarrow \quad -1 + \lambda = 0$

$\Rightarrow \quad \lambda = 1$ **Ans.**

18. *Find the volume of parallelopiped whose edges are represented by the vectors*

$$\vec{a} = 2\hat{i} - 3\hat{j} + 4\hat{k}$$

$$\vec{b} = \hat{i} + 2\hat{j} - \hat{k}$$

$$\vec{c} = 3\hat{i} - \hat{j} + 2\hat{k}.$$

Sol. Given,

$$\vec{a} = 2\hat{i} - 3\hat{j} + 4\hat{k}$$

$$\vec{b} = \hat{i} + 2\hat{j} - \hat{k}$$

$$\vec{c} = 3\hat{i} - \hat{j} + 2\hat{k}.$$

Volume of parallelopiped formed by $\vec{a}$, $\vec{b}$ and $\vec{c}$ is

$$V = (\vec{a} \times \vec{b}) \cdot \vec{c}$$

$$= \{(2\hat{i} - 3\hat{j} + 4\hat{k}) \times (\hat{i} + 2\hat{j} - \hat{k})\} \cdot (3\hat{i} - \hat{j} + 2\hat{k})$$

$$= \begin{vmatrix} \hat{i} & \hat{j} & \hat{k} \\ 2 & -3 & 4 \\ 1 & 2 & -1 \end{vmatrix} \cdot (3\hat{i} - \hat{j} + 2\hat{k})$$

$$= [\hat{i}(3 - 8) - \hat{j}(-2 - 4) + \hat{k}(4 + 3)].$$
$$(3\hat{i} - \hat{j} + 2\hat{k})$$

$$= (-5\hat{i} + 6\hat{j} + 7\hat{k}) \cdot (3\hat{i} - \hat{j} + 2\hat{k})$$

$$= -5 \times 3 + 6 \times (-1) + 7 \times (2)$$

$$= -15 - 6 + 14 = -21 + 14$$

$$= -7 \text{ cubic units.}$$

Hence, the volume of parallelopiped is 7 cubic units. We do not take (–) sign, we only take magnitude. **Ans.**

19. *Find the volume of the parallelopiped whose three co-terminus edges are represented by the vectors*

$$\hat{i} + \hat{j} + \hat{k},\ \hat{i} - \hat{j} + \hat{k} \text{ and } \hat{i} + 2\hat{j} - \hat{k}$$

Sol. Volume of parallelopiped

$$= \begin{vmatrix} 1 & 1 & 1 \\ 1 & -1 & 1 \\ 1 & 2 & -1 \end{vmatrix}$$

$$= 1(1-2) - 1(-1-1) + 1(2+1)$$

$$= -1 + 2 + 3$$

$$= 4 \text{ cubic units.}$$ **Ans.**

Chapter 12. Three Dimensional Geometry

1. **(a)** *Find the length of the perpendicular from origin to the plane* $\vec{r}\cdot(3\hat{i} - 4\hat{j} - 12\hat{k}) + 39 = 0$.

(b) *Find the angle between the two lines : $2x = 3y = -z$ and $6x = -y = -4z$.**

Sol. (a) Given, equation of the plane is

$$\vec{r}\cdot(3\hat{i}-4\hat{j}-12\hat{k}) + 39 = 0$$

Let $\vec{r} = (x\hat{i}+y\hat{j}+z\hat{k})$

$\therefore \quad (x\hat{i}+y\hat{j}+z\hat{k})\cdot(3\hat{i}-4\hat{j}-12\hat{k}) + 39 = 0$

$\Rightarrow \quad 3x - 4y - 12z + 39 = 0$

Perpendicular distance from the origin to plane

$$= \left|\frac{39}{\sqrt{9+16+144}}\right| = \left|\frac{39}{\sqrt{169}}\right|$$

$$= \left|\frac{39}{13}\right| = 3 \text{ units.}$$ **Ans.**

(b) Given equation of lines are

$$2x = 3y = -z$$

$$\therefore \quad \frac{x}{\frac{1}{2}} = \frac{y}{\frac{1}{3}} = \frac{z}{-1}$$

Direction Ratio of this line $= \langle a_1, b_1, c_1\rangle$

$$= \left\langle \frac{1}{2}, \frac{1}{3}, -1\right\rangle$$

$$= \langle 3, 2, -6\rangle$$

and $\quad 6x = -y = -4z$

$$\frac{x}{\frac{1}{6}} = \frac{y}{-1} = \frac{z}{-\frac{1}{4}}$$

Direction Ratio of this line $= \langle a_2, b_2, c_2\rangle$

$$= \left\langle \frac{1}{6}, -1, -\frac{1}{4}\right\rangle$$

$$= \langle 2, -12, -3\rangle$$

$$= \langle -2, 12, 3\rangle$$

Angle between the given lines :

$$\cos\theta = \frac{a_1a_2 + b_1b_2 + c_1c_2}{\sqrt{a_1^2+b_1^2+c_1^2}\sqrt{a_2^2+b_2^2+c_2^2}}$$

$$= \frac{3\times(-2)+2\times12+(-6)\times3}{\sqrt{9+4+36}\sqrt{4+144+9}}$$

$$= \frac{-6+24-18}{\sqrt{49}\sqrt{157}}$$

$$\cos\theta = 0$$

$$\cos\theta = \cos\frac{\pi}{2}$$

$$\theta = \frac{\pi}{2} \text{ or } 90°.$$ **Ans.**

2. *Find the cosine of the angle between the planes*

$$x + 2y - 2z + 6 = 0$$
$$2x + 2y + z + 8 = 0.$$

Sol. Given, $\quad x + 2y - 2z + 6 = 0$

$$2x + 2y + z + 8 = 0$$

If θ is the angle between the two planes,

$$a_1x + b_1y + c_1z + d_1 = 0$$

and $a_2x + b_2y + c_2z + d_2 = 0$, then

$$\cos\theta = \pm\frac{a_1a_2 + b_1b_2 + c_1c_2}{\sqrt{a_1^2+b_1^2+c_1^2}\cdot\sqrt{a_2^2+b_2^2+c_2^2}}$$

$$= \pm\frac{1\times2+2\times2+(-2)\times1}{\sqrt{1^2+2^2+(-2)^2}.\sqrt{2^2+2^2+1^2}}$$

$$= \frac{4}{3\times3} = \frac{4}{9}$$

$$\therefore \cos\theta = \frac{4}{9}$$ **Ans.**

3. **(a)** *Find the equation of the plane passing through the intersection of the planes $2x + 2y - 3z - 7 = 0$ and $2x + 5y + 3z - 9 = 0$, such that the intercepts made by the resulting plane on the X axis and the Z-axis are equal.**

OR

(b) *Find the equation of the lines passing through the point (2, 1, 3) and perpendicular to the lines $\frac{x-1}{1} = \frac{y-2}{2} = \frac{z-3}{3}$ and $\frac{x}{-3} = \frac{y}{2} = \frac{z}{5}$.*

---

* Frequently asked previous years Board Exam Questions.

Sol. (a) Equation of the required plane passing through the intersection of the given plane is :

$$(2x + 2y - 3z - 7) + \lambda\,(2x + 5y + 3z - 9) = 0$$

$$x(2 + 2\lambda) + y\,(2 + 5\lambda) + z\,(3\lambda - 3) - 7 - 9\lambda = 0 \qquad \text{...(i)}$$

For x-intercept, put $y = 0$ and $z = 0$.

we get

$$(2 + 2\lambda)\,x - 7 - 9\lambda = 0$$

$$x = \frac{7+9\lambda}{2+2\lambda}$$

For z- intercept, put $x = 0$ and $y = 0$

we get $z\,(3\lambda - 3) - 7 - 9\lambda = 0$

$$z = \frac{7+9\lambda}{3\lambda-3}$$

$\because$ x-intercept = z-intercept

$$\frac{7+9\lambda}{2+2\lambda} = \frac{7+9\lambda}{3\lambda-3}$$

$\Rightarrow$ $3\lambda - 3 = 2 + 2\lambda$

$\Rightarrow$ $3\lambda - 2\lambda = 2 + 3$

$\Rightarrow$ $\lambda = 5$

put $\lambda = 5$ in equation (i), we get

$$(2 + 10)\,x + (2 + 25)\,y + (15 - 3)\,z - 7 - 45 = 0$$

$\Rightarrow$ $12x + 27\,y + 12z - 52 = 0$

This is the required equation of the plane. **Ans.**

OR

(b) The given line is passing through the point (2, 1, 3) and perpendicular to the lines

$$\frac{x-1}{1} = \frac{y-2}{2} = \frac{z-3}{3} \text{ and } \frac{x}{-3} = \frac{y}{2} = \frac{z}{5}$$

Let $\vec{a} = \hat{i}+2\hat{j}+3\hat{k}$

$\vec{b} = -\,3\hat{i}+2\hat{j}+5\hat{k}$

$$\vec{a}\times\vec{b} = \begin{vmatrix} \hat{i} & \hat{j} & \hat{k} \\ 1 & 2 & 3 \\ -3 & 2 & 5 \end{vmatrix}$$

$$= \hat{i}\,(10 - 6) - \hat{j}\,(5 + 9) + \hat{k}\,(2 + 6)$$

$$= 4\hat{i} - 14\hat{j} + 8\hat{k}$$

$\therefore$ Direction ratios of required line

$= \langle 4, -14, 8\rangle$ or $\langle 2, -7, 4\rangle$

$\therefore$ Required equation of line is

$$\frac{x-2}{2} = \frac{y-1}{-7} = \frac{z-3}{4}$$ **Ans.**

* Frequently asked previous years Board Exam Questions.

4. (a) *Find the angle between the two lines :**

$$\frac{x+1}{2} = \frac{y-2}{5} = \frac{z+3}{4} \text{ and } \frac{x-1}{5} = \frac{y+2}{2} = \frac{z-1}{-5}$$

Sol. (a) Given lines are,

$$\frac{x+1}{2} = \frac{y-2}{5} = \frac{z+3}{4} \qquad \text{...(i)}$$

and $$\frac{x-1}{5} = \frac{y+2}{2} = \frac{z-1}{-5} \qquad \text{...(ii)}$$

$\therefore$ Direction ratios of the (i), $\langle a_1, b_1, c_1\rangle$ = $\langle 2, 5, 4\rangle$ and, direction ratios of line (ii), $\langle a_2, b_2, c_2\rangle = \langle 5, 2, -5\rangle$.

The angle between two lines is given as,

$$\cos\theta = \frac{a_1a_2 + b_1b_2 + c_1c_2}{\sqrt{a_1^2 + b_1^2 + c_1^2}.\sqrt{a_2^2 + b_2^2 + c_2^2}}$$

$$= \frac{2\times5+5\times2+4\times(-5)}{\sqrt{2^2+5^2+4^2}.\sqrt{5^2+2^2+(-5)^2}}$$

$$= \frac{10+10-20}{\sqrt{4+25+16}.\sqrt{25+4+25}}$$

$$= \frac{0}{\sqrt{45}.\sqrt{54}}$$

$\Rightarrow$ $\cos\theta = 0$

$\therefore$ $\theta = \frac{\neq}{2} = 90^\circ$

Hence, the angle between the two lines is $\frac{\neq}{2}$. **Ans.**

5. (a) *Find the image of the point (3, – 2, 1) in the plane $3x - y + 4z = 2$**

(b) *Determine the equation of the line passing through the point (– 1, 3, – 2) and perpendicular to the lines :*

$$\frac{x}{1} = \frac{y}{2} = \frac{z}{3} \text{ and } \frac{x+2}{-3} = \frac{y-1}{2} = \frac{z+1}{5}$$

Sol. (a) Given : Equation of plane is,

$$3x - y + 4z = 2 \qquad \text{...(i)}$$

Let Q (x_1, y_1, z_1) be the image of the point P (3, –2, 1) in the given plane.

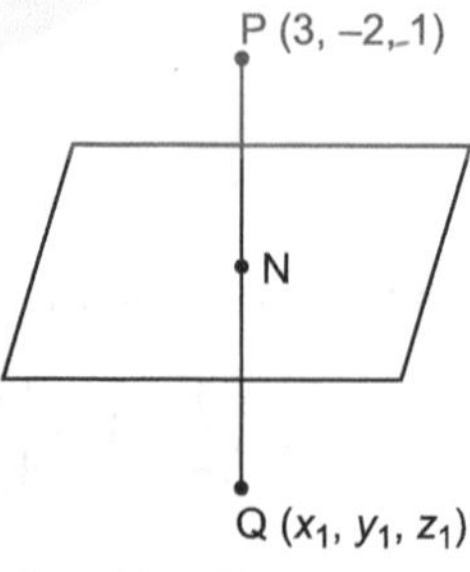

The equation of a line passing through (3, –2, 1) and perpendicular to the given plane is,

$$\frac{x-3}{3} = \frac{y+2}{-1} = \frac{z-1}{4} = k \text{ (say)}$$

So, the coordinates of a general point on this line are $(3k + 3, -k - 2, 4k + 1)$.

If N is the foot of the perpendicular from P to the given plane, then N lies on the plane.

Let the coordinates of N be

$$(3k + 3, -k - 2, 4k + 1).$$

$\therefore 3(3k + 3) - (-k - 2) + 4(4k + 1) - 2 = 0$

[Putting N in eq. (i)]

$$\Rightarrow \quad 9k + 9 + k + 2 + 16k + 4 - 2 = 0$$

$$\Rightarrow \quad 26k + 13 = 0$$

$$\Rightarrow \quad k = -\frac{13}{26} = -\frac{1}{2}$$

So, coordinates of,

$$N = \left[3\times\left(-\frac{1}{2}\right)+3, -\left(-\frac{1}{2}\right)-2, 4\times\left(-\frac{1}{2}\right)+1\right]$$

$$= \left(\frac{3}{2}, -\frac{3}{2}, -1\right)$$

Since, N is the mid-point of PQ,

$\therefore$ By mid-point formula,

$$\frac{3+x_1}{2} = \frac{3}{2}$$

$$\Rightarrow \quad x_1 = 3 - 3 = 0$$

Also,

$$\frac{-2+y_1}{2} = -\frac{3}{2}$$

$$\Rightarrow \quad y_1 = -3 + 2 = -1$$

and

$$\frac{1+z_1}{2} = -1$$

$$\Rightarrow \quad 1 + z_1 = -2$$

$$\Rightarrow \quad z_1 = -2 - 1 = -3$$

Hence, the required image of the point P is Q $(0, -1, -3)$. **Ans.**

(b) Let the equation of required line be,

$$\frac{x-x_1}{a} = \frac{y-y_1}{b} = \frac{z-z_1}{c}$$

$\therefore$ The line passes through $(-1, 3, -2)$.

$$\therefore \quad \frac{x+1}{a} = \frac{y-3}{b} = \frac{z+2}{c} \quad \text{...(i)}$$

Now, given lines are

$$\frac{x}{1} = \frac{y}{2} = \frac{z}{3}$$

and

$$\frac{x+2}{-3} = \frac{y-1}{2} = \frac{z+1}{5}$$

Since, the required line is perpendicular to these two lines

$$\therefore \quad a \times 1 + b \times 2 + 3 \times c = 0$$

$$\Rightarrow \quad a + 2b + 3c = 0 \quad \text{...(ii)}$$

$$\text{and} \quad a \times (-3) + b \times 2 + c \times 5 = 0$$

$$\Rightarrow \quad -3a + 2b + 5c = 0 \quad \text{...(iii)}$$

Solving equations (ii) and (iii) using cross-multiplication method,

$$\frac{a}{10-6} = \frac{b}{-9-5} = \frac{c}{2+6}$$

$$\Rightarrow \quad \frac{a}{4} = \frac{b}{-14} = \frac{c}{8} = k \text{ (say)}$$

$$\Rightarrow \quad a = 4k, b = -14k \text{ and } c = 8k.$$

Putting these values in equation (i), we get

$$\frac{x+1}{4k} = \frac{y-3}{-14k} = \frac{z+2}{8k}$$

$$\Rightarrow \quad \frac{x+1}{2} = \frac{y-3}{7} = \frac{z+2}{4},$$

Which is the required equation of line. **Ans.**

Chapter 14. Application of Calculus

1. *The total cost function of a firm for x units of commodity is given by*

$$C(x) = 4x^3 - 2x^2 + \frac{7}{2}x - 3.$$

*Find **(i)** Average cost function and **(ii)** Marginal cost function.*

Sol. We have,

$$C = C(x) = 4x^3 - 2x^2 + \frac{7}{2}x - 3$$

(i) Average cost function (AC)

$$= \frac{C}{x}$$

$$= \frac{4x^3 - 2x^2 + \frac{7}{2}x - 3}{x}$$

$$= 4x^2 - 2x + \frac{7}{2} - \frac{3}{x}$$

(ii) Marginal Cost function (MC) $= \dfrac{dC}{dx}$

Differentiating C (x) w.r.t. x

$$= \frac{d}{dx}\left(4x^3 - 2x^2 + \frac{7}{2}x - 3\right)$$

$$= 12x^2 - 4x + \frac{7}{2}.$$ **Ans.**

2. *The total cost C (x) of a firm is given by* $C(x) = 3x^3 + \frac{1}{3}x^2 - 7x + 4$, *where x is the output.*

Determine **(i)** *the average cost* **(ii)** *the marginal average cost,* **(iii)** *the marginal cost when x = 2.*

Sol. (i) We have :

$$C = C(x)$$
$$= 3x^3 + \frac{1}{3}x^2 - 7x + 4$$

$$\text{Average cost (AC)} = \frac{C}{x}$$
$$= \frac{3x^3 + \frac{1}{3}x^2 - 7x + 4}{x}$$
$$= 3x^2 + \frac{1}{3}x - 7 + \frac{4}{x}$$

when $x = 2$, $AC = 3(2)^2 + \frac{1}{3}(2) - 7 + \frac{4}{2}$
$$= 12 + \frac{2}{3} - 7 + 2 = \frac{23}{3}$$
$$= 7{\cdot}67$$

(ii) Marginal Average Cost (MAC)
$$= \frac{d}{dx}(AC)$$
$$= \frac{d}{dx}\left(3x^2 + \frac{1}{3}x - 7 + \frac{4}{x}\right)$$
$$= 6x + \frac{1}{3} - \frac{4}{x^2}$$

when $x = 2$, $MAC = 6(2) + \frac{1}{3} - \frac{4}{4}$
$$= 12 + \frac{1}{3} - 1 = \frac{34}{3}$$
$$= 11{\cdot}34$$

(iii) Marginal cost $(MC) = \frac{dC}{dx}$
$$= \frac{d}{dx}\left(3x^3 + \frac{1}{3}x^2 - 7x + 4\right)$$
$$= 9x^2 + \frac{2}{3}x - 7$$

when $x = 2$, $MC = 9(2)^2 + \frac{2}{3}(2) - 7$
$$= 36 + \frac{4}{3} - 7 = \frac{91}{3}$$
$$= 30{\cdot}34.$$ **Ans.**

3. *Given the total revenue received from the sale of x units of a product is given by* $R(x) = 24 - 5x + x^2$. *Find :*

   (i) The average revenue.

   (ii) The marginal revenue.

   (iii) The average revenue and marginal revenue at x = 5.

   (iv) The actual revenue from selling 40th item.

Sol. We have,

$$R = R(x) = 24 - 5x + x^2$$

(i) Average Revenue (AR)
$$= \frac{R}{x}$$
$$= \frac{24 - 5x + x^2}{x}$$
$$= \frac{24}{x} - 5 + x$$

(ii) Marginal Revenue (MR)
$$= \frac{dR}{dx}$$
$$= \frac{d}{dx}(24 - 5x + x^2).$$
$$= -5 + 2x$$

(iii) Average Revenue at $x = 5$
$$= \frac{24}{5} - 5 + 5$$
$$= \frac{24}{5} = 4{\cdot}8$$

and Marginal Revenue at $x = 5$
$$= -5 + 2(5)$$
$$= -5 + 10$$
$$= 5$$

(iv) Actual revenue on selling the 40th item

= Revenue received on selling 40th item – revenue received on 39th item.

$= [24 - 5(40) + 40)^2] - [24 - 5(39) + (39)^2]$

$= (24 - 200 + 1600) - (24 - 195 + 1521)$

$= 1424 - 1350$

= ₹ 74. **Ans.**

4. *The demand function is* $x = \frac{24 - 2p}{3}$ *where x is the number of units demanded and p is the price per unit. Find :*

   (i) The revenue function R in terms of p.

   (ii) The price and the number of units demanded for which the revenue is maximum.

Sol. Given, $x = \frac{24-2p}{3}$

$$p = 12 - \frac{3}{2}x$$

(i) Revenue function $R(p) = px$

$$= p\left(\frac{24-2p}{3}\right)$$

$$= 8p - \frac{2}{3}p^2$$

(ii) For maximum revenue, $\frac{dR(p)}{dp}$

$$\frac{dR(p)}{dp} = 8 - \frac{4}{3}p = 0$$

$\Rightarrow$ $p = 6$

and $\frac{d^2R(p)}{dp^2} = -\frac{4}{3}\ (<0)$

$\Rightarrow$ R (p) is maximum at $p = 6$

For maximum revenue, price per unit is 6

$\therefore$ $x = \frac{24-2\times6}{3}$

$$= \frac{24-12}{3} = \frac{12}{3} = 4$$

$\therefore$ Price is ₹ 6 per unit and number of units is 4 for which the revenue is maximum. **Ans.**

5. *The fixed cost of a new product is ₹ 30,000 and the variable cost per unit is ₹ 800. If the demand function is P (x) = 4500 – 100x. Find the break-even values.*

Sol. Given fixed cost = ₹ 30,000,

variable cost =₹ 800 per unit

Let x units of the product are produced.

The cost function C (x) is,

C (x) =fixed cost + variable cost = 30,000 + 800 x

and the Revenue function R (x) is

$$R(x) = xP(x)$$
$$= x(4500 - 100x)$$
$$= 4500x - 100x^2$$

For break-even values,

$$R(x) = C(x)$$
$$4500x - 100x^2 = 30000 + 800x$$
$$x^2 - 37x + 300 = 0$$
$$(x-25)(x-12) = 0$$
$$x = 12, 25$$

Hence, the break-even values are x = 12 and x = 25. **Ans.**

6. (a) *A company produces a commodity with ₹24,000 as fixed cost. The variable cost estimated to be 25% of the total revenue received on selling the product, is at the rate of ₹8 per unit. Find the break-even point.**

(b) *The total cost function for a production is given by*

$$C(x) = \frac{3}{4}x^2 - 7x + 27.$$

Find the number of units produced for which M. C. = A.C. (M.C. = Marginal Cost and A.C. = Average Cost)

Sol. (a) Given, fixed cost = ₹24,000

Let number of units be 'x'

$\therefore$ R (x) = 8x

[$\because$ Given selling price per unit, P = ₹8]

Variable cost = 25% of R (x)

$$= \frac{25}{100} \times 8x$$
$$= 2x$$
$$C(x) = 2x + 24{,}000$$

$\therefore$ Break-even point,

$$R(x) = C(x)$$
$$2x + 24{,}000 = 8x$$
$$6x = 24{,}000$$

x = 4,000 units. **Ans.**

(b) Given, $C(x) = \frac{3}{4}x^2 - 7x + 27$

$$\text{M. C.} = \frac{d}{dx}C(x) = \frac{3}{4}\times 2x - 7$$
$$= \frac{3}{2}x - 7$$

$$\text{A. C.} = \frac{C(x)}{x} = \frac{3}{4}x - 7 + \frac{27}{x}$$

$\because$ M. C. = A. C. (given)

$$\frac{3}{2}x - 7 = \frac{3}{4}x - 7 + \frac{27}{x}$$

$\Rightarrow$ $\frac{3}{2}x - \frac{3}{4}x = \frac{27}{x}$

$\Rightarrow$ $\frac{6x-3x}{4} = \frac{27}{x}$

$\Rightarrow$ $\frac{3x}{4} = \frac{27}{x}$

$\Rightarrow$ $x^2 = 36$

$\Rightarrow x = \pm 6$, but x cannot be negative

$\therefore$ Number of units = 6. **Ans.**

7. (a) *The cost function of a product is given by C (x)* $= \frac{x^3}{3} - 45x^2 - 900x + 36$ *where x is the number of units produced. How many units should be produced to minimise the marginal cost ?**

OR

(b) *The marginal cost function of x units of a product is given by MC =* $3x^2 - 10x + 3$. *The cost of producing one unit is ₹7. Find the total cost function and average cost function.*

Sol. (a) Given, $C(x) = \frac{x^3}{3} - 45x^2 - 900x + 36$

$$\text{M. C.} = \frac{d}{dx}C(x) = \frac{3x^2}{3} - 90x - 900$$

$$= x^2 - 90x - 900$$

Also, $\frac{d}{dx}(\text{MC}) = 2x - 90$

Now, the marginal cost is to be minimised *i.e.*

$$\frac{d}{dx}(\text{M. C.}) = 2x - 90$$

$$\frac{d}{dx}(\text{M. C.}) = 0$$

$$2x - 90 = 0$$

$$\Rightarrow \quad x = 45$$

$$\frac{d^2}{dx^2}(\text{M. C.}) = 2$$

$$\therefore \quad \frac{d^2}{dx^2}(\text{M.C.}) > 0$$

So, marginal cost is minimum.

∴ Number of units produced = 45 **Ans.**

OR

(b) Given, $\text{M. C.} = 3x^2 - 10x + 3$

$$C(x) = \int(\text{M. C.})\,dx$$

$$= \int(3x^2 - 10x + 3)\,dx$$

$$= 3\frac{x^3}{3} - 10\frac{x^2}{2} + 3x + k$$

where k is constant

$$= x^3 - 5x^2 + 3x + k$$

∵ Cost of one unit is ₹ 7 *i.e.*,

$$x = 1$$

$$\therefore \quad C(1) = 7$$

$$\Rightarrow \quad 1 - 5 + 3 + k = 7$$

$$\Rightarrow \quad -4 + 3 + k = 7$$

$$\Rightarrow \quad -1 + k = 7$$

$$\Rightarrow \quad k = 7 + 1$$

$$\Rightarrow \quad k = 8$$

$$\therefore \quad C(x) = x^3 - 5x^2 + 3x + 8$$

and

$$\text{Average cost, A. C.} = \frac{C(x)}{x}$$

$$= \frac{x^3 - 5x^2 + 3x + 8}{x}$$

$$= x^2 - 5x + 3 + \frac{8}{x}$$ **Ans**

7. (a) *The selling price of a commodity is fixed at ₹ 6[illegible] and its cost function is C (x) = 35x + 250**

(i) *Determine the profit function.*

(ii) *Find the break even points.*

(b) *The revenue function is given by* $R(x) = 100x - x^2 - x^3$. *Find*

(i) *The demand function.*

(ii) *Marginal revenue function.*

Sol. (a) Given : $C(x) = 35x + 250$

and $S(x) = 60x$

(i) $P(x) = S(x) - C(x)$

$$= 60x - 35x - 250$$

$$= 25x - 250$$ **Ans**

(ii) For break even points,

$$P(x) = 0$$

$$\Rightarrow \quad 25x - 250 = 0 \quad \text{[Putting P(x) from (i)]}$$

$$\Rightarrow \quad x = \frac{250}{25} = 10$$

$$\therefore \quad x = 10$$ **Ans**

(b) Given : $R(x) = 100x - x^2 - x^3$

(i) We know, $R(x) = P(x) \times x$

$$\therefore \quad P(x) = \frac{R(x)}{x}$$

$$= \frac{100x - x^2 - x^3}{x}$$

$$= 100 - x - x^2$$ **Ans**

(ii) We know $\text{M. R} = \frac{dR}{dx}$

$$= \frac{d}{dx}(100x - x^2 - x^3)$$

$$= 100 - 2x - 3x^2$$ **Ans**

9. (a) *The marginal cost of the production of the commodity is 30 + 2x, it is known that fixed costs are ₹ 200, find**

(i) *The total cost.*

(ii) *The cost of increasing output from 100 to 20[illegible] units.*

* Frequently asked previous years Board Exam Questions.

Formulae Based Questions

OR

(b) *The total cost function of a firm is given by $C(x) = \frac{1}{3}x^3 - 5x^2 + 30x - 15$ where the selling price per unit is given as ₹ 6. Find for what value of x will the profit be maximum.**

Sol. (a) Given : M. C. $= 30 + 2x$

(i) We know, M. C. $= \frac{dC}{dx}$

$$\Rightarrow \frac{dC}{dx} = 30 + 2x$$

Intergrating both sides w. r. t. x

$$\int \frac{dC}{dx}dx = \int (30+2x)dx$$

$$\Rightarrow C = 30x + \frac{2x^2}{2} + k$$

$$\Rightarrow C = 30x + x^2 + k$$

It is given that,

When $x = 0, C = 200$

$\therefore$ $200 = 0 + 0 + k$

$\Rightarrow$ $k = 200$

$\therefore$ Total cost $C(x) = 30x + x^2 + 200$. **Ans.**

(ii) Cost of increasing output from 100 to 200

$$= \int_{100}^{200} (30+2x)dx$$

$$= \left[30x + x^2\right]_{100}^{200}$$

$= [30 \times 200 + (200)^2] - [30 \times 100 + (100)^2]$

$= [6,000 + 40,000] - [3,000 + 10,000]$

$= 46,000 - 13,000$

$=$ ₹ 33,000 **Ans.**

OR

(b) Given : $C(x) = \frac{1}{3}x^3 - 5x^2 + 30x - 15$

[$\therefore$ P = 6 ... S(x) = 6x]

$\therefore$ $P(x) = S(x) - C(x)$

$$= 6x - \frac{1}{3}x^3 + 5x^2 - 30x + 15$$

$$= 5x^2 - \frac{1}{3}x^3 - 24x + 15$$

Now, $\frac{dP}{dx} = 10x - \frac{3x^2}{3} - 24$

$= 10x - x^2 - 24$

For maximum and minimum profit,

Put $\frac{dP}{dx} = 0$

$\Rightarrow$ $10x - x^2 - 24 = 0$

$\Rightarrow$ $x^2 - 10x + 24 = 0$

$\Rightarrow$ $(x - 6)(x - 4) = 0$

$\Rightarrow$ $x = 6$ or $x = 4$

Now, $\frac{d^2P}{dx^2} = 10 - 2x$

$\therefore$ $\left(\frac{d^2P}{dx^2}\right)_{x=6} = 10 - 2 \times 6 = -2 < 0$

So, at $x = 6$, P(x) is maximum.

and $\left(\frac{d^2P}{dx^2}\right)_{x=4} = 10 - 2 \times 4 = 2 > 0$

So, at $x = 4$, P(x) is minimum.

Hence, at $x = 6$, profit is maximum. **Ans.**

Chapter 15. Linear Regression

1. (c) *If $\bar{x} = 18$, $\bar{y} = 100$, $\sigma_x = 14$, $\sigma_y = 20$ and correlation coefficient $r_{xy} = 0{\cdot}8$, find the regression equation of y on x.*

Sol. (c) Given, $\bar{x} = 18$, $\bar{y} = 100$, $\sigma_x = 14$,

$\sigma_y = 20$, $r_{xy} = 0{\cdot}8$

$$b_{yx} = r_{xy}\left(\frac{\sigma_y}{\sigma_x}\right)$$

$$= 0{\cdot}8\left(\frac{20}{14}\right) = \frac{8}{7}$$

Now, Regression equation of y on x is

$$y - \bar{y} = b_{yx}(x - \bar{x})$$

$$\Rightarrow y - 100 = \frac{8}{7}(x - 18)$$

$$\Rightarrow 7y - 700 = 8x - 144$$

$$\Rightarrow 8x - 7y + 556 = 0$$ **Ans.**

2. (c) *For the lines of regression $4x - 2y = 4$ and $2x - 3y + 6 = 0$, find the mean of 'x' and the mean of 'y'.*

Sol. (c) Given : Lines of regression are,

$4x - 2y = 4$ ...(i)

and $2x - 3y + 6 = 0$

$\Rightarrow$ $2x - 3y = -6$ ...(ii)

Multiplying equation (ii) by 2 and then subtracting equation (i) from it,

$$4x - 6y = -12$$
$$4x - 2y = 4$$
$$- \quad + \quad -$$
$$-4y = -16$$

$$\Rightarrow y = \frac{16}{4} = 4$$

* Frequently asked previous years Board Exam Questions

Putting $y = 4$ in equation (i), we get

$$4x - 2 \times 4 = 4$$

$\Rightarrow \quad 4x = 4 + 8 = 12$

$\Rightarrow \quad x = \frac{12}{4} = 3$

$\therefore \quad \bar{x} = 3$ and $\bar{y} = 4$ **Ans.**

3. (a) *The correlation coefficient between x and y is 0.6. If the variance of x is 225, the variance of y is 400, mean of x is 10 and mean of y is 20, find*

(i) *the equations of two regression lines.*

(ii) *the expected value of y when x = 2.*

Sol. (a) Given : $r = 0{\cdot}6$, $\bar{x} = 10$, $\bar{y} = 20$,

var. $(x) = (\sigma_x^2) = 225$ and var. $(y) = \sigma_y^2 = 400$.

(i) $\sigma_x = \sqrt{225} = 15$

and $\sigma_y = \sqrt{400} = 20$

$$b_{yx} = r\frac{\sigma_y}{\sigma_x} = 0{\cdot}6 \times \frac{20}{15} = \frac{4}{5}$$

and $$b_{xy} = r\frac{\sigma_x}{\sigma_y} = 0{\cdot}6 \times \frac{15}{20} = \frac{9}{20}$$

So, line of regression of y on x is

$$y - \bar{y} = b_{yx}(x - \bar{x})$$

$\Rightarrow \quad y - 20 = \frac{4}{5}(x - 10)$

$\Rightarrow \quad 5y - 100 = 4x - 40$

$\Rightarrow \quad 4x - 5y + 60 = 0$ ...(i)

Line of regression of x on y is

$$x - \bar{x} = b_{xy}(y - \bar{y})$$

$\Rightarrow \quad x - 10 = \frac{9}{20}(y - 20)$

$\Rightarrow \quad 20x - 200 = 9y - 180$

$\Rightarrow \quad 20x - 9y - 20 = 0$ **Ans.**

(ii) At $x = 2$

$$4 \times 2 - 5y + 60 = 0$$

[Putting $x = 2$ in eq. (i)]

$\Rightarrow \quad 5y = 68$

$\therefore \quad y = \frac{68}{5}$ **Ans.**

□□

Formulae Based Questions

Find the Value Type Questions

Set 3

Chapter 2. Functions

1. *Determine whether the binary operation $*$ on R defined by $a * b = |a - b|$ is commutative. Also, find the value of $(-3) * 2$.**

Sol. We have,

$$\begin{aligned} a * b &= |a-b| \\ &= |-(b-a)| \\ &= b-a \\ &= |b-a| \\ &= b * a \end{aligned}$$

$\Rightarrow$ $a * b = b * a$

So, $*$ is commutative on R.

Now, $(-3) * 2 = |-3-2| = |-5| = 5$ **Ans.**

2. *Let '*' be a binary operation on N given by $a * b = a + b + 1$ for all $a, b \in N$.*

*Find : 6 * 5, 18 * 4.*

Sol. For any $a, b \in$ N, using definition of *, we get

$$6 * 5 = 6 + 5 + 1 = 12$$

and $18 * 4 = 18 + 4 + 1 = 23.$ **Ans.**

3. *Find all one-to-one functions from S to S given, S = {a, b}*

Sol. Let $f : S \to S$ be one-one function.

Then, $f(a)$ may be a or b.

i.e., $f(a) = a$ or $f(a) = b$

Case 1 : When $f(a) = a$

Since, $f : S \to S$ is one-one

$\therefore$ $f(b) = b$

Thus, we have $f(a) = a$ and $f(b) = b$

Case 2.: When $f(a) = b$

Since, $f : S \to S$ is one-one, therefore, $f(b) = a$

Thus, we have

$f(a) = b$ and $f(b) = a$

Hence, there are two functions say, f and g from S to S. **Ans.**

4. *Let A = {1, 2, 3}. Find all one-one functions from A to itself.*

Sol. All one-one functions from A to itself can easily be obtained by arranging the elements in the two row notation as follows :

$$f = \begin{pmatrix} 1 & 2 & 3 \\ 1 & 2 & 3 \end{pmatrix}, g = \begin{pmatrix} 1 & 2 & 3 \\ 1 & 3 & 2 \end{pmatrix}, h = \begin{pmatrix} 1 & 2 & 3 \\ 2 & 3 & 1 \end{pmatrix}$$

$$i = \begin{pmatrix} 1 & 2 & 3 \\ 2 & 1 & 3 \end{pmatrix}, j = \begin{pmatrix} 1 & 2 & 3 \\ 3 & 2 & 1 \end{pmatrix} k = \begin{pmatrix} 1 & 2 & 3 \\ 3 & 1 & 2 \end{pmatrix}$$

Hence, there are six functions from A to itself. **Ans.**

5. *Let A = {1, 2, 3 ...... n}. Find all onto functions from A to itself.*

Sol. We know that every onto function from any finite set to itself is one-one. Then, the number of onto functions from set A = {1, 2, 3...n} to itself is equal to the number of bijections from A to itself.

Here, no. of bijections (one-one onto) = $n\,!$

Hence, there are $n\,!$ onto functions from A to itself. **Ans.**

6. *Let R be the set of real numbers. If $f : R \to R$; $f(x) = \sin x$ and $g : R \to R$; $g(x) = x^2$. Then, find fog and gof.*

Sol. We know that range of f is a subset of domain of g and range of g is a subset of domain of f.

Therefore, fog and gof both exist.

Now, $$\begin{aligned} (fog)(x) &= f(g(x)) \ \forall\ x \in R \\ &= f(x^2) \\ &= \sin x^2 \end{aligned}$$

and $$\begin{aligned} (gof)(x) &= g(f(x)) \ \forall\ x \in R \\ &= g(\sin x) \\ &= (\sin x)^2 \\ &= \sin^2 x. \end{aligned}$$

Hence, $(fog)(x) = \sin x^2$ and $(gof)(x) = \sin^2 x$ **Ans.**

7. *If the function $f : R \to R$ is defined by $f(x) = x^2 + 3x + 1$. Find fof.*

Sol. We have

$$(fof)(x) = f(f(x))$$

$$f(f(x)) = f(x^2 + 3x + 1)$$

* Frequently asked previous years Board Exam Questions.

$fof(x) = f(f(x))$

$= (x^2 + 3x + 1)^2 + 3(x^2 + 3x + 1) + 1 \ \forall x \in R$

$= x^4 + 6x^3 + 14x^2 + 15x + 5$ **Ans.**

8. *If f, g : R → R are defined respectively by $f(x) = 4x - 7$, $g(x) = x^2 + 2x + 9$.*

*Find : **(i)** fog, **(ii)** gof, **(iii)** fof, **(iv)** gog.*

Sol. Here, range of f is a subset of domain of g and range of g is a subset of domain of f.

So, fog, gof, fof and gog all exist.

(i) $(fog)(x) = f(g(x))$

$= f(x^2 + 2x + 9)$

$= 4(x^2 + 2x + 9) - 7 \ \forall x \in R$

$= 4x^2 + 8x + 29$

(ii) $(gof)(x) = g(f(x))$

$= g(4x - 7)$

$= (4x - 7)^2 + 2(4x - 7) + 9 \ \forall x \in R$

$= 16x^2 - 48x + 44$

(iii) $(fof)(x) = f(f(x))$

$= f(4x - 7)$

$= 4(4x - 7) - 7 \ \forall x \in R$

$= 16x - 35$

(iv) $(gog)(x) = g(g(x))$

$= g(x^2 + 2x + 9)$

$= (x^2 + 2x + 9)^2 + 2(x^2 + 2x + 9) + 9$

$= x^4 + 4x^3 + 24x^2 + 40x + 108.$ **Ans.**

9. *If $f(x) = e^x$ and $g(x) = \log_e x \ (x > 0)$. Find fog and gof.*

Sol. Here, Range of f = Domain of $g = (0, \infty)$ and Range of g = Domain of f = R

So, fog and gof both exist.

For any $x \in R$,

$(fog)(x) = f(g(x))$

$= f(\log_e x)$

$= e^{\log_e x} = x$

and $(gof)(x) = g(f(x))$

$= g(e^x)$

$= \log_e e^x$

$= x \log_e e = x.$ **Ans.**

10. *Find gof and fog if $f(x) = |x|$ and $g(x) = |5x - 2|$.*

Sol. Given, $f(x) = |x|$ and $g(x) = |5x - 2|$

$(gof)(x) = g(f(x))$

$= g(|x|) = |5|x| - 2|$

$(fog)(x) = f(g(x))$

$= f(|5x - 2|)$

$= ||5x - 2||$

$= |5x - 2|$ **Ans**

11. *Find gof and fog if $f(x) = 8x^3$ and $g(x) = x^{1/3}$.*

Sol. Given, $f(x) = 8x^3$, $g(x) = x^{\frac{1}{3}}$

$(gof)(x) = g(f(x)) = g(8x^3) = (8x^3)^{\frac{1}{3}}$

$= [(2x)^3]^{\frac{1}{3}} = 2x$

$(fog)(x) = f(g(x)) = f(x^{\frac{1}{3}})$

$= 8.(x^{\frac{1}{3}})^3 = 8x$ **Ans**

12. *If $f : R \to R$ be given by $f(x) = (3 - x^3)^{1/3}$ find $fof(x)$.*

Sol. Given,

$f(x) = (3 - x^3)^{1/3}$

$(fof)(x) = f(f(x)) = f((3 - x^3)^{1/3})$

$= [3 - \{(3 - x^3)^{1/3}\}^3]^{1/3}$

$= (3 - 3 + x^3)^{1/3} = (x^3)^{1/3}$

$\Rightarrow (fof)(x) = x$ **Ans.**

13. *Let $f : R \to R$ be defined as $f(x) = 2x^2$.*
*Find : **(i)** $f^{-1}(8)$, **(ii)** $f^{-1}(-14)$*

Sol. **(i)** Let $f^{-1}(8) = x$

Then, $f(x) = 8$

$\Rightarrow 2x^2 = 8$

$\Rightarrow x^2 = 4$

$\Rightarrow x = \pm 2 \in R$

$\therefore f^{-1}(8) = \{-2, 2\}$

(ii) Let $f^{-1}(-14) = x$

Then, $f(x) = -14$

$\Rightarrow 2x^2 = -14$

$\Rightarrow x^2 = -7$

$\Rightarrow x = \pm\sqrt{-7} \notin R$

$\therefore f^{-1}(-14) = \phi.$ **Ans.**

14. *If f is an invertible function, defined as $f(x) = 3x + 1$, then find its inverse.*

Sol. Given, f is an invertible function.

So, let $y = 3x + 1$

$y - 1 = 3x$

$x = \dfrac{y-1}{3}$

$f^{-1}(y) = \dfrac{y-1}{3}$

Replacing y with x, we get

$f^{-1}(x) = \dfrac{x-1}{3}$ **Ans.**

15. *If $f : [-1, 1] \to R$ defined by $f(x) = \frac{x}{x+2}$ is invertible function then, find the inverse of the function $f : [-1, 1] \to$ Range (f).*

Sol. Since the function $f : [-1, 1] \to$ Range (f) is invertible, as $f : [-1, 1] \to R$ is invertible function.

Then, $fof^{-1}(x) = x \ \forall x \in$ Range (f)

$$f(f^{-1} x)) = x$$

$$\frac{f^{-1}(x)}{f^{-1}(x)+2} = x$$

$$f^{-1}(x) = xf^{-1}(x) + 2x$$

$$f^{-1}(x) - xf^{-1}(x) = 2x$$

$$f^{-1}(x)(1-x) = 2x$$

$$\therefore \quad f^{-1}(x) = \frac{2x}{1-x}.$$ **Ans**

Chapter 3. Inverse Trigonometric Functions

1. *Find x if : $\cos^{-1}[\sin(\cos^{-1} x)] = \frac{\pi}{3}$.*

Sol. Given,

$$\cos^{-1}[\sin(\cos^{-1} x)] = \frac{\pi}{3}$$

$$\Rightarrow \quad \sin\left(\sin^{-1}\sqrt{1-x^2}\right) = \cos\frac{\pi}{3}$$

$$\left[\because \cos^{-1} x = \sin^{-1}\sqrt{1-x^2}\right]$$

$$\Rightarrow \quad \sqrt{1-x^2} = \frac{1}{2}$$

$$\Rightarrow \quad 1-x^2 = \frac{1}{4}$$

$$\Rightarrow \quad x^2 = \frac{3}{4}$$

$$\Rightarrow \quad x = \pm\frac{\sqrt{3}}{2}$$ **Ans.**

2. *Find x if*

$$\sin^{-1}\frac{5}{x} + \sin^{-1}\left(\frac{12}{x}\right) = \frac{\pi}{2}.$$

Sol. Given, $\sin^{-1}\frac{5}{x} + \sin^{-1}\frac{12}{x} = \frac{\pi}{2}$

Let, $\sin^{-1}\frac{5}{x} = A$ and $\sin^{-1}\left(\frac{12}{x}\right) = B$ ... (i)

$$\therefore \quad A + B = \frac{\pi}{2}$$

$$\Rightarrow \quad A = \frac{\pi}{2} - B \qquad \text{... (ii)}$$

From (i), $\sin A = \frac{5}{x}$, $\sin B = \frac{12}{x}$

Since, $\cos A = \cos\left(\frac{\pi}{2} - B\right)$

$= \sin B$ [From (ii)]

$$\therefore \quad \cos A = \frac{12}{x}$$

$$\sin^2 A + \cos^2 A = \frac{25}{x^2} + \frac{144}{x^2} = 1$$

$$\Rightarrow \quad x^2 = 169$$

$$\therefore \quad x = \sqrt{169} = \pm 13$$ **Ans.**

3. *Find x if*

$$\sin^{-1}(6x) + \sin^{-1}(6\sqrt{3}x) = -\frac{\pi}{2}$$

Sol. Consider,

$$\sin^{-1} 6x = -\frac{\pi}{2} - \sin^{-1}(6\sqrt{3})x)$$

$$6x = \sin\left[-\frac{\pi}{2} - \sin^{-1}(6\sqrt{3}x)\right]$$

$$6x = -\cos[\sin^{-1}(6\sqrt{3}x)]$$

$$6x = -\cos[\cos^{-1}\sqrt{1-108x^2}]$$

$$[\because \sin^{-1} x = \cos^{-1}\sqrt{1-x^2}]$$

$$6x = -\sqrt{1-108x^2}$$

On squaring both sides, we get

$$36x^2 = 1 - 108x^2$$

$$144x^2 = 1$$

$$x^2 = \frac{1}{144}$$

$$x = \pm\frac{1}{12}$$

If $x = \frac{1}{12}$, we get

$$\sin^{-1}\left(6\times\frac{1}{12}\right) + \sin^{-1}\left(6\sqrt{3}\times\frac{1}{12}\right)$$

$$= \sin^{-1}\left(\frac{1}{2}\right) + \sin^{-1}\left(\frac{\sqrt{3}}{2}\right)$$

$$= \frac{\pi}{6} + \frac{\pi}{3}$$

$$= \frac{\pi}{2} \neq -\frac{\pi}{2}$$

$\Rightarrow$ Then, $x = -\frac{1}{12}$ satisfed the eqaution. **Ans.**

4. *Find the value of x :*

$$\sin^{-1}\left[\frac{2a}{1+a^2}\right] + \sin^{-1}\left[\frac{2b}{1+b^2}\right] = 2\tan^{-1} x.$$

Sol. Consider,

$$\text{L.H.S.} = \sin^{-1}\left[\frac{2a}{1+a^2}\right] + \sin^{-1}\left[\frac{2b}{1+b^2}\right]$$

$$= 2\tan^{-1} a + 2\tan^{-1} b$$

$$= 2[\tan^{-1} a + \tan^{-1} b]$$

$$= 2\tan^{-1}\left[\frac{a+b}{1-ab}\right]$$

Comparing L.H.S. with R.H.S.

$$\Rightarrow 2\tan^{-1}\left(\frac{a+b}{1-ab}\right) = 2\tan^{-1} x$$

$$\Rightarrow x = \frac{a+b}{1-ab}$$ **Ans.**

5. *Find the value of x if :*

$$\tan^{-1} 2x + \tan^{-1} 3x = \frac{\pi}{4}$$

Sol. Given, $\tan^{-1} 2x + \tan^{-1} 3x = \frac{\pi}{4}$

$$\Rightarrow \tan^{-1}\left[\frac{2x+3x}{1-2x\times 3x}\right] = \frac{\pi}{4}$$

$$\Rightarrow \tan^{-1}\left[\frac{5x}{1-6x^2}\right] = \frac{\pi}{4}$$

$$\Rightarrow \frac{5x}{1-6x^2} = \tan\frac{\pi}{4} = 1$$

$$\Rightarrow 5x = 1 - 6x^2$$

$$\Rightarrow 6x^2 + 5x - 1 = 0$$

$$\Rightarrow (6x-1)(x+1) = 0$$

$$\Rightarrow x = \frac{1}{6}, -1$$ **Ans.**

6. *Find the value of x :*

$$\sin^{-1} x + \sin^{-1}(1-x) = \cos^{-1} x.$$

Sol. The given equation is,

$$\sin^{-1} x + \sin^{-1}(1-x) = \cos^{-1} x$$

$$\Rightarrow \sin^{-1} x + \sin^{-1}(1-x) = \frac{\pi}{2} - \sin^{-1} x$$

$$\left[\because \sin^{-1} x + \cos^{-1} x = \frac{\pi}{2}\right]$$

$$\Rightarrow \sin^{-1}(1-x) = \frac{\pi}{2} - 2\sin^{-1} x \quad \text{...(i)}$$

Let, $\sin^{-1} x = y$

$\Rightarrow x = \sin y$

Therefore, from (i), we get

$$\sin^{-1}(1-x) = \frac{\pi}{2} - 2y$$

$$\Rightarrow 1 - x = \sin\left(\frac{\pi}{2} - 2y\right)$$

$$\Rightarrow 1 - x = \cos 2y$$

$$\Rightarrow 1 - x = 1 - 2\sin^2 y$$

$$\Rightarrow 1 - x = 1 - 2x^2 \quad [\because x = \sin y]$$

$$\Rightarrow 2x^2 - x = 0$$

$$\Rightarrow x(2x-1) = 0$$

$$\Rightarrow x = 0, \frac{1}{2}$$

Since both these values satisfy the given equation.

Hence, the solutions of the given equation are

$$x = 0, \frac{1}{2}.$$ **Ans.**

7. *Find the value of x in equation*

$$\tan^{-1}(2+x) + \tan^{-1}(2-x) = \tan^{-1}\frac{2}{3}.$$

Sol. Given,

$$\tan^{-1}(2+x) + \tan^{-1}(2-x) = \tan^{-1}\left(\frac{2}{3}\right)$$

We have to solve the equation for x

Consider,

$$\text{L.H.S.} = \tan^{-1}(2+x) + \tan^{-1}(2-x)$$

$$= \tan^{-1}\left[\frac{(2+x)+(2-x)}{1-(2+x)(2-x)}\right]$$

$$= \tan^{-1}\left[\frac{4}{1-(4-x^2)}\right]$$

$$= \tan^{-1}\left[\frac{4}{x^2-3}\right]$$

$$\therefore \tan^{-1}\left[\frac{4}{x^2-3}\right] = \tan^{-1}\left(\frac{2}{3}\right)$$

$$\Rightarrow \quad \frac{4}{x^2-3} = \frac{2}{3}$$

$$\Rightarrow \quad 2x^2 - 6 = 12$$

$$\Rightarrow \quad 2x^2 = 12 + 6 = 18$$

$$\Rightarrow \quad x^2 = 9$$

$$\Rightarrow \quad x = \sqrt{9} = \pm 3$$

$$\Rightarrow \quad x = 3 \text{ or } x = -3$$ **Ans.**

8. *Find the value of x : sin (2 tan⁻¹ x) = 1.*

Sol. Given,

$$\sin(2\tan^{-1}x) = 1$$

$$\Rightarrow \quad \sin\left[\sin^{-1}\left(\frac{2x}{1+x^2}\right)\right] = 1$$

$$\left[\because 2\tan^{-1}x = \sin^{-1}\left(\frac{2x}{1+x^2}\right)\right]$$

$$\Rightarrow \quad \frac{2x}{1+x^2} = 1$$

$$\Rightarrow \quad 2x = 1 + x^2$$

$$\Rightarrow \quad x^2 - 2x + 1 = 0$$

$$\Rightarrow \quad (x-1)^2 = 0$$

$$\therefore \quad x = 1, 1$$ **Ans.**

9. *Find x*

$$\sin[2\cos^{-1}(\cot(2\tan^{-1}x))] = 0$$

Sol. Since, $2\tan^{-1}x = \cot^{-1}\left[\frac{1-x^2}{2x}\right]$

$$\therefore \quad \cot(2\tan^{-1}x) = \cot\left[\cot^{-1}\left(\frac{1-x^2}{2x}\right)\right]$$

$$= \frac{1-x^2}{2x}$$

∴ Given expression reduces to

$$\sin\left[2\cos^{-1}\left[\frac{1-x^2}{2x}\right]\right] = 0$$

$$\Rightarrow \quad \sin\left[\cos^{-1}\left\{2\cdot\frac{(1-x^2)^2}{(2x)^2} - 1\right\}\right] = 0$$

$$\Rightarrow \quad \sin\left[\cos^{-1}\left\{\frac{(1-x^2)^2}{2x^2} - 1\right\}\right] = 0$$

$$\Rightarrow \quad \sin\left[\cos^{-1}\left\{\frac{x^4+1-2x^2-2x^2}{2x^2}\right\}\right] = 0$$

$$\Rightarrow \quad \sin\left[\cos^{-1}\left\{\frac{(x^4-4x^2+1)}{2x^2}\right\}\right] = 0$$

$$\Rightarrow \quad \sin\left[\sin^{-1}\sqrt{1-\frac{(x^4-4x^2+1)^2}{4x^4}}\right] = 0$$

$$\Rightarrow \quad 4x^4 - (x^4 - 4x^2 + 1)^2 = 0$$

$$\Rightarrow \quad (2x^2)^2 - (x^4 - 4x^2 + 1)^2 = 0$$

$$\Rightarrow \quad (x^4 - 4x^2 + 1 + 2x^2).(x^4 - 4x^2 + 1 - 2x^2) = 0$$

$$\Rightarrow \quad (x^4 - 2x^2 + 1)(x^4 - 6x^2 + 1) = 0$$

∴ Either $x^4 - 2x^2 + 1 = 0$

$$\Rightarrow \quad (x^2-1)^2 = 0$$

$$\Rightarrow \quad x = \pm 1$$ **Ans.**

Or $x^4 - 6x^2 + 1 = 0$

$$\Rightarrow \quad x^4 - 6x^2 + 1 + 8 = 8$$

$$\Rightarrow \quad (x^2-3)^2 = 8$$

$$\Rightarrow \quad x^2 - 3 = \pm 2\sqrt{2}$$

$$\Rightarrow \quad x^2 = 3 \pm 2\sqrt{2}$$

$$x = \pm\sqrt{3 \pm 2\sqrt{2}}$$ **Ans.**

3. *Solve for x :**

$$\tan^{-1}\left(\frac{x-1}{x-2}\right) + \tan^{-1}\left(\frac{x+1}{x+2}\right) = \frac{\pi}{4}$$

Sol. Given, $\tan^{-1}\left(\frac{x-1}{x-2}\right) + \tan^{-1}\left(\frac{x+1}{x+2}\right) = \frac{\pi}{4}$

$$\Rightarrow \quad \tan^{-1}\left[\frac{\left(\frac{x-1}{x-2}\right)+\left(\frac{x+1}{x+2}\right)}{1-\left(\frac{x-1}{x-2}\right)\left(\frac{x+1}{x+2}\right)}\right] = \frac{\pi}{4}$$

$$\Rightarrow \tan^{-1}\left[\frac{(x-1)(x+2)+(x+1)(x-2)}{(x-2)(x+2)-(x-1)(x+1)}\right] = \frac{\pi}{4}$$

$$\Rightarrow \quad \tan^{-1}\left[\frac{(x^2+x-1)+(x^2-x-2)}{(x^2-4)-(x^2-1)}\right] = \frac{\pi}{4}$$

$$\Rightarrow \quad \left[\frac{(x^2+x-2)+(x^2-x-2)}{(x^2-4)-(x^2-1)}\right] = \tan\frac{\pi}{4}$$

$$\Rightarrow \quad \frac{x^2+x-2+x^2-x-2}{x^2-4-x^2+1} = 1$$

$$\Rightarrow \quad \frac{2x^2-4}{-3} = 1$$

$$\Rightarrow \quad 2x^2 - 4 = -3$$

$$\Rightarrow \quad 2x^2 = -3 + 4$$

* Frequently asked previous years Board Exam Questions.

$\Rightarrow \qquad 2x^2 = 1$

$\Rightarrow \qquad x^2 = \dfrac{1}{2}$

$\Rightarrow \qquad x = \pm \dfrac{1}{\sqrt{2}}$

Ans.

3. *Evaluate : cos(2cos^{-1} x + sin^{-1} x) at x = $\dfrac{1}{5}$.**

Sol. $\cos(2\cos^{-1}x + \sin^{-1}x) = \cos[\cos^{-1}x + \cos^{-1}x + \sin^{-1}x]$

$= \cos[\cos^{-1}x + (\cos^{-1}x + \sin^{-1}x)]$

$= \cos\left[\cos^{-1}x + \dfrac{\pi}{2}\right]$

$= \cos\left(\dfrac{\pi}{2} + \cos^{-1}x\right)$

$= -\sin(\cos^{-1}x) \qquad \left[\because \cos\left(\dfrac{\pi}{2}+\theta\right) = -\sin\theta\right]$

$= -\sin\left(\sin^{-1}\sqrt{1-x^2}\right) \qquad \left[\because \cos^{-1}x = \sin^{-1}\sqrt{1-x^2}\right]$

$= -\sqrt{1-x^2}$

$= -\sqrt{1-\left(\dfrac{1}{5}\right)^2} \qquad \left[\because x = \dfrac{1}{5}\text{(Given)}\right]$

$= -\sqrt{1-\dfrac{1}{25}} = -\sqrt{\dfrac{25-1}{25}}$

$= -\sqrt{\dfrac{24}{25}} = -\dfrac{2\sqrt{6}}{5}$ **Ans.**

Chapter 4. Determinants

1. *If following system of equations is consistent. Find the value of λ.*

$2x + 3y - 8 = 0$

$7x - 5y + 3 = 0$

$4x - 6y + \lambda = 0.$

Sol. As the given system of equation are consistent

$$\begin{vmatrix} 2 & 3 & -8 \\ 7 & -5 & 3 \\ 4 & -6 & \lambda \end{vmatrix} = 0$$

$\Rightarrow 2(-5\lambda + 18) - 3(7\lambda - 12) - 8(-42 + 20) = 0$

$\Rightarrow \qquad -10\lambda + 36 - 21\lambda + 36 + 176 = 0$

$\Rightarrow \qquad 31\lambda = 248$

$\Rightarrow \qquad \lambda = \dfrac{248}{31} = 8$ **Ans.**

2. *Find the values of λ for which the determinant*

$$\begin{vmatrix} 3-\lambda & -1 & 1 \\ -1 & 5-\lambda & -1 \\ 1 & -1 & 3-\lambda \end{vmatrix} = 0$$

Sol. We have to find the value of λ, so that

$$\begin{vmatrix} 3-\lambda & -1 & 1 \\ -1 & 5-\lambda & -1 \\ 1 & -1 & 3-\lambda \end{vmatrix} = 0$$

Applying $C_1 \to C_1 + C_2 + C_3$, we obtain

$$\begin{vmatrix} 3-\lambda & -1 & 1 \\ 3-\lambda & 5-\lambda & -1 \\ 3-\lambda & -1 & 3-\lambda \end{vmatrix} = 0$$

Taking $(3 - \lambda)$ common from C_1, we get

$$(3-\lambda)\begin{vmatrix} 1 & -1 & 1 \\ 1 & 5-\lambda & -1 \\ 1 & -1 & 3-\lambda \end{vmatrix} = 0$$

Applying $R_2 \to R_2 - R_1$ and $R_3 \to R_3 - R_2$, we obtain

$$(3-\lambda)\begin{vmatrix} 1 & -1 & 1 \\ 0 & 6-\lambda & -2 \\ 0 & \lambda-6 & 4-\lambda \end{vmatrix} = 0$$

Expanding along C_1, we get

$(3-\lambda)\{1(6-\lambda)(4-\lambda) - (\lambda-6)(-2)\} = 0$

$\Rightarrow \qquad (3-\lambda)(6-\lambda)(4-\lambda-2) = 0$

$\Rightarrow \qquad (\lambda-2)(\lambda-6)(\lambda-3) = 0$

$\Rightarrow \qquad \lambda = 2, 3, 6.$

Ans.

3. *Using properties of determinant, find the value of the following determinant*

$$D = \begin{vmatrix} -a^2 & ab & ac \\ ba & -b^2 & bc \\ ac & bc & -c^2 \end{vmatrix}.$$

Sol. Given,

$$D = \begin{vmatrix} -a^2 & ab & ac \\ ba & -b^2 & bc \\ ac & bc & -c^2 \end{vmatrix}.$$

Taking a, b, c common from R_1, R_2 and R_3 respectively, we get

* Frequently asked previous years Board Exam Questions.

Find the Value Type Questions

$$D = abc\begin{vmatrix} -a & b & c \\ a & -b & c \\ a & b & -c \end{vmatrix}$$

Taking a, b, c common from C_1, C_2 and C_3 respectively, we get

$$D = a^2b^2c^2\begin{vmatrix} -1 & 1 & 1 \\ 1 & -1 & 1 \\ 1 & 1 & -1 \end{vmatrix}$$

Applying $C_1 \rightarrow C_1 + C_2$, we obtain

$$D = a^2b^2c^2\begin{vmatrix} 0 & 1 & 1 \\ 0 & -1 & 1 \\ 2 & 1 & -1 \end{vmatrix}$$

Expanding along C_1, we get

$= a^2b^2c^2\,[2\{1-(-1)\}]$

$= a^2b^2c^2\,(2 \times 2)$

$= 4a^2b^2c^2$ **Ans.**

4. *If $s = a + b + c$, then find the value of*

$$\begin{vmatrix} s+c & a & b \\ c & s+a & b \\ c & a & s+b \end{vmatrix}.$$

Sol. Let $\Delta = \begin{vmatrix} s+c & a & b \\ c & s+a & b \\ c & a & s+b \end{vmatrix}$.

Applying $C_1 \rightarrow C_1 + C_2 + C_3$, we obtain

$$\Delta = \begin{vmatrix} s+a+b+c & a & b \\ s+a+b+c & s+a & b \\ s+a+b+c & a & s+b \end{vmatrix}$$

Taking $(a + b + c + s)$, common from C_1, we get

$$= (s+a+b+c)\begin{vmatrix} 1 & a & b \\ 1 & s+a & b \\ 1 & a & s+b \end{vmatrix}$$

Now applying $R_1 \rightarrow R_1 - R_2$ and $R_2 \rightarrow R_2 - R_3$, we obtain

$$\Rightarrow \quad \Delta = (s+a+b+c)\begin{vmatrix} 0 & -s & 0 \\ 0 & s & -s \\ 1 & a & s+b \end{vmatrix}$$

$[\because a + b + c = s]$

Expanding along C_1, we get

$= 2s \times 1\,(s^2 - 0)$

$= 2s^3$ **Ans.**

5. *Using properties of determinants, find the value of the following determinant*

$$\begin{vmatrix} x^3 & x^2 & x \\ y^3 & y^2 & y \\ z^3 & z^2 & z \end{vmatrix}$$

Sol.

Let, $\begin{vmatrix} x^3 & x^2 & x \\ y^3 & y^2 & y \\ z^3 & z^2 & z \end{vmatrix} = D$

Taking x, y, z common from R_1, R_2, R_3 respectively, we get

$$\Rightarrow \quad D = xyz\begin{vmatrix} x^2 & x & 1 \\ y^2 & y & 1 \\ z^2 & z & 1 \end{vmatrix}$$

Applying $R_1 \rightarrow R_1 - R_2$, $R_2 \rightarrow R_2 - R_3$, we obtain

$$= xyz\begin{vmatrix} x^2-y^2 & x-y & 0 \\ y^2-z^2 & y-z & 0 \\ z^2 & z & 1 \end{vmatrix}$$

Taking $(x - y)$ common from R_1 and $(y - z)$ from R_2, we get

$$= (xyz)(x-y)(y-z)\begin{vmatrix} x+y & 1 & 0 \\ y+z & 1 & 0 \\ z^2 & z & 1 \end{vmatrix}$$

Expanding along C_3, we get

$D = (xyz)(x-y)(y-z)(x+y-y-z)$

$= (xyz)(x-y)(y-z)(x-z)$ **Ans.**

Chapter 5. Matrices

1. *If* $\begin{pmatrix} 2 & 3 \\ 5 & 7 \end{pmatrix}\begin{pmatrix} 1 & -3 \\ -2 & 4 \end{pmatrix} = \begin{pmatrix} -4 & 6 \\ -9 & x \end{pmatrix}$, *find x.**

Sol. Given : $\begin{pmatrix} 2 & 3 \\ 5 & 7 \end{pmatrix}\begin{pmatrix} 1 & -3 \\ -2 & 4 \end{pmatrix} = \begin{pmatrix} -4 & 6 \\ -9 & x \end{pmatrix}$

$$\Rightarrow \begin{pmatrix} 2\times1+3\times(-2) & 2\times(-3)+3\times4 \\ 5\times1+7\times(-2) & 5\times(-3)+7\times4 \end{pmatrix} = \begin{pmatrix} -4 & 6 \\ -9 & x \end{pmatrix}$$

$$\Rightarrow \begin{pmatrix} -4 & 6 \\ -9 & 13 \end{pmatrix} = \begin{pmatrix} -4 & 6 \\ -9 & x \end{pmatrix}$$

$\therefore$ $x = 13$ **Ans.**

2. *If the matrix* $\begin{bmatrix} 6 & -x^2 \\ 2x-15 & 10 \end{bmatrix}$ *is symmetric; find the value of x.*

* Frequently asked previous years Board Exam Questions.

Sol. Given, the matrix $\begin{bmatrix} 6 & -x^2 \\ 2x-15 & 10 \end{bmatrix}$ is symmetric

So, $\begin{bmatrix} 6 & -x^2 \\ 2x-15 & 10 \end{bmatrix}^T = \begin{bmatrix} 6 & -x^2 \\ 2x-15 & 10 \end{bmatrix}$

$\begin{bmatrix} 6 & 2x-15 \\ -x^2 & 10 \end{bmatrix} = \begin{bmatrix} 6 & -x^2 \\ 2x-15 & 10 \end{bmatrix}$

On comparing both sides, we get

$\Rightarrow \quad 2x - 15 = -x^2$

$\Rightarrow \quad x^2 + 2x - 15 = 0$

$\Rightarrow \quad x^2 + 5x - 3x - 15 = 0$

$\Rightarrow \quad x(x+5) - 3(x+5) = 0$

$(x+5)(x-3) = 0$

$x = -5$ or 3 **Ans.**

3. *Given*

$2\begin{bmatrix} 1 & 2 & 3 \\ -1 & -3 & 2 \end{bmatrix} + k\begin{bmatrix} 1 & 0 & 2 \\ 3 & 4 & 5 \end{bmatrix} = \begin{bmatrix} 4 & 4 & 10 \\ 4 & 2 & 14 \end{bmatrix}$

find the value of k (given $k \neq 0$).

Sol. Given,

$2\begin{bmatrix} 1 & 2 & 3 \\ -1 & -3 & 2 \end{bmatrix} + k\begin{bmatrix} 1 & 0 & 2 \\ 3 & 4 & 5 \end{bmatrix} = \begin{bmatrix} 4 & 4 & 10 \\ 4 & 2 & 14 \end{bmatrix}$

$\begin{bmatrix} 2 & 4 & 6 \\ -2 & -6 & 4 \end{bmatrix} + k\begin{bmatrix} 1 & 0 & 2 \\ 3 & 4 & 5 \end{bmatrix} = \begin{bmatrix} 4 & 4 & 10 \\ 4 & 2 & 14 \end{bmatrix}$

$\Rightarrow \quad k\begin{bmatrix} 1 & 0 & 2 \\ 3 & 4 & 5 \end{bmatrix} = \begin{bmatrix} 4 & 4 & 10 \\ 4 & 2 & 14 \end{bmatrix} - \begin{bmatrix} 2 & 4 & 6 \\ -2 & -6 & 4 \end{bmatrix}$

$= \begin{bmatrix} 4-2 & 4-4 & 10-6 \\ 4+2 & 2+6 & 14-4 \end{bmatrix}$

$= \begin{bmatrix} 2 & 0 & 4 \\ 6 & 8 & 10 \end{bmatrix}$

$= 2\begin{bmatrix} 1 & 0 & 2 \\ 3 & 4 & 5 \end{bmatrix}$

On comparing both sides, we get

$\therefore \quad k = 2$ **Ans.**

4. *If* $\begin{bmatrix} x^2 & 3 & 4 \\ 1 & 9 & 8 \end{bmatrix} + \begin{bmatrix} -3x & 1 & -5 \\ -3 & -2 & -6 \end{bmatrix} = \begin{bmatrix} 4 & 4 & -1 \\ -2 & 7 & 2 \end{bmatrix}$

find the value of x.

Sol.

Given, $\begin{bmatrix} x^2 & 3 & 4 \\ 1 & 9 & 8 \end{bmatrix} + \begin{bmatrix} -3x & 1 & -5 \\ -3 & -2 & -6 \end{bmatrix} =$

$\begin{bmatrix} 4 & 4 & -1 \\ -2 & 7 & 2 \end{bmatrix}$

$\Rightarrow \quad \begin{bmatrix} x^2-3x & 4 & -1 \\ -2 & 7 & 2 \end{bmatrix} = \begin{bmatrix} 4 & 4 & -1 \\ -2 & 7 & 2 \end{bmatrix}$

Comparing both sides, we get

$x^2 - 3x = 4$

$\Rightarrow \quad x^2 - 3x - 4 = 0$

$\Rightarrow \quad x^2 - 4x + x - 4 = 0$

$\Rightarrow \quad x(x-4) + 1(x-4) = 0$

$\Rightarrow \quad (x+1)(x-4) = 0$

$\Rightarrow \quad x = -1, x = 4$ **Ans.**

5. *If $(A - 2I)(A - 3I) = 0$, where*

$A = \begin{bmatrix} 4 & 2 \\ -1 & x \end{bmatrix}$ *and* $I = \begin{bmatrix} 1 & 0 \\ 0 & 1 \end{bmatrix}$ *find the value of x.*

Sol. Given, $(A - 2I)(A - 3I) = 0$

$\left\{\begin{bmatrix} 4 & 2 \\ -1 & x \end{bmatrix} - \begin{bmatrix} 2 & 0 \\ 0 & 2 \end{bmatrix}\right\} \cdot \left\{\begin{bmatrix} 4 & 2 \\ -1 & x \end{bmatrix} - \begin{bmatrix} 3 & 0 \\ 0 & 3 \end{bmatrix}\right\} = \begin{bmatrix} 0 & 0 \\ 0 & 0 \end{bmatrix}$

$\Rightarrow \quad \begin{bmatrix} 2 & 2 \\ -1 & x-2 \end{bmatrix}\begin{bmatrix} 1 & 2 \\ -1 & x-3 \end{bmatrix} = \begin{bmatrix} 0 & 0 \\ 0 & 0 \end{bmatrix}$

$\begin{bmatrix} 2\times1+2\times(-1) & 2\times2+2(x-3) \\ -1\times1+(x-2)\times(-1) & (-1)\times2+(x-2)(x-3) \end{bmatrix}$

$\Rightarrow \quad \begin{bmatrix} 0 & 2x-2 \\ -x+1 & x^2-5x+4 \end{bmatrix} = \begin{bmatrix} 0 & 0 \\ 0 & 0 \end{bmatrix}$

Equating corresponding elements of equal matrices, we get

$-x + 1 = 0$	$2x - 2 = 0$	$x^2 - 5x + 4 = 0$
$\Rightarrow x = 1$	$\Rightarrow x = 1$	$\Rightarrow x = 1, 4$

$\therefore \quad x = 1$ **Ans.**

6. *If* $A = \begin{bmatrix} 4 & 2 \\ 1 & 1 \end{bmatrix}$, *find $(A - 2I)(A - 3I)$, where I is a unit matrix.*

Sol. Given, $A = \begin{bmatrix} 4 & 2 \\ 1 & 1 \end{bmatrix}, I = \begin{bmatrix} 1 & 0 \\ 0 & 1 \end{bmatrix}$

$\therefore \quad A - 2I = \begin{bmatrix} 4 & 2 \\ 1 & 1 \end{bmatrix} - \begin{bmatrix} 2 & 0 \\ 0 & 2 \end{bmatrix}$

$= \begin{bmatrix} 2 & 2 \\ 1 & -1 \end{bmatrix}$

and $\quad A - 3I = \begin{bmatrix} 4 & 2 \\ 1 & 1 \end{bmatrix} - \begin{bmatrix} 3 & 0 \\ 0 & 3 \end{bmatrix}$

$= \begin{bmatrix} 1 & 2 \\ 1 & -2 \end{bmatrix}$

$$\therefore (A-2I)(A-3I) = \begin{bmatrix} 2 & 2 \\ 1 & -1 \end{bmatrix} \cdot \begin{bmatrix} 1 & 2 \\ 1 & -2 \end{bmatrix}$$

$$= \begin{bmatrix} 2+2 & 4-4 \\ 1-1 & 2+2 \end{bmatrix}$$

$$= \begin{bmatrix} 4 & 0 \\ 0 & 4 \end{bmatrix}$$

$= 4I$ **Ans.**

7. *If X is a 2 × 2 matrix given that*

$$\begin{bmatrix} 1 & 3 \\ 0 & 1 \end{bmatrix} X = \begin{bmatrix} 1 & -1 \\ 0 & 1 \end{bmatrix},$$ *find X.*

Sol. Let $X = \begin{bmatrix} a & b \\ c & d \end{bmatrix}$

$$\Rightarrow \begin{bmatrix} 1 & 3 \\ 0 & 1 \end{bmatrix}\begin{bmatrix} a & b \\ c & d \end{bmatrix} = \begin{bmatrix} 1 & -1 \\ 0 & 1 \end{bmatrix}$$

$$\Rightarrow \begin{bmatrix} 1\times a+3\times c & 1\times b+3\times d \\ 0\times a+1\times c & 0\times b+1\times d \end{bmatrix} = \begin{bmatrix} 1 & -1 \\ 0 & 1 \end{bmatrix}$$

$$\Rightarrow \begin{bmatrix} a+3c & b+3d \\ c & d \end{bmatrix} = \begin{bmatrix} 1 & -1 \\ 0 & 1 \end{bmatrix}$$

Equating corresponding elements of equal matrices, we get,

$c = 0, d = 1$

and $a + 3c = 1$

$\Rightarrow a + 0 = 1$

$\Rightarrow a = 1$

and $b + 3d = -1$

$\Rightarrow b + 3 \times 1 = -1$

$\Rightarrow b = -4$

Therefore, $X = \begin{bmatrix} 1 & -4 \\ 0 & 1 \end{bmatrix}$ **Ans.**

8. *Find the matrix X for which :*

$$\begin{bmatrix} 5 & 4 \\ 1 & 1 \end{bmatrix} X = \begin{bmatrix} 1 & -2 \\ 1 & 3 \end{bmatrix}$$

Sol. Let $X = \begin{bmatrix} a & b \\ c & d \end{bmatrix}$

Then, $\begin{bmatrix} 5 & 4 \\ 1 & 1 \end{bmatrix}\begin{bmatrix} a & b \\ c & d \end{bmatrix} = \begin{bmatrix} 1 & -2 \\ 1 & 3 \end{bmatrix}$

$$\Rightarrow \begin{bmatrix} 5a+4c & 5b+4d \\ a+c & b+d \end{bmatrix} = \begin{bmatrix} 1 & -2 \\ 1 & 3 \end{bmatrix}$$

$\Rightarrow 5a + 4c = 1, 5b + 4d = -2,$

$a + c = 1, \quad b + d = 3$

Solving these equations, we get

$a = -3, c = 4, b = -14,$

$d = 17$

Hence, $X = \begin{bmatrix} -3 & -14 \\ 4 & 17 \end{bmatrix}$ **Ans.**

9. *Find x such that*

$$[1\ x\ 1] \begin{bmatrix} 1 & 3 & 2 \\ 2 & 5 & 1 \\ 15 & 3 & 2 \end{bmatrix} \begin{bmatrix} 1 \\ 2 \\ x \end{bmatrix} = 0.$$

Sol. Given, $[1\ x\ 1] \begin{bmatrix} 1 & 3 & 2 \\ 2 & 5 & 1 \\ 15 & 3 & 2 \end{bmatrix} \begin{bmatrix} 1 \\ 2 \\ x \end{bmatrix} = 0$

$$\Rightarrow [1 + 2x + 15 \quad 3 + 5x + 3 \quad 2 + x + 2] \begin{bmatrix} 1 \\ 2 \\ x \end{bmatrix} = 0$$

$$\Rightarrow [2x + 16 \quad 5x + 6 \quad x + 4] \begin{bmatrix} 1 \\ 2 \\ x \end{bmatrix} = 0$$

$\Rightarrow [2x + 16 + 10x + 12 + x^2 + 4x] = 0$

$\Rightarrow x^2 + 16x + 28 = 0$

$\Rightarrow (x + 2)(x + 14) = 0$

$\Rightarrow x = -2, -14$ **Ans.**

10. *Find the value of k if* $M = \begin{bmatrix} 1 & 2 \\ 2 & 3 \end{bmatrix}$ *and*

$M^2 - kM - I^2 = 0.$

Sol. Given, $M = \begin{bmatrix} 1 & 2 \\ 2 & 3 \end{bmatrix}$

$M^2 = M \cdot M$

$$= \begin{bmatrix} 1 & 2 \\ 2 & 3 \end{bmatrix}\begin{bmatrix} 1 & 2 \\ 2 & 3 \end{bmatrix}$$

$$= \begin{bmatrix} 1\times1+2\times2 & 1\times2+2\times3 \\ 2\times1+3\times2 & 2\times2+3\times3 \end{bmatrix}$$

$$= \begin{bmatrix} 1+4 & 2+6 \\ 2+6 & 4+9 \end{bmatrix} = \begin{bmatrix} 5 & 8 \\ 8 & 13 \end{bmatrix}$$

Now, $M^2 - kM - I_2 = 0$

$\Rightarrow M^2 - I_2 = kM$

$$\Rightarrow \begin{bmatrix} 5 & 8 \\ 8 & 13 \end{bmatrix} - \begin{bmatrix} 1 & 0 \\ 0 & 1 \end{bmatrix} = k\begin{bmatrix} 1 & 2 \\ 2 & 3 \end{bmatrix}$$

$$\Rightarrow \begin{bmatrix} 4 & 8 \\ 8 & 12 \end{bmatrix} = k\begin{bmatrix} 1 & 2 \\ 2 & 3 \end{bmatrix}$$

$$\Rightarrow \quad 4\begin{bmatrix}1 & 2\\2 & 3\end{bmatrix} = k\begin{bmatrix}1 & 2\\2 & 3\end{bmatrix}$$

On comparing both sides, we get

$$\Rightarrow \quad k = 4.$$ **Ans.**

11. *If the matrix* $A = \begin{bmatrix}6 & x & 2\\2 & -1 & 2\\-10 & 5 & 2\end{bmatrix}$ *is a singular matrix, find the value of x.*

Sol. A is a singular matrix

$$\therefore \quad |A| = 0$$

$$\Rightarrow \quad 6(-12) - x(24) + 2(0) = 0$$

$$\Rightarrow \quad -24x = 72$$

$$\Rightarrow \quad x = -3$$ **Ans.**

12. *Given that* $A = \begin{bmatrix}\cos x & \sin x\\-\sin x & \cos x\end{bmatrix}$ *and*

$A\,(adj\,A) = K\begin{bmatrix}1 & 0\\0 & 1\end{bmatrix}$. *Find the value of K.*

Sol. Given,

$$A = \begin{bmatrix}\cos x & \sin x\\-\sin x & \cos x\end{bmatrix}$$

$$\therefore \quad |A| = \cos^2 x + \sin^2 x = 1 \neq 0$$

$$\therefore \quad \text{adj } A = \begin{bmatrix}\cos x & -\sin x\\\sin x & \cos x\end{bmatrix}$$

$$A\,(\text{adj } A) = \begin{bmatrix}\cos x & \sin x\\-\sin x & \cos x\end{bmatrix}\begin{bmatrix}\cos x & -\sin x\\\sin x & \cos x\end{bmatrix}$$

$$= \begin{bmatrix}\cos^2 x + \sin^2 x & -\sin x\cos x + \sin x\cos x\\-\sin x\cos x + \cos x\sin x & \sin^2 x + \cos^2 x\end{bmatrix}$$

$$\Rightarrow \quad A\,(\text{adj } A) = K\begin{bmatrix}1 & 0\\0 & 1\end{bmatrix}$$

$$\therefore \quad K\begin{bmatrix}1 & 0\\0 & 1\end{bmatrix} = A.\,(\text{adj } A)$$

$$\Rightarrow \quad K = 1$$ **Ans.**

13. *Find the inverse of matrix*

$$A = \begin{bmatrix}-1 & 1 & 2\\3 & -1 & 1\\-1 & 3 & 4\end{bmatrix}$$

Sol. $\therefore\ |A| = -1(-4-3) - 1(12+1) + 2(9-1)$
$= 7 - 13 + 16 = 10 \neq 0$

So, co-factors of A

$$A_{11} = (-1)^2\begin{vmatrix}-1 & 1\\3 & 4\end{vmatrix}$$

$$= -4 - 3 = -7$$

Similarly, $A_{12} = -13$, $A_{13} = 8$, $A_{21} = 2$
$A_{22} = -2$, $A_{23} = 2$, $A_{31} = 3$,
$A_{32} = 7$ and $A_{33} = -2$

$$\text{adj. (A)} = \begin{bmatrix}A_{11} & A_{12} & A_{13}\\A_{21} & A_{22} & A_{23}\\A_{31} & A_{32} & A_{33}\end{bmatrix}^T$$

$$= \begin{bmatrix}-7 & -13 & 8\\2 & -2 & 2\\3 & 7 & -2\end{bmatrix}^T$$

$$= \begin{bmatrix}-7 & 2 & 3\\-13 & -2 & 7\\8 & 2 & -2\end{bmatrix}$$

$$\Rightarrow \quad A^{-1} = \frac{\text{adj }(A)}{|A|}$$

$$\Rightarrow \quad = \frac{1}{10}\begin{bmatrix}-7 & 2 & 3\\-13 & -2 & 7\\8 & 2 & -2\end{bmatrix}$$ **Ans**

14. *Find the inverse of matrix* $\begin{bmatrix}a & b\\c & d\end{bmatrix}$ *given tha ad − bc* $\neq 0$.

Sol. Given, $A = \begin{bmatrix}a & b\\c & d\end{bmatrix}$

$$\therefore \quad |A| = ad - bc \neq 0$$

Hence, A^{-1} exists.

Now, cofactors of A are

$A_{11} = d$, $A_{12} = -c$
$A_{21} = -b$, $A_{22} = a$

$$\therefore \quad \text{adj }(A) = \begin{bmatrix}d & -b\\-c & a\end{bmatrix}$$

$$\therefore \quad A^{-1} = \frac{1}{|A|}.\text{adj }(A)$$

$$\Rightarrow \quad A^{-1} = \frac{1}{ad-bc}\begin{bmatrix}d & -b\\-c & a\end{bmatrix}$$ **Ans**

15. *Given the matrix* $A = \begin{bmatrix}1 & 0 & 2\\-2 & 1 & 0\\0 & -1 & 2\end{bmatrix}$, *compute* A^{-1}

Sol. Given $A = \begin{bmatrix}1 & 0 & 2\\-2 & 1 & 0\\0 & -1 & 2\end{bmatrix}$

Find the Value Type Questions

$\therefore \quad |A| = 1(2-0) + 2(2-0) = 6$

So, co-factors of A

$$A_{11} = \begin{vmatrix} 1 & 0 \\ -1 & 2 \end{vmatrix} = 2 + 0 = 2$$

$$A_{12} = -\begin{vmatrix} -2 & 0 \\ 0 & 2 \end{vmatrix} = -(-4) = 4$$

$$A_{13} = \begin{vmatrix} -2 & 1 \\ 0 & -1 \end{vmatrix} = (2-0) = 2$$

$$A_{21} = -\begin{vmatrix} 0 & 2 \\ -1 & 2 \end{vmatrix} = -(0+2) = -2$$

$$A_{22} = \begin{vmatrix} 1 & 0 \\ 0 & 2 \end{vmatrix} = 2 - 0 = 2$$

$$A_{23} = -\begin{vmatrix} 1 & 0 \\ 0 & -1 \end{vmatrix} = -(-1-0) = 1$$

$$A_{31} = \begin{vmatrix} 0 & 2 \\ 1 & 0 \end{vmatrix} = 0 - 2 = -2$$

$$A_{32} = -\begin{vmatrix} 1 & 2 \\ -2 & 0 \end{vmatrix} = -(0+4) = -4$$

$$A_{33} = \begin{vmatrix} 1 & 0 \\ -2 & 1 \end{vmatrix} = 1 - 0 = 1$$

$$\therefore \text{ adj } A = \begin{bmatrix} 2 & 4 & 2 \\ -2 & 2 & 1 \\ -2 & -4 & 1 \end{bmatrix}^T = \begin{bmatrix} 2 & -2 & -2 \\ 4 & 2 & -4 \\ 2 & 1 & 1 \end{bmatrix}$$

$$\therefore \quad A^{-1} = \frac{\text{adj } A}{|A|} = \frac{1}{6}\begin{bmatrix} 2 & -2 & -2 \\ 4 & 2 & -4 \\ 2 & 1 & 1 \end{bmatrix}$$ **Ans.**

Chapter 6. Differential Calculus

1. *$f(x) = \dfrac{x^2-9}{x-3}$ is not defined at $x = 3$. What value should be assigned to $f(3)$ for continuity of $f(x)$ at $x = 3$?**

Sol. Given, $f(x) = \dfrac{x^2-9}{x-3}$

$$\Rightarrow \quad f(x) = \frac{(x-3)(x+3)}{x-3}$$

$$\Rightarrow \quad \lim_{x\to3} f(x) = \lim_{x\to3} (x+3)$$

$$= \lim_{x\to3} (x+3)$$

$$f(3) = 6$$

Hence, $f(x)$ to be continuous at $x = 3$, $f(3) = 6$. **Ans.**

1. *Find $\dfrac{dy}{dx}$ when, $y = \dfrac{e^{2x}+e^{-2x}}{e^{2x}-e^{-2x}}$.*

Sol. Given,

$$y = \frac{e^{2x}+e^{-2x}}{e^{2x}-e^{-2x}}$$

$$= \frac{e^{2x}+\dfrac{1}{e^{2x}}}{e^{2x}-\dfrac{1}{e^{2x}}}$$

$$y = \frac{e^{4x}+1}{e^{4x}-1}$$

Differentiating w.r.t. x, we get

$$\frac{dy}{dx} = \frac{4e^{4x}(e^{4x}-1)-4e^{4x}(e^{4x}+1)}{(e^{4x}-1)^2}$$

$$= \frac{-8e^{4x}}{(e^{4x}-1)^2}$$ **Ans.**

3. *If $y = \log\sqrt{\dfrac{1-\cos x}{1+\cos x}}$, find $\dfrac{dy}{dx}$.*

Sol. Given, $y = \log\sqrt{\dfrac{1-\cos x}{1+\cos x}}$.

$$\Rightarrow \quad y = \frac{1}{2}\log\frac{1-\cos x}{1+\cos x}$$

$$\therefore \quad y = \frac{1}{2}[\log(1-\cos x) - \log(1+\cos x)]$$

Differentiating both sides w. r. t. x, we get

$$\therefore \quad \frac{dy}{dx} = \frac{1}{2}\left[\frac{\sin x}{1-\cos x} + \frac{\sin x}{1+\cos x}\right]$$

$$\Rightarrow \quad \frac{dy}{dx} = \frac{\sin x}{2}\left[\frac{1}{1-\cos x} + \frac{1}{1+\cos x}\right]$$

$$\frac{dy}{dx} = \frac{\sin x}{2}\left[\frac{1+\cos x+1-\cos x}{1-\cos^2 x}\right]$$

$$\Rightarrow \quad \frac{dy}{dx} = \frac{\sin x}{2}\left[\frac{2}{1-\cos^2 x}\right]$$

$$= \frac{\sin x}{2}\cdot\frac{2}{\sin^2 x}$$

* Frequently asked previous years Board Exam Questions.

$\Rightarrow \quad \frac{dy}{dx} = \frac{1}{\sin x}$

$\therefore \quad \frac{dy}{dx} = \text{cosec } x.$ **Ans.**

4. *If $y = e^x \log \tan 2x$, find $\frac{dy}{dx}$.*

Sol. Given, $y = e^x \log \tan 2x$

Differentiating both sides w.r.t. x, we get

$$\therefore \frac{dy}{dx} = e^x \frac{d}{dx}(\log \tan 2x) + \log \tan 2x \frac{d}{dx}(e^x)$$

$$= e^x \frac{1}{\tan 2x} . 2 \sec^2 2x + e^x \log \tan 2x$$

$$= e^x \left[\frac{2\cos 2x}{\sin 2x} \cdot \frac{1}{\cos^2 2x} + \log \tan 2x\right]$$

$$\left[\because \tan 2x = \frac{\sin 2x}{\cos 2x}\right]$$

$$= e^x \left[\frac{2}{\sin 2x \cos 2x} + \log \tan 2x\right]$$

$$= e^x \left[\frac{4}{\sin 4x} + \log \tan 2x\right]$$

$$[\because \sin 4x = 2 \sin 2x \cdot \cos 2x]$$

$$= e^x [4 \text{ cosec } 4x + \log \tan 2x]$$ **Ans.**

5. *Using a suitable substitution, find the derivative of $\tan^{-1} \sqrt{\frac{a-x}{a+x}}$ with respect to x.*

Sol. Given,

$$y = \tan^{-1} \sqrt{\frac{a-x}{a+x}}$$

Let, $x = a \cos 2\theta$

$$\Rightarrow \quad \theta = \frac{1}{2} \cos^{-1}\left(\frac{x}{a}\right)$$

Substituting value of x in the given equation, we get

$$y = \tan^{-1} \sqrt{\frac{a - a\cos 2\theta}{1 + a\cos 2\theta}}$$

$$\therefore y = \tan^{-1} \sqrt{\frac{1-\cos 2\theta}{1+\cos 2\theta}} = \tan^{-1} \sqrt{\tan^2 \theta}$$

$$\Rightarrow y = \tan^{-1} (\tan \theta)$$

$$\Rightarrow y = \theta = \frac{1}{2} \cos^{-1}\left(\frac{x}{a}\right)$$

Differentiating w.r.t. x, we get

$$\frac{dy}{dx} = -\frac{1}{2} \frac{1}{\sqrt{1 - x^2 / a^2}} \times \frac{1}{a}$$

$$= -\frac{1}{2\sqrt{a^2 - x^2}}$$ **Ans.**

6. *If $x = a (\cos \theta + \theta \sin \theta)$, $y = a (\sin \theta - \theta \cos \theta)$. Then find, $\frac{d}{dx}$ at $\theta = \frac{\pi}{4}$.*

Sol. Given,

$$x = a (\cos \theta + \theta \sin \theta)$$

Differentiating w.r.t. θ, we get

$$\therefore \frac{dx}{d\theta} = a (-\sin \theta + \sin \theta + \theta \cos \theta)$$

$$= a \theta \cos \theta$$

Also, $y = a (\sin \theta - \theta \cos \theta)$

Differentiating w.r.t. θ, we get

$$\therefore \frac{dy}{d\theta} = a (\cos \theta - \cos \theta + \theta \sin \theta)$$

$$= a \theta \sin \theta$$

$$\frac{dy}{dx} = \frac{dy / d\theta}{dx / d\theta} = \frac{a\theta \sin\theta}{a\theta \cos\theta} = \tan \theta$$

At $\theta = \frac{\pi}{4}, \frac{dy}{dx} = \tan \frac{\pi}{4} = 1.$ **Ans.**

7. *If $y = (\cos x)^{\cos x}$, find $\frac{dy}{dx}$.*

Sol. Given, $y = (\cos x)^{\cos x}$

Taking log on both sides, we get

$$\log y = \cos x \log (\cos x)$$

Differentiating both sides w. r. to x, we get

$$\frac{1}{y}\frac{dy}{dx} = - \sin x . \log (\cos x) + \cos x \cdot \frac{1}{\cos x} \cdot (-\sin x)$$

$$= - \sin x [1 + \log (\cos x)]$$

$$\Rightarrow \frac{dy}{dx} = - y . \sin x \; [1 + \log (\cos x)]$$

$$= - (\cos x)^{\cos x} . \sin x \; [1 + \log (\cos x)]$$ **Ans.**

8. *If $y = e^{\sin x^2}$, find $\frac{dy}{dx}$.*

Sol. Given, $y = e^{\sin x^2}$

Differentiating both sides w. r. t. x, we get

$$\frac{dy}{dx} = e^{\sin x^2} \frac{d}{dx} (\sin x^2)$$

$$= e^{\sin x^2} \times \cos x^2 \times \frac{d}{dx} (x^2)$$

$$= e^{\sin x^2} \cos x^2 (2x)$$

$$\frac{dy}{dx} = 2x \cos x^2 e^{\sin x^2}$$ **Ans.**

Find the Value Type Questions

9. *Using L'Hospital's rule, evaluate :* $\lim_{x\to 0}\frac{8^x-4^x}{4x}$.*

Sol. Applying L, Hospital's rule,

$$= \lim_{x\to 0}\frac{8^x-4^x}{4x} \qquad \left[\frac{0}{0}\text{ form}\right]$$

$$= \lim_{x\to 0}\frac{\frac{d}{dx}(8^x-4^x)}{\frac{d}{dx}(4x)}$$

$$= \lim_{x\to 0}\frac{8^x\log 8-4^x\log 4}{4}$$

$$\left\{\because \frac{d(a^x)}{dx}=a^x\log a\right\}$$

$$= \frac{8^0\log 8-4^0\log 4}{4}$$

$$= \frac{\log 8-\log 4}{4}$$

$$= \frac{1}{4}\log\frac{8}{4} = \frac{1}{4}\log 2.$$ **Ans.**

10. *Using L'Hospital's rule, evaluate :**

$$\lim_{x\to 0}\frac{xe^x-\log(1+x)}{x^2}$$

Sol. $\lim_{x\to 0}\frac{xe^x-\log(1+x)}{x^2}$ $\left[\frac{0}{0}\text{ form}\right]$

Differentiating w.r.t. x, on numerator and denominator

$$= \lim_{x\to 0}\frac{e^x+1.e^x+xe^x+\frac{1}{(1+x)^2}}{2} \qquad \left[\frac{0}{0}\text{ form}\right]$$

Again, differentiating w.r.t. x, on numerator and denominator,

$$= \frac{e^0+e^0+0.e^0+\frac{1}{(1+0)^2}}{2}$$

Using limits, $x = 0$

$$\frac{e^0+e^0+0.e^0+\frac{1}{(1+0)^2}}{2} = \frac{1+1+1}{2} = \frac{3}{2}$$ **Ans.**

Chapter 7. Applications of Derivatives

1. *Find the slope of the tangents to the following curves :**

(i) $y = x^2 - 4x + 3$ *at* $x = 5$

(ii) $y = x^3 - 12x + 5$ *at* $x = 0$

Sol. (i) The given curve is $y = x^2 - 4x + 3$
Differentiating w.r.t. x, we get

$$m_1 = \frac{dy}{dx} = 2x-4$$

At $x = 5$, $\frac{dy}{dx} = 2\times 5-4 = 6$ **Ans.**

(ii) The given curve is $y = x^3 - 12x + 5$
Differentiating w.r.t. x, we get

$$m_2 = \frac{dy}{dx} = 3x^2-12$$

At $x = 0$, $\frac{dy}{dx} = 3\times 0-12 = -12$ **Ans.**

2. *The equation of tangent at (2, 3) on the curve* $y^2 = px^3 + q$ *is* $y = 4x - 7$.*
Find the values of 'p' and 'q'.

Sol. Given : Equation of curve is,

$$y^2 = px^3 + q$$

Differentiating both sides w.r.t. x, we get

$$2y\frac{dy}{dx} = 3px^2$$

$$\Rightarrow \quad \frac{dy}{dx} = \frac{3px^2}{2y}$$

$$\therefore \quad \left(\frac{dy}{dx}\right)_{(2,3)} = \frac{3p\times 2^2}{2\times 3}$$

or $$m = \left(\frac{dy}{dx}\right)_{(2,3)} = \frac{12p}{6} = 2p$$

Since, $y = 4x - 7$ is the tangent to the curve at point (2, 3).

So, on comparing with $y = mx + c$, we get

$$m = 4$$

Now, $2p = 4$

$\Rightarrow$ $p = \frac{4}{2} = 2$

Since, point (2, 3) lies on the curve,

$\therefore$ $3^2 = p\times 2^3 + q$

$\Rightarrow$ $9 = 2\times 8 + q$ $[\because p = 2]$

$\Rightarrow$ $9 - 16 = q$

$\Rightarrow$ $q = -7$

Hence, $p = 2$ and $q = -7$ **Ans.**

3. *Find the approximate value of* f *(2·01), where* $f(x) = 4x^2 + 5x + 2$.

Sol. Let $y = f(x) = 4x^2 + 5x + 2$

Then, $\frac{dy}{dx} = 8x + 5$ ...(i)

Let $x = 2$ and $x + \delta x = 2{\cdot}01$

Then, $\delta x = 2{\cdot}01 - 2 = 0{\cdot}01$

$\therefore$ $y + \delta y = f(x + \delta x) = f(2{\cdot}01)$

* Frequently asked previous years Board Exam Questions.

$\Rightarrow \quad \delta y = f(2{\cdot}01) - y$

$\Rightarrow \quad \delta y = f(2{\cdot}01) - (4x^2 + 5x + 2)$

$\Rightarrow \quad \delta y = f(2{\cdot}01) - (4 \times 2 \times 2 + 5 \times 2 + 2) \quad (\because x = 2)$

$\Rightarrow \quad \delta y = f(2{\cdot}01) - 28$

$f(2{\cdot}01) = \delta y + 28 \quad \ldots(ii)$

$\Rightarrow \quad \delta y = \frac{dy}{dx} \cdot \delta x$

$\Rightarrow \quad \delta y = (8x + 5) \,.\, (0{\cdot}01)$

$\Rightarrow \quad \delta y = (16 + 5) \,.\, (0{\cdot}01) \quad (\because x = 2)$

$\Rightarrow \quad \delta y = 0{\cdot}21$

From (ii), $f(2{\cdot}01) = 0{\cdot}21 + 28 = 28{\cdot}21$ **Ans.**

4. *Using differential, find the approximate value of $(0{\cdot}999)^{1/5}$.*

Sol. Let $y = f(x) = x^{1/5}$

then, $\frac{dy}{dx} = \frac{1}{5x^{4/5}} \quad \ldots(i)$

Let $x = 1$ and $x + \delta x = 0{\cdot}999$

$\delta x = -0{\cdot}001$

and $y + \delta y = (x + \delta x)^{1/5} = (0{\cdot}999)^{1/5}$

$\delta y = (0{\cdot}999)^{1/5} - y$

$\delta y = (0{\cdot}999)^{1/5} - x^{1/5}$

$\delta y = (0{\cdot}999)^{1/5} - (1)^{1/5} \quad [\because x = 1]$

$\delta y = (0{\cdot}999)^{1/5} - 1$

$(0{\cdot}999)^{1/5} = \delta y + 1 \quad \ldots(ii)$

Now, $\delta y = \frac{dy}{dx} \delta x$

$= \frac{1}{5x^{4/5}} \cdot (-0{\cdot}001)$

$= \frac{-0{\cdot}001}{5(1)^{4/5}} \quad [\because x = 1]$

$= \frac{-0{\cdot}001}{5} = -0{\cdot}0002$

Putting the value of δy in (ii), we get

$(0{\cdot}999)^{1/5} = -0{\cdot}0002 + 1$

$= 0{\cdot}9998$ **Ans.**

5. *Find the values of x for which $f(x) = [x(x-2)]^2$ is increasing or decreasing.*

Sol. We have,

$f(x) = [x(x-2)]^2$

$= x^2 (x-2)^2, x \in R$

then, $f'(x) = x^2 \,.\, 2(x-2) + (x-2)^2 \cdot 2x$

$= (x-2)(2x^2 + 2x(x-2))$

$= (x-2)(2x^2 + 2x^2 - 4x)$

$= (x-2)(4x^2 - 4x)$

$= 4x(x-1)(x-2)$

Now, $f'(x) = 0$

$4x(x-1)(x-2) = 0$

$\Rightarrow \quad x = 0, 1, 2$

The points 0, 1, 2 split the interval $(-\infty, \infty)$ into disjoint sub-intervals.

$(-\infty, 0)$, $(0, 1)$, $(1, 2)$ and $(2, \infty)$

Interval	Sign of $f'(x)$ $= 4x(x-1)(x-2)$	Nature of function
$(-\infty, 0)$	– ve	strictly decreasing
$(0, 1)$	+ ve	strictly increasing
$(1, 2)$	– ve	strictly decreasing
$(2, \infty)$	+ ve	strictly increasing

It is clear from the table that $f(x)$ is increasing in $[0, 1]$ and $[2, \infty)$ *i.e.*, $0 \le x \le 1$ and $2 \le x < \infty$ and $f(x)$ is decreasing in $(-\infty, 0]$ and $[1, 2]$ *i.e.* $-\infty < x \le 0$ and $1 \le x \le 2$. **Ans.**

6. *Find the intervals in which the function f given by $f(x) = -2x^3 - 9x^2 - 12x + 1$ is strictly decreasing.*

Sol. Given,

$f(x) = -2x^3 - 9x^2 - 12x + 1$

$f'(x) = -6x^2 - 18x - 12$

$= -6(x^2 + 3x + 2)$

$= -6[x^2 + 2x + x + 2]$

$= -6[x(x+2) + 1(x+2)]$

$= -6(x+1)(x+2)$

$f'(x) = 0 \Rightarrow x = -2, -1$

$\therefore$ Intervals are $(-\infty, -2)$, $(-2, -1)$, $(-1, \infty)$

When $x \in (-\infty, -2)$,

$f'(x) = -6(-)(-) = -\text{ve} < 0$

$x \in (-2, -1)$,

$f'(x) = -6(-)(+) = +\text{ve} > 0$

$x \in (-1, \infty)$,

$f'(x) = -6(+)(+) = -\text{ve} < 0$

$f(x)$ is strictly decreasing if $f'(x) < 0$.

$\therefore f(x)$ is strictly decreasing in

$(-\infty, -2) \cup (-1, \infty)$. **Ans.**

7. *Find the interval for which the given function $f(x) = 10 - 6x - 2x^2$ is strictly increasing.*

Find the Value Type Questions

Sol. Given,

$$f(x) = 10 - 6x - 2x^2$$
$$f'(x) = -6 - 4x = -2(2x+3)$$
$$f'(x) = 0 \Rightarrow 2x + 3 = 0 \Rightarrow x = -\frac{3}{2}$$

$\therefore$ Intervals are $\left(-\infty, \frac{-3}{2}\right)$ and $\left(\frac{-3}{2}, \infty\right)$

When $x \in \left(-\infty, \frac{-3}{2}\right)$, $f'(x) = -2(-) = +\text{ve} > 0$

When $x \in \left(\frac{-3}{2}, \infty\right)$, $f'(x) = -2(+) = -\text{ve} < 0$

$\therefore f(x)$ is strictly increasing in the interval $\left(-\infty, -\frac{3}{2}\right)$. **Ans.**

8. *A rectangle is inscribed in a semicircle of radius r with one of its sides on the diameter of the semicircle. Find the dimensions of the rectangle to get maximum area. Also, find the maximum area.*

Sol. Let ABCD be the rectangle of sides x and $2y$ inscribed in the semicircle with centre O and radius r. Let OA = OB = y then AB = $2y$.

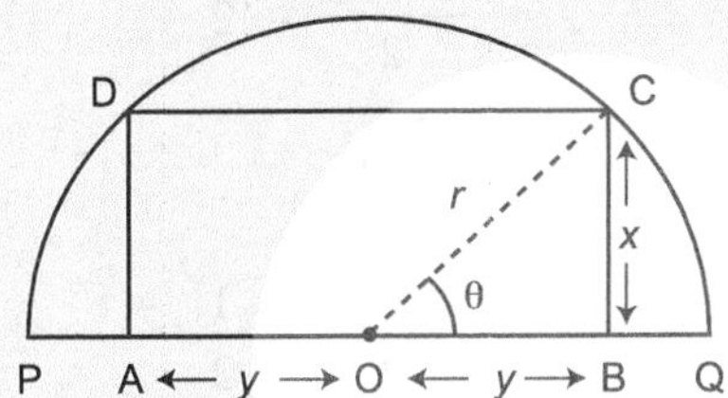

Let θ be the angle between radius of semicircle and side OB of rectangle, then $x = r\sin\theta$, $y = r\cos\theta$.

If A is the area of rectangle ABCD, then

$$A = 2y \cdot x = 2r\cos\theta . r\sin\theta$$

$$A = r^2 \sin 2\theta,$$

Differentiating w.r.t. θ, we get

$$\therefore \quad \frac{dA}{d\theta} = 2r^2 \cos 2\theta \quad \ldots(i)$$

For max. or min. $\frac{dA}{d\theta} = 0 \Rightarrow 2r^2\cos 2\theta = 0$

$$\cos 2\theta = 0 = \cos\frac{\pi}{2} \Rightarrow \theta = \frac{\pi}{4}$$

Again differentiating equation (i) w.r.t. θ, we get

$$\frac{d^2A}{d\theta^2} = -4r^2 \sin 2\theta$$

$$\left(\frac{d^2A}{d\theta^2}\right)_{\theta=\frac{\pi}{4}} = -4r^2 \sin\frac{\pi}{2} = -4r^2$$

Hence, area A is maximum when $\theta = \frac{\pi}{4}$

$\therefore$ For maximum area, $\theta = \frac{\pi}{4}$

and $x = r\sin\theta = r\sin\frac{\pi}{4} = \frac{r}{\sqrt{2}}$

$$2y = 2r\cos\theta$$
$$= 2r\cos\frac{\pi}{4} = 2r \times \frac{r}{\sqrt{2}} = r\sqrt{2}$$

Hence, the dimensions of the rectangle are $\frac{r}{\sqrt{2}}$ and $r\sqrt{2}$ and the area $= \frac{r}{\sqrt{2}} \times r\sqrt{2}$

$= r^2$ square units. **Ans.**

Chapter 8. Integral Calculus

1. *Evaluate :* $\int \frac{\sec^2 x}{\operatorname{cosec}^2 x} dx.$*

Sol. $\int \frac{\sec^2 x}{\operatorname{cosec}^2 x} dx = \int \frac{1}{\cos^2 x} \times \frac{\sin^2 x}{1} dx$

$$= \int \tan^2 x \, dx$$
$$= \int (\sec^2 x - 1)\, dx$$
$$= \int \sec^2 x \, dx - \int 1 \, dx$$
$$= \tan x - x + C$$
Ans.

2. (a) *Evaluate :* $\int \frac{x(1+x^2)}{1+x^4} dx.$*

(b) *Evaluate :* $\int_{-6}^{3} |x+3|\, dx$

Sol. (a) $\int \frac{x(1+x^2)}{1+x^4} dx = \int \frac{x}{1+x^4} dx + \int \frac{x^3}{1+x^4} dx$

Put $x^2 = t$ in I[st] Integral	Put $1 + x^4 = v$ in IInd integral
$2x\, dx = dt$	$4x^3\, dx = dv$
$x\, dx = \frac{dt}{2}$	$x^3\, dx = \frac{dv}{4}$

$$= \frac{1}{2}\int \frac{dt}{1+t^2} + \frac{1}{4}\int \frac{dv}{v}$$

$$= \frac{1}{2}\tan^{-1} t + \frac{1}{4}\log|v| + C$$

* Frequently asked previous years Board Exam Questions.

$$= \frac{1}{2}\tan^{-1}x^2 + \frac{1}{4}\log|1+x^4| + C$$

Ans.

(b) $$\int_{-6}^{3}|x+3|\,dx = \int_{-6}^{-3}|x+3|\,dx + \int_{-3}^{3}|x+3|\,dx$$

$$= -\int_{-6}^{-3}(x+3)dx + \int_{-3}^{3}(x+3)dx$$

$$= -\left[\frac{x^2}{2}+3x\right]_{-6}^{-3} + \left[\frac{x^2}{2}+3x\right]_{-3}^{3}$$

$$= -\left[\left(\frac{9}{2}-9\right)-\left(\frac{36}{2}-18\right)\right] + \left[\left(\frac{9}{2}+9\right)-\left(\frac{9}{2}-9\right)\right]$$

$$= \frac{9}{2}+0+18 = \frac{45}{2}$$ **Ans.**

3. *Evaluate :* $\int_0^{\pi}\frac{x\tan x}{\sec x+\tan x}dx$ *

Sol. Let $$I = \int_0^{\pi}\frac{x\tan x}{\sec x+\tan x}dx \quad \ldots(i)$$

Using $$\int_0^a f(x)dx = \int_0^a f(a-x)dx$$

$$I = \int_0^{\pi}\frac{(\pi-x)\tan(\pi-x)}{\sec(\pi-x)+\tan(\pi-x)}dx$$

$$I = \int_0^{\pi}\frac{(\pi-x)(-\tan x)}{(-\sec x)+(-\tan x)}dx$$

$$\Rightarrow \quad I = \int_0^{\pi}\frac{(\pi-x)\tan x}{\sec x+\tan x}dx \quad \ldots(ii)$$

Adding equations (i) and (ii), we get

$$2I = \int_0^{\pi}\frac{\pi\tan x}{\sec x+\tan x}dx$$

$$\Rightarrow 2I = \pi\int_0^{\pi}\frac{\tan x}{\sec x+\tan x}dx$$

$$\Rightarrow 2I = \pi\int_0^{\pi}\frac{\tan x}{\sec x+\tan x}\times\frac{(\sec x-\tan x)}{(\sec x-\tan x)}dx$$

$$\Rightarrow 2I = \pi\int_0^{\pi}\frac{\tan x(\sec x-\tan x)}{\sec^2 x-\tan^2 x}dx$$

$$\Rightarrow 2I = \pi\int_0^{\pi}\frac{\sec x\tan x-\tan^2 x}{1}dx$$

$[\because \sec^2 x - \tan^2 x = 1]$

$$\Rightarrow 2I = \pi\int_0^{\pi}\sec x\tan x - \sec^2 x + 1\,dx$$

$[\because \tan^2 x = \sec^2 x - 1]$

$$\Rightarrow 2I = \pi\left[\sec x - \tan x + x\right]_0^{\pi}$$

$$= \pi[-1-0+\pi-(1-0+0)]$$

$$\Rightarrow \quad = \pi[\pi-2] \Rightarrow I = \frac{\pi}{2}[\pi-2]$$ **Ans.**

4. *Evaluate :* $\int_4^5|x-5|\,dx$ *

Sol. $$\int_4^5|x-5|\,dx = \int_4^5-(x-5)dx$$

$$= \int_4^5(5-x)dx$$

$$= \left[5x-\frac{x^2}{2}\right]_4^5$$

$$= \left[5\times5-\frac{5\times5}{2}\right]-\left[5\times4-\frac{4\times4}{2}\right]$$

$$= \left[25-\frac{25}{2}\right]-\left[20-\frac{16}{2}\right]$$

$$= \frac{50-25}{2} - 12$$

$$= \frac{25}{2}-12$$

$$= \frac{25-24}{2} = \frac{1}{2}$$ **Ans.**

5. *(a) Evaluate :* $\int\frac{dx}{\sqrt{5x-4x^2}}$ *

(b) Evaluate : $\int\sin^3 x\cos^4 x\,dx$

Sol. (a) $$\int\frac{dx}{\sqrt{5x-4x^2}} = \frac{1}{2}\int\frac{dx}{\sqrt{\frac{5}{4}x-x^2}}$$

$$= \frac{1}{2}\int\frac{dx}{\sqrt{\frac{5}{4}x-x^2+\left(\frac{5}{8}\right)^2-\left(\frac{5}{8}\right)^2}}$$

$$= \frac{1}{2}\int\frac{dx}{\sqrt{\left(\frac{5}{8}\right)^2-\left(x^2-\frac{5}{4}x+\left(\frac{5}{8}\right)^2\right)}}$$

$$= \frac{1}{2}\int\frac{dx}{\sqrt{\left(\frac{5}{8}\right)^2-\left(x-\frac{5}{8}\right)^2}}$$

$$= \frac{1}{2}\sin^{-1}\left(\frac{x-\frac{5}{8}}{\frac{5}{8}}\right)+c$$

* Frequently asked previous years Board Exam Questions.

Find the Value Type Questions

$$\left[\because \int \frac{dx}{\sqrt{a^2-x^2}} = \sin^{-1}\frac{x}{a}+c\right]$$

$$= \frac{1}{2}\sin^{-1}\left(\frac{8x-5}{5}\right)+c \quad \textbf{Ans.}$$

(b) Let $I = \int \sin^3 x \cos^4 x\, dx$

$= \int \sin^2 x. \cos^4 x (\sin x\, dx)$

$= \int (1-\cos^2 x)\cos^4 x (\sin x\, dx)$

Let $\cos x = t$

$\Rightarrow \quad -\sin x\, dx = dt$

or $\quad \sin x\, dx = -dt$

$\therefore \quad I = \int (1-t^2)\, t^4 (-dt)$

$= -\int (t^4 - t^6)\, dt$

$= \int (t^6 - t^4)\, dt$

$= \frac{t^7}{7} - \frac{t^5}{5} + c$

$= \frac{\cos^7 x}{7} - \frac{\cos^5 x}{5} + c$ **Ans.**

Chapter 9. Differential Equations

1. *Find* $\frac{dy}{dx}$ *if* $x^3 + y^3 = 3axy$*

Sol. Given : $x^3 + y^3 = 3axy$

Differentiating w.r.t. x, we get

$$3x^2 + 3y^2\frac{dy}{dx} = 3a\left(1.y + x\frac{dy}{dx}\right)$$

$\Rightarrow \quad x^2 + y^2\frac{dy}{dx} = ay + ax\frac{dy}{dx}$

$\Rightarrow \quad \frac{dy}{dx}(y^2 - ax) = ay - x^2$

$\Rightarrow \quad \frac{dy}{dx} = \frac{ay - x^2}{y^2 - ax}$ **Ans.**

Chapter 11. Vectors

1. $\vec{a}$ *and* $\vec{b}$ *are unit vectors such that* $2\vec{a} - 4\vec{b}$ *and* $10\vec{a} + 8\vec{b}$ *are perpendicular to each other. Find the angle between the vectors* $\vec{a}$ *and* $\vec{b}$.

Sol. Given that the vectors $2\vec{a} - 4\vec{b}$ and $10\vec{a} + 8\vec{b}$ are perpendicular.

$\therefore \quad (2\vec{a} - 4\vec{b}).(10\vec{a} + 8\vec{b}) = 0$

$\Rightarrow \quad 20 + 16\,\vec{a}.\vec{b} - 40\,\vec{b}.\vec{a} - 32 = 0$

$\Rightarrow \quad -24\,\vec{a}\,\vec{b} = 12$

$\Rightarrow \quad \vec{a}.\vec{b} = -\frac{1}{2}$

$\because \quad \cos\theta = \frac{\vec{a}.\vec{b}}{|a||b|}$

and $\quad |a| = |b| = 1$

$\therefore \quad \cos\theta = -\frac{1}{2}$

where θ is angle between $\vec{a}$ and $\vec{b}$,

$\therefore \quad \cos\theta = -\cos\frac{\pi}{3}$

$\Rightarrow \quad \cos\theta = \cos(\pi - \pi/3)$

$\Rightarrow \quad \cos\theta = \cos 2\pi/3$

$\therefore \quad \theta = \frac{2\pi}{3}$ **Ans.**

2. *The vectors* $\vec{a} = 3\hat{i} + x\hat{j} - \hat{k}$ *and* $\vec{b} = 2\hat{i} + \hat{j} + y\hat{k}$ *are mutually perpendicular, given that* $|\vec{a}| = |\vec{b}|$, *find the value of x and y.*

Sol. Consider,

$|\vec{a}| = \sqrt{3^2 + x^2 + (-1)^2} = \sqrt{10 + x^2}$

$|\vec{b}| = \sqrt{2^2 + 1^2 + y^2} = \sqrt{5 + y^2}$

$|\vec{a}| = |\vec{b}|$

Squaring both sides, we get

$\sqrt{10 + x^2} = \sqrt{5 + y^2}$

$10 + x^2 = 5 + y^2$

$y^2 - x^2 = 5 \quad \ldots(i)$

Given, vectors $\vec{a}$ and $\vec{b}$ are mutually perpendicular, hence

$\vec{a}\cdot\vec{b} = 0$

$\Rightarrow (3\hat{i} + x\hat{j} - \hat{k})\cdot(2\hat{i} + \hat{j} + y\hat{k}) = 0$

$\Rightarrow \quad 3\cdot2 + x\cdot1 - 1\cdot y = 0$

$\Rightarrow \quad 6 + x - y = 0$

$\Rightarrow \quad x - y = -6$

$\Rightarrow \quad y - x = 6$

$\Rightarrow \quad y = 6 + x \quad \ldots(ii)$

* Frequently asked previous years Board Exam Questions.

By substituting the value of y in terms of x in equation (i), we get

$$(6 + x)^2 - x^2 = 5$$

$$\Rightarrow 36 + x^2 + 12x - x^2 = 5$$

$$\Rightarrow 36 + 12x = 5$$

$$\Rightarrow 12x = 5 - 36 = -31$$

$$\Rightarrow x = -\frac{31}{12}$$

$$\therefore y = 6 + \left(-\frac{31}{12}\right)$$

$$= \frac{72-31}{12}$$

$$\Rightarrow y = \frac{41}{12}$$ **Ans.**

3. *If $\vec{a}$ and $\vec{b}$ are perpendicular vectors,**
$|\vec{a}+\vec{b}| = 13$ and $|\vec{a}| = 5$, find the value of $|\vec{b}|$

Sol. Given, vectors $\vec{a}$ and $\vec{b}$ are perpendicular.

$$\therefore \vec{a}\cdot\vec{b} = 0$$

Given, $|\vec{a}+\vec{b}| = 13$

On squaring both sides, we get

$$|\vec{a}+\vec{b}|^2 = (13)^2$$

$$\Rightarrow \left(\vec{a}+\vec{b}\right).\left(\vec{a}+\vec{b}\right) = 169$$

$$\Rightarrow |\vec{a}|^2 + |\vec{b}|^2 + 2\vec{a}\cdot\vec{b} = 169$$

$$\Rightarrow 25 + |\vec{b}|^2 + 0 = 169$$

$[\because \text{Given } |\vec{a}| = 5]$

$$\Rightarrow |\vec{b}|^2 = 169 - 25$$

$$\Rightarrow |\vec{b}|^2 = 144$$

$$\Rightarrow |\vec{b}| = \sqrt{144}$$

$$\Rightarrow |\vec{b}| = 12$$ **Ans.**

4. *Find a unit vector perpendicular to each of the vectors $\vec{a} + \vec{b}$ and $\vec{a} - \vec{b}$ where $\vec{a} = 3\hat{i} + 2\hat{j} + 2\hat{k}$ and $\vec{b} = \hat{i} + 2\hat{j} - 2\hat{k}$.*

Sol. Given, $\vec{a} = 3\hat{i} + 2\hat{j} + 2\hat{k}$

and $\vec{b} = \hat{i} + 2\hat{j} - 2\hat{k}$

Now, $\vec{a} + \vec{b} = (3\hat{i} + 2\hat{j} + 2\hat{k}) + (\hat{i} + 2\hat{j} - 2\hat{k})$

$$= 4\hat{i} + 4\hat{j} + 0\hat{k}$$

$$\vec{a} - \vec{b} = (3\hat{i} + 2\hat{j} + 2\hat{k}) - (\hat{i} + 2\hat{j} - 2\hat{k})$$

$$= 2\hat{i} + 0\hat{j} + 4\hat{k}$$

* Frequently asked previous years Board Exam Questions.

We know that the unit vector perpendicular to both $\vec{a} + \vec{b}$ and $\vec{a} - \vec{b}$ is given by

$$\hat{n} = \frac{(\vec{a}+\vec{b})\times(\vec{a}-\vec{b})}{|(\vec{a}+\vec{b})\times(\vec{a}-\vec{b})|}$$

Here, $(\vec{a} + \vec{b}) \times (\vec{a} - \vec{b})$

$$= \begin{vmatrix} \hat{i} & \hat{j} & \hat{k} \\ 4 & 4 & 0 \\ 2 & 0 & 4 \end{vmatrix}$$

$$= \hat{i}\,(16 - 0) - \hat{j}\,(16 - 0) + \hat{k}\,(0 - 8)$$

$$= 16\hat{i} - 16\hat{j} - 8\hat{k}$$

Now, $|(\vec{a}+\vec{b})\times(\vec{a}-\vec{b})|$

$$= \sqrt{(16)^2 + (16)^2 + (8)^2}$$

$$= \sqrt{256 + 256 + 64}$$

$$= \sqrt{576} = 24$$

Hence, the required unit vector

$$\hat{n} = \frac{(\vec{a}+\vec{b})\times(\vec{a}-\vec{b})}{|(\vec{a}+\vec{b})\times(\vec{a}-\vec{b})|}$$

$$= \frac{16\hat{i} - 16\hat{j} - 8\hat{k}}{24}$$

$$= \frac{2}{3}\hat{i} - \frac{2}{3}\hat{j} - \frac{1}{3}\hat{k}$$ **Ans**

5. *The vectors $\hat{i} + 3\hat{j}$, $5\hat{k}$ and $\lambda\hat{i} - \hat{j}$ are coplanar. Find the value of λ.*

Sol. Given that the vectors $\hat{i} + 3\hat{j}$ and $5\hat{k}$, $\lambda\hat{i} - \hat{j}$ are coplanar. Condition when three vectors $\vec{a}$, $\vec{b}$ and $\vec{c}$ are coplanar is given as,

$$\vec{a}\cdot(\vec{b} \times \vec{c}) = 0$$

$$\therefore \begin{vmatrix} 1 & 3 & 0 \\ 0 & 0 & 5 \\ \lambda & -1 & 0 \end{vmatrix} = 0$$

Expanding along C_1,

$$\Rightarrow 1\,\{0 - (-5)\} + \lambda\,(15) = 0$$

$$\Rightarrow 5 + \lambda \times 15 = 0$$

$$\Rightarrow 15\lambda = -5$$

$$\Rightarrow \lambda = -\frac{1}{3}$$ **Ans**

6. *Find the value of λ, so that the vectors $2\hat{i} - 3\hat{j} + 3\hat{k}$, $\hat{i} + 2\hat{j} - 3\hat{k}$ and $\hat{j} + \lambda\hat{k}$ are coplanar.*

Sol. Let
$$\vec{a} = 2\hat{i} - 3\hat{j} + \hat{k}$$
$$\vec{b} = \hat{i} + 2\hat{j} - 3\hat{k}$$
$$\vec{c} = \hat{j} + \lambda\hat{k}$$

For $\vec{a}, \vec{b}$ and $\vec{c}$ to be coplanar,
$$\vec{a} \cdot (\vec{b} \times \vec{c}) = 0$$
$$\Rightarrow \begin{vmatrix} 2 & -3 & 1 \\ 1 & 2 & -3 \\ 0 & 1 & \lambda \end{vmatrix} = 0$$
$$\Rightarrow 2(2\lambda + 3) + 3(\lambda - 0) + 1(1 - 0) = 0$$
$$\Rightarrow 4\lambda + 6 + 3\lambda + 1 = 0$$
$$\Rightarrow 7\lambda = -7$$
$$\Rightarrow \lambda = -1$$
Ans.

7. *Find the volume of a parallelopiped whose edges are represented by the vectors :*
$\vec{a} = 2\hat{i} - 3\hat{j} - 4\hat{k}$, $\vec{b} = \hat{i} + 2\hat{j} - \hat{k}$ and $\vec{c} = 3\hat{i} + \hat{j} + 2\hat{k}$

Sol. Given,
$$\vec{a} = 2\hat{i} - 3\hat{j} - 4\hat{k}$$
$$\vec{b} = \hat{i} + 2\hat{j} - \hat{k}$$
$$\vec{c} = 3\hat{i} + \hat{j} + 2\hat{k}$$
$$[\vec{a}\,\vec{b}\,\vec{c}] = \begin{vmatrix} 2 & -3 & -4 \\ 1 & 2 & -1 \\ 3 & 1 & 2 \end{vmatrix}$$

Expanding along first row, we get
$$\Rightarrow 2\begin{vmatrix} 2 & -1 \\ 1 & 2 \end{vmatrix} - (-3)\begin{vmatrix} 1 & -1 \\ 3 & 2 \end{vmatrix} + (-4)\begin{vmatrix} 1 & 2 \\ 3 & 1 \end{vmatrix}$$
$$= 2(4 + 1) + 3(2 + 3) - 4(1 - 6)$$
$$= 2 \times 5 + 3 \times 5 - 4(-5)$$
$$= 10 + 15 + 20$$
$$= 45 \text{ cu. units.}$$
Ans.

8. *Find the value of λ for which the four points with position vectors $6\hat{i} - 7\hat{j}$, $16\hat{i} - 19\hat{j} - 4\hat{k}$, $\lambda\hat{j} - 6\hat{k}$ and $2\hat{i} - 5\hat{j} + 10\hat{k}$ are coplanar.*

Sol. Let A, B, C, D be the given points. Then,
$$\vec{AB} = \text{Position vector of B} - \text{Position vector of A}$$
$$= (16\hat{i} - 19\hat{j} - 4\hat{k}) - (6\hat{i} - 7\hat{j})$$
$$= 10\hat{i} - 12\hat{j} - 4\hat{k}$$
$$\vec{AC} = \text{Position vector of C} - \text{Position vector of A}$$
$$= (\lambda\hat{j} - 6\hat{k}) - (6\hat{i} - 7\hat{j})$$
$$= -6\hat{i} + (\lambda + 7)\hat{j} - 6\hat{k}$$
$$\vec{AD} = \text{Position vector of D} - \text{Position vector of A}$$
$$= (2\hat{i} - 5\hat{j} + 10\hat{k}) - (6\hat{i} - 7\hat{j})$$
$$= -4\hat{i} + 2\hat{j} + 10\hat{k}$$

Given points are coplanar, if vectors $\vec{AB}, \vec{AC}, \vec{AD}$ are coplanar.

i.e., $[\vec{AB}, \vec{AC}, \vec{AD}] = 0$

Now,
$$[\vec{AB}, \vec{AC}, \vec{AD}] = \begin{vmatrix} 10 & -12 & -4 \\ -6 & \lambda+7 & -6 \\ -4 & 2 & 10 \end{vmatrix}$$
$$\Rightarrow 10(10\lambda + 70 + 12) + 12(-60 - 24) - 4(-12 + 4\lambda + 28) = 0$$
$$\Rightarrow 10(10\lambda + 82) + 12(-84) - 4(16 + 4\lambda) = 0$$
$$\Rightarrow 100\lambda + 820 - 1008 - 64 - 16\lambda = 0$$
$$\Rightarrow 84\lambda = 252$$
$$\Rightarrow \lambda = \frac{252}{84}$$
$$\Rightarrow \lambda = 3$$
Ans.

Chapter 12. Three Dimensional Geometry

1. *Find the angle between the pairs of lines*
$$\frac{x-2}{2} = \frac{y-1}{5} = \frac{z+3}{-3} \text{ and } \frac{x+2}{-1} = \frac{y-4}{8} = \frac{z-5}{4}$$

Sol. Let θ be the angle between two lines.

The direction ratios of given lines are 2, 5, – 3 and – 1, 8, 4.

The angle between two lines is
$$\cos\theta = \frac{a_1a_2 + b_1b_2 + c_1c_2}{\sqrt{a_1^2 + b_1^2 + c_1^2}\sqrt{a_2^2 + b_2^2 + c_2^2}}$$
$$= \frac{2 \times (-1) + 5 \times 8 + (-3) \times 4}{\sqrt{2^2 + 5^2 + (-3)^2}\sqrt{(-1)^2 + 8^2 + 4^2}}$$

$$\cos\theta = \frac{-2+40-12}{\sqrt{4+25+9}\sqrt{1+64+16}}$$

$$= \frac{26}{\sqrt{38}\sqrt{81}}$$

$$= \frac{26}{9\sqrt{38}}$$

$$\Rightarrow \quad \theta = \cos^{-1}\left(\frac{26}{9\sqrt{38}}\right)$$ **Ans.**

2. *Find the equation of a line passing through the points P (– 1, 3, 2) and Q (– 4, 2, – 2). Also, if the point R (5, 5, λ) is collinear with the points P and Q, then find the value of λ.*

Sol. Equation of the line passing through the points P (– 1, 3, 2) and Q (– 4, 2, – 2) is

$$\Rightarrow \quad \frac{x+1}{-4+1} = \frac{y-3}{2-3} = \frac{z-2}{-2-2}$$

$$\Rightarrow \quad \frac{x+1}{-3} = \frac{y-3}{-1} = \frac{z-2}{-4}$$

∵ Point R (5, 5, λ) lies on it,

$$\frac{5+1}{3} = \frac{5-3}{1} = \frac{\lambda-2}{4}$$

$$\Rightarrow \quad \frac{\lambda-2}{4} = 2$$

$$\Rightarrow \quad \lambda = 10$$ **Ans.**

3. *Find the image of the point (2, – 1, 5) in the line*

$$\frac{x-11}{10} = \frac{y+2}{-4} = \frac{z+8}{-11}.$$

Also, find the length of the perpendicular from the point (2, – 1, 5) to the line.

Sol. Let P (2, – 1, 5) be the given point and AB we the given line

$$\frac{x-11}{10} = \frac{y+2}{-4} = \frac{z+8}{-11}. = r \text{ (say)}$$

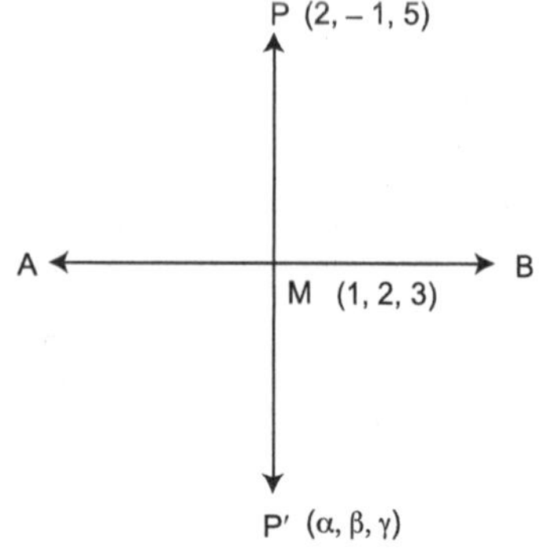

Draw PM ⊥ AB. Produce PM to P′ such that PM = MP′. Then P′ is image of P in AB.

Any point on AB is M (10*r* + 11, – 4*r* – 2, – 11*r* – 8) ...(i)

Then, DR's of MP are 10*r* + 11 – 2, –4*r* – 2 – (–1), –11*r* – 8 – 5

i.e., 10*r* + 9, – 4*r* – 1, – 11*r* – 13.

and the D.C.'s of AB are proportional to 10, – 4, –11.

$$\therefore \quad MP \perp AB$$

$$\therefore 10(10r+9) - 4(-4r-1) - 11(-11r-13) = 0$$

$$\Rightarrow \quad r = -1$$

From equation (i), M is (1, 2, 3)

Let P′ be the point (α, β, γ).

Since M (1, 2, 3) is the mid point of PP′

$$\therefore \quad 1 = \frac{\alpha+2}{2}, 2 = \frac{\beta-1}{2}, 3 = \frac{\gamma+5}{2}$$

$$\Rightarrow \quad \alpha = 0, \beta = 5, \gamma = 1$$

$$\therefore \quad P' \equiv (0, 5, 1)$$

Thus, P′ ≡ (0, 5, 1) is the image of P (2, – 1, 5) in the line AB.

Also, $$MP = \sqrt{(1-2)^2 + (2+1)^2 + (3-5)^2}$$

$$= \sqrt{1+9+4} = \sqrt{14} \text{ units.}$$ **Ans.**

4. *Find the shortest distance between the lines whose vector equations are :*

$$\vec{r} = (\hat{i} + 2\hat{j} + 3\hat{k}) + \lambda(\hat{i} - 3\hat{j} + 2\hat{k})$$

$$\vec{r} = (4\hat{i} + 5\hat{j} + 6\hat{k}) + \mu(2\hat{i} + 3\hat{j} + \hat{k})$$

Sol. The vector equations of given lines are

$$\vec{r} = (\hat{i} + 2\hat{j} + 3\hat{k}) + \lambda(\hat{i} - 3\hat{j} + 2\hat{k}) \quad \text{...(i)}$$

$$\vec{r} = (4\hat{i} + 5\hat{j} + 6\hat{k}) + \mu(2\hat{i} + 3\hat{j} + \hat{k}) \quad \text{...(ii)}$$

Comparing equations (i) and (ii) with

$\vec{r} = \vec{a}_1 + \lambda\vec{b}_1$ and $\vec{r} = \vec{a}_2 + \mu\vec{b}_2$, we get

$$\vec{a}_1 = \hat{i} + 2\hat{j} + 3\hat{k},$$

$$\vec{b}_1 = \hat{i} - 3\hat{j} + 2\hat{k}$$

$$\vec{a}_2 = 4\hat{i} + 5\hat{j} + 6\hat{k},$$

$$\vec{b}_2 = 2\hat{i} + 3\hat{j} + \hat{k}$$

$$\vec{a_2} - \vec{a_1} = (4\hat{i} + 5\hat{j} + 6\hat{k}) - (\hat{i} + 2\hat{j} + 3\hat{k})$$

$$= 3\hat{i} + 3\hat{j} + 3\hat{k}$$

$$\vec{b_1} \times \vec{b_2} = \begin{vmatrix} \hat{i} & \hat{j} & \hat{k} \\ 1 & -3 & 2 \\ 2 & 3 & 1 \end{vmatrix}$$

$$= \hat{i}(-3-6) - \hat{j}(1-4) + \hat{k}(3+6)$$

$$= -9\hat{i} + 3\hat{j} + 9\hat{k}$$

$$|\vec{b_1} \times \vec{b_2}| = \sqrt{(-9)^2 + 3^2 + 9^2}$$

$$= \sqrt{81+9+81} = \sqrt{171}$$

$$(\vec{b_1} \times \vec{b_2}).(\vec{a_2} - \vec{a_1})$$

$$= (-9\hat{i} + 3\hat{j} + 9\hat{k}).(3\hat{i} + 3\hat{j} + 3\hat{k})$$

$$= -27 + 9 + 27 = 9$$

Shortest distance between lines

$$d = \left| \frac{(\vec{b_1} \times \vec{b_2}).(\vec{a_2} - \vec{a_1})}{|\vec{b_1} \times \vec{b_2}|} \right|$$

$$= \left| \frac{9}{\sqrt{171}} \right| = \frac{9}{\sqrt{171}} \text{ units.}$$ **Ans.**

5. *Find the shortest distance between the lines*

$$\vec{r} = (\hat{i} + 2\hat{j} + \hat{k}) + \lambda(\hat{i} - \hat{j} + \hat{k})$$

and $\vec{r} = (2\hat{i} - \hat{j} - \hat{k}) + \mu(2\hat{i} + \hat{j} + 2\hat{k})$.

Sol. Comparing the given equations with

$$\vec{r} = \vec{a_1} + \lambda \vec{b_1}$$

and $\vec{r} = \vec{a_2} + \mu \vec{b_2}$

$$\vec{a_1} = \hat{i} + 2\hat{j} + \hat{k}$$

$$\vec{b_1} = \hat{i} - \hat{j} + \hat{k}$$

$$\vec{a_2} = 2\hat{i} - \hat{j} - \hat{k}$$

$$\vec{b_2} = 2\hat{i} + \hat{j} + 2\hat{k}$$

$$\vec{a_2} - \vec{a_1} = (2\hat{i} - \hat{j} - \hat{k}) - (\hat{i} + 2\hat{j} + \hat{k})$$

$$= \hat{i} - 3\hat{j} - 2\hat{k}$$

$$\vec{b_1} \times \vec{b_2} = \begin{vmatrix} \hat{i} & \hat{j} & \hat{k} \\ 1 & -1 & 1 \\ 2 & 1 & 2 \end{vmatrix}$$

$$= \hat{i}(-2-1) - \hat{j}(2-2) + \hat{k}(1+2)$$

$$= -3\hat{i} - 0\hat{j} + 3\hat{k}$$

$$= -3\hat{i} + 3\hat{k}$$

$$|\vec{b_1} \times \vec{b_2}| = \sqrt{(-3)^2 + (3)^2}$$

$$= \sqrt{9+9} = \sqrt{18} = 3\sqrt{2}$$

$$(\vec{b_1} \times \vec{b_2}).(\vec{a_2} - \vec{a_1})$$

$$= (-3\hat{i} + 3\hat{k}).(\hat{i} - 3\hat{j} - 2\hat{k})$$

$$= -3 - 6 = -9$$

Shortest distance between lines

$$d = \left| \frac{(\vec{b_1} \times \vec{b_2}).(\vec{a_2} - \vec{a_1})}{|\vec{b_1} \times \vec{b_2}|} \right|$$

$$= \left| \frac{-9}{3\sqrt{2}} \right| = \frac{3}{\sqrt{2}} \text{ units.}$$ **Ans**

6. *Find the shortest distance between the lines*

$$\frac{x+1}{7} = \frac{y+1}{-6} = \frac{z+1}{1} \text{ and } \frac{x-3}{1} = \frac{y-5}{-2} = \frac{z-7}{1}.$$

Sol. Comparing the given equations with

$$\frac{x-x_1}{a_1} = \frac{y-y_1}{b_1} = \frac{z-z_1}{c_1}$$

and $\frac{x-x_2}{a_2} = \frac{y-y_2}{b_2} = \frac{z-z_2}{c_2}$

we get $(x_1, y_1, z_1) = (-1, -1, -1)$,

$$(a_1, b_1, c_1) = (7, -6, 1)$$

$$(x_2, y_2, z_2) = (3, 5, 7),$$

$$(a_2, b_2, c_2) = (1, -2, 1)$$

Re-writing the equation in vector form

$$\vec{r} = -\hat{i} - \hat{j} - \hat{k} + \lambda(7\hat{i} - 6\hat{j} + \hat{k})$$

$$\vec{r} = 3\hat{i} + 5\hat{j} + 7\hat{k}, + \mu(\hat{i} - 2\hat{j} + \hat{k})$$

Comparing with

$$\vec{r} = \vec{a_1} + \lambda\vec{b_1} \text{ and } \vec{r} = \vec{a_2} + \mu\vec{b_2}$$

$$\vec{a_1} = -\hat{i} - \hat{j} - \hat{k} \text{ and } \vec{b_1} = 7\hat{i} - 6\hat{j} + \hat{k}$$

$$\vec{a_2} = 3\hat{i} + 5\hat{j} + 7\hat{k} \text{ and } \vec{b_2} = \hat{i} - 2\hat{j} + \hat{k}$$

$$\vec{a_2} - \vec{a_1} = (3\hat{i} + 5\hat{j} + 7\hat{k}) - (-\hat{i} - \hat{j} - \hat{k})$$

$$= 4\hat{i} + 6\hat{j} + 8\hat{k}$$

$$\vec{b}_1 \times \vec{b}_2 = \begin{vmatrix} \hat{i} & \hat{j} & \hat{k} \\ 7 & -6 & 1 \\ 1 & -2 & 1 \end{vmatrix}$$

$$= \hat{i}\,(-6+2) - \hat{j}\,(7-1) + \hat{k}\,(-14+6)$$

$$= -4\hat{i} - 6\hat{j} - 8\hat{k}$$

$$|\vec{b}_1 \times \vec{b}_2| = \sqrt{(-4)^2 + (-6)^2 + (-8)^2}$$

$$= \sqrt{16+36+64}$$

$$= \sqrt{116}$$

$$(\vec{b}_1 \times \vec{b}_2).(\vec{a}_2 - \vec{a}_1) = -16 - 36 - 64 = -116$$

Shortest distance between lines

$$d = \left| \frac{(\vec{b}_1 \times \vec{b}_2).(\vec{a}_2 - \vec{a}_1)}{|\vec{b}_1 \times \vec{b}_2|} \right|$$

$$= \left| \frac{-116}{\sqrt{116}} \right|$$

$$= \sqrt{116}$$

$$= 2\sqrt{29} \text{ units}$$ **Ans.**

7. *Find the shortest distance between the lines*

$\vec{r} = \hat{i} + 2\hat{j} + 3\hat{k} + \lambda\,(2\hat{i} + 3\hat{j} + 4\hat{k})$ *and*

$\vec{r} = 2\hat{i} + 4\hat{j} + 5\hat{k} + \mu\,(4\hat{i} + 6\hat{j} + 8\hat{k})$

Sol. Let l_1 and l_2 be the given lines whose equations are

$$\vec{r} = \vec{a}_1 + \lambda \vec{b}_1$$

and $$\vec{r} = \vec{a}_2 + \mu \vec{b}_2$$

Since the lines are parallel to each other

$$\vec{b}_1 = 2\hat{i} + 2\hat{j} + 4\hat{k}$$

$$\vec{b}_2 = 4\hat{i} + 6\hat{j} + 8\hat{k}$$

$$\vec{a}_2 - \vec{a}_1 = (2\hat{i} + 4\hat{j} + 5\hat{k}) - (\hat{i} + 2\hat{j} + 3\hat{k})$$

$$= \hat{i} + 2\hat{j} + 2\hat{k}$$

Hence, the distance between the lines using the formula :

$$\frac{|\vec{b}(a_2 - a_1)|}{|\vec{b}|} = \frac{|(2\hat{i}+3\hat{j}+4\hat{k}) \times (\hat{i}+2\hat{j}+2\hat{k})|}{|\vec{b}|}$$

$$= \frac{|(2\hat{i}+3\hat{j}+4\hat{k}) \times (\hat{i}+2\hat{j}+2\hat{k})|}{|2\hat{i}+3\hat{j}+4\hat{k}|}$$

$$= \frac{\begin{vmatrix} \hat{i} & \hat{j} & \hat{k} \\ 2 & 3 & 4 \\ 1 & 2 & 2 \end{vmatrix}}{\sqrt{4+9+16}} = \frac{|-2\hat{i}+\hat{k}|}{\sqrt{29}}$$

$$= \frac{\sqrt{5}}{\sqrt{29}} = \frac{\sqrt{5}}{\sqrt{29}} = \sqrt{\frac{5}{29}}$$ **Ans.**

[**Note :** b_1 and b_2 are parallel as $b_2 = 2b_1$ and hence in the aforesoid formula b can be substituted by either of the two and answer remain unaltered.]

8 *Find the Cartesian equation of the plane, passing through the line of intersection of the plane* $\vec{r} \cdot (2\hat{i} + 3\hat{j} - 4\hat{k}) + 5 = 0$ *and* $\vec{r} \cdot (\hat{i} - 5\hat{j} + 7\hat{k}) + 2 = 0$ *and intersecting y-axis at (0, 3).*

Sol. Any plane passing through the intersection of the given planes is

$$[\vec{r}.(2\hat{i} + 3\hat{j} - 4\hat{k}) + 5] + \lambda\,[\vec{r}.(\hat{i} - 5\hat{j} + 7\hat{k}) + 2] = 0$$

$$\vec{r}\,[(2+\lambda)\,\hat{i} + (3-5\lambda)\,\hat{j} + (-4+7\lambda)\hat{k}]$$
$$= -5 - 2\lambda \quad \text{...(i)}$$

Given equation (i) intersects the y-axis at (0, 3) so at y axis

coordinate of $x = 0$

coordinate of $y = 3$

coordinate of $z = 0$

As at y axis coordinate of x and z will be 0.

$$\therefore \quad \vec{r} = 0\hat{i} + 3\hat{j} + 0\hat{k}$$

$$(0\hat{i} + 3\hat{j} + 0\hat{k}).\,[(2+\lambda)\,\hat{i} + (3-5\lambda)\,\hat{j} + (-4+7\lambda)\,\hat{k}]$$
$$= -5 - 2\lambda$$

$$0\,(2+\lambda) + 3\,(3-5\lambda) + 0\,(-4+7\lambda) + 5 + 2\lambda$$
$$= 0$$

$$9 - 15\lambda + 5 + 2\lambda = 0$$

$$\lambda = \frac{14}{13}$$

$\therefore$ The required plane is

$$\left(2+\frac{14}{13}\right)x + \left(3-5\times\frac{14}{13}\right)y + \left(-4+7\times\frac{14}{13}\right)z$$

$$= -5 - 2 \times \left(\frac{14}{13}\right)$$

$$\frac{(26+14)}{13}x + \left(\frac{(39-70)}{13}\right)y + \frac{(-52+98)}{13}z$$

$$+\frac{65+28}{13} = 0$$

$40x - 31y + 46z + 93 = 0$ is the required equation.

Ans.

9. *Find the equation of the plane passing through the points (2, –3, 1) and (– 1, 1, – 7) and perpendicular to the plane $x - 2y + 5z + 1 = 0$.*

Sol. Required plane is perpendicular to the given plane

$$x - 2y + 5z + 1 = 0$$

∵ Required plane is parallel to the line which is perpendicular to the given plane. Direction ratio of line, $a = 1, b = -2, c = 5$.

Hence, required plane is

$$\begin{vmatrix} x-x_1 & y-y_1 & z-z_1 \\ x_2-x_1 & y_2-y_1 & z_2-z_1 \\ a & b & c \end{vmatrix} = 0$$

$$\Rightarrow \begin{vmatrix} x-2 & y+3 & z-1 \\ -1-2 & 1+3 & -7-1 \\ 1 & -2 & 5 \end{vmatrix} = 0$$

$$\Rightarrow \begin{vmatrix} x-2 & y+3 & z-1 \\ -3 & 4 & -8 \\ 1 & -2 & 5 \end{vmatrix} = 0$$

$\Rightarrow (x-2)(20-16) - (y+3)(-15+8)$
$+ (z-1)(6-4) = 0$

$\Rightarrow (x-2)4 - (y+3)(-7) + (z-1)2 = 0$

$\Rightarrow \quad 4x + 7y + 2z + 11 = 0$ **Ans.**

10. *Find the equation of the plane passing through the point (1, – 2, 1) and perpendicular to the line joining the points A(3, 2, 1) and B(1, 4, 2).*

Sol. Equation of line AB is

$$\frac{x-3}{1-3} = \frac{y-2}{4-2} = \frac{z-1}{2-1}$$

$$\frac{x-3}{-2} = \frac{y-2}{2} = \frac{z-1}{1} = r \text{ (say)}$$

Direction ratios of the line are – 2, 2, 1. Since the line is perpendicular to the required plane. Therefore, direction ratios of a normal to the plane are proportional to – 2, 2, 1.

Hence, the equation of the plane passing through the point (1, – 2, 1) and having the line as normal is,

$$-2(x-1) + 2(y+2) + 1(z-1) = 0$$
$$-2x + 2 + 2y + 4 + z - 1 = 0$$
$$-2x + 2y + z + 5 = 0$$

or $2x - 2y - z - 5 = 0$ is the required equation of plane. **Ans.**

11. *Find the equation of the plane through the intersection of the planes $3x - y + 2z - 4 = 0$ and $x + y + z - 2 = 0$ and the point (2, 2, 1).*

Sol. The equation of plane through the intersection of planes

$3x - y + 2z - 4 = 0$ and $x + y + z - 2 = 0$ is

$3x - y + 2z - 4 + k(x + y + z - 2) = 0$ ...(i)

It passes through the point (2, 2, 1)

$(6 - 2 + 2 - 4) + k(2 + 2 + 1 - 2) = 0$

$$2 + 3k = 0 \Rightarrow k = -\frac{2}{3}$$

Put value of k in equation (i), we get

$$(3x - y + 2z - 4) - \frac{2}{3}(x + y + z - 2) = 0$$

$\Rightarrow 3(3x - y + 2z - 4) - 2(x + y + z - 2) = 0$

$\Rightarrow 9x - 3y + 6z - 12 - 2x - 2y - 2z + 4 = 0$

$\Rightarrow \quad 7x - 5y + 4z - 8 = 0$ **Ans.**

12. *Find the equation of plane through the line of intersection of plane $x + y + z = 1$ and $2x + 3y + 4z = 5$ which is perpendicular to the plane $x - y + z = 0$.*

Sol. The equation of plane passing through the intersection of planes $x + y + z - 1 = 0$ and $2x + 3y + 4z - 5 = 0$ is

$(x + y + z - 1) + k(2x + 3y + 4z - 5) = 0$ ...(i)

$(1 + 2k)x + (1 + 3k)y + (1 + 4k)z + (-1 - 5k) = 0$

It is perpendicular to the plane

$$x - y + z = 0$$

Then, $a_1a_2 + b_1b_2 + c_1c_2 = 0$

$\Rightarrow (1 + 2k).1 + (1 + 3k)(-1) + (1 + 4k).1 = 0$

$\therefore \quad (1 + 2k).1 - (1 + 3k) + (1 + 4k).1 = 0$

$\Rightarrow \quad 1 + 2k - 1 - 3k + 1 + 4k = 0$

$\Rightarrow \quad 3k + 1 = 0$

$$\Rightarrow \quad k = -\frac{1}{3}$$

Put value of k in equation (i), we get

$$(x + y + z - 1) - \frac{1}{3}(2x + 3y + 4z - 5) = 0$$

$\Rightarrow \quad 3(x + y + z - 1) - 2x - 3y - 4z + 5 = 0$

$\Rightarrow \quad x - z + 2 = 0$

Ans.

13. *Find the angle between the two planes $3x - 6y + 2z = 7$ and $2x + 2y - 2z = 5$.*

Sol. Let θ be the angle between two planes

Comparing given planes with $a_1x + b_1y + c_1z + d_1 = 0$ and $a_2x + b_2y + c_2z + d_2 = 0$

We get, $a_1 = 3, b_1 = -6, c_1 = 2$

$a_2 = 2, b_2 = 2, c_2 = -2$

The angle between two planes is

$$\cos\theta = \frac{a_1a_2 + b_1b_2 + c_1c_2}{\sqrt{a_1^2 + b_1^2 + c_1^2}\sqrt{a_2^2 + b_2^2 + c_2^2}}$$

$$= \frac{3\times2+(-6)\times2+2\times(-2)}{\sqrt{9+36+4}\sqrt{4+4+4}}$$

$$= \frac{6-12-4}{\sqrt{49}\sqrt{12}} = \frac{-10}{7\times2\sqrt{3}} = \frac{-5}{7\sqrt{3}}$$

$\therefore \quad \theta = \cos^{-1}\left(\frac{-5}{7\sqrt{3}}\right)$. **Ans.**

14. *Find the equation of the plane which contains the line of intersection of the planes $\vec{r}\cdot(\hat{i} + 2\hat{j} + 3\hat{k}) - 4 = 0$, $\vec{r}\cdot(2\hat{i} + \hat{j} - \hat{k}) + 5 = 0$ and which is perpendicular to the plane $\vec{r}\cdot(5\hat{i} + 3\hat{j} - 6\hat{k}) + 8 = 0$.*

Sol. The equation of plane containing the line of intersection of the planes

$$\vec{r}\cdot(\hat{i} + 2\hat{j} + 3\hat{k}) = 4$$

and $\quad \vec{r}\cdot(2\hat{i} + \hat{j} - \hat{k}) = -5$

$$\vec{r}\cdot(\vec{r_1} + \lambda\vec{r_2}) = d_1 + \lambda d_2$$

$\vec{r}\cdot[(\hat{i} + 2\hat{j} + 3\hat{k}) + \lambda(2\hat{i} + \hat{j} - \hat{k})] = 4 + \lambda(-5)$ ...(i)

$\Rightarrow \vec{r}\cdot[(1 + 2\lambda)\hat{i} + (2 + \lambda)\hat{j} + (3 - \lambda)\hat{k}] = 4 - 5\lambda$

It is given that plane (i) is perpendicular to the plane

$$\vec{r}\cdot(5\hat{i} + 3\hat{j} - 6\hat{k}) + 8 = 0$$

$$a_1a_2 + b_1b_2 + c_1c_2 = 0$$

$\therefore 5 \times (1 + 2\lambda) + 3 \times (2 + \lambda) + (-6) \times (3 - \lambda) = 0$

$\Rightarrow \quad 5 + 10\lambda + 6 + 3\lambda - 18 + 6\lambda = 0$

$\Rightarrow \quad 19\lambda - 7 = 0$

$\Rightarrow \quad \lambda = \frac{7}{19}$

Put value of λ in equation (i)

$$\vec{r}\cdot[(\hat{i} + 2\hat{j} + 3\hat{k}) + \frac{7}{19}(2\hat{i} + \hat{j} - \hat{k})] = 4 - 5\left(\frac{7}{19}\right)$$

$$\Rightarrow \vec{r}\cdot\frac{[19(\hat{i}+2\hat{j}+3\hat{k})+7(2\hat{i}+2\hat{j}-\hat{k})]}{19} = 4 - \frac{35}{19}$$

$$\Rightarrow \vec{r}\cdot\frac{(19\hat{i}+38\hat{j}+57\hat{k}+14\hat{i}+7\hat{j}-7\hat{k})}{196} = \frac{41}{19}$$

$\Rightarrow \quad \vec{r}\cdot(33\hat{i} + 45\hat{j} + 50\hat{k}) = 41$. **Ans.**

15. *Find the equation of the plane passing through the intersection of the planes :*

$x + y + z + 1 = 0$ and $2x - 3y + 5z - 2 = 0$ and the point $(-1, 2, 1)$.

Sol. The equations of the given planes are

$x + y + z + 1 = 0$ ...(i)

$2x - 3y + 5z - 2 = 0$ ...(ii)

The equation of any plane passing through the line of intersection of the planes (i) and (ii) is

$x + y + z + 1 + k(2x - 3y + 5z - 2) = 0$ ...(iii)

where k is a parameter.

It passes through the point $(-1, 2, 1)$ if

$-1 + 2 + 1 + 1 + k(2 \times (-1) - 3(2) + 5 \times 1 - 2) = 0$

$\Rightarrow \quad 3 + k(-10 + 5) = 0$

$\Rightarrow \quad 3 - 5k = 0$

$\Rightarrow \quad 5k = 3$

$\Rightarrow \quad k = \frac{3}{5}$

Substituting this value of k in (iii),

$\Rightarrow x + y + z + 1 + \frac{3}{5}(2x - 3y + 5z - 2) = 0$

$\Rightarrow 5x + 5y + 5z + 5 + 6x - 9y + 15z - 6 = 0$

$\Rightarrow \quad 11x - 4y + 20z - 1 = 0$

which is the equation of the required plane.

Ans.

Chapter 13. Application of Integrals

1. *Find the area of the region bounded by the ellipse* $\frac{x^2}{4}+\frac{y^2}{9}=1.$

Sol. The given ellipse is

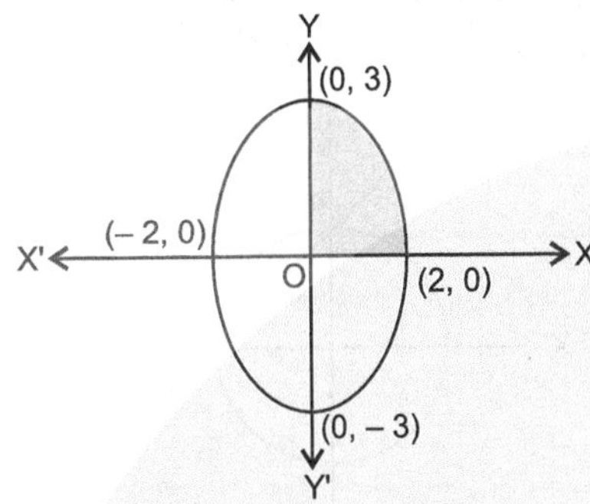

$$\frac{x^2}{4}+\frac{y^2}{9}=1$$

$$\frac{x^2}{2^2}+\frac{y^2}{3^2}=1$$

$\therefore a = 2, b = 3$

$$\frac{y^2}{9}=1-\frac{x^2}{4}=\frac{4-x^2}{4}$$

$$y^2=\frac{9}{4}[4-x2]$$

$$y=\frac{3}{2}\sqrt{4-x^2}=\frac{3}{2}\sqrt{2^2-x^2}$$

Required Area = 4 (Area of shaded region)

$$=4\int_0^2 y\,dx=4\times\frac{3}{2}\int_0^2\sqrt{2^2-x^2}\,dx$$

$$=6\left[\frac{x}{2}\sqrt{4-x^2}+\frac{4}{2}\sin^{-1}\frac{x}{2}\right]_0^2$$

$$=6\,[(0+2\sin^{-1}1)-(0+2.0)]$$

$$=6\left(2\times\frac{\pi}{2}\right)$$

$= 6\pi$ square units. **Ans.**

2. *Using integration find the area of region bounded by triangle whose vertices are A (– 1, 0) B (1, 3) and C (3, 2).*

Sol. The vertices of ΔABC are A (– 1, 0), B (1, 3) and C (3, 2).

Equation of side AB is

$$y-0=\frac{3-0}{1+1}(x+1)$$

$$\Rightarrow \quad y=\frac{3}{2}(x+1)$$

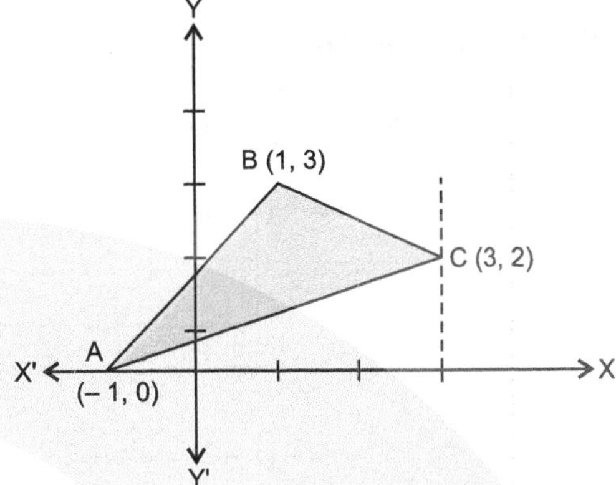

Equation of side BC is

$$y-3=\frac{2-3}{3-1}(x-1)$$

$$\Rightarrow \quad y-3=\frac{-1}{2}(x-1)$$

$$\Rightarrow \quad 2y-6=-x+1$$

$$\Rightarrow \quad 2y=7-x$$

$$\Rightarrow \quad y=\frac{7-x}{2}$$

Equation of side AC is

$$y-0=\frac{2-0}{3+1}(x+1)$$

$$y=\frac{1}{2}(x+1)$$

Area of ΔABC = Area under line AB + Area under line BC – Area under line AC

$$=\int_{-1}^{1}\frac{3}{2}(x+1)\,dx+\int_{1}^{3}\frac{7-x}{2}\,dx-\int_{-1}^{3}\frac{x+1}{2}\,dx$$

$$=\frac{3}{2}\left[\frac{x^2}{2}+x\right]_{-1}^{1}+\frac{1}{2}\left[7x-\frac{x^2}{2}\right]_{1}^{3}-\frac{1}{2}\left[\frac{x^2}{2}+x\right]_{-1}^{3}$$

$$=\frac{3}{2}\left[\left(\frac{1}{2}+1\right)-\left(\frac{1}{2}-1\right)\right]+\frac{1}{2}\left[\left(21-\frac{9}{2}\right)-\left(7-\frac{1}{2}\right)\right]$$

$$-\frac{1}{2}\left[\left(\frac{9}{2}+3\right)-\left(\frac{1}{2}-1\right)\right]$$

$$=\frac{3}{2}(2)+\frac{1}{2}(14-4)-\frac{1}{2}(4+4)$$

$= 3 + 5 – 4 = 4$ square units. **Ans.**

3. *Using integration find the area of region bounded by the triangle whose vertices are (1, 0), (2, 2) and (3, 1).*

Sol. Let the vertices of ABC

A (1, 0), B (2, 2) and C (3, 1)

Equation of side AB is

$$y - 0 = \frac{2-0}{2-1}(x-1)$$

$$\Rightarrow \quad y = 2(x-1)$$

Equation of side BC is

$$y - 2 = \frac{1-2}{3-2}(x-2)$$

$$= -(x-2) = -x+2$$

$$\Rightarrow \quad y = 4 - x$$

Equation of side AC is

$$y - 0 = \frac{1-0}{3-1}(x-1)$$

$$\Rightarrow \quad y = \frac{1}{2}(x-1)$$

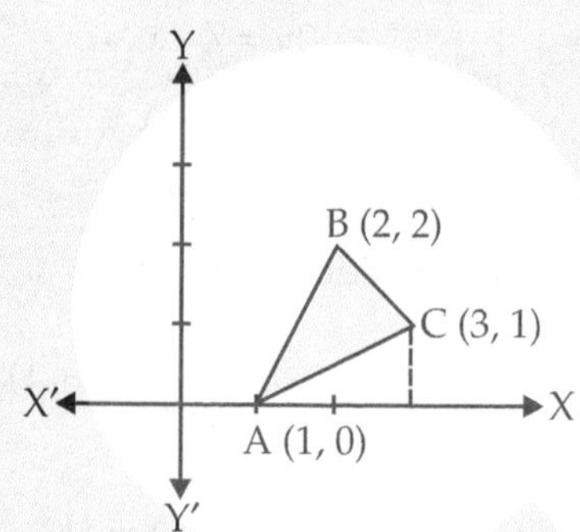

Area of ΔABC = Area under line AB + Area under line BC – Area under line AC

$$= \int_1^2 2(x-1)dx + \int_2^3 (4-x)dx - \int_1^3 \frac{1}{2}(x-1)dx$$

$$= 2\left[\frac{x^2}{2} - x\right]_1^2 + \left[4x - \frac{x^2}{2}\right]_2^3 - \frac{1}{2}\left[\frac{x^2}{2} - x\right]_1^3$$

$$= 2\left[(2-2) - \left(\frac{1}{2} - 1\right)\right] + \left[\left(12 - \frac{9}{2}\right) - (8-2)\right] - \frac{1}{2}\left[\left(\frac{9}{2} - 3\right) - \left(\frac{1}{2} - 1\right)\right]$$

$$= 2\left(0 + \frac{1}{2}\right) + 12 - \frac{9}{2} - 6 - \frac{1}{2}\left(\frac{3}{2} + \frac{1}{2}\right)$$

$$= 1 + 12 - \frac{9}{2} - 6 - 1$$

$$= 6 - \frac{9}{2} = \frac{12-9}{2}$$

$$= \frac{3}{2} \text{ sq. units.}$$ **Ans**

4. *Find the smaller area enclosed by the circle $x^2 + y^2 = 4$ and line $x + y = 2$.*

Sol. The given circle is $x^2 + y^2 = 4 = 2^2$ with centre (0, 0) and radius = 2 and line is $x + y = 2$

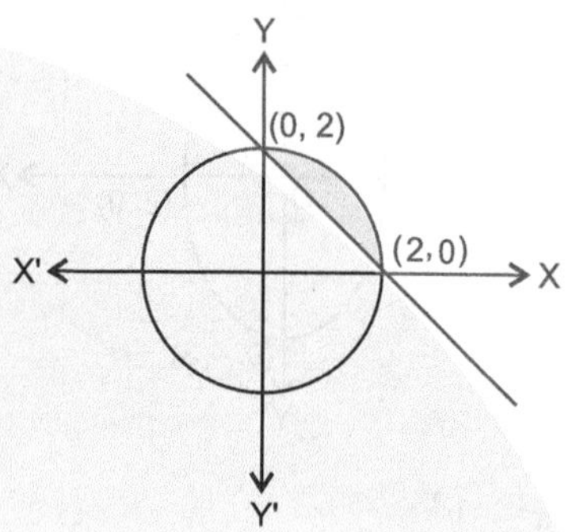

which passes through points (2, 0) and (0, 2)

Required area

= Area of shaded region

= Area under circle – Area under line

$$= \int_0^2 y \text{ circle } dx - \int_0^2 y \text{ line } dx$$

$$= \int_0^2 \sqrt{4-x^2}\,dx - \int_0^2 (2-x)\,dx$$

$$= \left[\frac{x}{2}\sqrt{4-x^2} + \frac{4}{2}\sin^{-1}\frac{x}{2}\right]_0^2 - \left[2x - \frac{x^2}{2}\right]_0^2$$

$$= [(0 + 2\sin^{-1} 1) - (0 + 2 \times 0)] - [(4-2) - 0]$$

$$= 2 \times \frac{\pi}{2} - 2 = (\pi - 2) \text{ square units.}$$ **Ans**

5. *Find the area of the smaller region bounded by the ellipse $\frac{x^2}{9} + \frac{y^2}{4} = 1$ and the line $\frac{x}{3} + \frac{y}{2} = 1$.*

Sol. The given curves are

$$\frac{x^2}{9} + \frac{y^2}{4} = 1 \quad \text{...(i)}$$

and

$$\frac{x}{3} + \frac{y}{2} = 1 \quad \text{...(ii)}$$

(i) is equation of ellipse with $a = 3$, $b = 2$.

(ii) is equation of line with x-intercept = 3 and y intercept = 2.

i.e., it passes through points (3, 0) and (0, 2)

Find the Value Type Questions

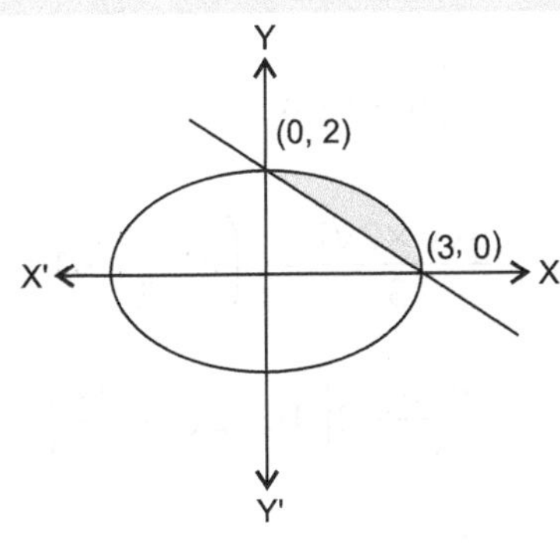

From (i) $\frac{y^2}{4} = 1 - \frac{x^2}{9} = \frac{9-x^2}{9}$

$$y = \frac{2}{3}\sqrt{9-x^2}$$

From (ii) $\frac{y}{2} = 1 - \frac{x}{3} = \frac{3-x}{3}$

$$y = \frac{2}{3}(3-x)$$

Required area

= Area of shaded region

= Area under ellipse – Area under line

$$= \int_0^3 y \text{ ellipse } dx - \int_0^3 y \text{ line } dx$$

$$= \int_0^3 \frac{2}{3}\sqrt{3^2-x^2}\, dx - \int_0^3 \frac{2}{3}(3-x)\, dx$$

$$= \frac{2}{3}\left[\frac{x}{2}\sqrt{9-x^2} + \frac{9}{2}\sin^{-1}\frac{x}{3} - 3x + \frac{x^2}{2}\right]_0^3$$

$$= \frac{2}{3}\left[\left(0 + \frac{9}{2}\sin^{-1}1 - 9 - \frac{9}{2}\right) - 0\right]$$

$$= \frac{2}{3}\left(\frac{9}{2}\times\frac{\pi}{2} - \frac{27}{2}\right) = \frac{2}{3}\left(\frac{9}{4}\pi - \frac{27}{2}\right)$$

$$= \frac{2}{3}\times\frac{9}{4}(\pi - 6) = \frac{3}{2}(\pi - 6) \text{ sq. units. } \textbf{Ans.}$$

6. *Draw a rough sketch of the curve $y^2 = 4x$ and find the area of the region enclosed by the curve and the line $y = x$.*

Sol. Given equation of the curves are

$$y^2 = 4x \qquad \ldots\text{(i)}$$

$$y = x \qquad \ldots\text{(ii)}$$

Solving (i) and (ii), we get $x = 0, 4$

Equation (i) gives

$$y = 2\sqrt{x}$$

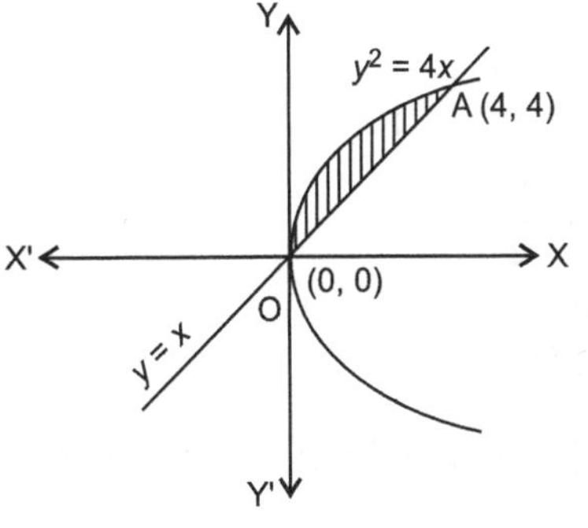

Point of inter-section of (i) and (ii) are (0, 0) and (4, 4)

$$\therefore \text{Required area} = \int_0^4 2\sqrt{x}\, dx - \int_0^4 x\, dx$$

$$= 2\frac{\left[x^{3/2}\right]_0^4}{3/2} - \frac{\left[x^2\right]_0^4}{2}$$

$$= \frac{4}{3}(4^{3/2} - 0^{3/2}) - \frac{1}{2}(4^2 - 0^2)$$

$$= \frac{4}{3}\times 8 - \frac{1}{2}\times 16$$

$$= \frac{32}{3} - 8$$

$$= \frac{32-24}{3} = \frac{8}{3} \text{ sq. units. } \textbf{Ans}$$

7. *Find the area of the region bounded by the curves $y = x^2 + 2$, $y = x$, $x = 0$ and $x = 3$. Also sketch the region bounded by these curves.*

Sol. The shaded area is the requireed area. It is enclosed by the two curves

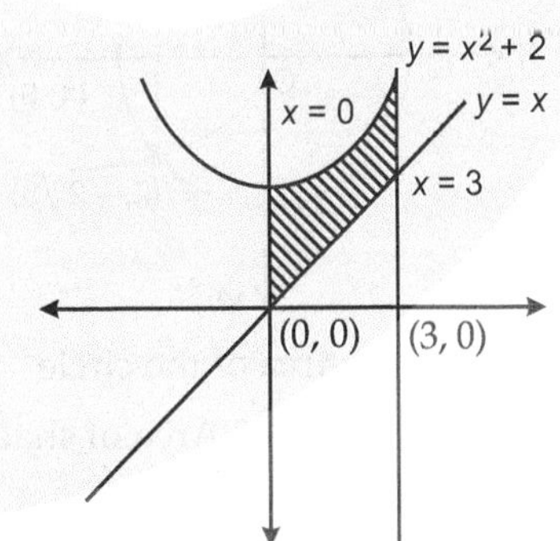

$$= \int_0^3 (y \text{ of parabola})dx - \int_0^3 (y \text{ of line})dx$$

$$= \int_0^3 (x^2 + 2)dx - \int_0^3 x\, dx$$

$$= \left[\frac{x^3}{3} + 2x\right]_0^3 - \left[\frac{x^2}{2}\right]_0^3$$

$= [9+6] - \left[\frac{9}{2}\right]$

$= 15 - \frac{9}{2}$

$= \frac{21}{2}$ sq. units **Ans.**

8. *Find the area of the circle $x^2 + y^2 = 16$ exterior to the parabola $y^2 = 6x$.*

Sol. The equations of circle and parabola are

$$x^2 + y^2 = 16 \quad \text{...(i)}$$

and

$$y^2 = 6x \quad \text{...(ii)}$$

Circle has centre (0, 0) and radius 4. The given parabola is right hand parabola with vertex at origin. To find point of intersection of circle and parabola, put value of y^2 from eq. (ii) in eq. (i), we get

$$x^2 + 6x = 16$$

$\Rightarrow \quad x^2 + 6x - 16 = 0$

$\Rightarrow \quad x^2 + 8x - 2x - 16 = 0$

$\Rightarrow \quad x(x + 8) - 2(x + 8) = 0$

$\Rightarrow \quad (x - 2)(x + 8) = 0$

$\Rightarrow \quad x - 2 = 0$ or $x + 8 = 0$

$\Rightarrow \quad x = 2$ or $x = -8$

$\therefore \quad y = \sqrt{6x} = \sqrt{12} = \pm 2\sqrt{3}$

[not possible as $x \geq 0$]

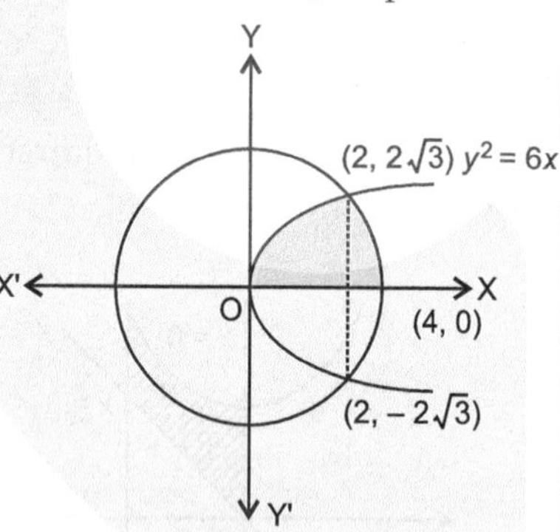

Required Area = Area of the circle − 2 Area of shaded region

$= \pi r^2 - 2\left\{\int_0^2 y \text{ parabola } dx + \int_2^4 y \text{ circle } dx\right\}$

$= \pi(4)^2 - 2\left\{\int_0^2 \sqrt{6}\sqrt{x}\, dx + \int_2^4 \sqrt{16 - x^2}\, dx\right\}$

$= 16\pi - 2\sqrt{6}\left[\frac{x^{3/2}}{3/2}\right]_0^2 - 2\left[\frac{x}{2}\sqrt{16 - x^2} + \frac{16}{2}\sin^{-1}\frac{x}{4}\right]_2^4$

$= 16\pi - \frac{4\sqrt{6}}{3}(2^{3/2} - 0)$

$- 2\left[(0 + 8\sin^{-1}1) - \left(\sqrt{16-4} + 8\sin^{-1}\frac{1}{2}\right)\right]$

$= 16\pi - \frac{4\sqrt{6} \times 2\sqrt{2}}{3} - 2\left(8 \times \frac{\pi}{2} - \sqrt{12} - 8 \times \frac{\pi}{6}\right)$

$= 16\pi - \frac{16\sqrt{3}}{3} - 2\left(4\pi - 2\sqrt{3} - \frac{4\pi}{3}\right)$

$= 16\pi - \frac{16\sqrt{3}}{3} - 2\left(\frac{8\pi}{3} - 2\sqrt{3}\right)$

$= 16\pi - \frac{16\sqrt{3}}{3} - \frac{16\pi}{3} + 4\sqrt{3}$

$= \frac{48\pi - 16\pi}{3} + \sqrt{3}\left(4 - \frac{16}{3}\right)$

$= \frac{32\pi}{3} + \sqrt{3}\left(\frac{-4}{3}\right) = \frac{32\pi}{3} - \frac{4\sqrt{3}}{3}$

$= \frac{4}{3}(8\pi - \sqrt{3})$ sq. units. **Ans.**

9. *Find the area of the region bound by the curves $y = 6x - x^2$ and $y = x^2 - 2x$.*

Sol. Given equations to the curves are

$$y = 6x - x^2 \quad \text{...(i)}$$

$$y = x^2 - 2x \quad \text{...(ii)}$$

Solving equations (i) and (ii), we get $x = 0, 4$

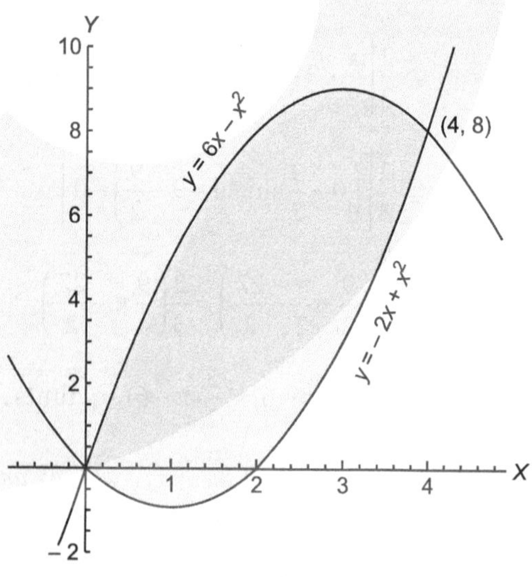

Required area $= \int_0^4 (6x - x^2)dx - \int_0^4 (x^2 - 2x)dx$

$= \int_0^4 (6x - x^2 - x^2 + 2x)\,dx$

$= \int_0^4 (8x - 2x^2)\,dx$

$$= \left[4x^2 - \frac{2x^3}{3}\right]_0^4$$

$$= \left(64 - \frac{128}{3}\right)$$

$$= \frac{64}{3} = 21\frac{1}{3} \text{ sq. units}$$ **Ans.**

10. *Sketch the graphs of the curves $y^2 = x$ and $y^2 = 4 - 3x$ and find the area enclosed between them.*

Sol.

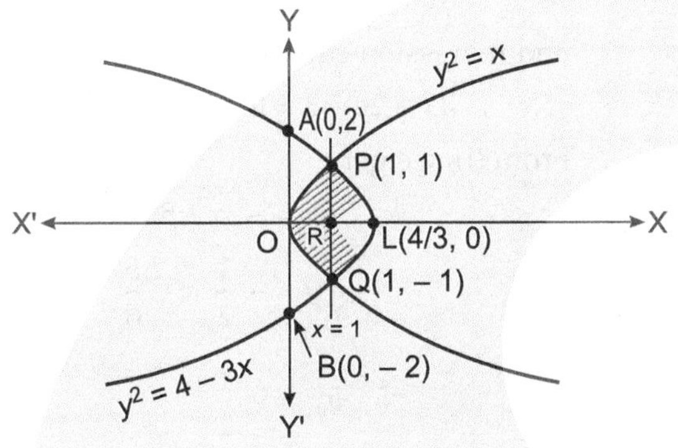

Given curves are $y^2 = x$ and $y^2 = -3x + 4$

or $$y^2 = -3\left(x - \frac{4}{3}\right)$$

The vertex L of this parabola is $\left(\frac{4}{3}, 0\right)$. It cuts the y-axis at A (0, 2) and B (0, – 2).

The points of intersection of these two parabolas are given by the equation

$$x = -3x + 4 \Rightarrow x = 1 \qquad [\because y^2 = x]$$

Then $$y^2 = 1 \Rightarrow y = \pm 1$$

Thus, the points of intersection are P (1, 1) and Q(1, – 1). Let PQ cut the x-axis at R

∴ Total area of POQLP

= 2 area of OPLRO

$$= 2\left[\int_0^1 \sqrt{x}\,dx + \int_1^{4/3} \sqrt{4-3x}\,dx\right]$$

$$= 2\left[\left(\frac{x^{3/2}}{3/2}\right)_0^1 + \left(\frac{2(4-3x)^{3/2}}{(-3)\times 3}\right)_1^{4/3}\right]$$

$$= 2\left[\left(\frac{2}{3} - 0\right) - \frac{2}{9}(0-1)\right]$$

$$= 2\left[\frac{2}{3} + \frac{2}{9}\right] = 2\left[\frac{6+2}{9}\right]$$

$$= 2\left(\frac{8}{9}\right) = \frac{16}{9} \text{ sq. units}$$ **Ans.**

Chapter 15. Linear Regression

1. *Find the line of best fit for the following data, treating x as dependent variable (Regression equation x on y) :*

X	14	12	13	14	16	10	13	12
Y	14	23	17	24	18	25	23	24

Hence, estimate the value of x when $y = 16$.

Sol. Let us construct the table :

x	y	xy	y^2
14	14	196	196
12	23	276	529
13	17	221	289
14	24	336	576
16	18	288	324
10	25	250	625
13	23	299	529
12	24	288	576
$\Sigma x = 104$	$\Sigma y = 168$	$\Sigma xy = 2154$	$\Sigma y^2 = 3644$

$$\bar{x} = \frac{\Sigma x}{n} = \frac{104}{8} = 13$$

$$\bar{y} = \frac{\Sigma y}{n} = \frac{168}{8} = 21$$

Now, $$b_{xy} = \frac{\Sigma xy - \frac{\Sigma x\,\Sigma y}{n}}{\Sigma y^2 - \frac{1}{n}(\Sigma y)^2}$$

$$= \frac{2154 - \frac{104\times 168}{8}}{3644 - \frac{(168)^2}{8}}$$

$$b_{xy} = \frac{2154 - 2184}{3644 - 3528}$$

$$b_{xy} = \frac{-30}{116}$$

The regression equation of x on y is

$$x - \bar{x} = b_{xy}(y - \bar{y})$$

$$x - 13 = \frac{-30}{116}(y - 21)$$

$$\Rightarrow \quad 116x - 1508 = -30y + 630$$

$\Rightarrow \quad 116x + 30y = 1508 + 630$

$\Rightarrow \quad 116x + 30y = 2138$

$\Rightarrow \quad 58x + 15y = 1069$

Value of x when $y = 16$

$58x + 15 \times 16 = 1069$

$\Rightarrow \quad 58x = 1069 - 240$

$\Rightarrow \quad 58x = 829$

$\Rightarrow \quad x = \frac{829}{58}$

$x = 14{\cdot}2931$ **Ans.**

2. *If the regression equation of x on y is given by $mx - y + 10 = 0$ and the equation of y on x is given by $-2x + 5y = 14$, determine the value of 'm' if the coefficient of correlation between x and y is $\frac{1}{\sqrt{10}}$.*

Sol. The regression equation of x on y is

$$mx - y + 10 = 0$$

i.e., $\quad x = \frac{y}{m} - \frac{10}{m}$

$\therefore \quad b_{xy} = \frac{1}{m}$

and the regression equation of y on x is
$-2x + 5y = 14.$

i.e., $\quad y = \frac{2x}{5} + \frac{14}{5}$

$\therefore \quad b_{yx} = \frac{2}{5}$

Now $\quad r^2 = b_{xy}.\, b_{yx}$

$\Rightarrow \quad \frac{1}{10} = \frac{2}{5} \times \frac{1}{m} \quad \left(\because r = \frac{1}{\sqrt{10}}\right)$

$\Rightarrow \quad m = \frac{10 \times 2}{5} = 4$

$\Rightarrow \quad m = 4.$ **Ans.**

3. *The two lines of regressions are $4x + 2y - 3 = 0$ and $3x + 6y + 5 = 0$. Find the correlation co-efficient between x and y.*

Sol. Let $4x + 2y - 3 = 0$ be the regression equation of y on x and $3x + 6y + 5 = 0$ the regression equation of x on y then

$2y = -4x + 3$

$y = -2x + \frac{3}{2}$

$\therefore \quad b_{yx} = -2$

and, $\quad 3x = -6y - 5$

$x = -\frac{6}{3}y - \frac{5}{3}$

$x = -2y - \frac{5}{3}$

$\therefore \quad b_{xy} = -2$

$r^2 = b_{yx} \times b_{xy}$

$= -2 \times -2 = 4 > 1$

which is impossible.

So, regression equation of y on x is

$3x + 6y + 5 = 0 \quad$ ...(i

and regression equation of x on y is

$4x + 2y - 3 = 0 \quad$ ...(ii

From (i), we get

$6y = -3x - 5$

$y = -\frac{1}{2}x - \frac{5}{6}$

$\therefore \quad b_{yx} = -\frac{1}{2}$

From (ii), we get

$4x = -2y + 3$

$x = -\frac{1}{2}y + \frac{3}{4}$

$\therefore \quad b_{xy} = -\frac{1}{2}$

$\therefore$ Correlation coefficient is given by

$| r | = \sqrt{b_{yx}.b_{xy}}$

$= \sqrt{-\frac{1}{2} \times -\frac{1}{2}} = \frac{1}{2}$

$\Rightarrow \quad r = \pm \frac{1}{2}$

But r has the same sign as regression coefficien

$\therefore \quad r = -\frac{1}{2}.$ **Ans**

4. *Given that the observations are :*
(9, – 4), (10, – 3), (11, – 1), (12, 0), (13, 1), (14, 3)
(15, 5), (16, 8).

Find the two lines of regression and estimate th value of y when x = 13·5.

Sol.

x	x^2	y	y^2	xy
9	81	– 4	16	– 36
10	100	– 3	9	– 30
11	121	– 1	1	– 11
12	144	0	0	0

13	169	1	1	13
14	196	3	9	42
15	225	5	25	75
16	256	8	64	128
$\Sigma x = 100$	$\Sigma x^2 = 1292$	$\Sigma y = 9$	$\Sigma y^2 = 125$	$\Sigma xy = 181$

$$\bar{x} = \frac{\Sigma x}{N} = \frac{100}{8} = 12{\cdot}5 \qquad (\because N = 8)$$

$$\bar{y} = \frac{\Sigma y}{N} = \frac{9}{8} = 1{\cdot}125$$

$$b_{yx} = \frac{\Sigma xy - \frac{1}{N}\Sigma x \Sigma y}{\Sigma x^2 - \frac{1}{N}(\Sigma x)^2}$$

$$= \frac{181 - \frac{9 \times 100}{8}}{1292 - \frac{10000}{8}}$$

$$= \frac{181 - 112{\cdot}5}{1292 - 1250}$$

$$= \frac{181 - 112{\cdot}5}{42} = \frac{68{\cdot}5}{42} = 1{\cdot}63$$

$$b_{xy} = \frac{\Sigma xy - \frac{1}{N}\Sigma x \Sigma y}{\Sigma y^2 - \frac{1}{N}(\Sigma x)^2}$$

$$= \frac{181 - \frac{900}{8}}{125 - \frac{81}{8}} = \frac{181 - 112{\cdot}5}{125 - 10{\cdot}125}$$

$$= \frac{68{\cdot}5}{114{\cdot}875} = 0{\cdot}596$$

Regression line of y on x

$$y - \bar{y} = b_{yx}(x - \bar{x})$$

$$y - 1{\cdot}125 = 1{\cdot}63\,(x - 12{\cdot}5)$$

$$\Rightarrow \quad y - 1{\cdot}125 = 1{\cdot}63x - 20{\cdot}375$$

$$\Rightarrow \quad y = 1{\cdot}63x - 19{\cdot}25$$

Regression line of x on y

$$x - \bar{x} = b_{xy}(y - \bar{y})$$

$$x - 12{\cdot}5 = 0{\cdot}596\,(y - 1{\cdot}125)$$

$$x - 12{\cdot}5 = 0{\cdot}596y - 0{\cdot}67$$

$$x = 0{\cdot}596y + 11{\cdot}83$$

Now, $y = 1{\cdot}63x - 19{\cdot}25$

For $x = 13{\cdot}5$

The estimated value of

$$y = 1{\cdot}63 \times 13{\cdot}5 - 19{\cdot}25$$

$$= 22{\cdot}005 - 19{\cdot}25$$

$$= 2{\cdot}755 \cong 2{\cdot}8$$

Ans.

Prove the Following Type Questions

Set 4

Chapter 1. Relations

1. *Show that the relation R in the set A of real numbers defined as R = {(a, b) : a ≤ b} is reflexive and transitive but not symmetric.*

Sol. Given,

R ={(a, b) : a ≤ b} where $a, b \in R$

Reflexive : $a \le a$ for $a \in R$

$\Rightarrow (a, a) \in R \ \forall \ a \in R$

i.e., R is reflexive.

Symmetric : $1 \le 2$, for any, $1, 2 \in R$

but $2 \nleq 1$

$\Rightarrow (1, 2) \in R$ but $(2, 1) \notin R$

i.e., R is not symmetric

Transitive : Let $(a, b) \in R$ and $(b, c) \in R$

Then, $a \le b$ and $b \le c$

$\Rightarrow a \le c$

$\Rightarrow (a, c) \in R$

i.e., R is transitive. **Hence Proved.**

2. *Show that the relation R on the set R of real numbers, defined as R = {(a, b) : a ≤ b²} is neither reflexive nor symmetric nor transitive.*

Sol. We have, R = $\{(a, b) : a \le b^2\}$,

where $a, b \in R$

Reflexivity : We know that, $\frac{1}{2}$ is a real number

and $\frac{1}{2} \le \left(\frac{1}{2}\right)^2$ is not true.

Therefore, R is not reflexive.

Symmetry : Consider the real numbers 1 and 2.

Now, $1 \le 2^2$

$\Rightarrow (1, 2) \in R$

But, $2 \le (1)^2$ is not true and so, $(2, 1) \notin R$.

Thus, $(1, 2) \in R$ but $(2, 1) \notin R$

Hence, R is not symmetric.

Transitivity : By taking real numbers 2, – 2 and 1.

We have, $2 \le (-2)^2$ and $-2 \le (1)^2$ but $2 \le (1)^2$ is not true.

Thus, $(2, -2) \in R$ and $(-2, 1) \in R$, but $(2, 1) \notin R$.

Hence, R is not transitive. **Hence Proved.**

3. *Let the relation R in the set of real numbers be defined as a R b if and only if 1 + ab > 0. Show that this relation is reflexive and symmetric, but not transitive.*

Sol. Let S denote the set of all real numbers. Let R be a relation on S defined as aRb if $1 + ab > 0$.

(i) R is reflexive

Let a be any real number.

Then $1 + a\,a = 1 + a^2 > 0$, since $a^2 \ge 0$.

Thus, $a \, R \, a \ \forall \ a \in S$.

Therefore, R is reflexive.

(ii) R is symmetric

Let a, b be any two real numbers. Then

$a \, R \, b$ is

$\Rightarrow 1 + ab > 0$

$\Rightarrow 1 + ba > 0$ $[\because ab = ba]$

$\Rightarrow b \, R \, a$

$\therefore$ R is symmetric.

(iii) R is not transitive

Consider three real numbers $1, -\frac{1}{2}, -4$, we have

$$1+1\left(-\frac{1}{2}\right) = \frac{1}{2} > 0, \ \therefore \left(1, \frac{-1}{2}\right) \in R$$

Further $1+\left(-\frac{1}{2}\right)(-4) = 3 > 0, \ \therefore \left(-\frac{1}{2}, -4\right) \in R$

But $1 + 1(-4) = -3$, which is not greater than 0,

$\therefore$ 1 is not R-related to – 4.

Thus, $\left(1, \frac{1}{2}\right) \in R, \left(-\frac{1}{2'} -4\right) \in R$. But $(1, -4) \notin R$

and 1 is not related to – 4.

$\therefore$ R is not transitive. **Hence Proved.**

4. *Show that the relation R defined on the set A of all triangles in a plane as R = {(T_1, T_2) : T_1 is congruent to T_2} is an equivalence relation.*

Sol. Given,

$R = \{(T_1, T_2) : T_1 \text{ is congruent to } T_2\}$

Reflexivity : $(T, T) \in R \ \forall \ T \in A$

Since every triangle is congruent to itself

i.e., R is reflexive.

Symmetric : Let, $(T_1, T_2) \in R$

Then, T_1 is congruent to T_2

⇒ T_2 is congruent to T_1

⇒ $(T_2, T_1) \in R$

So, R is symmetric.

Transitivity : Let $T_1, T_2, T_3 \in A$ such that $(T_1, T_2) \in R$ and $(T_2, T_3) \in R$. Then,

⇒ T_1 is congruent to T_2 and T_2 is congruent to T_3

⇒ T_1 is congruent to T_3

⇒ $(T_1, T_3) \in R$

So, R is transitive.

Hence, R is an equivalence relation on set A.

Hence Proved.

5. *Show that the relation R on the set Z of all integers numbers defined by $(x, y) \in R$ if and only if $x - y$ is divisible by 3 is an equivalence relation on Z.*

Sol. Given,

$R = \{(x, y) : x, y \in Z \text{ and } (x - y) \text{ is divisible by } 3\}$

Reflexivity : For $x \in Z$, $(x, x) \in R$

⇒ $x - x = 0$ which is divisible by 3

⇒ R is reflexive.

Symmetric : For $x, y \in Z$; $(x, y) \in R$

⇒ $(x - y)$ is divisible by 3

⇒ $-(x - y)$ *i.e.*, $(y - x)$ is divisible by 3.

⇒ $(y, x) \in R$

$(x, y) \in R \Rightarrow (y, x) \in R \ \forall \ x, y \in Z$

∴ R is symmetric.

Transitivity : For $x, y, z \in Z$ such that

$(x, y) \in R$ and $(y, z) \in R$

⇒ $(x - y)$ is divisible by 3 and $(y - z)$ is divisible by 3.

⇒ $(x - z) = (x - y) + (y - z)$ which is divisible by 3.

⇒ $(x - z)$ is divisible by 3.

$(x, z) \in R$

So, R is transitive.

Thus, R is reflexive, symmetric and transitive, is an equivalence relation on Z. **Hence Proved.**

6. *Let A be the set of real numbers and R be the relation on A defined as $R = \{(a, b) : a, b \in A$ and $|a - b|$ is even$\}$. Prove that R is an equivalence relation.*

Sol. Given,

$R = \{(a, b) : a, b \in A \text{ and } |a - b| \text{ is even}\}$

Reflexivity : For $a \in A$, such that $(a, a) \in R$

⇒ $|a - a| = 0$, an even integer

So, R is reflexive.

Symmetric : For $a, b \in A$ such that $(a, b) \in R$

⇒ $|a - b|$ is an even integer

then $|b - a| = |-(a - b)|$ is an even integer

$(a, b) \in R$

⇒ $(b, a) \in R$

So, R is symmetric

Transitivity : For $a, b, c \in A$ such that

$(a, b) \in R$ and $(b, c) \in R$

if $|a - b|$ is an even integer and $|b - c|$ is an even integer.

then $|a - b| + |b - c| = |a - c|$ is an even integer.

Hence, R is transitive.

Thus, R is an equivalence relation on A.

Hence Proved.

7. *Show that the relation R on the set A of all the books in a library of a college given by $R = \{(x, y) : x$ and y have the same number of pages$\}$, is an equivalence relation.*

Sol. Given, $R = \{(x, y) : x$ and y have the same number of pages$\}$

Reflexivity : For any books, $x \in A$

$(x, x) \in R$

Since, x and x have the same number of pages.

⇒ $(x, x) \in R \ \forall \ x \in A$

So, R is reflexive.

Symmetric : For any book $x, y \in A$

Let $(x, y) \in R$

⇒ x and y have the same number of pages.

⇒ y and x have the same number of pages.

$(y, x) \in R$

∴ $(x, y) \in R$

⇒ $(y, x) \in R$

So, R is symmetric.

Transitivity : For book $x, y, z \in A$

Let $(x, y) \in R$ and $(y, z) \in R$.

⇒ x and y have the same number of pages.

and, y and z have the same number of pages.

$\Rightarrow$ x and z have the same number of pages

$\Rightarrow (x, z) \in R$

So, R is transitive.

Hence, R is reflexive, symmetric and transitive, is an equivalence relation. **Hence Proved.**

8. *Prove that the relation R on the set N × N defined by $(a, b)\ R\ (c\ d) \Leftrightarrow a + d = b + c\ \forall\ (a, b), (c, d) \in N \times N$ is an equivalence relation.*

Sol. Given,

$(a, b)\ R\ (c, d) \Leftrightarrow a + d = b + c\ \ \forall\ (a, b), (c, d) \in N \times N.$

Reflexivity : Let $(a, b) \in N \times N$

then, $a, b \in N$

$$a + b = b + a$$

(by commutativity of addition on N)

$(a, b)\ R\ (a, b)$

So, R is reflexive.

Symmetric : Let $(a, b), (c, d) \in N \times N$ and $(a, b)\ R\ (c, d)$ then, $a + d = b + c$.

$$c + b = d + a$$

(By commutativity of addition on N)

$(c, d)\ R\ (a, b)$

Thus, $(a, b)\ R\ (c, d) \Rightarrow (c, d)\ R\ (a, b)\ \forall\ (a, b)\ (c, d) \in N \times N$

So, R is symmetric on N × N.

Transitivity : Let $(a, b), (c, d), (e, f) \in N \times N$ such that

$(a, b)\ R\ (c, d)$ and $(c, d)\ R\ (e, f)$

Then, $(a, b)\ R\ (c, d)$

$\Rightarrow$ $a + d = b + c$

$(c, d)\ R\ (e, f)$

$\Rightarrow$ $c + f = d + e$

then, $(a + d) + (c + f) = (b + c) + (d + e)$

$\Rightarrow$ $a + f = b + e$

$(a, b)\ R\ (e, f)$

Thus, $(a, b)\ R\ (c, d)$ and $(c, d)\ R\ (e, f)$

$\Rightarrow (a, b)\ R\ (e, f)\ \forall\ (a, b), (c, d), (e, f) \in N \times N.$

So, R is transitive on N × N

Hence, R is an equivalence relation on N × N.

Hence Proved.

9. *Show that the relation R defined on the set A of all polygons as $R = \{(P_1, P_2) : P_1$ and P_2 have same number of sides$\}$ is an equivalence relation. What is the set of all elements in A related to the right angle triangle T with sides 3, 4 and 5 ?*

Sol. Given, $R = \{(P_1, P_2) : P_1$ and P_2 have same number of sides$\}$

Reflexivity : Let P be any polygon in A

Then, P and P have same number of sides.

$\Rightarrow$ $(P, P) \in R$

Thus, $(P, P) \in R\ \forall\ P \in A$

So, R is reflexive.

Symmetric : Let P_1 and P_2 be two polygon in A such that $(P_1, P_2) \in R$.

Then, $(P_1, P_2) \in R$.

Now, P_1 and P_2 have same number of sides

$\Rightarrow$ P_2 and P_1 have same number of sides

$(P_2, P_1) \in R$

So, R is symmetric on A.

Transitivity : Let P_1, P_2 and P_3 be three polygons in A such that $(P_1, P_2) \in R$ and $(P_2, P_3) \in R$.

Then, $(P_1, P_2) \in R$

$\Rightarrow$ P_1 and P_2 have same number of sides.

and $(P_2, P_3) \in R$

$\Rightarrow$ P_2 and P_3 have same number of sides.

$\therefore$ P_1 and P_3 have same number of sides

$\Rightarrow (P_1, P_3) \in R.$

Thus, $(P_1, P_2) \in R$ and $(P_2, P_3) \in R$

$\Rightarrow (P_1, P_3) \in R.$

So, R is transitive.

Hence, R is an equivalence relation on A.

Let P be a polygon in A such that $(P, T) \in R$. Then, polygon P and triangle T have same number of sides.

Thus, P is any triangle in A with sides 3, 4 and 5. **Hence Proved.**

10. *Prove that the relation R on the set Z of integers defined by $R = \{(a, b) : 2$ divides $a - b\}$, is an equivalence relation.*

Sol. Given, $R = \{(a, b) : 2 \text{ divides } a - b\}$

Reflexivity : Let $a \in Z$

$$a - a = 0$$
$$= 0 \times 2$$

2 divides $a - a$

$(a, a) \in R$

So, R is a reflexive relation on Z.

Symmetric : Let $a, b \in Z$ such that $(a, b) \in R$

$\Rightarrow$ 2 divides $a - b$

then, $a - b = 2k$ where, $k \in Z$

$b - a = 2(-k)$ where, $-k \in Z$

2 divides $b - a$

$\Rightarrow (b, a) \in R.$

Prove the Following Type Questions

Thus, $(a, b) \in R$

$\Rightarrow (b, a) \in R$

So, R is symmetric.

Transitivity : Let $a, b, c \in Z$ such that $(a, b) \in R$ and $(b, c) \in R$

Then, $(a, b) \in R$

$\Rightarrow$ 2 divides $b - a$

$\Rightarrow b - a = 2k$ where, $k \in Z$

and $(b, c) \in R$

$\Rightarrow$ 2 divides $c - b$

$\Rightarrow c - b = 2l$ where, $l \in Z$

$\therefore \quad b - a + c - b = 2(k + l)$

$c - a = 2(k + l)$ where, $(k + l) \in Z$

$\therefore$ 2 divides $c - a$

Thus, $(a, b) \in R$ and $(b, c) \in R$

$\Rightarrow (a, c) \in R$

So, R is transitive.

Hence, R is an equivalence relation on Z.

Hence Proved.

11. *Show that the relation R on the set A $\{x \in Z : 0 \le x \le 12\}$, given by $R = \{(a, b) : |a - b|$ is a multiple of 4$\}$ is an equivalence relation.*

Sol. We have the given relation

$R = \{(a, b) : |a - b|$ is a multiple of 4$\}$, where $a, b \in A$ and $A = \{x \in Z : 0 \le x \le 12\} = \{0, 1, 2, \ldots 12\}$. We discuss the following properties of relation R on set A.

Reflexivity : For any $a \in A$, we have

$|a - a| = 0$, which is multiple of 4

$\Rightarrow (a, b) \in R$ for all $a \in R$.

So, R is reflexive.

Symmetric : Let $(a, b) \in R$

$\Rightarrow |a - b|$ is divisible by 4

$\Rightarrow \quad |a - b| = 4k$ [where $k \in Z$]

$\Rightarrow \quad a - b = \pm 4k$

$\Rightarrow \quad b - a = \pm 4k$

$\Rightarrow \quad |b - a| = 4k$

$\Rightarrow |b - a|$ is divisible by 4

$\Rightarrow \quad (b - a) \in R$

So, R is symmetric

Transitivity : Let $a, b, c, \in A$ such that $(a, b) \in R$ and $(b, c) \in R$

$\Rightarrow |a - b|$ is multiple of 4 and $|b - c|$ is multiple of 4.

$\Rightarrow \quad |a - b| = 4m$ and

$|b - c| = 4n, m, n \in N$

$\Rightarrow \quad a - b = \pm 4m$ and $b - c = \pm 4n$

$\therefore \quad (a - b) + (b - c) = \pm 4(m + n)$

$\Rightarrow \quad a - c = \pm 4(m + n)$

$\Rightarrow \quad |a - c| = 4(m + n)$

$\Rightarrow |a - b|$ is multiple of 4

$\Rightarrow (a, c) \in R$

Thus, if $(a, b) \in R$ and $(b, c) \in R$

Then $(a, c) \in R$

So, R is transitive

Hence, R is equivalence relation.

Hence Proved.

12. *Show that the relation R on the set A of points in a plane is given by*

R = {(P, Q) : Distance of the point P from the origin = Distance of point Q from origin} is an equivalence relation.

Further, show that the set of all points related to a point P ≠ (0, 0) is the circle passing through P with origin as centre.

Sol. If O is the origin, then

R = {(P, Q) : OP = OQ}

Reflexivity : For all point $P \in A$

OP = OP

$\Rightarrow \quad (P, P) \in R$

i.e., R is reflexive.

Symmetry : Let $P, Q \in A$, such that $(P, Q) \in R$

$\Rightarrow \quad$ OP = OQ

$\Rightarrow \quad$ OQ = OP

$\Rightarrow \quad (Q, P) \in R$

i.e., R is symmetric.

Transitivity : Let P, Q, S $\in$ A, such that $(P, Q) \in R$ and $(Q, S) \in R$

$\Rightarrow \quad$ OP = OQ and OQ = OS

$\Rightarrow \quad$ OP = OS

$\Rightarrow \quad (P, S) \in R$

i. e., R is transitive.

Now, we have R is reflexive, symmetric and transitive.

Therefore, R is an equivalence relation.

Hence Proved.

Let P, Q, R... be points in the set A, such that

$(P, Q), (P, R) \ldots \in R$

$\Rightarrow \quad$ OP = OQ; OP = OR;...

[where O is origin]

$\Rightarrow$ OP = OQ = OR = ...

i.e., All points P, Q, R... ∈ A, which are related to P are equidistant from origin 'O'.

Hence, set of all points of A related to P is the circle passing through P, having origin as centre. **Hence Proved.**

Chapter 2. Functions

1. *Discuss the commutativity and associativity of the binary operation * on R defined by*

$$a * b = \frac{ab}{4} \ \forall \ a, b \in R.$$

Sol. We have,

$$a * b = \frac{ab}{4} \ \forall \ a, b \in \text{R}$$

Commutativity : For $a, b \in$ R, we have

$$a * b = \frac{ab}{4}$$

and $$b * a = \frac{ba}{4}$$

Since, multiplication on R is a commutative binary operation

$$\therefore \quad ab = ba \ \forall \ a, b \in \text{R}$$

$$\frac{ab}{4} = \frac{ba}{4} \ \forall \ a, b \in \text{R}$$

So, '*' is a commutative binary operation on R

Associativity : Let $a, b, c \in$ R. Then

$$(a * b) * c = \left(\frac{ab}{4}\right) * c$$

$$= \frac{\left(\frac{ab}{4}\right)c}{4} = \frac{(ab)c}{16}$$

and $$a * (b * c) = a * \left(\frac{bc}{4}\right)$$

$$= \frac{a\left(\frac{bc}{4}\right)}{4} = \frac{a(bc)}{16}$$

Since, multiplication on R is an associative binary operation

$$\therefore \quad (ab)\,c = a\,(bc) \ \forall \ a, b, c \in \text{R}$$

$$\frac{(ab)c}{16} = \frac{a(bc)}{16} \ \forall \ a, b, c \in \text{R}.$$

$$(a * b) * c = a * (b * c) \ \forall \ a, b, c \in \text{R}$$

So, '*' is an associative binary operation on R.

Hence Proved.

2. *Let '*' be a binary operation on N given by a * b = HCF (a, b) for all a, b ∈ N.*

*(i) Find : 12 * 4, 18 * 24, 7 * 5.*

(ii) Check the commutativity and associativity of '' on N.*

Sol. (i) We have,

$$a * b = \text{HCF}\,(a, b)$$
$$12 * 4 = \text{HCF}\,(12, 4) = 4$$
$$18 * 24 = \text{HCF}\,(18, 24) = 6$$
$$7 * 5 = \text{HCF}\,(7, 5) = 1$$

(ii) Commutativity : For $a, b \in$ N.

We have $a * b = \text{HCF}\,(a, b)$

$= \text{HCF}\,(b, a) = b * a$

So, '*' is commutative on N.

(iii) Associativity : For $a, b, c \in$ N

$$(a * b) * c = \text{HCF}\,(a, b) * c$$
$$= \text{HCF}\,(a, b, c)$$

and $$a * (b * c) = a * \text{HCF}\,(a, b)$$
$$= \text{HCF}\,(a, b, c)$$

$$\therefore \quad (a * b) * c = a * (b * c) \text{ for all } a, b, c \in \text{N}$$

So, '*' is associative on N. **Ans.**

3. *Show that the number of binary operations on [1, 2] having 1 as identity and having 2 as inverse of 2 is exactly one.*

Sol. It is known that a binary operation on a set A is a function from A × A to A.

So, a binary operation on set A = [1, 2] is a function from A × A → A

i.e., {(1, 1) (1, 2), (2, 1), (2, 2)} to {1, 2}

Let * be the binary operation. Given, 1 as identity and 2 as inverse of 2 then,

$$1 * 1 = 1, 1 * 2 = 2 * 1 = 2$$

and $$2 * 2 = 1$$

Thus, * associates every element of A × A be elements of A.

Hence, '*' can be defined uniquely and the number of binary operation is 1.

Hence Proved.

4. *A binary operation * on the set {0, 1, 2, 3, 4, 5} is defined as :*

$$a * b = \begin{cases} a + b, & \text{if } a + b < 6 \\ a + b - 6, & \text{if } a + b \geq 6 \end{cases}$$

Prove that zero is the identity for this operation and each element 'a' of the set is invertible with 6 – a, being the inverse of 'a'.

Sol. Identity : Let a be an arbitrary element of set {0, 1, 2, 3, 4, 5}

Now, $a * 0 = a + 0 = a$ ...(i)

$0 * a = 0 + a = a$ ...(ii)

$[\because a + 0 = 0 + a < 6 \ \forall \ a \in \{0, 1, 2, 3, 4, 5\}]$

From equation (i) and (ii)

$$a * 0 = 0 * a = a \ \ \forall \ a \in \{0, 1, 2, 3, 4, 5\}$$

Hence, 0 is identity for binary operation *.

Inverse : Let a be an arbitrary element of set {0, 1, 2, 3, 4, 5}.

Now, $a * (6 - a) = a + (6 - a) - 6$

$[\because a + (6 - a) \geq 6]$

$= a + 6 - a - 6$

$= 0$ (identity) ...(iii)

Also $(6 - a) * a = (6 - a) + a - 6$

$= 6 - a + a - 6$

$= 0$ (identity) ...(iv)

From equations (iii) and (iv)

$$a * (6 - a) = (6 - a) * a = 0 \text{ (identity)}$$

$$\forall \ a \in \{0, 1, 2, 3, 4, 5\}$$

Hence, each element 'a' of given set is invertible with inverse $(6 - a)$. **Hence Proved.**

5. *Let A = N × N and * be a binary operation on A defined by (a, b) * (c, d) = (a + c, b + d). Prove that * is commutative and associative. Also, find the identity element for * on A, if any.*

Sol. Given, A = N × N and * is a binary operation on A defined by

$$(a, b) * (c, d) = (a + c, b + d)$$

Commutativity : Let $(a, b), (c, d) \in N \times N$

Then $(a, b) * (c, d) = (a + c, b + d)$

$= (c + a, d + b)$

$(\because a, b, c, d \in N, a + c = c + a \text{ and } b + d = d + b)$

$= (c, d) * (a, b)$

Hence, $(a, b) * (c, d) = (c, d) * (a, b)$

∴ * is commutative.

Associativity : Let $(a, b), (c, d), (e, f) \in N \times N$

Then $[(a, b) * (c, d)] * (e, f)$

$= [a + c, b + d] * (e, f)$

$= [(a + c) + e, (b + d) + f]$

$= [a + (c + e), b + (d + f)]$

(∵ set N is associative)

$= (a, b) * (c + e, d + f)$

$= (a, b) * [(c, d) * (e, f)]$

Hence, $[(a, b) * (c, d)] * (e, f) = (a, b) * [(c, d) * (e, f)]$

∴ * is associative. **Hence Proved.**

Identity : Let (x, y) be identity element for * on A,

Then $(a, b) * (x, y) = (a, b)$

⇒ $(a + x, b + y) = (a, b)$

⇒ $a + x = a, b + y = b$

⇒ $x = 0, y = 0$

But $(0, 0) \notin A = N \times N$

∴ For *, there is no identity element. **Ans.**

6. *Consider the binary operations * : R × R → R and o : R × R → R defined as a * b = |a – b| and aob = a for all a, b ∈ R. Prove that '*' is commutative but not associative, 'o' is associative but not commutative.*

Sol. For operation '*'

'*' : R × R → R such that $a * b = |a - b| \ \forall \ a, b \in R$

Commutativity :

$\forall \ a, b \in R, a * b = |a - b|$

$= |b - a| = b * a$

i.e., '*' is commutative

Associativity :

$\forall \ a, b, c \in R, (a * b) * c = |a - b| * c = ||a - b| - c|$

and $a * (b * c) = a * |b - c|$

$= |a - |b - c||$

But $||a - b| - c| \neq |a - |b - c||$

⇒ $(a * b) * c \neq a * (b * c)$

⇒ * is not associative.

Hence, '*' is commutative but not associative.

For Operation 'o'

o : R × R → R such that $aob = a$

Commutativity :

$\forall \ a, b \in R, aob = a$ and $boa = b$

∴ $a \neq b$

⇒ $aob \neq boa$

⇒ 'o' is not commutative.

Associativity :

$\forall \ a, b, c \in R, (aob) oc = aoc = a$

⇒ $ao (boc) - aob = a$

⇒ $(aob) oc = ao(boc)$

⇒ 'o' is associative.

Hence, 'o' is not commutative but associative.

Hence Proved.

7. *Prove that the function f : N → N given by f (x) = 2x, is one-one.*

Sol. Let $x_1, x_2 \in N$ such that $f(x_1) = f(x_2)$

$$\Rightarrow \quad 2x_1 = 2x_2$$

$$\Rightarrow \quad x_1 = x_2$$

Thus, f is one-one. **Hence Proved.**

8. *Prove that the function $f : R \to R$ given by $f(x) = 2x$ is one-one and onto.*

Sol. Let $x_1, x_2 \in R$ such that $f(x_1) = f(x_2)$ then,

$$\Rightarrow \quad 2x_1 = 2x_2$$

$$\Rightarrow \quad x_1 = x_2$$

Thus, f is one-one.

Now, for onto, let $y \in R$,

then, $f(x) = y$

$$2x = y$$

$$x = y/2$$

Clearly, $y/2 \in R$ for $y \in R$, such that

$$f\left(\frac{y}{2}\right) = 2\left(\frac{y}{2}\right) = y$$

Thus, for $y \in R\ \exists\ x = \frac{y}{2} \in R$ such that $f(x) = y$.

Hence, $f : R \to R$ is onto. **Hence Proved.**

9. *Show that $f : N \to N$, given by*

$$f(x) = \begin{cases} x+1, \text{ if } x \text{ is odd} \\ x-1, \text{ if } x \text{ is even} \end{cases}$$

is both one-one and onto.

Sol. One-one :

Case I : When x_1, x_2 are odd natural number.

$$\therefore \quad f(x_1) = f(x_2)$$

$$\Rightarrow \quad x_1 + 1 = x_2 + 1 \quad \forall\ x_1, x_2 \in N$$

$$\Rightarrow \quad x_1 = x_2$$

i.e., f is one-one.

Case II : When x_1, x_2 are even natural number.

$$\therefore \quad f(x_1) = f(x_2)$$

$$\Rightarrow \quad x_1 - 1 = x_2 - 1$$

$$\Rightarrow \quad x_1 = x_2$$

i.e., f is one-one.

Case III : When x_1 is odd and x_2 is even natural number.

$$f(x_1) = f(x_2)$$

$$\Rightarrow \quad x_1 + 1 = x_2 - 1$$

$\Rightarrow$ $x_2 - x_1 = 2$ which is never possible as the difference of odd and even number is alway odd number.

Hence in this case $f(x_1) \neq f(x_2)$ *i.e.*, f is one-one

Case IV : When x_1 is even and x_2 is odd natura number (Similar as case III). So, f is one-one.

Onto : $f(x) = \begin{cases} x+1 \text{ if } x \text{ is odd} \\ x-1 \text{ if } x \text{ is even} \end{cases}$

$\Rightarrow$ For every even number, 'y' of codomai $\exists$ odd number $y - 1$ in domain and for ever odd number y of codomain $\exists$ even number y + 1 in domain.

i.e., f is onto function.

Hence, f is one-one onto function.

Hence Proved

10. *Prove that the function $f : R \to R$ given b $f(x) = ax + b$ where, $a, b \in R$, $a \neq 0$ is a bijection.*

Sol. Let x, y be any two real numbers

Then, $f(x) = f(y)$,

$$\Rightarrow \quad ax + b = ay + b$$

$$ax = ay$$

$$\Rightarrow \quad x = y$$

Thus, if $f(x) = f(y)$

$$\Rightarrow \quad x = y\ \forall\ x, y \in R$$

So, f is one-one.

Now, let $y \in R$. Then, $f(x) = y$

$$ax + b = y$$

$$\Rightarrow \quad x = \frac{y-b}{a}$$

Clearly, $x = \frac{y-b}{a} \in R\ \forall\ y \in R$.

Thus, for all $y \in R\ \exists\ x = \frac{y-b}{a} \in R$ such that

$$f(x) = f\left(\frac{y-b}{a}\right)$$

$$= a\left(\frac{y-b}{a}\right) + b = y$$

So, f is onto.

Hence, f is one-one and onto *i.e.*, bijective.

Hence Proved.

11. *Let S be any non-empty set. Then, prove that th identity function on set S is a bijection.*

Sol. Let identity function on set S is

$I : S \to S$ defined as

$$I(x) = x \text{ for all } x \in S$$

Injectivity : Let x, y be any two elements of S then, $I(x) = I(y)$

$\Rightarrow \quad x = y$

So, I is injective.

Surjectivity : Let $y \in S$, then there exists $x = y \in S$ Such that

$$I(x) = x = y$$

So, I is surjective.

Hence, $I : S \to S$ is bijective. **Hence Proved.**

12. *Let $A = R - \{3\}$ and $B = R - \left\{\frac{2}{3}\right\}$. If $f : A \to B : f(x) = \frac{2x-4}{3x-9}$, then prove that f is a bijective function.*

Sol. Let x_1, x_2 be any two elements of A, then

$$f(x_1) = f(x_2)$$

$$\Rightarrow \frac{2x_1-4}{3x_1-9} = \frac{2x_2-4}{3x_2-9}$$

$$\Rightarrow 6x_1x_2 - 18x_1 - 12x_2 + 36 = 6x_1x_2 - 12x_1 - 18x_2 + 36$$

$$\Rightarrow -18x_1 - 12x_2 = -12x_1 - 18x_2$$

$$\Rightarrow -18x_1 + 12x_1 = -18x_2 + 12x_2$$

$$\Rightarrow -6x_1 = -6x_2$$

$$\Rightarrow x_1 = x_2$$

Hence, f is one-one function, ...(i)

Onto : Let $y = \frac{2x-4}{3x-9}$

$$\Rightarrow 3xy - 9y = 2x - 4$$

$$\Rightarrow 3xy - 2x = 9y - 4$$

$$\Rightarrow x(3y-2) = 9y - 4$$

$$\Rightarrow x = \frac{9y-4}{3y-2}$$

From above, it is clear that $\forall\, y \neq \frac{2}{3}$ *i.e.*, $\forall\, y \in B$, $\exists\, x \in A$

Hence, f is onto function. ...(ii)

From equation (i) and (ii)

$\Rightarrow f$ is one-one onto *i.e.*, bijective function.

Hence Proved.

13. *Let $f : x \to y$ be an invertible function. Show that f has unique inverse.*

Sol. Given, $f : x \to y$

Suppose $\quad y = f(x)$

Let $g_1(y)$ and $g_2(y)$ be two inverses of f

$$\Rightarrow fog_1(y) = I = y \quad (\because y \in I) \ldots\text{(i)}$$

$$fog_2(y) = I = y \quad \ldots\text{(ii)}$$

From (i) and (ii), we get

$$fog_1(y) = fog_2(y)$$

$$g_1(y) = g_2(y) \quad [\because f \text{ is one-one}]$$

Hence, f has a unique inverse. **Hence Proved.**

14. *Let A be a non-empty set and let * be a binary operation on P(A), the power set of A defined by*

$$X * Y = X \cap Y \text{ for } X, Y \in P(A)$$

(i) *Find the identity element with respect to * in P(A).*

(ii) *Show that A is the only invertible element of P(A).*

Sol. (i) Let E be the identity element with respect to * in P(A) then,

$$X * E = X = E * X \;\forall \in P(A)$$

$$X \cap E = X = E \cap X \;\forall X \in A \quad \ldots\text{(i)}$$

$$[\because X * E = X \cap E]$$

Also, $X \cap A = X = A \cap X$ ...(ii)

From equations (i) and (ii)

$$E = A$$

Hence, A is the identity element with respect to * in P (A).

(ii) Let X be an invertible element and B be the inverse of X

Then, $\quad X * B = E = B * X$

$$X * B = A = B * X \quad [\because E = A]$$

$$X \cap B = A = B \cap X$$

$$\Rightarrow X = B = A$$

Thus, A is the only invertible element of P (A) and it is the inverse of itself. **Hence Proved.**

15. *Consider $f : R_+ \to (4, \infty)$, given by $f(x) = x^2 + 4$. Show that f is invertible with the inverse f^{-1} of f given by $f^{-1}(y) = \sqrt{y-4}$, where R_+ is the set of all non-negative real numbers.*

Sol. $f(x) = x^2 + 4$, $D_f = R_+ [0, \infty)$

One-one function : Let x_1, x_2 be any two points on R_+ such that

$$f(x_1) = f(x_2)$$

$$\Rightarrow x_1^2 + 4 = x_2^2 + 4$$

$$\Rightarrow x_1^2 = x_2^2$$

$$\Rightarrow |x_1| = |x_2|$$

$$\Rightarrow x_1 - x_2$$

$[\because$ both x^1 and x^2 are non – negative$]$

$\Rightarrow f$ is one-one function

Onto function : Let $y \in (4, \infty)$ then

$$y = f(x) = x^2 + 4$$

$$\Rightarrow x^2 = 4 - y$$

$\Rightarrow \quad x = \sqrt{y-4}$

[– ve sign not taken $\because x \geq 0$]

Corresponding to every element y of $(4, \infty)$ $\exists$ a unique element $\sqrt{y-4}$ of R_+ s.t. $f(\sqrt{y-4}) = y$

$\Rightarrow f$ is onto

$\therefore f$ is both one-one and onto function.

$\Rightarrow f^{-1}$ exists

Inverse : since $f\left(\sqrt{y-4}\right) = y$

$\Rightarrow \quad \sqrt{y-4} = f^{-1}(y)$ **Ans.**

16. *Consider $f : R \to$ given by $f(x) = 4x + 3$. Show that f is invertible. Find the inverse of f.*

Sol. $f(x) = 4x + 3$

Injectivity : Let $x_1, x_2 \in R$

$\therefore \quad f(x_1) = f(x_2)$

$\Rightarrow \quad 4x_1 + 3 = 4x_2 + 3$

$\Rightarrow \quad 4x_1 = 4x_2$

$\Rightarrow \quad x_1 = x_2$

$\therefore f$ is one-one

Surjectivity : Let $y \in R_f$ then

$y = f(x)$

$\Rightarrow \quad y = 4x + 3$

$\Rightarrow \quad x = \dfrac{y-3}{4} \in R$

$\therefore$ Corresponding to each element y of R $\exists$ a unique element $\dfrac{y-3}{4}$ of R such that

$$f\left(\frac{y-3}{4}\right) = y$$

$\Rightarrow f$ is onto

$\Rightarrow f^{-1}$ exists

Inverse : As

$$f\left(\frac{y-3}{4}\right) = y$$

$\Rightarrow \quad \dfrac{y-3}{4} = f^{-1}(y)$

$\therefore \quad f^{-1}(x) = \dfrac{x-3}{4}$ **Ans.**

17. *Let $A = R - \{3\}$ and $B = R - \{1\}$. Consider the function $f : A \to B$ defined by $f(x) = \left(\dfrac{x-2}{x-3}\right)$. Show that f is one-one and onto and hence find f^{-1}.*

Sol. One-one :

Let $x_1, x_2 \in A$.

Now, $\quad f(x_1) = f(x_2)$

$\Rightarrow \quad \dfrac{x_1-2}{x_1-3} = \dfrac{x_2-2}{x_2-3}$

$\Rightarrow \quad (x_1 - 2)(x_2 - 3) = (x_1 - 3)(x_2 - 2)$

$\Rightarrow \quad x_1x_2 - 3x_1 - 2x_2 + 6 = x_1x_2 - 2x_1 - 3x_2 + 6$

$\Rightarrow \quad -3x_1 - 2x_2 = -2x_1 - 3x_2$

$\Rightarrow \quad -x_1 = -x_2$

$\Rightarrow \quad x_1 = x_2$

Hence, f is one-one function.

Onto :

Let, $\quad y = \dfrac{x-2}{x-3}$

$\Rightarrow \quad xy - 3y = x - 2$

$\Rightarrow \quad xy - x = 3y - 2$

$\Rightarrow \quad x(y - 1) = 3y - 2$

$\Rightarrow \quad x = \dfrac{3y-2}{y-1}$

From above it is obvious that $\forall\ y$ except 1, *i.e.*, $\forall\ y \in B = R - \{1\}\ \exists\ x \in A$

Hence, f is onto function.

Thus, f is one-one onto function.

Hence Proved.

If f^{-1} is inverse function of f then

$f^{-1}(y) = \dfrac{3y-2}{y-1}$ [From (i)]

Ans.

18. *Show that function f in $A = R - \left\{\dfrac{2}{3}\right\}$ defined as $f(x) = \dfrac{4x+3}{6x-4}$ is one-one and onto. Hence, find f^{-1}.*

Sol. One-one : Let $x_1, x_2 \in A$

Now, $\quad f(x_1) = f(x_2)$

$\Rightarrow \quad \dfrac{4x_1+3}{6x_1-4} = \dfrac{4x_2+3}{6x_2-4}$

$\Rightarrow 24x_1x_2 + 18x_2 - 16x_1 - 12$

$= 24x_1x_2 + 18x_1 - 16x_2 - 12$

$\Rightarrow \quad -34x_1 = -34x_2$

$\Rightarrow \quad x_1 = x_2$

Hence, f is one-one function.

Onto :

Let $\quad y = \dfrac{4x+3}{6x-4}$

$\Rightarrow \quad 6xy - 4y = 4x + 3$

$\Rightarrow \quad 6xy - 4x = 4y + 3$

$\Rightarrow \quad x(6y - 4) = 4y + 3$

$\Rightarrow \quad x = \dfrac{4y+3}{6y-4}$

$\Rightarrow \forall\, y \in$ codomain $\exists\, x \in$ domain $\left[\because x \neq \dfrac{2}{3}\right]$

$\Rightarrow$ f is onto function.

Thus, f is one-one onto function.

Also, $\quad f^{-1}(x) = \dfrac{4x+3}{6x-4}$ **Hence Proved.**

19. *If $f : A \to A$ and $A = R - \left\{\dfrac{8}{5}\right\}$, show that the function $f(x) = \dfrac{8x+3}{5x-8}$ is one-one onto. Hence, find f^{-1}.**

Sol. Given, $\quad f(x) = \dfrac{8x+3}{5x-8}$

For one-one, Let $x_1, x_2 \in$ A such that,

$$f(x_1) = f(x_2)$$

$\therefore \quad \dfrac{8x_1+3}{5x_1-8} = \dfrac{8x_2+3}{5x_2-8}$

$\Rightarrow \quad (8x_1 + 3)(5x_2 - 8) = (8x_2 + 3)(5x_1 - 8)$

$\Rightarrow 40x_1x_2 - 64x_1 + 15x_2 - 24 = 40x_1x_2 - 64x_2 + 15x_1 - 24$

$\Rightarrow \quad -64x_1 - 15x_1 = -64x_2 - 15x_2$

$\Rightarrow \quad -79x_1 = -79x_2$

$\Rightarrow \quad x_1 = x_2$

$\therefore f(x)$ is one-one **Hence Proved.**

For onto,

Let $\quad f(x) = y$

$\Rightarrow \quad y = \dfrac{8x+3}{5x-8}$

$\Rightarrow \quad y(5x - 8) = 8x + 3$

$\Rightarrow \quad 5xy - 8y = 8x + 3$

$\Rightarrow \quad 5xy - 8x = 8y + 3$

$\Rightarrow \quad x(5y - 8) = 8y + 3$

$\Rightarrow \quad x = \dfrac{8y+3}{5y-8}$, when $y \in$ A

$\therefore \quad 5y - 8 \neq 0$

$\quad y \neq \dfrac{8}{5}$

Range of $\quad f = \text{R} - \left\{\dfrac{8}{5}\right\} = \text{A}$

$\therefore \quad$ Codomain = Range

$\therefore f(x)$ is onto function. **Hence Proved**

$\therefore f$ is one one onto

$\Rightarrow f$ is invertible

$\Rightarrow \quad f(x) = y$

$\Rightarrow \quad f^{-1}(y) = x$

$\Rightarrow \quad f^{-1}(y) = \dfrac{8y+3}{5y-8}$

i.e., $\quad f^{-1}(x) = \dfrac{8x+3}{5x-8}$ **Ans.**

20. *If the function $f : R \to R$ be defined as** *$f(x) = \dfrac{3x+4}{5x-7}, \left(x \neq \dfrac{7}{5}\right)$ and $g : R \to R$ be defined as $g(x) = \dfrac{7x+4}{5x-3}, \left(x \neq \dfrac{3}{5}\right)$ show that $(g \circ f)(x) = (f \circ g)(x)$.*

Sol. Given : $\quad f(x) = \dfrac{3x+4}{5x-7}, x \neq \dfrac{7}{5}$

and $\quad g(x) = \dfrac{7x+4}{5x-3}, 3x \neq \dfrac{3}{5}$

Now, $\quad (g \circ f)(x) = g\{f(x)\}$

$$= g\left(\frac{3x+4}{5x-7}\right)$$

$$= \frac{7\left(\frac{3x+4}{5x-7}\right)+4}{5\left(\frac{3x+4}{5x-7}\right)-3}$$

$$= \frac{\frac{21x+28+20x-28}{5x-7}}{\frac{15x+20-15x+21}{5x-7}}$$

$$= \frac{41x}{41} = x$$

Also, $\quad (f \circ g)(x) = f\{g(x)\}$

$$= f\left(\frac{7x+4}{5x-3}\right)$$

$$= \frac{3\left(\frac{7x+4}{5x-3}\right)+4}{5\left(\frac{7x+4}{5x-3}\right)-7}$$

$$= \frac{\frac{21x+12+20x-12}{5x-3}}{\frac{35x+20-35x+21}{5x-3}}$$

$$= \frac{41x}{41} = x$$

$\therefore \quad (g \circ f)(x) = (f \circ g)(x) = x$

Thus, $\quad (g \circ f)(x) = (f \circ g)(x)$ **Hence Proved.**

* Frequently asked previous years Board Exam Questions.

Chapter 3. Inverse Trigonometric Functions

1. *Prove that :**

$\tan^2(\sec^{-1} 2) + \cot^2(\mathrm{cosec}^{-1} 3) = 11.$

Sol. To prove : $\tan^2(\sec^{-1} 2) + \cot^2(\mathrm{cosec}^{-1} 3) = 11$

Let $\sec^{-1} 2 = \theta$

$\Rightarrow \sec\theta = 2$

$\therefore \tan^2\theta = \sec^2\theta - 1 = 2^2 - 1 = 3$

Also, let $\mathrm{cosec}^{-1} 3 = \alpha$

$\Rightarrow \mathrm{cosec}\,\alpha = 3$

$\therefore \cot^2\alpha = \mathrm{cosec}^2\alpha - 1 = 3^2 - 1 = 8$

Now, L.H.S. $= \tan^2(\sec^{-1} 2) + \cot^2(\mathrm{cosec}^{-1} 3)$

$= \tan^2\theta + \cot^2\alpha$

$= 3 + 8 = 11 =$ R.H.S. **Hence Proved.**

2. *If* $\cos^{-1}\frac{x}{2} + \cos^{-1}\frac{y}{3} = \theta$, *then prove that*
$9x^2 - 12xy\cos\theta + 4y^2 = 36\sin^2\theta$

Sol. Given : $\cos^{-1}\frac{x}{2} + \cos^{-1}\frac{y}{3} = \theta$

$$\Rightarrow \cos^{-1}\left[\frac{x}{2}\times\frac{y}{3} - \sqrt{1-\frac{x^2}{4}}.\sqrt{1-\frac{y^2}{9}}\right] = \theta$$

$$\left[\because \cos^{-1}a + \cos^{-1}b = \cos^{-1}\left(ab - \sqrt{1-a^2}.\sqrt{1-b^2}\right)\right]$$

$$\Rightarrow \frac{xy}{6} - \sqrt{1-\frac{x^2}{4}}.\sqrt{1-\frac{y^2}{9}} = \cos\theta$$

$$\Rightarrow \frac{xy}{6} - \cos\theta = \sqrt{1-\frac{x^2}{4}}.\sqrt{1-\frac{y^2}{9}}$$

Squaring both sides, we get

$$\Rightarrow \left(\frac{xy}{6} - \cos\theta\right)^2 = \left(1-\frac{x^2}{4}\right)\left(1-\frac{y^2}{9}\right)$$

$$\Rightarrow \frac{x^2y^2}{36} + \cos^2\theta - \frac{2xy}{6}\cos\theta = 1 - \frac{x^2}{4} - \frac{y^2}{9} + \frac{x^2y^2}{36}$$

$$\Rightarrow \frac{x^2}{4} + \frac{y^2}{9} - \frac{xy\cos\theta}{3} = 1 - \cos^2\theta$$

$$\Rightarrow \frac{9x^2 + 4y^2 - 12xy\cos\theta}{36} = \sin^2\theta$$

$$\Rightarrow 9x^2 + 4y^2 - 12xy\cos\theta = 36\sin^2\theta$$

$$\Rightarrow 9x^2 - 12xy\cos\theta + 4y^2 = 36\sin^2\theta$$

Hence Proved.

3. *Show that*

$$\sin^{-1}\left(\frac{\sqrt{3}}{2}\right) + 2\tan^{-1}\left(\frac{1}{\sqrt{3}}\right) = \frac{2\pi}{3}.$$

Sol. Consider, L.H.S. $= \sin^{-1}\left(\frac{\sqrt{3}}{2}\right) + 2\tan^{-1}\left(\frac{1}{\sqrt{3}}\right)$

$$= \frac{\pi}{3} + 2\times\frac{\pi}{6} = \frac{\pi}{3} + \frac{\pi}{3} = \frac{2\pi}{3}$$

= R.H.S. **Hence Proved**

4. *Prove that*

$\sec^2(\tan^{-1} 2) + \mathrm{cosec}^2(\cot^{-1} 3) = 15.$

Sol. Consider,

L.H.S. $= \sec^2(\tan^{-1} 2) + \mathrm{cosec}^2(\cot^{-1} 3)$

$$= \sec^2\left(\sec^{-1}\sqrt{5}\right) + \mathrm{cosec}^2\left(\mathrm{cosec}^{-1}\sqrt{10}\right)$$

$$= \left\{\sec\left(\sec^{-1}\sqrt{5}\right)\right\}^2 + \left\{\mathrm{cosec}\left(\mathrm{cosec}^{-1}\sqrt{10}\right)\right\}^2$$

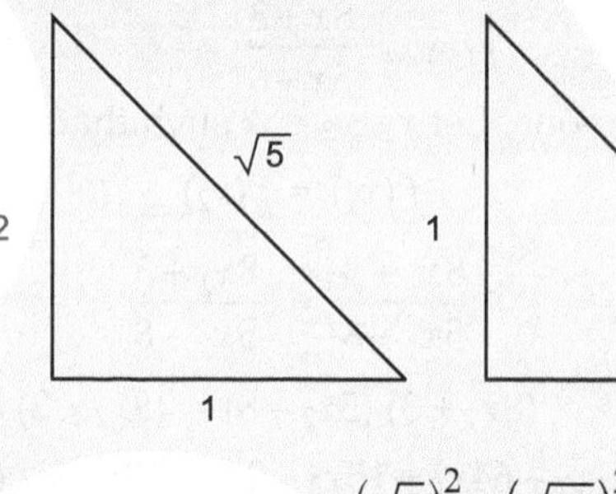

$$= \left(\sqrt{5}\right)^2 + \left(\sqrt{10}\right)^2$$

$= 5 + 10 = 15$

= R.H.S. **Hence Proved**

5. *Show that*

$2\cos^{-1}x = \cos^{-1}(2x^2 - 1).$

Sol. Let $\cos^{-1}x = \theta$

$\Rightarrow x = \cos\theta$

$\because \cos 2\theta = 2\cos^2\theta - 1 = 2x^2 - 1$

$\Rightarrow 2\theta = \cos^{-1}(2x^2 - 1)$

$\Rightarrow 2\cos^{-1}x = \cos^{-1}(2x^2 - 1)$

$\therefore$ L. H. S. = R. H. S. **Hence Proved.**

6. *Show that*

$$\frac{1}{2}\tan^{-1}x = \cos^{-1}\left[\sqrt{\frac{1+\sqrt{1+x^2}}{2\sqrt{1+x^2}}}\right]$$

Sol. Let $x = \tan\theta$

$\Rightarrow \theta = \tan^{-1}x.$

$$\text{R.H.S.} = \cos^{-1}\left[\sqrt{\frac{1+\sqrt{1+x^2}}{2\sqrt{1+x^2}}}\right]$$

$$= \cos^{-1}\left[\sqrt{\frac{1+\sqrt{1+\tan^2\theta}}{2\sqrt{1+\tan^2\theta}}}\right]$$

* Frequently asked previous years Board Exam Questions.

Prove the Following Type Questions

$$= \cos^{-1}\left[\sqrt{\frac{1+\sqrt{\sec^2\theta}}{2\sqrt{\sec^2\theta}}}\right]$$

$$= \cos^{-1}\left[\sqrt{\frac{1+\sec\theta}{2\sec\theta}}\right]$$

$$= \cos^{-1}\left[\sqrt{\frac{\cos\theta+1}{2}}\right]$$

$$= \cos^{-1}\left\{\sqrt{\frac{2\cos^2\frac{\theta}{2}-1+1}{2}}\right\}$$

$$= \cos^{-1}\left\{\frac{\sqrt{2\cos^2\frac{\theta}{2}}}{\sqrt{2}}\right\}$$

$$= \cos^{-1}\left(\cos\frac{\theta}{2}\right)$$

$$= \frac{\theta}{2} = \frac{\tan^{-1}x}{2} = \frac{1}{2}\tan^{-1}x$$

= L.H.S. **Hence Proved.**

7. *If* $\sin^{-1}x + \tan^{-1}x = \frac{\pi}{2}$, *prove that :*

$$2x^2 + 1 = \sqrt{5}$$

Sol. Given,

$$\sin^{-1}x + \tan^{-1}x = \frac{\pi}{2},$$

$$\Rightarrow \quad \tan^{-1}x = \frac{\pi}{2} - \sin^{-1}x$$

$$\Rightarrow \quad x = \tan\left(\frac{\pi}{2} - \sin^{-1}x\right)$$

$$\Rightarrow \quad x = \cot(\sin^{-1}x)$$

$$\Rightarrow \quad x = \cot\left(\cot^{-1}\frac{\sqrt{1-x^2}}{x}\right)$$

$$\left[\because \sin^{-1}x = \cot^{-1}\frac{\sqrt{1-x^2}}{x}\right]$$

$$\Rightarrow \quad x = \frac{\sqrt{1-x^2}}{x}$$

$$\Rightarrow \quad x^2 = \sqrt{1-x^2}$$

$$\Rightarrow \quad x^4 = 1 - x^2$$

$$\Rightarrow \quad x^4 + x^2 - 1 = 0$$

On solving, we get

$$x^2 = \frac{-1 \pm \sqrt{1-4(-1)}}{2}$$

$$\Rightarrow \quad x^2 = \frac{-1 \pm \sqrt{5}}{2}$$

$$\Rightarrow \quad 2x^2 + 1 = \sqrt{5}$$

Hence Proved.

8. *Show that :*

$$4\tan^{-1}\left(\frac{1}{5}\right) - \tan^{-1}\left(\frac{1}{70}\right) + \tan^{-1}\left(\frac{1}{99}\right) = \frac{\pi}{4}.$$

Sol. Consider,

$$\text{L.H.S.} = 4\tan^{-1}\frac{1}{5} - \tan^{-1}\frac{1}{70} + \tan^{-1}\frac{1}{99}$$

$$= 2\left(2\tan^{-1}\frac{1}{5}\right) - \left(\tan^{-1}\frac{1}{70} - \tan^{-1}\frac{1}{99}\right)$$

$$= 2\tan^{-1}\left[\frac{\frac{2}{5}}{1-\left(\frac{1}{5}\right)^2}\right] - \tan^{-1}\left[\frac{\left(\frac{1}{70}-\frac{1}{99}\right)}{\left(1+\frac{1}{70}\cdot\frac{1}{99}\right)}\right]$$

$$= 2\tan^{-1}\left(\frac{5}{12}\right) - \tan^{-1}\left(\frac{29}{6931}\right)$$

$$\left[\because \tan^{-1}x - \tan^{-1}y = \tan^{-1}\left(\frac{x-y}{1+xy}\right)\right.$$

$$\left.\because 2\tan^{-1}x = \tan^{-1}\left(\frac{2x}{1-x^2}\right)\right]$$

$$= 2\tan^{-1}\left(\frac{5}{12}\right) - \tan^{-1}\left(\frac{1}{239}\right)$$

$$= \tan^{-1}\left[\frac{2+\frac{5}{12}}{1-\left(\frac{5}{12}\right)^2}\right] - \tan^{-1}\left(\frac{1}{239}\right)$$

$$= \tan^{-1}\left(\frac{120}{119}\right) - \tan^{-1}\left(\frac{1}{239}\right)$$

$$= \tan^{-1}\left[\frac{\frac{120}{119} - \frac{1}{239}}{1+\frac{120}{119}\times\frac{1}{239}}\right]$$

$$= \tan^{-1}\left[\frac{28561}{28561}\right] = \tan^{-1}(1)$$

$$= \frac{\pi}{4} = \text{R.H.S.}$$

Hence Proved.

9. *If* $\cos^{-1}x + \cos^{-1}y + \cos^{-1}z = \pi$, *prove that* $x^2 + y^2 + z^2 + 2xyz = 1$.

Sol. Given,

$$\Rightarrow \quad \cos^{-1} x + \cos^{-1} y = \pi - \cos^{-1} z$$

$$\cos^{-1} x + \cos^{-1} y + \cos^{-1} z = \pi$$

$$\Rightarrow \quad \cos^{-1} x + \cos^{-1} y = \pi - \cos^{-1} z$$

$$\Rightarrow \cos^{-1}\left(xy - \sqrt{1-x^2}\sqrt{1-y^2}\right) = \pi - \cos^{-1} z$$

$$\Rightarrow \quad xy - \sqrt{1-x^2}\sqrt{1-y^2} = \cos(\pi - \cos^{-1} z)$$

$$\Rightarrow \quad xy - \sqrt{1-x^2}\sqrt{1-y^2} = -\cos(\cos^{-1} z)$$

$$\Rightarrow \quad xy - \sqrt{1-x^2}\sqrt{1-y^2} = -z$$

$$\Rightarrow \quad xy + z = \sqrt{1-x^2}\sqrt{1-y^2}$$

$$\Rightarrow \quad (xy+z)^2 = (1-x^2)(1-y^2)$$

$$\Rightarrow \quad x^2y^2 + z^2 + 2xyz = 1 - x^2 - y^2 + x^2y^2$$

$$\Rightarrow \quad x^2 + y^2 + z^2 + 2xyz = 1.$$

Hence Proved.

10. *Prove that :*

$$\tan^{-1}\left(\frac{1}{4}\right) + \tan^{-1}\left(\frac{2}{9}\right) = \frac{1}{2}\sin^{-1}\frac{4}{5}.$$

Sol. Consider,

$$\text{L.H.S.} = \tan^{-1}\left(\frac{1}{4}\right) + \tan^{-1}\left(\frac{2}{9}\right)$$

$$= \tan^{-1}\left(\frac{\frac{1}{4}+\frac{2}{9}}{1-\frac{1}{4}.\frac{2}{9}}\right)$$

$$= \tan^{-1}\left(\frac{\frac{17}{36}}{\frac{17}{18}}\right) = \tan^{-1}\left(\frac{1}{2}\right) \quad \text{...(i)}$$

$$\text{R.H.S.} = \frac{1}{2}\sin^{-1}\left(\frac{4}{5}\right) = \alpha \text{ (Say)}$$

$$\Rightarrow \quad 2\alpha = \sin^{-1}\left(\frac{4}{5}\right)$$

$$\Rightarrow \quad \sin 2\alpha = \frac{4}{5}$$

$$\Rightarrow \frac{2\tan\alpha}{1+\tan^2\alpha} = \frac{4}{5}$$

$$\Rightarrow \quad 4\tan^2\alpha - 10\tan\alpha + 4 = 0$$

$$\Rightarrow \quad 4\tan^2\alpha - 8\tan\alpha - 2\tan\alpha + 4 = 0$$

$$\Rightarrow 4\tan\alpha(\tan\alpha - 2) - 2(\tan\alpha - 2) = 0$$

$$\Rightarrow \quad (\tan\alpha - 2)(4\tan\alpha - 2) = 0$$

$$\Rightarrow \quad \tan\alpha = 2, \tfrac{1}{2}$$

$$\therefore \quad \tan\alpha = \frac{1}{2}$$

$$\left(\because -\frac{\pi}{2} \le 2\alpha \le \frac{\pi}{2}\right)$$

$$\Rightarrow \quad -\frac{\pi}{4} \le \alpha \le \frac{\pi}{4}$$

$$\therefore \quad \tan\alpha = \frac{1}{2}$$

$$\Rightarrow \quad \alpha = \tan^{-1}\frac{1}{2} \quad \text{...(ii)}$$

From equations (i) and (ii),

$$\tan^{-1}\frac{1}{4} + \tan^{-1}\frac{2}{9} = \frac{1}{2}\sin^{-1}\frac{4}{5}$$ **Hence Proved.**

11. *Prove that*

$$\sin^{-1}\left[\frac{x}{\sqrt{1+x^2}}\right] + \cos^{-1}\left[\frac{x+1}{\sqrt{x^2+2x+2}}\right] = \tan^{-1}(x^2+x+1).$$

Sol. Let $\sin^{-1}\left(\frac{x}{\sqrt{1+x^2}}\right) = \theta$ ...(i)

$$\Rightarrow \quad \sin\theta = \frac{x}{\sqrt{1+x^2}}$$

$$\Rightarrow \quad \tan\theta = \frac{x}{1}$$

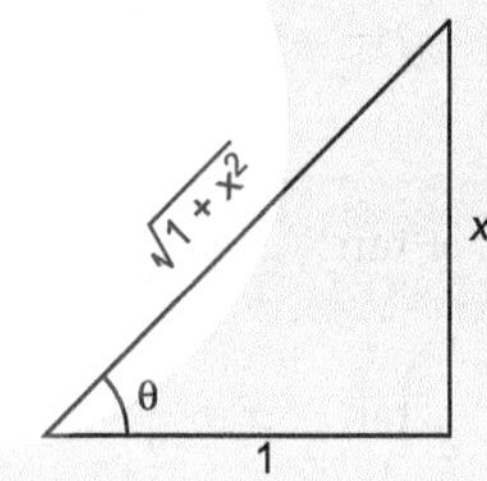

$$\Rightarrow \quad \theta = \tan^{-1}(x) \quad \text{...(ii)}$$

From equations (i) and (ii),

$$\sin^{-1}\left(\frac{x}{\sqrt{1+x^2}}\right) = \tan^{-1}(x) \quad \text{...(iii)}$$

Also let,

$$\cos^{-1}\left(\frac{x+1}{\sqrt{x^2+2x+2}}\right) = \phi \quad \text{...(iv)}$$

$$\Rightarrow \quad \cos\phi = \frac{x+1}{\sqrt{x^2+2x+2}}$$

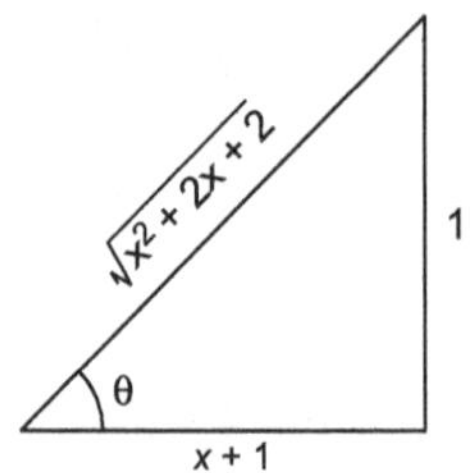

$$\Rightarrow \qquad \tan\phi = \frac{1}{x+1}$$

$$\Rightarrow \qquad \phi = \tan^{-1}\left(\frac{1}{x+1}\right) \qquad \ldots(v)$$

From equations (iv) and (v),

$$\cos^{-1}\left(\frac{x+1}{\sqrt{x^2+2x+2}}\right) = \tan^{-1}\left(\frac{1}{x+1}\right) \qquad \ldots(vi)$$

Now consider,

$$\text{L.H.S.} = \sin^{-1}\left(\frac{x}{\sqrt{1+x^2}}\right) + \cos^{-1}\left(\frac{x+1}{\sqrt{x^2+2x+2}}\right)$$

$$= \tan^{-1}(x) + \tan^{-1}\left(\frac{1}{x+1}\right)$$

[from equations (iii) and (vi)]

$$= \tan^{-1}\left(\frac{x+\frac{1}{x+1}}{1-\frac{x}{x+1}}\right) \qquad \text{[Using identity]}$$

$$= \tan^{-1}\left(\frac{\frac{x^2+x+1}{x+1}}{\frac{x+1-x}{x+1}}\right)$$

$$= \tan^{-1}(x^2+x+1)$$

$= \text{R.H.S.}$ **Hence Proved.**

12. *Prove that* $\cot\left(\frac{\pi}{4}-2\cos^{-1}(3)\right) = 7.$

Sol. Consider,

$$\text{L.H.S} = \cot\left[\frac{\pi}{4}-2\tan^{-1}\left(\frac{1}{3}\right)\right]$$

$$= \cot\left[\tan^{-1}(1)-\tan^{-1}\left(\frac{\frac{2}{3}}{1-\frac{1}{9}}\right)\right]$$

$$= \cot\left[\tan^{-1}1-\tan^{-1}\left(\frac{\frac{2}{3}}{\frac{8}{9}}\right)\right]$$

$$= \cot\left[\tan^{-1}1-\tan^{-1}\frac{3}{4}\right]$$

$$= \cot\left[\tan^{-1}\left(\frac{1-\frac{3}{4}}{1+\frac{3}{4}}\right)\right]$$

[Using identity]

$$= \cot\left[\tan^{-1}\frac{1}{7}\right] = \cot(\cot^{-1}7)$$

$= 7 = \text{R.H.S.}$ **Hence Proved.**

13. *Prove that :*

$$\cos^{-1}\left(\frac{63}{65}\right)+2\tan^{-1}\left(\frac{1}{5}\right)=\sin^{-1}\left(\frac{3}{5}\right).$$

Sol. We have to prove that

$$\cos^{-1}\left(\frac{63}{65}\right)+2\tan^{-1}\left(\frac{1}{5}\right) = \sin^{-1}\left(\frac{3}{5}\right)$$

Taking L.H.S., we have

$$\sin^{-1}\sqrt{1-\left(\frac{63}{65}\right)^2} + \sin^{-1}\left[\frac{2\cdot\frac{1}{5}}{1+\left(\frac{1}{5}\right)^2}\right]$$

$$\left[\begin{array}{l}\text{Using,}\\ \cos^{-1}x = \sin^{-1}\left[\sqrt{1-x^2}\right]\\ 2\tan^{-1}x = \sin^{-1}\left[\frac{2x}{1+x^2}\right]\end{array}\right]$$

$$= \sin^{-1}\left[\sqrt{1-\frac{3969}{4225}}\right]+\sin^{-1}\left(\frac{\frac{2}{5}}{\frac{26}{25}}\right)$$

$$= \sin^{-1}\left[\sqrt{\frac{4225-3969}{4225}}\right]+\sin^{-1}\left(\frac{5}{13}\right)$$

$$= \sin^{-1}\left[\sqrt{\frac{256}{4225}}\right]+\sin^{-1}\left(\frac{5}{13}\right)$$

$$= \sin^{-1}\left(\frac{16}{65}\right)+\sin^{-1}\left(\frac{5}{13}\right)$$

$$= \sin^{-1}\left\{\frac{16}{65}\sqrt{1-\left(\frac{5}{13}\right)^2}+\frac{5}{13}\sqrt{1-\left(\frac{16}{65}\right)^2}\right\}$$

$$\Big[\text{Using } \sin^{-1}x+\sin^{-1}y$$

$$= \sin^{-1}\left(x\left\{\sqrt{1-y^2}\right\}+y\sqrt{1-x^2}\right)\Big]$$

$$= \sin^{-1}\left(\frac{16}{65}\times\frac{12}{13}+\frac{5}{13}\times\frac{63}{65}\right)$$

$$= \sin^{-1}\left(\frac{16\times12+315}{13\times65}\right)$$

$$= \sin^{-1}\left(\frac{507}{65 \times 13}\right)$$

$$= \sin^{-1}\frac{3}{5} = \text{R.H.S.}$$ **Hence Proved.**

14. *Prove that :*

$$2\tan^{-1}\left(\frac{1}{5}\right) + \cos^{-1}\left(\frac{7}{5\sqrt{2}}\right) + 2\tan^{-1}\left(\frac{1}{8}\right) = \frac{\pi}{4}.$$

Sol. Consider,

$$\text{L.H.S.} = 2\tan^{-1}\left(\frac{1}{5}\right) + \cos^{-1}\left(\frac{7}{5\sqrt{2}}\right) + 2\tan^{-1}\left(\frac{1}{8}\right)$$

$$= 2\left(\tan^{-1}\frac{1}{5} + \tan^{-1}\frac{1}{8}\right) + \cos^{-1}\left(\frac{7}{5\sqrt{2}}\right)$$

$$= 2\tan^{-1}\left[\frac{\frac{1}{5}+\frac{1}{8}}{1-\frac{1}{5}\times\frac{1}{8}}\right] + \cos^{-1}\left(\frac{7}{5\sqrt{2}}\right)$$

$$\left[\because \tan^{-1}x + \tan^{-1}y = \tan^{-1}\left(\frac{x+y}{1-xy}\right)\right]$$

$$= 2\tan^{-1}\left[\frac{\frac{13}{40}}{\frac{39}{40}}\right] + \tan^{-1}\left[\frac{\sqrt{1-\frac{49}{50}}}{\frac{7}{5\sqrt{2}}}\right]$$

$$\left[\because \cos^{-1}x = \tan^{-1}\left(\frac{\sqrt{1-x^2}}{x}\right)\right]$$

$$= 2\tan^{-1}\left(\frac{1}{3}\right) + \tan^{-1}\frac{1}{7}$$

$$= \tan^{-1}\left(\frac{1}{3}\right) + \left(\tan^{-1}\frac{1}{3} + \tan^{-1}\frac{1}{7}\right)$$

$$= \tan^{-1}\left(\frac{1}{3}\right) + \tan^{-1}\left[\frac{\left(\frac{1}{3}+\frac{1}{7}\right)}{\left(1-\frac{1}{3}.\frac{1}{7}\right)}\right]$$

$$= \tan^{-1}\left(\frac{1}{3}\right) + \tan^{-1}\left(\frac{1}{2}\right)$$

$$= \tan^{-1}\left(\frac{\frac{1}{3}+\frac{1}{2}}{1-\frac{1}{3}.\frac{1}{2}}\right)$$

$$= \tan^{-1}\left(\frac{\frac{5}{6}}{\frac{5}{6}}\right) = \tan^{-1}1$$

$$= \frac{\pi}{4} = \text{R.H.S.}$$ **Hence Proved.**

15. *Show that :*

$$\sin^{-1}\left(\frac{1}{\sqrt{17}}\right) + \cos^{-1}\left(\frac{9}{\sqrt{85}}\right) = \tan^{-1}\left(\frac{1}{2}\right).$$

Sol.

Let, $\theta = \sin^{-1}\left(\frac{1}{\sqrt{17}}\right)$

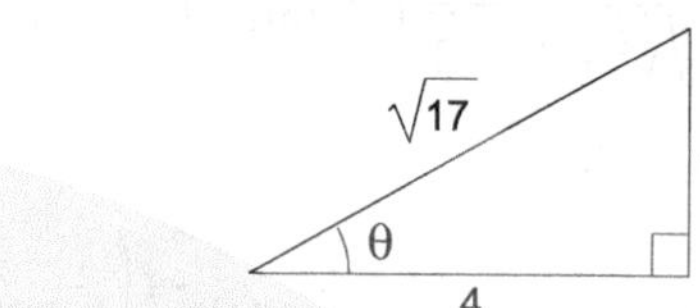

$\therefore$ $\sin\theta = \frac{1}{\sqrt{17}}$

$\therefore$ $\tan\theta = \frac{1}{4}$

or $\theta = \tan^{-1}\left(\frac{1}{4}\right)$

Also, let $\alpha = \cos^{-1}\left(\frac{9}{\sqrt{85}}\right)$

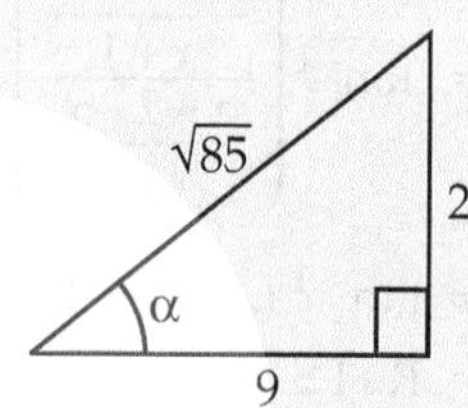

$\therefore$ $\cos\alpha = \frac{9}{\sqrt{85}}$

$\therefore$ $\tan\alpha = \frac{2}{9}$

or $\alpha = \tan^{-1}\left(\frac{2}{9}\right)$

Now, L.H.S. $= \tan^{-1}\left(\frac{1}{4}\right) + \tan^{-1}\left(\frac{2}{9}\right)$

$$= \tan^{-1}\left[\frac{\frac{1}{4}+\frac{2}{9}}{1-\frac{1}{4}\times\frac{2}{9}}\right]$$

$$= \tan^{-1}\left[\frac{\frac{9+8}{36}}{\frac{36-2}{36}}\right]$$

$$= \tan^{-1}\left(\frac{17}{34}\right) = \tan^{-1}\left(\frac{1}{2}\right)$$

= R.H.S. **Hence Proved.**

Prove the Following Type Questions

16. *Prove that*

$$tan^{-1}\left(\frac{1}{3}\right)+tan^{-1}\left(\frac{1}{5}\right)+tan^{-1}\left(\frac{1}{7}\right)+tan^{-1}\left(\frac{1}{8}\right) = \frac{\pi}{4}$$

Sol. Consider,

$$\text{L.H.S.} = \tan^{-1}\frac{1}{3}+\tan^{-1}\frac{1}{5}+\tan^{-1}\frac{1}{7}+\tan^{-1}\frac{1}{8}$$

$$= \tan^{-1}\left[\frac{\frac{1}{3}+\frac{1}{5}}{1-\frac{1}{3}\times\frac{1}{5}}\right]+\tan^{-1}\left[\frac{\frac{1}{7}+\frac{1}{8}}{1-\frac{1}{7}\times\frac{1}{8}}\right]$$

$$= \tan-1\left[\frac{\frac{8}{15}}{\frac{14}{15}}\right]+\tan^{-1}\left[\frac{\frac{15}{56}}{\frac{55}{56}}\right]$$

$$= \tan^{-1}\left(\frac{8}{14}\right)+\tan^{-1}\left(\frac{15}{55}\right)$$

$$= \tan^{-1}\left(\frac{4}{7}\right)+\tan^{-1}\left(\frac{3}{11}\right)$$

$$= \tan^{-1}\left[\frac{\frac{4}{7}+\frac{3}{11}}{1-\frac{4}{7}\times\frac{3}{11}}\right]$$

$$= \tan^{-1}\left[\frac{44+21}{77-12}\right] = \tan^{-1}\left(\frac{65}{65}\right)$$

$$= \tan^{-1}(1)$$

$$= \frac{\pi}{4}$$

= R.H.S. **Hence Proved.**

17. *Prove that*

$$tan^{-1}\left(\frac{1}{2}tan\ 2A\right)+tan^{-1}(cot\ A)+tan^{-1}(cot^3\ A) = 0.$$

Sol. Consider,

$$\text{L.H.S.} = \tan^{-1}\left(\frac{1}{2}\tan 2A\right)+\tan^{-1}\left[\frac{\cot A+\cot^3 A}{1-\cot^4 A}\right]$$

$$= \tan^{-1}\left(\frac{1}{2}\tan 2A\right)$$

$$+ \tan^{-1}\left[\frac{\cot A\,(1+\cot^2 A)}{(1+\cot^2 A)\,(1-\cot^2 A)}\right]$$

$$= \tan^{-1}\left(\frac{1}{2}\cdot\frac{2\tan A}{1-\tan^2 A}\right)+\tan^{-1}\left(\frac{\cot A}{1-\cot^2 A}\right)$$

$$= \tan^{-1}\left(\frac{\tan A}{1-\tan^2 A}\right)-\tan^{-1}\left(\frac{\tan A}{1-\tan^2 A}\right)$$

$$= 0$$

= R.H.S. **Hence Proved.**

18. *Prove that*

$$2\left(tan^{-1}1+tan^{-1}\frac{1}{2}+tan^{-1}\frac{1}{3}\right) = \pi.$$

Sol. Consider,

$$\text{L.H.S.} = 2\left[\tan^{-1}1+\tan^{-1}\frac{1}{2}+\tan^{-1}\frac{1}{3}\right]$$

$$= 2\left[\tan^{-1}\left[\frac{1+\frac{1}{2}}{1-\frac{1}{2}}\right]+\tan^{-1}\frac{1}{3}\right]$$

$$= 2\left[\tan^{-1}\left[\frac{\frac{3}{2}}{\frac{1}{2}}\right]+\tan^{-1}\left(\frac{1}{3}\right)\right]$$

$$= 2\left[\tan^{-1}(3)+\tan^{-1}\left(\frac{1}{3}\right)\right]$$

$$= 2\left[\tan^{-1}\left(\frac{3+\frac{1}{3}}{1-3\times\frac{1}{3}}\right)\right] = 2\left[\tan^{-1}\left(\frac{\frac{10}{3}}{0}\right)\right]$$

$$= 2\tan^{-1}(\infty) = 2\times\frac{\pi}{2}$$

$$= \pi$$

= R.H.S. **Hence Proved.**

19. *Prove that*

$$tan^{-1}x + cot^{-1}(x+1) = tan^{-1}(x^2+x+1).$$

Sol. Consider,

$$\text{L.H.S.} = \tan^{-1}x+\cot^{-1}(x+1)$$

$$= \tan^{-1}x+\tan^{-1}\left(\frac{1}{x+1}\right)$$

$$= \tan^{-1}\left\{\frac{x+\frac{1}{x+1}}{1-x\times\frac{1}{x+1}}\right\}$$

$$= \tan^{-1}\left[\frac{x^2+x+1}{x+1-x}\right]$$

$$= \tan^{-1}(x^2+x+1)$$

= R.H.S. **Hence Proved.**

20. *Prove that :*

$$\sin\left[2\tan^{-1}\left(\frac{3}{5}\right)-\sin^{-1}\left(\frac{7}{25}\right)\right]=\frac{304}{425}.$$

Sol.

L.H.S.

$$=\sin\left[2\tan^{-1}\left(\frac{3}{5}\right)-\sin^{-1}\left(\frac{7}{25}\right)\right]$$

$$=\sin\left[\sin^{-1}\frac{2\times\frac{3}{5}}{1+\left(\frac{3}{5}\right)^2}-\sin^{-1}\left(\frac{7}{25}\right)\right]$$

$$=\sin\left[\sin^{-1}\left(\frac{\frac{6}{5}}{\frac{34}{25}}\right)-\sin^{-1}\left(\frac{7}{25}\right)\right]$$

$$=\sin\left[\sin^{-1}\left(\frac{15}{17}\right)-\sin^{-1}\left(\frac{7}{25}\right)\right]$$

$$=\sin\left[\sin^{-1}\left\{\left(\frac{15}{17}\right)\sqrt{1-\frac{49}{625}}-\frac{7}{25}\sqrt{1-\frac{225}{289}}\right\}\right]$$

$$=\frac{15}{17}\times\frac{24}{25}-\frac{7}{25}\times\frac{8}{17}$$

$$=\frac{360-56}{425}=\frac{304}{425}$$

= R.H.S. **Hence Proved.**

21. *Prove that*

$$\tan^{-1}\left(\frac{1}{4}\right)+\tan^{-1}\left(\frac{2}{9}\right)=\left(\frac{1}{2}\right)\cos^{-1}\left(\frac{3}{5}\right).$$

Sol. Consider, L.H.S.

$$=\tan^{-1}\left(\frac{1}{4}\right)+\tan^{-1}\left(\frac{2}{9}\right)$$

$$=\tan^{-1}\left[\frac{\frac{1}{4}+\frac{2}{9}}{1-\frac{1}{4}\times\frac{2}{9}}\right]$$

$$=\tan^{-1}\left[\frac{\frac{17}{36}}{\frac{34}{36}}\right]=\tan^{-1}\left(\frac{1}{2}\right)$$

Multiplying L.H.S. by 2, we get

$$2\times\text{L.H.S.}=2\tan^{-1}\frac{1}{2}$$

$$=\tan^{-1}\left[\frac{2\times\frac{1}{2}}{1-\left(\frac{1}{2}\right)^2}\right]$$

$$=\tan^{-1}\left[\frac{\frac{1}{3}}{\frac{3}{4}}\right]=\tan^{-1}\left(\frac{4}{3}\right)$$

$$=\cos^{-1}\left(\frac{3}{5}\right)$$

Hence, $\text{L.H.S.}=\frac{1}{2}\cos^{-1}\frac{3}{5}$

= R.H.S. **Hence Proved.**

22. *If $\tan^{-1}a+\tan^{-1}b+\tan^{-1}c=\pi$, then prove that $a+b+c=abc$.*

Sol. Given, $\tan^{-1}a+\tan^{-1}b+\tan^{-1}c=\pi$

$$\Rightarrow \tan^{-1}\left[\frac{a+b}{1-ab}\right]+\tan^{-1}c=\pi$$

$$\Rightarrow \tan^{-1}\left\{\frac{\frac{a+b}{1-ab}+c}{1-\frac{(a+b)}{(1-ab)}\times c}\right\}=\pi$$

$$\Rightarrow \tan^{-1}\left\{\frac{a+b+c(1-ab)}{(1-ab)-(a+b)c}\right\}=\pi$$

$$\Rightarrow \frac{a+b+c-abc}{1-ab-ac-bc}=\tan\pi$$

$$\Rightarrow a+b+c-abc=0$$

$$\Rightarrow a+b+c=abc$$

Hence Proved.

23. *Show that*

$$\sin^{-1}\left(\frac{4}{5}\right)+\cos^{-1}\left(\frac{2}{\sqrt{5}}\right)=\cot^{-1}\left(\frac{2}{11}\right).$$

Sol. Consider,

$$\text{L.H.S.}=\sin^{-1}\left(\frac{4}{5}\right)+\cos^{-1}\left(\frac{2}{\sqrt{5}}\right)$$

As, $\sin^{-1}\left(\frac{4}{5}\right)=\tan^{-1}\left(\frac{4}{3}\right)$

and $\cos^{-1}\left(\frac{2}{\sqrt{5}}\right)=\tan^{-1}\left(\frac{1}{2}\right)$

$$\text{L.H.S.}=\tan^{-1}\left(\frac{4}{3}\right)+\tan^{-1}\left(\frac{1}{2}\right)$$

$$=\tan^{-1}\left[\frac{\frac{4}{3}+\frac{1}{2}}{1-\frac{4}{3}\times\frac{1}{2}}\right]$$

$$=\tan^{-1}\left[\frac{\frac{11}{6}}{\frac{6-4}{6}}\right]$$

Prove the Following Type Questions

$$= \tan^{-1}\left[\frac{11}{2}\right] = \cot^{-1}\left[\frac{2}{11}\right]$$

= R.H.S. **Hence Proved.**

24. *If $\sec^{-1} x = \text{cosec}^{-1} y$, show that* $\frac{1}{x^2}+\frac{1}{y^2} = 1$.*

Sol. Given, $\sec^{-1} x = \text{cosec}^{-1} y$

$$\Rightarrow \cos^{-1}\frac{1}{x} = \sin^{-1}\frac{1}{y}$$

$$\Rightarrow \cos^{-1}\frac{1}{x} = \frac{\pi}{2} - \cos^{-1}\frac{1}{y}$$

$$\Rightarrow \cos^{-1}\frac{1}{x} + \cos^{-1}\frac{1}{y} = \frac{\pi}{2}$$

$$\Rightarrow \cos^{-1}\left(\frac{1}{x}\times\frac{1}{y} - \sqrt{1-\frac{1}{x^2}}\sqrt{1-\frac{1}{y^2}}\right) = \frac{\pi}{2}$$

$$\Rightarrow \frac{1}{xy} - \sqrt{\frac{x^2-1}{x^2}}\sqrt{\frac{y^2-1}{y^2}} = \cos\frac{\pi}{2}$$

$$\Rightarrow \frac{1-\sqrt{x^2-1}\sqrt{y^2-1}}{xy} = 0$$

$$\Rightarrow 1-\sqrt{x^2-1}\sqrt{y^2-1} = 0$$

$$\Rightarrow 1 = \sqrt{x^2-1}\sqrt{y^2-1}$$

On squaring both sides, we get

$$(x^2-1)(y^2-1) = 1$$

$$\Rightarrow x^2y^2 - x^2 - y^2 + 1 = 1$$

$$\Rightarrow x^2 + y^2 = x^2y^2$$

$$\Rightarrow \frac{x^2}{x^2y^2} + \frac{y^2}{x^2y^2} = 1$$

[Dividing by x^2y^2 on both sides]

$$\Rightarrow \frac{1}{x^2}+\frac{1}{y^2} = 1$$ **Hence Proved.**

Alternatively,

$$\sec^{-1} x = \text{cosec}^{-1} y$$

$$\Rightarrow \cos^{-1}\frac{1}{x} = \sin^{-1}\frac{1}{y}$$

$$\Rightarrow \cos^{-1}\frac{1}{x} = \cos^{-1}\sqrt{1-\frac{1}{y^2}}$$

$$\left[\because \sin^{-1}\alpha = \cos^{-1}\sqrt{1-\alpha^2}\right]$$

$$\Rightarrow \frac{1}{x} = \sqrt{1-\frac{1}{y^2}}$$

On squaring both sides, we get

$$\frac{1}{x^2} = 1-\frac{1}{y^2}$$

$$\Rightarrow \frac{1}{x^2}+\frac{1}{y^2} = 1$$ **Hence Proved.**

Chapter 4. Determinants

1. *Using properties of determinants, show that**

$$\begin{vmatrix} x & p & q \\ p & x & q \\ q & q & x \end{vmatrix} = (x-p)(x^2+px-2q^2)$$

Sol. To prove :

$$\begin{vmatrix} x & p & q \\ p & x & q \\ q & q & x \end{vmatrix} = (x-p)(x^2+px-2q^2)$$

Let $\Delta = \text{LHS} = \begin{vmatrix} x & p & q \\ p & x & q \\ q & q & x \end{vmatrix}$

Applying $C_1 \to C_1 - C_2$, we get

$$\Delta = \begin{vmatrix} x-p & p & q \\ p-x & x & q \\ q-q & q & x \end{vmatrix}$$

Taking $(x-p)$ as common from C_1, we get

$$\Delta = (x-p)\begin{vmatrix} 1 & p & q \\ -1 & x & q \\ 0 & q & x \end{vmatrix}$$

Now, expanding along R_1, we get

$$\Delta = (x-p)[1(x^2-q^2) - p(-x-0) + q(-q-0)]$$

$$= (x-p)[x^2-q^2+px-q^2]$$

$= (x-p)(x^2+px-2q^2) = \text{RHS}$. **Hence Proved.**

2. *Using properties of determinants, prove that :*

$$\begin{vmatrix} a & a+b & a+b+c \\ 2a & 3a+2b & 4a+3b+2c \\ 3a & 6a+3b & 10a+6b+3c \end{vmatrix} = a^3$$

Sol. Consider,

$$\text{L.H.S.} = \begin{vmatrix} a & a+b & a+b+c \\ 2a & 3a+2b & 4a+3b+2c \\ 3a & 6a+3b & 10a+6b+3c \end{vmatrix}$$

* Frequently asked previous years Board Exam Questions.

Prove the Following Type Questions

Operating $R_2 \to R_2 - 2R_1$ and $R_3 \to R_3 - 3R_1$, we get

$$= \begin{vmatrix} a & a+b & a+b+c \\ 0 & a & 2a+b \\ 0 & 3a & 7a+3b \end{vmatrix}$$

Expanding along C_1, we get

$$= a\begin{vmatrix} a & 2a+b \\ 3a & 7a+3b \end{vmatrix}$$

$$= a\,[a\,(7a+3b) - 3a\,(2a+b)]$$

$$= a\,[7a^2 + 3ab - 6a^2 - 3ab]$$

$$= a \cdot a^2$$

$$= a^3$$ **Hence Proved.**

3. *Using properties of determinants, prove that*

$$\begin{vmatrix} x^2 & y^2 & z^2 \\ x^3 & y^3 & z^3 \\ xyz & yzx & zxy \end{vmatrix} = xyz\,(x-y)\,(y-z)\,(z-x)\,(xy+yz+zx)$$

Sol. Consider,

$$\text{L.H.S} = \begin{vmatrix} x^2 & y^2 & z^2 \\ x^3 & y^3 & z^3 \\ xyz & yzx & zxy \end{vmatrix}$$

$$= xyz\begin{vmatrix} x & y & z \\ x^2 & y^2 & z^2 \\ yz & xz & xy \end{vmatrix}$$

[We have taken x, y, z common from columns 1, 2, 3 respectively]

Operating $C_1 \to C_1 - C_2$, we obtain

$$= xyz\begin{vmatrix} x-y & y & z \\ x^2-y^2 & y^2 & z^2 \\ yz-xz & xz & xy \end{vmatrix}$$

$$= xyz\begin{vmatrix} (x-y) & y & z \\ (x-y)\,(x+y) & y^2 & z^2 \\ z\,(y-x) & xz & xy \end{vmatrix}$$

$$= xyz\,(x-y)\begin{vmatrix} 1 & y & z \\ x+y & y^2 & z^2 \\ -z & xz & xy \end{vmatrix}$$

Operating $C_2 \to C_2 - C_3$, we obtain

$$= xyz\,(x-y)\begin{vmatrix} 1 & y-z & z \\ x+y & y^2-z^2 & z^2 \\ -z & xz-xy & xy \end{vmatrix}$$

$$= xyz\,(x-y)\,(y-z)\begin{vmatrix} 1 & 1 & z \\ x+y & y+z & z^2 \\ -z & -x & xy \end{vmatrix}$$

Operating $C_2 \to C_2 - C_1$, we obtain

$$= xyz\,(x-y)\,(y-z)\begin{vmatrix} 1 & 0 & z \\ x+y & z-x & z^2 \\ -z & z-x & xy \end{vmatrix}$$

$$= xzy\,(x-y)\,(y-z)\,(z-x)\begin{vmatrix} 1 & 0 & z \\ x+y & 1 & z^2 \\ -z & 1 & xy \end{vmatrix}$$

Expanding along R_1, we get

$$= xyz\,(x-y)\,(y-z)\,(z-x)\,[1\,(xy-z^2) - 0\,\{(x^2y+xy^2 + z^3\} + z(x+y+z)]$$

$$= xyz\,(x-y)\,(y-z)\,(z-x)\,[xy - z^2 + xz + yz + z^2]$$

$$= xyz\,(x-y)\,(y-z)\,(z-x)\,(xy+yz+xz).$$

Hence Proved

4. *Using properties of determinants prove that*

$$\begin{vmatrix} y+z & x+y & x \\ z+x & y+z & y \\ x+y & z+x & z \end{vmatrix} = (x^3+y^3+z^3-3xyz).$$

Sol.

$$\text{L.H.S.} = \begin{vmatrix} y+z & x+y & x \\ z+x & y+z & y \\ x+y & z+x & z \end{vmatrix}$$

Operating $C_1 \to C_1 + C_3$ and $C_2 \to C_2 - C_3$, we obtain

$$= \begin{vmatrix} x+y+z & y & x \\ x+y+z & z & y \\ x+y+z & x & z \end{vmatrix}$$

Taking $(x+y+z)$ common from C_1

$$= (x+y+z)\begin{vmatrix} 1 & y & x \\ 1 & z & y \\ 1 & x & z \end{vmatrix}$$

Operating $R_2 \to R_2 - R_1$ and $R_3 \to R_3 - R_2$

$$= (x+y+z)\begin{vmatrix} 1 & y & x \\ 0 & z-y & y-x \\ 0 & x-z & z-y \end{vmatrix}$$

Expanding along C_1, we obtain

$$= (x+y+z)\,[1\,(z-y)\,(z-y) - (y-x)\,(x-z)]$$

$$= (x+y+z)\,[(z^2-yz-yz+y^2) - (yx-x^2-zy+xz)]$$

$$= (x+y+z)\,(x^2+y^2+z^2-xy-yz-zx)$$

$$= x^3+y^3+z^3-3xyz.$$ **Hence Proved.**

5. *Using properties of determinants, prove that*

$$\begin{vmatrix} a & b-c & c-b \\ a-c & b & c-a \\ a-b & b-a & c \end{vmatrix} = (a+b-c)\,(b+c-a)$$

Sol. Consider,

$$\text{L.H.S.} = \begin{vmatrix} a & b-c & c-b \\ a-c & b & c-a \\ a-b & b-a & c \end{vmatrix}$$

Operating $C_1 \to C_1 + C_2$ and $C_2 \to C_2 + C_3$, we obtain

$$= \begin{vmatrix} a+b-c & 0 & c-b \\ a+b-c & b+c-a & c-a \\ 0 & b+c-a & c \end{vmatrix}$$

Taking $(a + b - c)$ common from C_1 and $(b + c - a)$ common from C_2, we get

$$= (a+b-c)(b+c-a)\begin{vmatrix} 1 & 0 & c-b \\ 1 & 1 & c-a \\ 0 & 1 & c \end{vmatrix}$$

Expanding along R_1, we get

$= (a + b - c) (b + c - a) [(c - c + a) + (c - b) (1)]$

$= (a + b - c) (b + c - a) (a + c - b)$. **Hence Proved.**

6. *Using properties of determinants, show that $p\alpha^2 + 2q\alpha + r = 0$, given that :*

p, q and r are not in G. P. and

$$\begin{vmatrix} 1 & \frac{q}{p} & \alpha + \frac{q}{p} \\ 1 & \frac{r}{q} & \alpha + \frac{r}{q} \\ p\alpha + q & q\alpha + r & 0 \end{vmatrix} = 0$$

Sol. As p, q and r are not in G. P.

$$\Rightarrow \frac{q}{p} \neq \frac{r}{q} \Rightarrow \left(\frac{q}{p} - \frac{r}{q}\right) \neq 0 \qquad \text{..(i)}$$

Given, $$\begin{vmatrix} 1 & \frac{q}{p} & \alpha + \frac{q}{p} \\ 1 & \frac{r}{q} & \alpha + \frac{r}{q} \\ p\alpha + q & q\alpha + r & 0 \end{vmatrix} = 0$$

Applying $R_1 \to R_1 - R_2$, we obtain

$$\begin{vmatrix} 0 & \frac{q}{p} - \frac{r}{q} & \frac{q}{p} - \frac{r}{q} \\ 1 & \frac{r}{q} & \alpha + \frac{r}{q} \\ p\alpha + q & q\alpha + r & 0 \end{vmatrix} = 0$$

Taking $\left(\frac{q}{p} - \frac{r}{q}\right)$ common from R_1, we get

$$\Rightarrow \left(\frac{q}{p} - \frac{r}{q}\right)\begin{vmatrix} 0 & 1 & 1 \\ 1 & \frac{r}{q} & \alpha + \frac{r}{q} \\ p\alpha + q & q\alpha + r & 0 \end{vmatrix} = 0$$

But by equation (i), $\left(\frac{q}{p} - \frac{r}{q}\right) \neq 0$

$$\therefore \begin{vmatrix} 0 & 1 & 1 \\ 1 & \frac{r}{q} & \alpha + \frac{r}{q} \\ p\alpha + q & q\alpha + r & 0 \end{vmatrix} = 0$$

Applying $C_2 \to C_2 - C_3$, we obtain

$$\Rightarrow \begin{vmatrix} 0 & 0 & 1 \\ 1 & -\alpha & \alpha + \frac{r}{q} \\ p\alpha + q & q\alpha + r & 0 \end{vmatrix} = 0$$

Expanding along R_1, we get

$\Rightarrow \quad (q\alpha + r) + \alpha (p\alpha + q) = 0$

$\Rightarrow \quad q\alpha + r + p\alpha^2 + q\alpha = 0$

$\Rightarrow \quad p\alpha^2 + 2q\alpha + r = 0$. **Hence Proved.**

7. *Prove that*

$$\begin{vmatrix} y+z & z & y \\ z & z+x & x \\ y & x & x+y \end{vmatrix} = 4xyz.$$

Sol. $$\text{L. H. S.} = \begin{vmatrix} y+z & z & y \\ z & z+x & x \\ y & x & x+y \end{vmatrix}$$

Applying $C_1 \to C_1 - C_2 - C_3$, we obtain

$$\text{L. H. S.} = \begin{vmatrix} 0 & z & y \\ -2x & z+x & x \\ -2x & x & x+y \end{vmatrix}$$

Now applying $R_2 \to R_2 - R_3$, we obtain

$$\text{L. H. S.} = \begin{vmatrix} 0 & z & y \\ 0 & z & -y \\ -2x & x & x+y \end{vmatrix}$$

Expanding along C_1, we get

$= -2x(-zy - zy)$

$= -2x(-2zy)$

$= 4xyz$

$=$ R.H.S. **Hence Proved.**

8. *If x, y and z are all different and*

$$\begin{vmatrix} x & x^2 & 1+x^3 \\ y & y^2 & 1+y^3 \\ z & z^2 & 1+z^3 \end{vmatrix} = 0$$

then show that $xyz = -1$.

Sol. Given,

$$\begin{vmatrix} x & x^2 & 1+x^3 \\ y & y^2 & 1+y^3 \\ z & z^2 & 1+z^3 \end{vmatrix} = 0$$

Consider, $\text{L.H.S.} = \begin{vmatrix} x & x^2 & x^3 \\ y & y^2 & y^3 \\ z & z^2 & z^3 \end{vmatrix} + \begin{vmatrix} x & x^2 & 1 \\ y & y^2 & 1 \\ z & z^2 & 1 \end{vmatrix}$

Taking x common from R_1, y common from R_2 and z common from R_3, we get

$$= xyz\begin{vmatrix} 1 & x & x^2 \\ 1 & y & y^2 \\ 1 & z & z^2 \end{vmatrix} + \begin{vmatrix} x & x^2 & 1 \\ y & y^2 & 1 \\ z & z^2 & 1 \end{vmatrix}$$

$$= xyz\begin{vmatrix} 1 & x & x^2 \\ 1 & y & y^2 \\ 1 & z & z^2 \end{vmatrix} + (-1)^2\begin{vmatrix} 1 & x & x^2 \\ 1 & y & y^2 \\ 1 & z & z^2 \end{vmatrix}$$

$$\Rightarrow \begin{vmatrix} 1 & x & x^2 \\ 1 & y & y^2 \\ 1 & z & z^2 \end{vmatrix}(1+xyz) = 0 \qquad \text{...(i)}$$

Consider, $\begin{vmatrix} 1 & x & x^2 \\ 1 & y & y^2 \\ 1 & z & z^2 \end{vmatrix}$

Now, applying $R_2 \to R_2 - R_1$ and $R_3 \to R_3 - R_1$, we obtain

$$= \begin{vmatrix} 1 & x & x^2 \\ 0 & y-x & y^2-x^2 \\ 0 & z-x & z^2-x^2 \end{vmatrix}$$

Taking $(y - x)$ common from R_2 and $(z - x)$ common from R_3, we get

$$= (y-x)(z-x)\begin{vmatrix} 1 & x & x^2 \\ 0 & 1 & y+x \\ 0 & 1 & z+x \end{vmatrix}$$

Expanding along C_1, we get

$= (y - x)(z - x)\,1\,[(z + x) - (y + x)]$

$= (x - y)(y - z)(z - x)$ ...(ii)

Therefore, from equation (i) and (ii),

$(x - y)(y - z)(z - x)(1 + xyz) = 0$

$\because \quad x \neq y \neq z$

$\Rightarrow$ So, $\quad (1 + xyz) = 0$

$xyz = -1$ **Hence Proved.**

9. *Show that*

$$\begin{vmatrix} \sin^2 A & \sin A & \cos^2 A \\ \sin^2 B & \sin B & \cos^2 B \\ \sin^2 C & \sin C & \cos^2 C \end{vmatrix} = -(\sin A - \sin B)(\sin B - \sin C)(\sin C - \sin A).$$

Sol. Consider,

$$\text{L.H.S.} = \begin{vmatrix} \sin^2 A & \sin A & \cos^2 A \\ \sin^2 B & \sin B & \cos^2 B \\ \sin^2 C & \sin C & \cos^2 C \end{vmatrix}$$

Applying $C_3 \to C_3 + C_1$, we obtain

$$= \begin{vmatrix} \sin^2 A & \sin A & \sin^2 A + \cos^2 A \\ \sin^2 B & \sin B & \sin^2 B + \cos^2 B \\ \sin^2 C & \sin C & \sin^2 C + \cos^2 C \end{vmatrix}$$

$$= \begin{vmatrix} \sin^2 A & \sin A & 1 \\ \sin^2 B & \sin B & 1 \\ \sin^2 C & \sin C & 1 \end{vmatrix}$$

Applying $R_1 \to R_1 - R_2$ and $R_2 \to R_2 - R_3$, we obtain

$$= \begin{vmatrix} \sin^2 A - \sin^2 B & \sin A - \sin B & 0 \\ \sin^2 B - \sin^2 C & \sin B - \sin C & 0 \\ \sin^2 C & \sin C & 1 \end{vmatrix}$$

$$= \begin{vmatrix} (\sin A - \sin B)(\sin A + \sin B) & \sin A - \sin B & 0 \\ (\sin B - \sin C)(\sin B + \sin C) & \sin B - \sin C & 0 \\ \sin^2 C & \sin C & 1 \end{vmatrix}$$

Taking $(\sin A - \sin B)$ common from R_1 and $(\sin B - \sin C)$ common from R_2, we get

$$= (\sin A - \sin B)(\sin B - \sin C)\begin{vmatrix} \sin A + \sin B & 1 & 0 \\ \sin B + \sin C & 1 & 0 \\ \sin^2 C & \sin C & 1 \end{vmatrix}$$

Now, expanding along C_3, we get

$= (\sin A - \sin B)(\sin B - \sin C) \times 1 \{(\sin A + \sin B) - (\sin B + \sin C)\}$

$= (\sin A - \sin B)(\sin B - \sin C)\{\sin A - \sin C\}$

$= -(\sin A - \sin B)(\sin B - \sin C)(\sin C - \sin A)$

= R. H. S. **Hence Proved.**

10. *By using properties of determinants, prove that the determinant*

$$\begin{vmatrix} a & \sin x & \cos x \\ -\sin x & -a & 1 \\ \cos x & 1 & a \end{vmatrix}$$ *is independent of x.*

Sol.

Let $$\Delta = \begin{vmatrix} a & \sin x & \cos x \\ -\sin x & -a & 1 \\ \cos x & 1 & a \end{vmatrix}$$

Applying $C_1 \to C_1 + C_2$, we obtain

$$\Delta = \begin{vmatrix} a+\sin x & \sin x & \cos x \\ -(a+\sin x) & -a & 1 \\ 1+\cos x & 1 & a \end{vmatrix}$$

Applying $R_1 \to R_1 + R_2$, we obtain

$$\Delta = \begin{vmatrix} 0 & -a+\sin x & 1+\cos x \\ -(a+\sin x) & -a & 1 \\ 1+\cos x & 1 & a \end{vmatrix}$$

Now expanding along R_1, we get

$$\Delta = -(-a+\sin x)[-a^2 - a\sin x - 1 - \cos x]$$
$$+ (1+\cos x)[-a - \sin x + a + a\cos x]$$
$$= -[a^3 + a^2\sin x + a + a\cos x - a^2\sin x - a\sin^2 x$$
$$- \sin x - \sin x\cos x] + [a\cos x - \sin x$$
$$- \sin x\cos x + a\cos^2 x]$$
$$= -a^3 - a + a(\sin^2 x + \cos^2 x) = -a^3 + a - a$$
$$= -a^3, \text{ which is independent of } x.$$

Hence Proved.

11. *Using properties of determinants show that*

$$\begin{vmatrix} b^2c^2 & bc & b+c \\ c^2a^2 & ca & c+a \\ a^2b^2 & ab & a+b \end{vmatrix} = 0.$$

Sol. Consider,

$$\text{L.H.S} = \begin{vmatrix} b^2c^2 & bc & b+c \\ c^2a^2 & ca & c+a \\ a^2b^2 & ab & a+b \end{vmatrix}$$

Applying $R_1 \to aR_1$, $R_2 \to bR_2$ and $R_3 \to cR_3$, we obtain

$$\text{L.H.S.} = \frac{1}{abc}\begin{vmatrix} ab^2c^2 & abc & ab+ac \\ bc^2a^2 & abc & bc+ab \\ a^2b^2c & abc & ac+bc \end{vmatrix}$$

Taking abc common from C_1 and C_2, we get

$$= \frac{(abc)^2}{abc}\begin{vmatrix} bc & 1 & ab+ac \\ ca & 1 & bc+ab \\ ab & 1 & ac+bc \end{vmatrix}$$

Applying $C_3 \to C_3 + C_1$, we obtain

$$= abc\begin{vmatrix} bc & 1 & ab+bc+ca \\ ca & 1 & ab+bc+ca \\ ab & 1 & ab+bc+ca \end{vmatrix}$$

Taking $(ab + bc + ca)$ common from C_3, we get

$$= abc\,(ab+bc+ac)\begin{vmatrix} bc & 1 & 1 \\ ca & 1 & 1 \\ ab & 1 & 1 \end{vmatrix}$$

As C_2 and C_3 are identical,

$\therefore$ L.H.S. $= abc\,(ab+bc+ac).0$

$= 0$

$=$ R.H.S. **Hence Proved.**

12. *Show that* $\begin{vmatrix} 1 & 1 & 1 \\ \alpha^2 & \beta^2 & \gamma^2 \\ \alpha^3 & \beta^3 & \gamma^3 \end{vmatrix}$

$$= (\alpha-\beta)(\beta-\gamma)(\gamma-\alpha)(\alpha\beta+\beta\gamma+\alpha\gamma).$$

Sol. Consider,

$$\text{L.H.S.} = \begin{vmatrix} 1 & 1 & 1 \\ \alpha^2 & \beta^2 & \gamma^2 \\ \alpha^3 & \beta^3 & \gamma^3 \end{vmatrix}$$

Applying $C_1 \to C_1 - C_2$ and $C_2 \to C_2 - C_3$, we obtain

$$= \begin{vmatrix} 0 & 0 & 1 \\ \alpha^2-\beta^2 & \beta^2-\gamma^2 & \gamma^2 \\ \alpha^3-\beta^3 & \beta^3-\gamma^3 & \gamma^3 \end{vmatrix}$$

$$= \begin{vmatrix} 0 & 0 & 1 \\ (\alpha-\beta)(\alpha+\beta) & (\beta-\gamma)(\beta+\gamma) & \gamma^2 \\ (\alpha-\beta)(\alpha^2+\alpha\beta+\beta^2) & (\beta-\gamma)(\beta^2+\beta\gamma+\gamma^2) & \gamma^2 \end{vmatrix}$$

Taking $(\alpha - \beta)$ common from C_1 and $(\beta - \gamma)$ common from C_2, we get

$$= (\alpha-\beta)(\beta-\gamma)\begin{vmatrix} 0 & 0 & 1 \\ \alpha+\beta & \beta+\gamma & \gamma^2 \\ \alpha^2+\alpha\beta+\beta^2 & \beta^2+\beta\gamma+\gamma^2 & \gamma^3 \end{vmatrix}$$

Applying $C_2 \to C_2 - C_1$, we obtain

$$= (\alpha-\beta)(\beta-\gamma)\begin{vmatrix} 0 & 0 & 1 \\ \alpha+\beta & (\gamma-\alpha) & \gamma^2 \\ \alpha^2+\alpha\beta+\beta^2 & (\gamma-\alpha)(\gamma+\alpha+\beta) & \gamma^3 \end{vmatrix}$$

Taking $(\gamma - \alpha)$ common from C_2, we get

$$= (\alpha-\beta)(\beta-\gamma)(\gamma-\alpha)\begin{vmatrix} 0 & 0 & 1 \\ \alpha+\beta & 1 & \gamma^2 \\ \alpha^2+\alpha\beta+\beta^2 & \gamma+\alpha+\beta & \gamma^3 \end{vmatrix}$$

Now expanding along R_1 we get

$$= (\alpha-\beta)(\beta-\gamma)(\gamma-\alpha)\,1\begin{vmatrix} \alpha+\beta & 1 \\ \alpha^2+\alpha\beta+\beta^2 & \alpha+\beta+\gamma \end{vmatrix}$$

$$= (\alpha-\beta)(\beta-\gamma)(\gamma-\alpha)\times 1\,\{(\alpha+\beta)(\alpha+\beta+\gamma) - 1(\alpha^2+\beta^2+\alpha\beta)\}$$

$$= (\alpha-\beta)(\beta-\gamma)(\gamma-\alpha)\times 1\,\{\alpha^2+\beta^2+2\alpha\beta+\alpha\gamma+\beta\gamma - \alpha^2-\beta^2-\alpha\beta)$$

$$= (\alpha-\beta)(\beta-\gamma)(\gamma-\alpha)(\alpha\beta+\beta\gamma+\gamma\alpha)$$

= R.H.S. **Hence Proved.**

13. *Using properties of determinants, prove that*

$$\begin{vmatrix} x & y & z \\ x^2 & y^2 & z^2 \\ y+z & z+x & x+y \end{vmatrix} = (x-y)(y-z)(z-x)(x+y+z)$$

Sol. Let $\Delta = \begin{vmatrix} x & y & z \\ x^2 & y^2 & z^2 \\ y+z & z+x & x+y \end{vmatrix}$

Applying $R_3 \to R_3 + R_1$, we get

$$\Delta = \begin{vmatrix} x & y & z \\ x^2 & y^2 & z^2 \\ x+y+z & x+y+z & x+y+z \end{vmatrix}$$

Taking $(x+y+z)$ common from R_3, we get

$$= (x+y+z)\begin{vmatrix} x & y & z \\ x^2 & y^2 & z^2 \\ 1 & 1 & 1 \end{vmatrix}$$

Applying $C_1 \to C_1 - C_2$ and $C_2 \to C_2 - C_3$, we get

$$\Delta = (x+y+z)\begin{vmatrix} x-y & y-z & z \\ x^2-y^2 & y^2-z^2 & z^2 \\ 0 & 0 & 1 \end{vmatrix}$$

Taking $(x-y)$ and $(y-z)$ common from C_1 and C_2 respectively, we get

$$= (x+y+z)(x-y)(y-z)\begin{vmatrix} 1 & 1 & z \\ x+y & y+z & z^2 \\ 0 & 0 & 1 \end{vmatrix}$$

Expanding along R_3, we get

$$= (x+y+z)(x-y)(y-z)(y+z-x-y)$$

$$= (x-y)(y-z)(z-x)(x+y+z).$$

Hence Proved.

14. *Prove that*

$$\begin{vmatrix} 1+a & 1 & 1 \\ 1 & 1+b & 1 \\ 1 & 1 & 1+c \end{vmatrix} = abc\left[1+\frac{1}{a}+\frac{1}{b}+\frac{1}{c}\right].$$

Sol. Consider,

$$\text{L.H.S} = \begin{vmatrix} 1+a & 1 & 1 \\ 1 & 1+b & 1 \\ 1 & 1 & 1+c \end{vmatrix}$$

Applying $C_1 \to C_1 - C_2$ and $C_2 \to C_2 - C_3$, we get

$$\text{L.H.S.} = \begin{vmatrix} a & 0 & 1 \\ -b & b & 1 \\ 0 & -c & 1+c \end{vmatrix}$$

Expanding along R_1, we get

$$\Rightarrow \quad \text{L.H.S.} = a\,[b(1+c)+c] + 1\,(bc - b\times 0)$$

$$= a\,[b+bc+c] + bc$$

$$= abc + ab + bc + ac$$

$$= abc\left[1+\frac{1}{a}+\frac{1}{b}+\frac{1}{c}\right]$$

= R.H.S. **Hence Proved**

15. *Show that*

$$\begin{vmatrix} a & b & c \\ a^2 & b^2 & c^2 \\ bc & ca & ab \end{vmatrix} = (a-b)(b-c)(c-a)(ab+bc+ca)$$

Sol. Consider,

$$\text{L.H.S.} = \begin{vmatrix} a & b & c \\ a^2 & b^2 & c^2 \\ bc & ca & ab \end{vmatrix}$$

Applying $C_1 \to C_1 - C_2$ and $C_2 \to C_2 - C_3$, we obtain

$$\text{L.H.S.} = \begin{vmatrix} a-b & b-c & c \\ a^2-b^2 & b^2-c^2 & c^2 \\ bc-ca & ca-ab & ab \end{vmatrix}$$

$$= \begin{vmatrix} (a-b) & (b-c) & c \\ (a-b)(a+b) & (b-c)(b+c) & c^2 \\ -c(a-b) & -a(b-c) & ab \end{vmatrix}$$

Taking $(a-b)$ common from C_1 and $(b-c)$ common from C_2, we get

$$= (a-b)(b-c)\begin{vmatrix} 1 & 1 & c \\ a+b & b+c & c^2 \\ -c & -a & ab \end{vmatrix}$$

Now applying $C_2 \to C_2 - C_1$, we obtain

$$\text{L.H.S.} = (a-b)(b-c)\begin{vmatrix} 1 & 0 & c \\ a+b & c-a & c^2 \\ -c & c-a & ab \end{vmatrix}$$

Prove the Following Type Questions

Taking $(c-a)$ common from C_2, we get

$$\Rightarrow \quad \text{L.H.S.} = (a-b)(b-c)(c-a)\begin{vmatrix} 1 & 0 & c \\ a+b & 1 & c^2 \\ -c & 1 & ab \end{vmatrix}$$

Now applying $R_2 \to R_2 - R_3$, we obtain

$$\Rightarrow \quad \text{L.H.S.} = (a-b)(b-c)(c-a)$$

$$\begin{vmatrix} 1 & 0 & c \\ a+b+c & 0 & c^2-ab \\ -c & 1 & ab \end{vmatrix}$$

Expanding along C_2, we get

$$\text{L.H.S.} = (a-b)(b-c)(c-a)(-1)$$

$$[(c^2-ab)-c(a+b+c)]$$

$$= -(a-b)(b-c)(c-a)[-ab-bc-ac]$$

$$= (a-b)(b-c)(c-a)(ab+bc+ac)$$

$= \text{R.H.S.}$ **Hence Proved.**

16. *Show that*

$$\begin{vmatrix} a & b & c \\ a-b & b-c & c-a \\ b+c & c+a & a+b \end{vmatrix} = a^3+b^3+c^3-3abc.$$

Sol. Consider,

$$\text{L.H.S.} = \begin{vmatrix} a & b & c \\ a-b & b-c & c-a \\ b+c & c+a & a+b \end{vmatrix}$$

Applying $R_1 \to R_2 + R_3$, we get

$$= \begin{vmatrix} a+b+c & b+c+a & a+b+c \\ a-b & b-c & c-a \\ b+c & c+a & a+b \end{vmatrix}$$

Taking $(a+b+c)$ common from R_1, we get

$$= (a+b+c)\begin{vmatrix} 1 & 1 & 1 \\ a-b & b-c & c-a \\ b+c & c+a & a+b \end{vmatrix}$$

Applying $C_2 \to C_2 - C_1$ and $C_3 \to C_3 - C_1$, we obtain

$$= (a+b+c)\begin{vmatrix} 1 & 0 & 0 \\ a-b & 2b-a-c & b+c-2a \\ b+c & a-b & a-c \end{vmatrix}$$

Expanding along R_1, we get

$$= (a+b+c)\begin{vmatrix} 2b-a-c & b+c-2a \\ a-b & a-c \end{vmatrix}$$

Applying $R_1 \to R_1 + R_2$

$$= (a+b+c)\begin{vmatrix} b-c & b-a \\ a-b & a-c \end{vmatrix}$$

$$= (a+b+c)[(b-c)(a-c)-(a-b)(b-a)]$$

$$= (a+b+c)[(ab-ac-bc+c^2)-(ab-b^2-a^2+ab)]$$

$$= (a+b+c)[a^2+b^2+c^2-ab-bc-ac]$$

$$= a^3+b^3+c^3-3abc.$$

17. *Show that*

$$\begin{vmatrix} a-b-c & 2a & 2a \\ 2b & b-c-a & 2b \\ 2c & 2c & c-a-b \end{vmatrix} = (a+b\ c)^3.$$

Sol. Consider,

$$\text{L.H.S.} = \begin{vmatrix} a-b-c & 2a & 2a \\ 2b & b-c-a & 2b \\ 2c & 2c & c-a-b \end{vmatrix}$$

Applying $R_1 \to R_1 + R_2 + R_3$, we obtain

$$\Rightarrow \text{L.H.S.} = \begin{vmatrix} a+b+c & a+b+c & a+b+c \\ 2b & b-c-a & 2b \\ 2c & 2c & c-a-b \end{vmatrix}$$

Taking $(a+b+c)$ common from R_1, we get

$$= (a+b+c)\begin{vmatrix} 1 & 1 & 1 \\ 2b & b-c-a & 2b \\ 2c & 2c & c-a-b \end{vmatrix}$$

Applying $C_3 \to C_3 - C_2$ and $C_2 \to C_2 - C_1$, we obtain

$$= (a+b+c)\begin{vmatrix} 1 & 0 & 0 \\ 2b & -a-b-c & a+b+c \\ 2c & 0 & -a-b-c \end{vmatrix}$$

Taking $(a+b+c)$ common from C_3, we get

$$= (a+b+c)^2\begin{vmatrix} 1 & 0 & 0 \\ 2b & -a-b-c & 1 \\ 2c & 0 & -1 \end{vmatrix}$$

Expanding along R_1, we get

$$\text{L. H. S.} = (a+b+c)^2 \times 1\,(a+b+c-0)$$

$$= (a+b+c)^3$$

$= \text{R. H. S.}$ **Hence Proved.**

18. *Prove that* $\begin{vmatrix} 1 & \omega & \omega^2 \\ \omega & \omega^2 & 1 \\ \omega^2 & 1 & \omega \end{vmatrix} = 0$

(ω is the cube root of unity)

Sol. Consider,

$$\text{L.H.S.} = \begin{vmatrix} 1 & 1 & 1 \\ \omega & \omega^2 & 1 \\ \omega^2 & 1 & \omega \end{vmatrix}$$

Applying $R_1 \to R_1 + R_2 + R_3$, we obtain

$$= \begin{vmatrix} 1+\omega+\omega^2 & 1+\omega+\omega^2 & 1+\omega+\omega^2 \\ \omega & \omega^2 & 1 \\ \omega^2 & 1 & \omega \end{vmatrix}$$

Taking $(1 + \omega + \omega^2)$ common from R_1, we get

$$= (1+\omega+\omega^2)\begin{vmatrix} 1 & 1 & 1 \\ \omega & \omega^2 & 1 \\ \omega^2 & 1 & \omega \end{vmatrix}$$

$= 0$ (as $1 + \omega + \omega^2 = 0$) **Hence Proved.**

19. *Using properties of determinants show that*

$$\begin{vmatrix} a^2+1 & ab & ac \\ ba & b^2+1 & bc \\ ca & cb & c^2+1 \end{vmatrix} = a^2 + b^2 + c^2 + 1.$$

Sol. Consider,

$$\text{L.H.S.} = \begin{vmatrix} a^2+1 & ab & ac \\ ba & b^2+1 & bc \\ ca & cb & c^2+1 \end{vmatrix}$$

Multiplying by a, b and c in R_1, R_2 and R_3 respectively, we get

$$= \frac{1}{abc}\begin{vmatrix} a^3+a & a^2b & a^2c \\ b^2a & b^3+b & b^2c \\ c^2a & c^2b & c^3+c \end{vmatrix}$$

Taking a, b, c common from C_1, C_2, C_3 respectively, we get

$$= \frac{abc}{abc}\begin{vmatrix} a^2+1 & a^2 & a^2 \\ b^2 & b^2+1 & b^2 \\ c^2 & c^2 & c^2+1 \end{vmatrix}$$

Applying $R_1 \to R_1 + R_2 + R_3$, we obtain

$$= \begin{vmatrix} a^2+b^2+c^2+1 & a^2+b^2+c^2+1 & a^2+b^2+c^2+1 \\ b^2 & b^2+1 & b^2 \\ c^2 & c^2 & c^2+1 \end{vmatrix}$$

Taking $(a^2 + b^2 + c^2 + 1)$ common from R_1, we get

$$= (a^2+b^2+c^2+1)\begin{vmatrix} 1 & 1 & 1 \\ b^2 & b^2+1 & b^2 \\ c^2 & c^2 & c^2+1 \end{vmatrix}$$

Applying $C_1 \to C_1 - C_3$ and $C_2 \to C_2 - C_3$, we obtain

$$= (a^2+b^2+c^2+1)\begin{vmatrix} 0 & 0 & 1 \\ 0 & 1 & b^2 \\ -1 & -1 & c^2+1 \end{vmatrix}$$

Expanding along R_1, we get

$= (a^2 + b^2 + c^2 + 1)(1)$

$= a^2 + b^2 + c^2 + 1$

= R. H. S. **Hence Proved.**

20. *By using properties of determinants, prove that :*

$$\begin{vmatrix} 1+\sin^2 x & \cos^2 x & 4\sin 2x \\ \sin^2 x & 1+\cos^2 x & 4\sin 2x \\ \sin^2 x & \cos^2 x & 1+4\sin 2x \end{vmatrix} = 2 + 4\sin 2x.$$

Sol. Consider,

$$\text{L. H. S.} = \begin{vmatrix} 1+\sin^2 x & \cos^2 x & 4\sin 2x \\ \sin^2 x & 1+\cos^2 x & 4\sin 2x \\ \sin^2 x & \cos^2 x & 1+4\sin 2x \end{vmatrix}$$

Applying $C_1 \to C_1 + C_2$, we obtain

$$= \begin{vmatrix} 2 & \cos^2 x & 4\sin 2x \\ 2 & 1+\cos^2 x & 4\sin 2x \\ 1 & \cos^2 x & 1+4\sin 2x \end{vmatrix}$$

Applying $R_2 \to R_2 - R_1$ and $R_3 \to R_3 - R_1$; we obtain

$$= \begin{vmatrix} 2 & \cos^2 x & 4\sin 2x \\ 0 & 1 & 0 \\ -1 & 0 & 1 \end{vmatrix}$$

Expanding along C_1, we get

$= 2(1 - 0) - 1(0 - 4\sin 2x)$

$= 2 + 4\sin 2x.$ **Hence Proved.**

21. *Prove that :*

$$\begin{vmatrix} a & b & ax+by \\ b & c & bx+cy \\ ax+by & bx+cy & 0 \end{vmatrix} = (b^2 - ac)(ax^2 + 2bx + cy^2).$$

Sol. $$\text{L.H.S.} = \begin{vmatrix} a & b & ax+by \\ b & c & bx+cy \\ ax+by & bx+cy & 0 \end{vmatrix}$$

Applying $R_3 \to R_3 - xR_1 - yR_2$, we obtain

$$= \begin{vmatrix} a & b & ax+by \\ b & c & bx+cy \\ 0 & 0 & -ax^2-bxy-bxy-cy^2 \end{vmatrix}$$

Expanding along R_3, we get

$$= -(ax^2+2bxy+cy^2)(ac-b^2)$$

$$= (b^2-ac)(ax^2+2bxy+cy^2).$$

Hence Proved.

22. *Using properties of determinants, prove that :*

$$\begin{vmatrix} 1+a^2-b^2 & 2ab & -2b \\ 2ab & 1-a^2+b^2 & 2a \\ 2b & -2a & 1-a^2-b^2 \end{vmatrix} = (1+a^2+b^2)^3$$

Sol. $\Delta = \begin{vmatrix} 1+a^2-b^2 & 2ab & -2b \\ 2ab & 1-a^2+b^2 & 2a \\ 2b & -2a & 1-a^2-b^2 \end{vmatrix}$

Operating $R_1 \to R_1 + b\,R_3$,

$$= \begin{vmatrix} 1+a^2+b^2 & 0 & -b(1+a^2+b^2) \\ 2ab & 1-a^2+b^2 & 2a \\ 2b & -2a & 1-a^2-b^2 \end{vmatrix}$$

Taking $(1+a^2+b^2)$ common from R_1,

$$= (1+a^2+b^2)\begin{vmatrix} 1 & 0 & -b \\ 2ab & 1-a^2+b^2 & 2a \\ 2b & -2a & 1-a^2-b^2 \end{vmatrix}$$

Operating $R_2 \to R_2 - aR_3$,

$$= (1+a^2+b^2)\begin{vmatrix} 1 & 0 & -b \\ 0 & 1+a^2+b^2 & a(1+a^2+b^2) \\ 2b & -2a & 1-a^2-b^2 \end{vmatrix}$$

Taking $(1+a^2+b^2)$ common from R_2,

$$= (1+a^2+b^2)^2\begin{vmatrix} 1 & 0 & -b \\ 0 & 1 & a \\ 2b & -2a & 1-a^2-b^2 \end{vmatrix}$$

Now, on expanding along R_1, we get

$$= (1+a^2+b^2)^2\,[(1-a^2-b^2+2a^2)+(2b^2)]$$

$$= (1+a^2+b^2)^2\cdot(1+a^2+b^2)$$

$$= (1+a^2+b^2)^3$$

Hence Proved.

23. *Using properties of determinants, prove that :*

$$\begin{vmatrix} b+c & a & a \\ b & a+c & b \\ c & c & a+b \end{vmatrix} = 4\,abc$$

Sol. Let $\Delta = \begin{vmatrix} b+c & a & a \\ b & a+c & b \\ c & c & a+b \end{vmatrix}$

Operating $R_1 \to R_1 + R_2 + R_3$, we get

$$\Delta = \begin{vmatrix} 2(b+c) & 2(a+c) & 2(a+b) \\ b & a+c & b \\ c & c & a+b \end{vmatrix}$$

Taking 2 as common from R_1, we get

$$\Delta = 2\begin{vmatrix} b+c & a+c & a+b \\ b & a+c & b \\ c & c & a+b \end{vmatrix}$$

Applying $R_2 \to R_2 - R_1$ and $R_3 \to R_3 - R_1$

$$\Delta = 2\begin{vmatrix} b+c & a+c & a+b \\ -c & 0 & -a \\ -b & -a & 0 \end{vmatrix}$$

Now expanding along R_1, we get

$$\Delta = 2\,[(b+c)(-a^2) - (a+c)(0-ab) + (a+b)(ac)]$$

$$= 2\,[-a^2b - a^2c + a^2b + abc + a^2c + abc]$$

$$= 2\cdot 2abc$$

$$= 4abc$$

Hence Proved.

24. *Using properties of determinants, prove that :*

$$\begin{vmatrix} a & b & b+c \\ c & a & c+a \\ b & c & a+b \end{vmatrix} = (a+b+c)(a-c)^2$$

Sol.

Let $\Delta = \begin{vmatrix} a & b & b+c \\ c & a & c+a \\ b & c & a+b \end{vmatrix}$ = L.H.S.

Applying $R_1 \to R_1 + R_2 + R_3$

$$\Delta - \begin{vmatrix} a+b+c & a+b+c & 2(a+b+c) \\ c & a & c+a \\ b & c & a+b \end{vmatrix}$$

$$\Delta = (a+b+c)\begin{vmatrix} 1 & 1 & 2 \\ c & a & c+a \\ b & c & a+b \end{vmatrix}$$

[Taking out $(a+b+c)$ as common]

Applying $C_1 \to C_1 - \frac{1}{2}C_3$ and $C_2 \to C_2 - \frac{1}{2}C_3$

$$\Delta = (a+b+c)\begin{vmatrix} 0 & 0 & 2 \\ (c-a)/2 & (a-c)/2 & c+a \\ (b-a)/2 & c-\frac{(a+b)}{2} & a+b \end{vmatrix}$$

Now, expanding along R_1

$$\Delta = (a+b+c)\cdot 2\left[\frac{(c-a)}{2}\cdot\frac{(2c-a-b)}{2} - \frac{(a-c)}{2}\frac{(b-a)}{2}\right]$$

$$= (a+b+c)\cdot\frac{1}{2}[2c^2 - ac - bc - 2ac + a^2 + ab - ab + a^2 + bc - ac]$$

$$= (a+b+c)\frac{1}{2}(2a^2 + 2c^2 - 4ac)$$

$$= (a+b+c)(a-c)^2$$

= R.H.S. **Hence Proved.**

25. *Using properties of determinants prove that :**

$$\begin{vmatrix} x & x(x^2+1) & x+1 \\ y & y(y^2+1) & y+1 \\ z & z(z^2+1) & z+1 \end{vmatrix} = (x-y)(y-z)(z-x)(x+y+z)$$

Sol. L.H.S. = $\begin{vmatrix} x & x(x^2+1) & x+1 \\ y & y(y^2+1) & y+1 \\ z & z(z^2+1) & z+1 \end{vmatrix}$

Applying $R_2 \to R_2 - R_1$ and $R_3 \to R_3 - R_1$

$$\Rightarrow \begin{vmatrix} x & x(x^2+1) & x+1 \\ y-x & y^3+y-x^3-x & y-x \\ z-x & z^3+z-x^3-x & z-x \end{vmatrix}$$

$$\Rightarrow \begin{vmatrix} x & x(x^2+1) & x+1 \\ (y-x) & (y-x)(y^2+x^2+xy+1) & y-x \\ (z-x) & (z-x)(z^2+x^2+zx+1) & z-x \end{vmatrix}$$

Taking $(y-x)$ and $(z-x)$ common from R_2 and R_3, respectively.

$$\Rightarrow (y-x)(z-x)\begin{vmatrix} x & x(x^2+1) & x+1 \\ 1 & y^2+x^2+xy+1 & 1 \\ 1 & z^2+x^2+zx+1 & 1 \end{vmatrix}$$

Now, applying $R_2 \to R_2 - R_3$,

$$\Rightarrow (y-x)(z-x)\begin{vmatrix} x & x(x^2+1) & x+1 \\ 0 & (y-z)(x+y+z) & 0 \\ 1 & z^2+x^2+zx+1 & 1 \end{vmatrix}$$

Taking $(y-z)$ common from R_2,

$$\Rightarrow (y-z)(y-x)(z-x)\begin{vmatrix} x & x(x^2+1) & x+1 \\ 0 & x+y+z & 0 \\ 1 & z^2+x^2+zx+1 & 1 \end{vmatrix}$$

Now, Expanding along R_2,

$$\Rightarrow (y-x)(y-z)(z-x)(x+y+z)\begin{vmatrix} x & x+1 \\ 1 & 1 \end{vmatrix}$$

$$\Rightarrow (y-x)(y-z)(z-x)(x+y+z)(x-x-1)$$

$\Rightarrow (x-y)(y-z)(z-x)(x+y+z)$ **Hence Proved**

Chapter 5. Matrices

1. *Show that $(A + A')$ is symmetric matrix, if $A = \begin{bmatrix} 2 & 4 \\ 3 & 5 \end{bmatrix}$.* *

Sol. Given, $A = \begin{bmatrix} 2 & 4 \\ 3 & 5 \end{bmatrix}$

Then, $A' = \begin{bmatrix} 2 & 3 \\ 4 & 5 \end{bmatrix}$

$\therefore$ $A + A' = \begin{bmatrix} 2 & 4 \\ 3 & 5 \end{bmatrix} + \begin{bmatrix} 2 & 3 \\ 4 & 5 \end{bmatrix}$

$= \begin{bmatrix} 4 & 7 \\ 7 & 10 \end{bmatrix}$

Now, $(A + A')' = \begin{bmatrix} 4 & 7 \\ 7 & 10 \end{bmatrix}$

$\because$ $(A + A') = (A + A')'$

Hence, $(A + A')$ is a symmetric matrix.

Hence Proved

2. *Let A be a square matrix, show that $\frac{1}{2}(A + A')$ is symmetric matrix and that $\frac{1}{2}(A - A')$ is a skew symmetric matrix. Hence, prove that every square matrix may be expressed as the sum of symmetric and skew symmetric matrices.*

Sol. Let $B = \frac{1}{2}(A + A')$

$\Rightarrow$ $B' = \left(\frac{1}{2}(A + A')\right)'$

$= \frac{1}{2}(A + A')'$

* Frequently asked previous years Board Exam Questions.

Prove the Following Type Questions

$$= \frac{1}{2}(A' + (A)')'$$

$$= \frac{1}{2}(A' + A)$$

$$= \frac{1}{2}(A + A') = B$$

Therefore B′ = B, *i.e.*, B is symmetric

Let $C = \frac{1}{2}(A - A')$

$$C' = \left(\frac{1}{2}(A - A')\right)'$$

$$= \frac{1}{2}(A - A')'$$

$$= \frac{1}{2}(A' - (A)')'$$

$$= \frac{1}{2}(A' - A)$$

$$= -C$$

Therefore C′ = – C *i.e.*, is skew symmetric.

Now consider,

$$\frac{1}{2}(A + A') + \frac{1}{2}(A - A')$$

$$= \frac{A}{2} + \frac{A'}{2} + \frac{A}{2} - \frac{A'}{2}$$

$$= \frac{A}{2} + \frac{A}{2}$$

$$= A$$

∴ We can say that

$$A = \frac{1}{2}(A + A') + \frac{1}{2}(A - A')$$

i.e., a square matrix can be expressed as the sum of symmetric and skew symmetric matrix.

Hence Proved.

3. *A matrix is given by*

$$R(t) = \begin{bmatrix} \cos t & \sin t \\ -\sin t & \cos t \end{bmatrix}$$

then show that R (s) R (t) = R (s + t).

Sol. Given,

$$R(t) = \begin{bmatrix} \cos t & \sin t \\ -\sin t & \cos t \end{bmatrix} \quad ...(i)$$

$$\text{Similarly, } R(s) = \begin{bmatrix} \cos s & \sin s \\ -\sin s & \cos s \end{bmatrix} \quad ...(ii)$$

$$\therefore \quad R(s)\,R(t) = \begin{bmatrix} \cos s & \sin s \\ -\sin s & \cos s \end{bmatrix}\begin{bmatrix} \cos t & \sin t \\ -\sin t & \cos t \end{bmatrix}$$

$$= \begin{bmatrix} \cos s \cos t - \sin s \sin t & \cos s \sin t + \sin s \cos t \\ -\sin s \cos t - \cos s \sin t & -\sin s \sin t + \cos s \cos t \end{bmatrix}$$

$$= \begin{bmatrix} \cos(s+t) & \sin(s+t) \\ -\sin(s+t) & \cos(s+t) \end{bmatrix}$$

$= R(s + t)$ **Hence Proved.**

4. *If* $A = \begin{bmatrix} 0 & -\tan\frac{\theta}{2} \\ \tan\frac{\theta}{2} & 0 \end{bmatrix}$ *and I is a unit matrix, then prove that*

$$(I + A) = (I - A)\begin{bmatrix} \cos\theta & -\sin\theta \\ \sin\theta & \cos\theta \end{bmatrix}$$

Sol. Let,

$$\tan\frac{\theta}{2} = t$$

$$\Rightarrow \quad \cos\theta = \frac{1-t^2}{1+t^2}$$

$$\text{And, } \sin\theta = \frac{2t}{1+t^2}$$

$$\text{Now, R.H.S.} = (I - A)\begin{bmatrix} \cos\theta & -\sin\theta \\ \sin\theta & \cos\theta \end{bmatrix}$$

$$= \left\{\begin{bmatrix} 1 & 0 \\ 0 & 1 \end{bmatrix} - \begin{bmatrix} 0 & -t \\ t & 0 \end{bmatrix}\right\}\begin{bmatrix} \frac{1-t^2}{1+t^2} & \frac{-2t}{1+t^2} \\ \frac{2t}{1+t^2} & \frac{1-t^2}{1+t^2} \end{bmatrix}$$

$$= \begin{bmatrix} 1 & t \\ -t & 1 \end{bmatrix}\begin{bmatrix} \frac{1-t^2}{1+t^2} & \frac{-2t}{1+t^2} \\ \frac{2t}{1+t^2} & \frac{1-t^2}{1+t^2} \end{bmatrix}$$

Taking $\left(\frac{1}{1+t^2}\right)$ common from C_1 and C_2, we get

$$= \frac{1}{(1+t^2)^2}\begin{bmatrix} 1 & t \\ -t & 1 \end{bmatrix}\begin{bmatrix} 1-t^2 & -2t \\ 2t & 1-t^2 \end{bmatrix}$$

$$= \frac{1}{(1+t^2)^2}\begin{bmatrix} 1-t^2+2t^2 & -2t+t-t^3 \\ -t+t^3+2t & 2t^2+1-t^2 \end{bmatrix}$$

$$= \frac{1}{(1+t^2)^2}\begin{bmatrix} 1+t^2 & -t-t^3 \\ t+t^3 & t^2+1 \end{bmatrix}$$

$$= \frac{1}{(1+t^2)^2}\begin{bmatrix} 1+t^2 & -t(1+t^2) \\ t(1+t^2) & 1+t^2 \end{bmatrix}$$

Taking $(1 + t^2)$ common from C_1 and C_2, we get

$$= \begin{bmatrix} 1 & -t \\ t & 1 \end{bmatrix}$$

Now, L.H.S. = I + A

$$= \begin{bmatrix} 1 & 0 \\ 0 & 1 \end{bmatrix} + \begin{bmatrix} 0 & -t \\ t & 0 \end{bmatrix}$$

$$= \begin{bmatrix} 1 & -t \\ t & 1 \end{bmatrix}$$

= R.H.S. **Hence proved.**

5. *If* $A = \begin{bmatrix} 1 \\ -5 \\ 7 \end{bmatrix}$, $B = [3 \ 1 \ -2]$ *prove that* $(AB)' = B'A'$.

Sol. Given,

$$A = \begin{bmatrix} 1 \\ -5 \\ 7 \end{bmatrix}$$

and $B = [3 \ 1 \ -2]$

$$\therefore \quad AB = \begin{bmatrix} 1 \\ -5 \\ 7 \end{bmatrix} [3 \ 1 \ -2]$$

$$= \begin{bmatrix} 3 & 1 & -2 \\ -15 & -5 & 10 \\ 21 & 7 & -14 \end{bmatrix}$$

L.H.S. = (AB)′

$$= \begin{bmatrix} 3 & -15 & 21 \\ 1 & -5 & 7 \\ -2 & 10 & -14 \end{bmatrix}$$

$$A' = [1 \ -5 \ 7]$$

$$B' = \begin{bmatrix} 3 \\ 1 \\ -2 \end{bmatrix}$$

R.H.S. = B′ A′

$$= \begin{bmatrix} 3 \\ 1 \\ -2 \end{bmatrix} [1 \ -5 \ 7]$$

$$= \begin{bmatrix} 3 & -15 & 21 \\ 1 & -5 & 7 \\ -2 & 10 & -14 \end{bmatrix}$$

= L.H.S. **Hence proved.**

6. *Show that*

$$\begin{bmatrix} 1 & 1 \\ 0 & 1 \end{bmatrix}^3 = \begin{bmatrix} 1 & 3 \\ 0 & 1 \end{bmatrix}.$$

Sol. Consider,

$$\begin{bmatrix} 1 & 1 \\ 0 & 1 \end{bmatrix}^2 = \begin{bmatrix} 1 & 1 \\ 0 & 1 \end{bmatrix}\begin{bmatrix} 1 & 1 \\ 0 & 1 \end{bmatrix}$$

$$= \begin{bmatrix} 1\times1+1\times0 & 1\times1+1\times1 \\ 0\times1+1\times0 & 0\times1+1\times1 \end{bmatrix}$$

$$= \begin{bmatrix} 1 & 2 \\ 0 & 1 \end{bmatrix}$$

Now, $\begin{bmatrix} 1 & 1 \\ 0 & 1 \end{bmatrix}^3 = \begin{bmatrix} 1 & 1 \\ 0 & 1 \end{bmatrix}^2 \begin{bmatrix} 1 & 1 \\ 0 & 1 \end{bmatrix}$

$$= \begin{bmatrix} 1 & 2 \\ 0 & 1 \end{bmatrix}\begin{bmatrix} 1 & 1 \\ 0 & 1 \end{bmatrix}$$

$$= \begin{bmatrix} 1\times1+2\times0 & 1\times1+2\times1 \\ 0\times1+1\times0 & 0\times1+1\times1 \end{bmatrix}$$

$$= \begin{bmatrix} 1 & 3 \\ 0 & 1 \end{bmatrix}$$ **Hence Proved.**

7. $A = \begin{bmatrix} 1 & 1 & 2 \\ 2 & 1 & 0 \end{bmatrix}$ *and* $B = \begin{bmatrix} 1 & 2 \\ 2 & 0 \\ -1 & 1 \end{bmatrix}$

verify that $B^T A^T = (AB)^T$.

Sol. Given,

$$A = \begin{bmatrix} 1 & 1 & 2 \\ 2 & 1 & 0 \end{bmatrix}, \ A^T = \begin{bmatrix} 1 & 2 \\ 1 & 1 \\ 2 & 0 \end{bmatrix}$$

Also, $B = \begin{bmatrix} 1 & 2 \\ 2 & 0 \\ -1 & 1 \end{bmatrix}, \ B^T = \begin{bmatrix} 1 & 2 & -1 \\ 2 & 0 & 1 \end{bmatrix}$

$$\therefore \quad B^T A^T = \begin{bmatrix} 1 & 2 & -1 \\ 2 & 0 & 1 \end{bmatrix}\begin{bmatrix} 1 & 2 \\ 1 & 1 \\ 2 & 0 \end{bmatrix} = \begin{bmatrix} 1 & 4 \\ 4 & 4 \end{bmatrix}$$

Now, $AB = \begin{bmatrix} 1 & 1 & 2 \\ 2 & 1 & 0 \end{bmatrix}\begin{bmatrix} 1 & 2 \\ 2 & 0 \\ -1 & 1 \end{bmatrix} = \begin{bmatrix} 1 & 4 \\ 4 & 4 \end{bmatrix}$

$$\therefore \quad (AB)^T = \begin{bmatrix} 1 & 4 \\ 4 & 4 \end{bmatrix} = B^T A^T$$ **Hence Proved.**

8. *If a matrix* $A = \begin{bmatrix} 3 & -4 \\ 1 & -1 \end{bmatrix}$ *prove that*

$A^k = \begin{bmatrix} 1+2K & -4K \\ K & 1-2K \end{bmatrix}$, *where K is any positive integer.*

Sol. Given,

$$A = \begin{bmatrix} 3 & -4 \\ 1 & -1 \end{bmatrix}$$

$$\therefore \quad A^2 = \begin{bmatrix} 3 & -4 \\ 1 & -1 \end{bmatrix}\begin{bmatrix} 3 & -4 \\ 1 & -1 \end{bmatrix} = \begin{bmatrix} 5 & -8 \\ 2 & -3 \end{bmatrix}$$

If $K = 2$,

we find

$$\begin{bmatrix} 1+2K & -4K \\ K & 1-2K \end{bmatrix} = \begin{bmatrix} 5 & -8 \\ 2 & -3 \end{bmatrix}$$

i.e., the proposition is true for $K = 2$.

Let it be true for $K = n$.

i.e., Let $A^n = \begin{bmatrix} 1+2n & -4n \\ n & 1-2n \end{bmatrix}$

We will prove that it is true for $K = n + 1$

$$A^{n+1} = \begin{bmatrix} 1+2n & -4n \\ n & 1-2n \end{bmatrix}\begin{bmatrix} 3 & -4 \\ 1 & -1 \end{bmatrix}$$

$$= \begin{bmatrix} 3+2n & -4-4n \\ n+1 & -2n-1 \end{bmatrix}$$

$$= \begin{bmatrix} 1+2(n+1) & -4(n+1) \\ n+1 & 1-2(n+1) \end{bmatrix}$$

i.e., the proposition is true for $K = n + 1$ if it is true for $K = n$.

Hence, by principle of mathematical induction it is true for all $K \in I^+$.

Hence Proved.

9. *Show that*

$$\begin{bmatrix} 1 & -\tan\theta \\ \tan\theta & 1 \end{bmatrix} \times \begin{bmatrix} 1 & -\tan\theta \\ \tan\theta & 1 \end{bmatrix}^{-1} = \begin{bmatrix} \cos 2\theta & -\sin 2\theta \\ \sin 2\theta & \cos 2\theta \end{bmatrix}$$

Sol. Let

$$A = \begin{bmatrix} 1 & -\tan\theta \\ \tan\theta & 1 \end{bmatrix}$$

and $$B = \begin{bmatrix} 1 & \tan\theta \\ -\tan\theta & 1 \end{bmatrix}$$

$$\therefore \quad |B| = 1 + \tan^2\theta = \sec^2\theta \neq 0$$

Now cofactors,

$$B_{11} = 1, \quad B_{12} = \tan\theta$$

$$B_{21} = -\tan\theta, \quad B_{22} = 1$$

$$\therefore \quad B^{-1} = \frac{1}{\sec^2\theta}\begin{bmatrix} 1 & -\tan\theta \\ \tan\theta & 1 \end{bmatrix}$$

$$\Rightarrow \quad \text{L.H.S.} = A \times B^{-1}$$

$$= \begin{bmatrix} 1 & -\tan\theta \\ \tan\theta & 1 \end{bmatrix} \times \frac{1}{\sec^2\theta}\begin{bmatrix} 1 & -\tan\theta \\ \tan\theta & 1 \end{bmatrix}$$

$$= \cos^2\theta \begin{bmatrix} 1-\tan^2\theta & -\tan\theta-\tan\theta \\ \tan\theta+\tan\theta & -\tan^2\theta+1 \end{bmatrix}$$

$$= \cos^2\theta \begin{bmatrix} 1-\tan^2\theta & -2\tan\theta \\ 2\tan\theta & 1-\tan^2\theta \end{bmatrix}$$

$$= \cos^2\theta \begin{bmatrix} \dfrac{\cos^2\theta-\sin^2\theta}{\cos^2\theta} & \dfrac{-2\sin\theta}{\cos\theta} \\ \dfrac{2\sin\theta}{\cos\theta} & \dfrac{\cos^2\theta-\sin^2\theta}{\cos^2\theta} \end{bmatrix}$$

$$\left[\because \tan\theta = \frac{\sin\theta}{\cos\theta}\right]$$

$$= \begin{bmatrix} \cos^2\theta-\sin^2\theta & -2\sin\theta.\cos\theta \\ 2\sin\theta.\cos\theta & \cos^2\theta-\sin^2\theta \end{bmatrix}$$

$$= \begin{bmatrix} \cos 2\theta & -\sin 2\theta \\ \sin 2\theta & \cos 2\theta \end{bmatrix}$$

= R.H.S. **Hence Proved.**

10. *If the matrix* $A = \begin{bmatrix} 2 & 3 \\ 5 & -2 \end{bmatrix}$, *show that* $A^{-1} = \frac{1}{19}A$.

Sol. Given,

$$A = \begin{bmatrix} 2 & 3 \\ 5 & -2 \end{bmatrix}$$

$$\therefore \quad |A| = \begin{bmatrix} 2 & 3 \\ 5 & -2 \end{bmatrix}$$

$$= -4 - 15 = -19 \neq 0$$

$$\Rightarrow \quad \text{adj } A = \begin{bmatrix} -2 & -3 \\ -5 & 2 \end{bmatrix}$$

$$\therefore \quad A^{-1} = \frac{1}{-19}\begin{bmatrix} -2 & -3 \\ -5 & 2 \end{bmatrix}$$

$$= \frac{1}{19}\begin{bmatrix} 2 & 3 \\ 5 & -2 \end{bmatrix}$$

$$= \frac{1}{19}A$$

Hence Proved.

Chapter 6. Differential Calculas

1. *(a) Show that the function f(x) = |x – 4|, x ∈ R is continuous, but not differentiable at x = 4.*

(b) Verify the Lagrange's mean value theorem for the function :

$f(x) = x + \frac{1}{x}$ *in the interval [1, 3].**

Sol. (a) $f(x) = |x-4| = \begin{cases} x-4, x \geq 4 \\ -(x-4), x < 4 \end{cases}$

$$\text{R.H.L.} = \lim_{x\to 4^+} f(4) = \lim_{h\to 0} f(4+h) = \lim_{h\to 0} (4+h-4) = \lim_{h\to 0} h = 0$$

$$\text{L.H.L.} = \lim_{x\to 4^-} f(x) = \lim_{h\to 0} f(4-h) = \lim_{h\to 0} -(4-h-4) = \lim_{h\to 0} h = 0$$

$$f(4) = 4-4 = 0$$

$\because$ R.H.L. = L.H.L. = $f(4) = 0$

$\therefore f(x)$ is continuous at $x = 4$

Now, R.H.D.= $\lim_{x\to 4^+} f'(x)$

$$= \lim_{h\to 0}\frac{f(4+h)-f(4)}{h} = \lim_{h\to 0}\frac{(4+h-4)-(4-4)}{h} = \lim_{h\to 0}\frac{h}{h} = \lim_{h\to 0} 1 = 1$$

and L.H.D. = $\lim_{x\to 4^-} f'(x)$

$$= \lim_{h\to 0} f\frac{(4-h)-f(4)}{-h} = \lim_{h\to 0}\frac{-(4-h-4)-(4-4)}{-h} = \lim_{h\to 0}\frac{h}{-h} = -1$$

$\therefore$ R.H.D. $\neq$ L.H.D.

$\therefore f(x)$ is not differentiable at $x = 4$.

Hence, $f(x)$ is continuous at $x = 4$ but not differentiable at $x = 4$. **Hence Proved.**

OR

(b) Given, $f(x) = x + \frac{1}{x}, x \in [1, 3]$

(i) $f(x)$ is continuous in [1, 3]

(ii) $f(x)$ is differentiable for $x \in (1, 3)$.

$\therefore$ Lagrange's mean value theorem is applicable.

$\therefore$ There must atleast be one value of x, say c, in the open interval (1, 3), such that

$$f'(c) = \frac{f(b)-f(a)}{b-a} \quad \text{...(i)}$$

$$\therefore f(x) = x + \frac{1}{x}$$

$$\therefore f'(x) = 1 - \frac{1}{x^2} \quad \text{...(ii)}$$

$$\Rightarrow 1 - \frac{1}{c^2} = \frac{\left(3+\frac{1}{3}\right)-\left(1+\frac{1}{1}\right)}{3-1}$$

[from eq. (i) and (ii)]

$$\Rightarrow 1 - \frac{1}{c^2} = \frac{\frac{10}{3}-2}{2}$$

$$\Rightarrow 1 - \frac{1}{c^2} = \frac{10-6}{6}$$

$$\Rightarrow 1 - \frac{1}{c^2} = \frac{4}{6} = \frac{2}{3}$$

$$\Rightarrow 1 - \frac{2}{3} = \frac{1}{c^2}$$

$$\Rightarrow \frac{1}{3} = \frac{1}{c^2}$$

$$\Rightarrow c = \pm\sqrt{3}$$

= 1·73 lies in the open interval (1, 3)

$$\therefore 1 < \sqrt{3} < 3.$$

Hence, Lagrange's mean value theorem is verified. **Hence Proved**

2. *If* $y = e^{\sin^{-1}x}$ *and* $z = e^{-\cos^{-1}x}$, *prove that* $\frac{dy}{dz} = e^{\pi/2}$.

Sol. Given, $y = e^{\sin^{-1}x}$ and $z = e^{-\cos^{-1}x}$

$$\frac{dy}{dx} = e^{\sin^{-1}x} \times \frac{1}{\sqrt{1-x^2}}$$

and $$\frac{dz}{dx} = e^{-\cos^{-1}x} \times \frac{1}{\sqrt{1-x^2}}$$

$$\Rightarrow \frac{dy}{dz} = \frac{dy}{dx} \times \frac{dx}{dz}$$

$$\Rightarrow \frac{dy}{dz} = \frac{e^{\sin^{-1}x}}{\sqrt{1-x^2}} \times \frac{\sqrt{1-x^2}}{e^{-\cos^{-1}x}}$$

$$\Rightarrow \frac{dy}{dz} = \frac{e^{\sin^{-1}x}}{e^{-\cos^{-1}x}} = e^{\sin^{-1}x} \cdot e^{\cos^{-1}x}$$

$$\Rightarrow \frac{dy}{dz} = e^{\sin^{-1}x + \cos^{-1}x}$$

$$\Rightarrow \frac{dy}{dz} = e^{\pi/2}$$

Hence Proved

* Frequently asked previous years Board Exam Questions.

3. *If* $x^y = e^{x-y}$, *prove that*

$$\frac{dy}{dx} = \frac{\log_e x}{(1+\log_e x)^2}.$$

Sol. Given, $x^y = e^{x-y}$

Taking log on both sides, we get

$$y \log_e x = (x-y)\log_e e$$

$$\Rightarrow \quad y \log_e x = x - y \quad \ldots(i)$$

$$\Rightarrow \quad y + y\log_e x = x$$

$$\Rightarrow \quad y(1+\log_e x) = x$$

$$\Rightarrow \quad \frac{y}{x} = \frac{1}{1+\log_e x} \quad \ldots(ii)$$

Differentiating equation (i) w.r.t. x, we get

$$y \times \frac{1}{x} + (\log_e x)\cdot\frac{dy}{dx} = 1 - \frac{dy}{dx}$$

$$\Rightarrow \quad (1+\log_e x)\frac{dy}{dx} = 1 - \frac{y}{x}$$

$$\Rightarrow \quad (1+\log_e x)\frac{dy}{dx} = 1 - \frac{1}{1+\log_e x}$$

[from (ii)]

$$\therefore \quad (1+\log_e x)\frac{dy}{dx} = \frac{1+\log_e x - 1}{1+\log_e x}$$

$$\Rightarrow \quad \frac{dy}{dx} = \frac{\log_e x}{(1+\log_e x)^2}$$

Hence Proved.

4. *If* $x^y y^x = 5$, *then show that*

$$\frac{dy}{dx} = -\left[\frac{\log y + \frac{y}{x}}{\log x + \frac{x}{y}}\right].$$

Sol. Given, $x^y y^x = 5$

Taking log on both sides, we get

$$\log x^y y^x = \log 5$$

$$\log(x^y) + \log(y^x) = \log 5$$

$$y\log x + x\log y = \log 5$$

Differentiating *w. r.* to x, we get

$$\frac{y}{x} + \log x\frac{dy}{dx} + \frac{x}{y}\frac{dy}{dx} + \log y = 0$$

$$\Rightarrow \quad \frac{dy}{dx} = -\left[\frac{\frac{y}{x} + \log y}{\frac{x}{y} + \log x}\right]$$

Hence Proved.

5. *If* $x^p y^q = (x+y)^{p+q}$, *then prove that*

$$\frac{dy}{dx} = \frac{y}{x}.$$

Sol. Given,

$$x^p y^q = (x+y)^{p+q}$$

Taking log on both sides,

$$\therefore \quad \log(x^p y^q) = \log(x+y)^{(p+q)}$$

$$\Rightarrow \quad \log x^p + \log y^q = (p+q)\log(x+y)$$

$$\Rightarrow \quad p\log x + q\log y = (p+q)\log(x+y)$$

Differentiating both sides w.r.t. x, we get

$$\frac{p}{x} + \frac{q}{y}\frac{dy}{dx} = \frac{(p+q)}{(x+y)}\frac{d}{dx}(x+y)$$

$$\Rightarrow \quad \frac{p}{x} + \frac{q}{y}\frac{dy}{dx} = \frac{(p+q)}{(x+y)}\left(1 + \frac{dy}{dx}\right)$$

$$\Rightarrow \quad \frac{p}{x} + \frac{q}{y}\frac{dy}{dx} = \frac{(p+q)}{(x+y)} + \frac{(p+q)}{(x+y)}\frac{dy}{dx}$$

$$\Rightarrow \quad \left(\frac{q}{y} - \frac{p+q}{x+y}\right)\frac{dy}{dx} = \frac{p+q}{x+y} - \frac{p}{x}$$

$$\Rightarrow \quad \frac{(qx+qy-py-qy)}{y(x+y)}\frac{dy}{dx} = \frac{(px+qx-px-py)}{x(x+y)}$$

$$\Rightarrow \quad \frac{(qx-py)}{y}\frac{dy}{dx} = \frac{(qx-py)}{x}$$

$$\therefore \quad \frac{dy}{dx} = \frac{y}{x}$$

Hence Proved.

6. *If* $y = x^y$, *prove that*

$$x.\frac{dy}{dx} = \frac{y^2}{1 - y\log x}.$$

Sol. Given, $y = x^y$

Taking log on both sides, we get

$$\log y = \log x^y = y\log x$$

Differentiating both sides w.r.t. x, we get

$$\frac{1}{y}\cdot\frac{dy}{dx} = y\cdot\frac{1}{x} + \frac{dy}{dx}.\log x$$

$$\Rightarrow \quad \frac{1}{y}\frac{dy}{dx} - \log x.\frac{dy}{dx} = \frac{y}{x}$$

$$\Rightarrow \quad \frac{dy}{dx}\left[\frac{1}{y} - \log x\right] = \frac{y}{x}$$

$$\Rightarrow \quad x\cdot\frac{dy}{dx} = \frac{y}{\frac{1}{y} - \log x}$$

$$\Rightarrow \quad = \frac{y}{\frac{1 - y\log x}{y}}$$

$$\Rightarrow \quad x \cdot \frac{dy}{dx} = \frac{y^2}{1 - y\log x}$$

Hence Proved.

7. *If* $y = \frac{\sin^{-1} x}{\sqrt{1-x^2}}$, *then prove that*

$$(1 - x^2)\frac{dy}{dx} - xy = 1.$$

Sol. Given,

$$y = \frac{\sin^{-1} x}{\sqrt{1-x^2}} \quad \text{...(i)}$$

$$y = \frac{\sqrt{1-x^2} \times \sin^{-1} x}{(\sqrt{1-x^2})^2}$$

Differentiating w.r.t. x, we get

$$\frac{dy}{dx} = \frac{\sqrt{1-x^2}\frac{d}{dx}(\sin^{-1} x) - \sin^{-1} x\frac{d}{dx}(\sqrt{1-x^2})}{(\sqrt{1-x^2})^2}$$

$$= \frac{\sqrt{1-x^2} \times \frac{1}{\sqrt{1-x^2}} - \sin^{-1} x \times \frac{1}{2\sqrt{1-x^2}}(-2x)}{(1-x^2)}$$

$$\Rightarrow \quad (1 - x^2)\frac{dy}{dx} = 1 + \frac{x\sin^{-1} x}{\sqrt{1-x^2}}$$

$$\Rightarrow \quad (1 - x^2)\frac{dy}{dx} = 1 + xy \qquad \text{[From (i)]}$$

$$\therefore \quad (1 - x^2)\frac{dy}{dx} - xy = 1$$

Hence Proved.

8. *if* $\sqrt{1-x^4} + \sqrt{1-y^4} = \lambda(x^2 - y^2)$, *then show that*

$$y\sqrt{1-x^4}\frac{dy}{dx} = x\sqrt{1-y^4}.$$

Sol. Let $x^2 = \sin A$ and $y^2 = \sin B$

Then, the given expression becomes,

$$\sqrt{1-\sin^2 A} + \sqrt{1-\sin^2 B} = \lambda(\sin A - \sin B)$$

$$\Rightarrow \quad \cos A + \cos B = \lambda(\sin A - \sin B)$$

$$\Rightarrow 2\cos\frac{A+B}{2}\cos\frac{A-B}{2} = \lambda . 2\cos\frac{A+B}{2}\sin\frac{A-B}{2}$$

$$\Rightarrow \quad \text{A} - \text{B} = 2\tan^{-1}\frac{1}{\lambda}$$

$$\Rightarrow \quad \sin^{-1} x^2 - \sin^{-1} y^2 = 2\tan^{-1}\frac{1}{\lambda}$$

Differentiating both sides w.r.t. x, we get

$$\Rightarrow \quad \frac{1}{\sqrt{1-x^4}} . 2x - \frac{1}{\sqrt{1-y^4}} \cdot 2y\frac{dy}{dx} = 0$$

$$\Rightarrow \quad \frac{x}{\sqrt{1-x^4}} = \frac{y}{\sqrt{1-y^4}}\frac{dy}{dx}$$

$$\Rightarrow \quad y\sqrt{1-x^4}\frac{dy}{dx} = x\sqrt{1-y^4}$$

Hence Proved.

9. *If* $\sin(xy) + \cos(xy) = 1$ *and* $\tan(xy) \neq 1$, *then show that* $\frac{dy}{dx} = -\frac{y}{x}$.

Sol. Given, $\sin(xy) + \cos(xy) = 1$

On differentiating w.r.t. x, we get

$$\cos xy . \frac{d}{dx}(xy) - \sin xy . \frac{d}{dx}(xy) = 0$$

$$\Rightarrow \quad \frac{d}{dx}(xy)(\cos xy - \sin xy) = 0$$

$$\Rightarrow \quad \left(x \times \frac{dy}{dx} + y\right) = 0 \qquad [\because \tan(xy) \neq 1]$$

$$\Rightarrow \quad x\frac{dy}{dx} = -y$$

$$\Rightarrow \quad \frac{dy}{dx} = -\frac{y}{x}$$

Hence Proved.

10. *If* $y = (x + \sqrt{x^2 - 1})^m$, *prove that*

$$(x^2 - 1)\left(\frac{dy}{dx}\right)^2 = m^2y^2.$$

Sol. Given, $y = (x + \sqrt{x^2 - 1})^m$

Let $\quad t = x + \sqrt{x^2 - 1} \quad$ ...(i)

$\Rightarrow \quad y = t^m \quad$ ...(ii)

Differentiating w.r.t. x, we get

$$\frac{dy}{dt} = mt^{m-1}$$

$$\because \quad t = x + \sqrt{x^2 - 1}$$

Differentiating w.r.t. x, we get

$$\frac{dt}{dx} = 1 + \frac{1}{2} \times \frac{2x}{\sqrt{x^2 - 1}}$$

$$= 1 + \frac{x}{\sqrt{x^2 - 1}}$$

$$= \frac{\sqrt{x^2-1}+x}{\sqrt{x^2-1}}$$

$$\therefore \quad \frac{dy}{dx} = \frac{dy}{dt}\times\frac{dt}{dx}$$

$$= mt^{m-1}\times\frac{\sqrt{x^2-1}+x}{\sqrt{x^2-1}}$$

$$= m.(x+\sqrt{x^2-1})^{m-1}.\frac{(x+\sqrt{x^2-1})}{\sqrt{x^2-1}}$$

$$= m\frac{(x+\sqrt{x^2-1})^m}{\sqrt{x^2-1}} = \frac{my}{\sqrt{x^2-1}}$$

(From (i) & (ii))

Squaring both sides, we get

$$\left(\frac{dy}{dx}\right)^2 = \frac{m^2y^2}{x^2-1}$$

$$\Rightarrow \quad (x^2-1).\left(\frac{dy}{dx}\right)^2 = m^2y^2$$ **Hence Proved.**

11. *If $e^y(x+1) = 1$, then show that :*

$$\frac{d^2y}{dx^2} = \left(\frac{dy}{dx}\right)^2.$$

Sol. Given, $e^y(x+1) = 1$

Taking log on both sides, we get

$$y\log e + \log(x+1) = \log 1$$

$$\Rightarrow \quad y + \log(x+1) = 0$$

$$\Rightarrow \quad y = -\log(x+1)$$

Differentiating w.r.t. x, we get

$$\frac{dy}{dx} = -\frac{1}{x+1} \quad \ldots(i)$$

Differentiating again w.r.t. x, we get

$$\frac{d^2y}{dx^2} = \frac{1}{(x+1)^2}$$

$$= \left(\frac{-1}{x+1}\right)^2$$

$$\Rightarrow \quad \frac{d^2y}{dx^2} = \left(\frac{dy}{dx}\right)^2 \quad \text{[From (i)]}$$

Hence Proved.

12. *If $y = x^x$, prove that :*

$$\frac{d^2y}{dx^2} - \frac{1}{y}\left(\frac{dy}{dx}\right)^2 - \frac{y}{x} = 0.$$

Sol. Given, $y = x^x$

Taking log on both sides, we get

$$\log y = x\log x$$

Differentiating w.r. t. x, we get

$$\frac{1}{y}\frac{dy}{dx} = x\cdot\frac{1}{x} + \log x$$

$$\frac{dy}{dx} = y(1+\log x) \quad \ldots(i)$$

Differentiating again w.r. t. x, we get

$$\Rightarrow \quad \frac{d^2y}{dx^2} = y\left(0+\frac{1}{x}\right) + (1+\log x)\frac{dy}{dx}$$

$$\Rightarrow \quad \frac{d^2y}{dx^2} = \frac{y}{x} + \frac{1}{y}\frac{dy}{dx}\cdot\frac{dy}{dx} \quad \text{[From (i)]}$$

$$\Rightarrow \quad \frac{d^2y}{dx^2} - \frac{1}{y}\left(\frac{dy}{dx}\right)^2 - \frac{y}{x} = 0$$ **Hence Proved.**

13. *If $y = \cos(\sin x)$, show that :*

$$\frac{d^2y}{dx^2} + \tan x\frac{dy}{dx} + y\cos^2 x = 0$$

Sol. Given, $y = \cos(\sin x)$ ...(i)

On differentiating both sides, we get

$$\Rightarrow \quad \frac{dy}{dx} = -\sin(\sin x)\cos x \quad \ldots(ii)$$

Again differentiating both sides

$$\Rightarrow \quad \frac{d^2y}{dx^2} = -\cos(\sin x)\cos^2 x + \sin(\sin x)\sin x$$

$$\Rightarrow \quad \frac{d^2y}{dx^2} = -y\cos^2 x + \frac{\sin x}{(-\cos x)}\frac{dy}{dx}$$

[Using (i) and (ii)]

$$\Rightarrow \quad \frac{d^2y}{dx^2} + y\cos^2 x = -\tan x\cdot\frac{dy}{dx}$$

$$\Rightarrow \quad \frac{d^2y}{dx^2} + \tan x\frac{dy}{dx} + y\cos^2 x \text{ s}$$

Hence Proved.

14. *If $y = e^{m\cos^{-1}x}$, prove that*

$$(1-x^2)\frac{d^2y}{dx^2} - x\frac{dy}{dx} = m^2y.$$

Sol. Given, $y = e^{m\cos^{-1}x}$

Differentiating both sides w.r.t. x

$$\therefore \quad \frac{dy}{dx} = e^{m\cos^{-1}x}m.\left(\frac{-1}{\sqrt{1-x^2}}\right)$$

$$(\sqrt{1-x^2})\frac{dy}{dx} = -m.e^{m\cos^{-1}x} = -m.y$$

Squaring both sides, we get

$$(1-x^2)\left(\frac{dy}{dx}\right)^2 = m^2y^2$$

Differentiating both sides w.r.t. x, we get

$$-2x\left(\frac{dy}{dx}\right)^2 + (1-x^2).\frac{2dy}{dx}\cdot\frac{d^2y}{dx^2} = m^2.2y\frac{dy}{dx}$$

Dividing both sides by $2\frac{dy}{dx}$, we get

$$-x\frac{dy}{dx} + (1-x^2)\frac{d^2y}{dx^2} = m^2y$$

$$\Rightarrow \quad (1-x^2)\frac{d^2y}{dx^2} - x\frac{dy}{dx} = m^2y.$$

Hence Proved.

15. *If $y = (\cot^{-1}x)^2$, show that*

$$(1+x^2)^2\frac{d^2y}{dx^2} + 2x(1+x^2)\frac{dy}{dx} = 2.$$

Sol. Given,

$$y = (\cot^{-1}x)^2$$

Differentiating w.r.t. x, we get

$$\frac{dy}{dx} = 2\cot^{-1}x\frac{d}{dx}(\cot^{-1}x)$$

$$\Rightarrow \quad \frac{dy}{dx} = 2\cot^{-1}x\left(\frac{-1}{1+x^2}\right)$$

$$\Rightarrow \quad (1+x^2)\frac{dy}{dx} = -2\cot^{-1}x$$

Differentiating once again w.r.t. x, we get

$$(1+x^2)\frac{d^2y}{dx^2} + 2x\frac{dy}{dx} = -2\cdot\frac{(-1)}{1+x^2}$$

$$\Rightarrow \quad (1+x^2)\frac{d^2y}{dx^2} + 2x\frac{dy}{dx} = \frac{2}{1+x^2}$$

$$\Rightarrow \quad (1+x^2)^2\frac{d^2y}{dx^2} + 2x(1+x^2)\frac{dy}{dx} = 2.$$

Hence Proved

16. *If $\log y = \tan^{-1}x$, prove that*

$$(1+x^2)\frac{d^2y}{dx^2} + (2x-1)\frac{dy}{dx} = 0$$

Sol. Given,

$$\log y = \tan^{-1}x$$

Differentiating both sides, w.r.t x, we get

$$\frac{1}{y}\frac{dy}{dx} = \frac{1}{1+x^2}$$

$$(1+x^2)\frac{dy}{dx} - y = 0$$

Again differentiating w.r.t x, we get

$$2x\frac{dy}{dx} + (1+x^2)\frac{d^2y}{dx^2} - \frac{dy}{dx} = 0$$

$$\Rightarrow \quad (1+x^2)\frac{d^2y}{dx^2} + (2x-1)\frac{dy}{dx} = 0$$

Hence Proved

17. *Verify Rolle's theorem for the function, $f(x) = -1 + \cos x$ in the interval $[0, 2\pi]$.**

Sol. Given : $f(x) = -1 + \cos x$ in $[0, 2\pi]$

(i) Since, $\cos x$ is continuous in $[0, 2\pi]$, so, $f(x)$ is a continuous function in $[0, 2\pi]$.

(ii) $f(x)$ is differentiable on $(0, 2\pi)$

(iii) $\because \quad f(0) = -1 + \cos 0 = -1 + 1 = 0$

and $f(2\pi) = -1 + \cos(2\pi)$

$= -1 + \cos(2\pi - 0)$

$= -1 + \cos 0$

$= -1 + 1 = 0$

$\therefore \quad f(0) = f(2\pi)$

Thus, $f(x)$ satisfies all conditions of Rolle's theorem. So, there exists a value of x, say $c \in (0, 2\pi)$ such that

$$f'(c) = 0$$

Now, $f(x) = -1 + \cos x$

$\therefore \quad f'(x) = 0 - \sin x$

$\Rightarrow \quad f'(c) = 0 - \sin c$

$\therefore \quad f'(c) = 0$

$\therefore \quad -\sin c = 0$

or $\quad \sin c = 0$

$\therefore \quad c = 0, \pi, 2\pi, ...$

Thus, $c = \pi \in (0, 2\pi)$, such that $f'(c) = 0$

Hence, Rolle's theorem is verified. **Hence Proved**

18. *If $y = e^{m\sin^{-1}x}$, prove that*

$$(1-x^2)\frac{d^2y}{dx^2} - x\frac{dy}{dx} = m^2y$$

Sol. Given : $\quad y = e^{m\sin^{-1}x}$

Differentiating w.r.t. x, we get

$$\frac{dy}{dx} = e^{m\sin^{-1}x}\frac{d}{dx}(m\sin^{-1}x)$$

$$\Rightarrow \quad \frac{dy}{dx} = y \times \frac{m}{\sqrt{1-x^2}} \quad ...(i)$$

$$\text{or} \quad \sqrt{1-x^2}\frac{dy}{dx} = my$$

* Frequently asked previous years Board Exam Questions.

Squaring both LHS and RHS,

$$\sqrt{1-x^2}\left(\frac{dy}{dx}\right)^2 = m^2y^2$$

Differentiating w.r.t. x,

$$\Rightarrow (1-x^2)\times 2\left(\frac{dy}{dx}\right)\left(\frac{d^2y}{dx^2}\right)+\left(\frac{dy}{dx}\right)^2(-2x) = m^2y\frac{dy}{dx}$$

$$\Rightarrow \quad (1-x^2)\frac{d^2y}{dx^2}+\frac{dy}{dx}(-x) = m^2y$$

$$\Rightarrow \quad (1-x^2)\frac{d^2y}{dx^2}-x\frac{dy}{dx} = m^2y$$

Hence Proved.

19. *If* $y=\dfrac{x\sin^{-1}x}{\sqrt{1-x^2}}$, *prove that*

$$(1-x^2)\frac{dy}{dx} = x+\frac{y}{x}$$

Sol. Given, $y=\dfrac{x\sin^{-1}x}{\sqrt{1-x^2}}$

$$y\sqrt{1-x^2} = x\sin^{-1}x \quad \text{...(i)}$$

Differentiating both sides w.r.t. x, we get

$$y\cdot\frac{(-2x)}{2\sqrt{1-x^2}}+\sqrt{1-x^2}\frac{dy}{dx} = x\cdot\frac{1}{\sqrt{1-x^2}}+\sin^{-1}x$$

$$\Rightarrow \quad -xy+(1-x^2)\frac{dy}{dx} = x+\sqrt{1-x^2}\sin^{-1}x$$

$$\Rightarrow \quad -xy+(1-x^2)\frac{dy}{dx} = x+\sqrt{1-x^2}\cdot\frac{y}{x}\sqrt{1-x^2}$$

[From equation (i)]

$$\Rightarrow \quad -xy+(1-x^2)\frac{dy}{dx} = x+\frac{y}{x}(1-x^2)$$

$$\Rightarrow \quad -xy+(1-x^2)\frac{dy}{dx} = x+\frac{y}{x}-yx$$

$$\Rightarrow \quad (1-x^2)\frac{dy}{dx} = x+\frac{y}{x}$$ **Hence Proved.**

20. *A function $f(x)$ is defined on an interval (0, 2) as following :*

$$f(x) = x \quad \textit{when } 0\le x\le 1$$
$$= 2x-1 \quad \textit{when } 1<x\le 2.$$

Prove that the function is not differentiable at $x = 1$.

Sol. Given,

$$f(x) = x \quad \text{when } 0 \le x \le 1$$
$$= 2x-1 \quad \text{when } 1 \le x \le 2$$

Then, function is differentiable at $x = 1$, if at $x = 1$,

Left hand derivative = Right hand derivative

Now, $\text{LHD} = \lim\limits_{h\to 0^-}\dfrac{f(1-h)-f(1)}{h}$

$$= \lim_{h\to 0^-}\frac{1-h-1}{h} = -1$$

where h is a negative quantity

And, $\text{RHD} = \lim\limits_{h\to 0^+}\dfrac{f(1+h)-f(1)}{h}$

$$= \lim_{h\to 0^+}\frac{2(1+h)-1-1}{h} = \frac{2h}{h} = 2$$

where h is a positive quantity

$\Rightarrow$ LHD $\neq$ RHD

Hence, the function is not differentiable at $x = 1$. **Hence Proved.**

21. *Show that the function $f(x) = x^2 - 6x + 1$ satisfies the Lagrange's Mean Value Theorem. Also find the co-ordinate of a point at which the tangent to the curve represented by the above function is parallel to the chord joining A (1, – 4) and B (3, – 8).*

Sol. Given, $f(x) = x^2 - 6x + 1$

As it is a quadratic polynomial so it is continuous.

Differentiating $f(x)$ w.r. t. x, we get

$$f'(x) = 2x-6$$

Gradient m of line AB is given as

$$m = \frac{-8-(-4)}{3-1} = \frac{-4}{2} = -2$$

$\Rightarrow$ $2x-6 = -2$ $\quad$ [$\because$ gradient $= f'(x)$]

$\Rightarrow$ $2x = 4$

$\Rightarrow$ $x = 2$ and $y = (2)^2 - 6\times 2 + 1 = -7$

$\therefore$ The required point is (2, –7). **Ans.**

22. *Given $f(x) = (x - 3)\log x$. Prove that there is atleast one value of x in the interval [1, 3] which satisfies the equation $x \log x = 3 - x$.*

Sol. Given, $f(x) = (x - 3)\log x$ in the interval [1, 3].

$f(x)$ is continuous in [1, 3] as it is a product of polynomial and logarithmic functions in x.

$$\therefore \quad f'-(x) = (x-3)\times\frac{1}{x}+\log x$$

$$\Rightarrow \quad f'(x) = 1-\frac{3}{x}+\log x \quad \text{...(i)}$$

Hence, it is differentiable in open interval (1, 3).

Now, $f(1) = 0 = f(3)$ $\quad$ ($\because \log 1 = 0$)

Hence, Rolle's theorem is applicable.

$\therefore$ There exists x in the interval $(1, 3)$

such that $(-) f'(x) = 0$

$$\therefore \quad 1-\frac{3}{x}+\log x = 0 \qquad \text{...(From (i))}$$

$$\Rightarrow \quad x + x\log x = 3$$

$$\Rightarrow \quad x\log x = 3 - x.$$

Hence Proved.

Chapter 7. Applications of Derivatives

1. *Prove that the function $f(x) = x^3 - 6x^2 + 12x + 5$ is increasing on R.**

Sol. Given, $\quad f(x) = x^3 - 6x^2 + 12x + 5$

$$\therefore \quad f'(x) = 3x^2 - 12x + 12$$

$$= 3(x^2 - 4x + 4)$$

$$= 3(x-2)^2$$

$$\Rightarrow \quad 3(x-2)^2 \geq 0, \text{ for all } x \in R$$

$$\Rightarrow \quad f'(x) \geq 0, \text{ for all } x \in R.$$

$\therefore f(x)$ is increasing for $x \in R$. **Hence Proved.**

2. (a) *Show that the radius of a closed right circular cylinder of given surface area and maximum volume is equal to half of its height.**

OR

(b) *Prove that the area of right-angled triangle of given hypotenuse is maximum when the triangle is isosceles.**

Sol. (a) Let S be the given surface area of a closed right circular cylinder whose radius is r and height is h.

$$\therefore \quad S = 2\pi rh + 2\pi r^2$$

$$\Rightarrow \quad h = \frac{S - 2\pi r^2}{2\pi r} \qquad \text{...(i)}$$

Now, volume of cylinder $(V) = \pi r^2 h$

$$\therefore \quad V = \pi r^2\left(\frac{S-2\pi r^2}{2\pi r}\right) \quad \text{[From equation (i)]}$$

$$\Rightarrow \quad V = \frac{Sr - 2\pi r^3}{2}$$

$$\therefore \quad \frac{dV}{dr} = \frac{S - 6\pi r^2}{2}$$

[differentiating w.r.t. r.] ...(ii)

$$\text{and} \quad \frac{d^2V}{dr^2} = \frac{-12\pi r}{2} = -6\pi r$$

[Again diff. w.r.t. r.] ...(iii)

For maximum and minimum value,

$$\text{Put} \quad \frac{dV}{dr} = 0 \qquad \text{[from eq. (ii)]}$$

$$\Rightarrow \quad \frac{S-6\pi r^2}{2} = 0$$

$$\Rightarrow \quad S = 6\pi r^2$$

Putting the value of S in equation (i), we get

$$h = \frac{6\pi r^2 - 2\pi r^2}{2\pi r}$$

$$= \frac{4\pi r^2}{2\pi r} = 2r$$

$$\text{or} \quad r = \frac{h}{2}$$

$$\text{and} \quad \left[\frac{d^2V}{dr^2}\right]_{r=h/2} = -6\pi \times \frac{h}{2} \quad \text{[from eq. (iii)]}$$

$$= -3\pi h < 0$$

[$\therefore$ h cannot be negative]

So, volume is maximum when $r = \dfrac{h}{2}$.

OR

(b) Let h be the hypotenuse of the right-angled triangle and x be its altitude.

So, base $= \sqrt{h^2 - x^2}$ [using Pythagoras theorem]

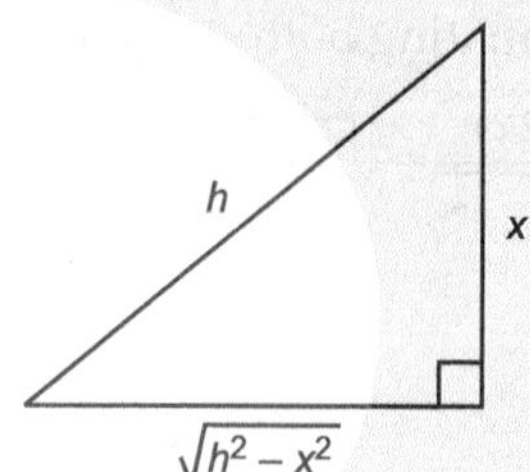

Now, area of triangle (A) $= \dfrac{1}{2} \times \sqrt{h^2 - x^2} \times x$

$$\therefore \quad \frac{dA}{dx} = \frac{1}{2}\left[x\left\{\frac{1}{2}(h^2-x^2)^{-1/2} \times (-2x)\right\} + \sqrt{h^2-x^2}\right]$$

$$= \frac{1}{2}\left[\frac{-x^2}{\sqrt{h^2-x^2}} + \sqrt{h^2-x^2}\right]$$

$$= \frac{1}{2}\left[\frac{-x^2 + h^2 - x^2}{\sqrt{h^2-x^2}}\right]$$

$$= \frac{1}{2}\left[\frac{h^2 - 2x^2}{\sqrt{h^2-x^2}}\right]$$

$$\text{and} \quad \frac{d^2A}{dx^2} = \frac{1}{2}\left[\frac{\sqrt{h^2-x^2}.(-4x) - (h^2-2x^2)\left\{\frac{1}{2}(h^2-x^2)^{-1/2}(-2x)\right\}}{\left(\sqrt{h^2-x^2}\right)^2}\right]$$

* Frequently asked previous years Board Exam Questions.

$$= \frac{1}{2}\left[\frac{-4x\left(\sqrt{h^2-x^2}\right)+\frac{\left(h^2-2x^2\right)x}{\sqrt{h^2-x^2}}}{\left(h^2-x^2\right)}\right]$$

$$= \frac{1}{2}\left[\frac{-4x\left(h^2-x^2\right)+\left(h^2-2x^2\right)x}{\left(h^2-x^2\right).\sqrt{h^2-x^2}}\right]$$

$$= \frac{1}{2}\left[\frac{-4xh^2+4x^3+h^2x-2x^3}{\left(h^2-x^2\right)^{3/2}}\right]$$

$$= \frac{1}{2}\left[\frac{2x^3-3xh^2}{\left(h^2-x^2\right)^{3/2}}\right]$$

For maximum and minimum value,

Put $\frac{dA}{dx} = 0$

$$\Rightarrow \quad \frac{1}{2}\left[\frac{h^2-2x^2}{\sqrt{h^2-x^2}}\right] = 0$$

$$\Rightarrow \quad h^2-2x^2 = 0$$

$$\Rightarrow \quad x = \frac{h}{\sqrt{2}}$$

$$\therefore \quad \left(\frac{d^2A}{dx^2}\right)_{x=\frac{h}{\sqrt{2}}} = \frac{1}{2}\left[\frac{2\times\frac{h^3}{2\sqrt{2}}-3\times\frac{h}{\sqrt{2}}\times h^2}{\left(h^2-\frac{h^2}{2}\right)^{3/2}}\right]$$

$$= \frac{1}{2}\left[\frac{h^3-3h^3}{\left(\frac{h^2}{2}\right)^{3/2}\times\sqrt{2}}\right]$$

$$= \frac{-2h^3}{2\times\frac{h^3}{2\sqrt{2}}\times\sqrt{2}}$$

$$= -2 < 0$$

Thus, A is maximum at $x = \frac{h}{\sqrt{2}}$

Now, Base $= \sqrt{h^2-x^2}$

$$= \sqrt{h^2-\frac{h^2}{2}} = \sqrt{\frac{h^2}{2}} = \frac{h}{\sqrt{2}}$$

and Altitude $= x = \frac{h}{\sqrt{2}}$

Since, Base = Altitude $= \frac{h}{\sqrt{2}}$

Hence, the triangle is isosceles. **Hence Proved.**

3. *A rectangle is given whose area is constant. Prove that the sum of the length of its sides is least when it is a square.*

Sol. Let the length be x and breadth be y.

$\therefore$ Perimeter $(P) = 2x + 2y$,

and Area $(A) = xy$

$$\Rightarrow \quad P = 2x + 2\cdot\frac{A}{x} \qquad \left[\because y = \frac{A}{x}\right]$$

Differentiating w.r.t. x, we get

$$\frac{dP}{dx} = 2 - \frac{2A}{x^2} \qquad \ldots(i)$$

For maxima or minima,

$$\frac{dP}{dx} = 0$$

$$\Rightarrow \quad 2 - \frac{2A}{x^2} = 0$$

$$\Rightarrow \quad 2x^2 = 2A$$

$$\Rightarrow \quad x^2 = A$$

$$\Rightarrow \quad x = \sqrt{A}$$

Now differentiating again equation (i) w.r.t. x, we get

$$\frac{d^2P}{dx^2} = 0 + \frac{4A}{x^3}$$

which is positive when $x = \sqrt{A}$

$\therefore$ P is minimum when $x = \sqrt{A}$

Since, $y = \frac{A}{x}$

$$= \frac{A}{\sqrt{A}} = \sqrt{A}$$

$$\therefore \quad y = x$$

Thus, it is square. **Hence Proved.**

4. *If the sum of the lengths of the hypotenuse and a side of a right angled triangle is given, show that the area of the triangle is maximum when the angle between them is* $\frac{\pi}{3}$.

Sol. Let 'l' be the length of the hypotenuse of the given right angled Δ ABC at B and $\angle CAB = \theta$ (in radian measure) and $0 < \theta < \pi/2$.

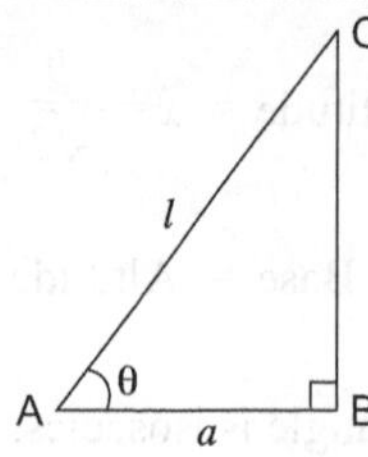

Then, $a = AB = l\cos\theta$ and $BC = l\sin\theta$.

Given, $S = l + a$

$S = l + l\cos\theta = l(1+\cos\theta)$

$\Rightarrow \quad l = \dfrac{S}{1+\cos\theta}$ ...(i)

Now, Area of the $\Delta ABC = \dfrac{1}{2}\cdot AB\cdot BC$

$= \dfrac{1}{2}(l\cos\theta)\cdot l\sin\theta$

$= \dfrac{1}{2}l^2\sin\theta\cos\theta$

$= \dfrac{1}{4}l^2(2\sin\theta\cos\theta)$

$\Rightarrow \quad A = \dfrac{\sin 2\theta}{4}\cdot\left[\dfrac{S^2}{(1+\cos\theta)^2}\right]$

[From (i)]

Differentiating w.r.t. θ, we get

$$\frac{dA}{d\theta} = \frac{S^2}{4}$$

$$\text{X}\ \frac{\left[(1+\cos\theta)^2\frac{d}{d\theta}\sin 2\theta - \sin 2\theta\frac{d}{d\theta}(1+\cos\theta)^2\right]}{(1+\cos\theta)^4}$$

For maxima or minima

$$\frac{dA}{d\theta} = 0$$

$\Rightarrow (1+\cos\theta)^2\cdot 2\cos 2\theta + 2\sin 2\theta\,(1+\cos\theta)\cdot\sin\theta = 0$

$\Rightarrow 2(1+\cos\theta)\,[(1+\cos\theta)\cdot\cos 2\theta + 2\sin\theta\cos\theta\cdot\sin\theta] = 0$

$\Rightarrow (1+\cos\theta)\,[(1+\cos\theta)(1-2\sin^2\theta) + 2\sin^2\theta\cos\theta] = 0$

$\Rightarrow (1+\cos\theta)\,[1-2\sin^2\theta+\cos\theta - 2\sin^2\theta\cos\theta + 2\sin^2\theta\cos\theta] = 0$

$\Rightarrow (1+\cos\theta)\,[1-2(1-\cos^2\theta)+\cos\theta] = 0$

$\Rightarrow (1+\cos\theta)\,[2\cos^2\theta+\cos\theta-1] = 0$

$\Rightarrow (1+\cos\theta)\,[2\cos^2\theta+2\cos\theta-\cos\theta-1] = 0$

$\Rightarrow (1+\cos\theta)\,[2\cos\theta(\cos\theta+1) - 1(\cos\theta+1)] = 0$

$\Rightarrow (1+\cos\theta)(1+\cos\theta)(2\cos\theta-1) = 0$

$\Rightarrow \cos\theta = -1$ or $\cos\theta = \dfrac{1}{2}$

$\Rightarrow \theta = \pi$ (not possible) or $\theta = \dfrac{\pi}{3}$

i.e., $\theta = \dfrac{\pi}{3}$ **Hence Proved**

5. *An open tank with a square base of side 'x' metre and vertical height 'h' metres is to be constructed s as to contain 'c' cubic metres of water. Show that th expenses on lining the inside of the tank with lea would be least if*

$$h = \frac{x}{2}.$$

Sol. Given side of square base = 'x' metres and height of tank = 'h' metres

$\therefore$ Volume $= c = x^2h$

$\Rightarrow \quad h = \dfrac{c}{x^2}$ ...(i

Now, cost of lining from inside is directly proportional to surface area.

$\therefore \quad S = x^2 + 4xh$

$= x^2 + 4x\cdot\dfrac{c}{x^2}$ [From eq. (i)

$= x^2 + \dfrac{4c}{x}$

Differentiating w.r.t. x, we get

$$\frac{dS}{dx} = 2x - \frac{4c}{x^2} \quad \text{...(ii}$$

For S to be minimum or maximum, $\dfrac{dS}{dx} = 0.$

$\Rightarrow \quad 2x - \dfrac{4c}{x^2} = 0$ [From (ii)

$\Rightarrow \quad 2x^3 = 4c$

$\Rightarrow \quad c = \dfrac{x^3}{2}$

$\Rightarrow \quad x = (2c)^{1/3}$

Differentiating eq. (ii) w.r.t. x, we get

$$\frac{d^2S}{dx^2} = 2 + \frac{8c}{x^3} > 0$$

It is always positive. Hence, it is point of minima

By equation (i), $h = \dfrac{c}{x^2}$

$\Rightarrow \quad h = \dfrac{x^3}{2x^2}$

$\Rightarrow \quad h = \dfrac{x}{2}$ **Hence Proved**

6. *A right-angled triangle ABC with constant area S is given. Prove that the hypotenuse of the triangle is least when the triangle is isosceles.*

Sol. Given, area of triangle $= \dfrac{x \cdot y}{2} = S$

$$\Rightarrow \quad y = \frac{2S}{x} \quad \ldots\text{(i)}$$

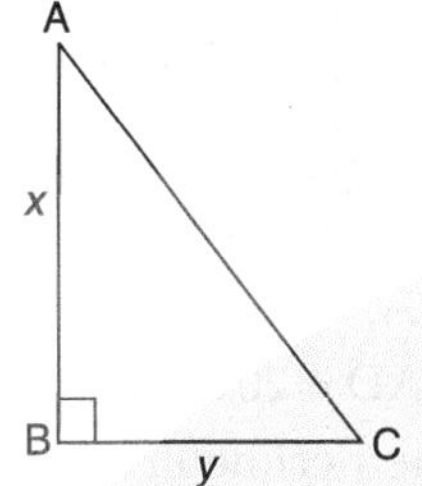

Hypotenuse,

$$z = \sqrt{x^2 + y^2}$$

$$\Rightarrow \quad z^2 = x^2 + y^2$$

$$\Rightarrow \quad z^2 = x^2 + \left(\frac{2S}{x}\right)^2$$

Differentiating both sides w.r.t. x, we get

$$2z\frac{dz}{dx} = 2x + 2\left(\frac{2S}{x}\right)\cdot\frac{-2S}{x^2}$$

$$= 2x - \frac{8S^2}{x^3} \quad \ldots\text{(ii)}$$

For maxima and minima, $\dfrac{dz}{dx} = 0$

$$\Rightarrow \quad 2z\frac{dz}{dx} = 0$$

$$\Rightarrow \quad 2x - \frac{8S^2}{x^3} = 0$$

$$\Rightarrow \quad 2x = \frac{8S^2}{x^3}$$

Again, $\quad x^4 = 4S^2$

Taking square root on both sides

$$\Rightarrow \quad x^2 = 2S$$

$$\Rightarrow \quad x = \sqrt{2S}$$

Differentiating equation (ii) w.r.t. x, we get

$$2z.\frac{d^2z}{dx^2} + 2\left(\frac{dz}{dx}\right)^2 = 2 + 24\cdot\frac{S^2}{x^4}$$

We can see that $\left(\dfrac{dz}{dx}\right)^2$, 2 and $\dfrac{S^2}{x^4}$ will be always positive

$$\Rightarrow \quad 2z\frac{d^2z}{dx^2} > 0$$

$$\Rightarrow \quad \frac{d^2z}{dx^2} > 0$$

Hence, it is point of minima.

At $\quad x = \sqrt{2S}$,

$$y = \frac{2S}{x} = \frac{2S}{\sqrt{2S}} = \sqrt{2S}$$

$$\Rightarrow \quad x = y$$

Hence, for minimum hypotenuse in right-angled triangle, the triangle should be isosceles. **Hence Proved.**

7. *Prove that the right circular cone of maximum volume which can be inscribed in a sphere of radius 'a' has a height of $\frac{4}{3}$ a.*

Sol. Let OD = x and DC = r. Then h = AD = AO + OD $= a + x$ and $a^2 = r^2 + x^2$.

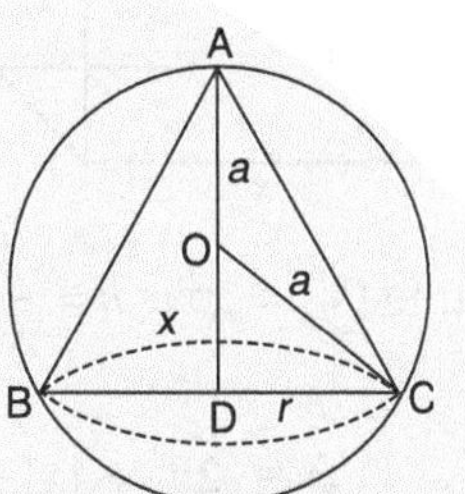

The volume of the cone is given by,

$$V = \frac{1}{3}\pi r^2 h$$

$$= \frac{1}{3}\pi\,(a^2 - x^2)\,(a + x)$$

Differentiating w.r.t. x, we get

$$\Rightarrow \quad \frac{dV}{dx} = \frac{1}{3}\pi\,[-2x\,(a + x) + 1\,(a^2 - x^2)]$$

$$= \frac{1}{3}\pi\,[-2x\,(a + x) + (a + x)\,(a - x)]$$

$$= \frac{1}{3}\pi\,(a + x)\,(a - 3x)$$

For maxima and minima, $\dfrac{dV}{dx} = 0$

$$\Rightarrow \quad \frac{1}{3}\pi\,(a + x)\,(a - 3x) = 0$$

$$\Rightarrow \quad x = -a, \frac{a}{3}$$

$$\Rightarrow \quad x = \frac{a}{3}, \text{ because } x \text{ cannot be negative}$$

Also, $\quad \dfrac{d^2V}{dx^2} = \dfrac{1}{3}\pi\,[1\,(a - 3x) - 3\,(a + x)]$

$$= \frac{1}{3}\pi\,(-2a - 6x)$$

Differentiating again w.r.t. x, we get

$$\Rightarrow \left(\frac{d^2V}{dx^2}\right)_{x = a/3} = \frac{1}{3}\pi\,(-2a - 2a) = \frac{-4a\pi}{3} < 0$$

Therefore, volume is maximum when $x = \dfrac{a}{3}$

$$\Rightarrow \quad h = a + x = a + \frac{a}{3} = \frac{4a}{3}.$$

Hence Proved.

8. *Show that the surface area of a closed cuboid with square base and given volume is minimum when it is a cube.*

Sol. Let h be the height and x be the side of the square base of the closed cuboid. Then,

Surface Area (A) = $2(x^2 + 2hx)$

$$A = 2x^2 + 4hx \quad \text{...(i)}$$

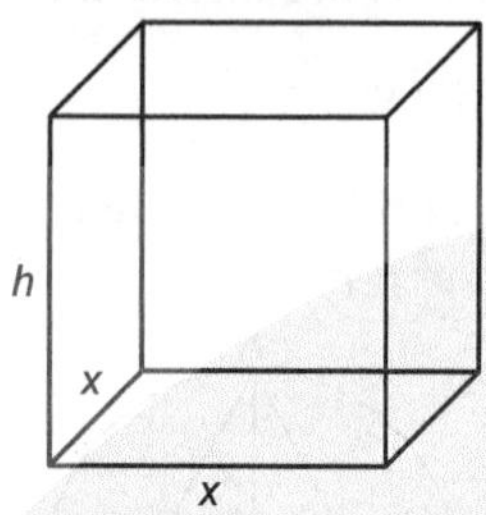

and Volume (V) = x^2h, $h = \dfrac{V}{x^2}$ ...(ii)

Now, $$A = 2x^2 + 4x\left(\frac{V}{x^2}\right)$$

$$A = 2x^2 + \frac{4V}{x}$$

On differentiating w.r.t. x, we get

$$\frac{dA}{dx} = 4x - \frac{4V}{x^2} \quad \text{...(iii)}$$

Putting $\dfrac{dA}{dx} = 0$, we get

$$4x - \frac{4V}{x^2} = 0$$

$$4x = \frac{4V}{x^2}$$

$$V = x^3$$

$$x^2h = x^3$$

$$h = x$$

Differentiating (iii) w.r.t. x, we get

$$\frac{d^2A}{dx^2} = 4 + \frac{8V}{x^3}$$

$$= 4 + \frac{8x^2h}{x^3} \quad \text{[From (ii)]}$$

$$= 4 + \frac{8h}{x}$$

$$\Rightarrow \left(\frac{d^2A}{dx^2}\right)_{h=x} = 4 + 8 = 12 > 0$$

Hence, the surface area is minimum when length = breadth = height = x.

i.e, when it is a cube. **Hence Proved.**

9. *Show that the rectangle of maximum perimeter which can be inscribed in a circle of radius 10 cm is a square of side $10\sqrt{2}$ cm.*

Sol. Let ABCD be a rectangle inscribed in a circle of radius 10 cm with centre at O, then DB = 20 cm.

Let ∠OBA = θ, (0 < θ < π/2) and θ is in radian.

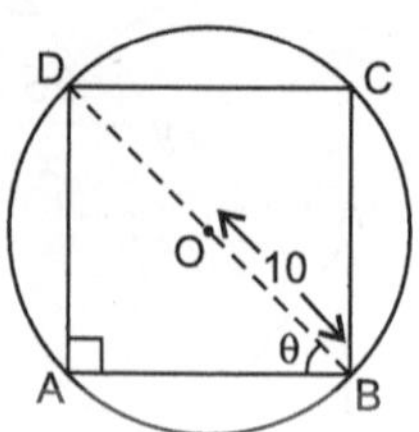

Then, AB = 20 cos θ

AD = 20 sin θ

Let p be the perimeter of the rectangle ABCD, then

$$p = 2AB + 2AD$$

$$= 2(20\cos\theta + 20\sin\theta)$$

$$p = 40(\cos\theta + \sin\theta) \quad \text{...(i)}$$

Differentiating (i) w.r.t. θ, we get

$$\frac{dp}{d\theta} = 40(-\sin\theta + \cos\theta) \quad \text{...(ii)}$$

Again differentiating, we get

$$\frac{d^2p}{d\theta^2} = -40(\cos\theta + \sin\theta) \quad \text{...(iii)}$$

Now, for maximum value

$$\frac{dp}{d\theta} = 0$$

$$\Rightarrow 40(-\sin\theta + \cos\theta) = 0$$

$$\Rightarrow \cos\theta = \sin\theta$$

$$\Rightarrow \tan\theta = 1$$

$$\theta = \tan^{-1}(1)$$

$$\Rightarrow \theta = \frac{\pi}{4}$$

Also, $$\left(\frac{d^2p}{d\theta^2}\right)_{\theta=\pi/4} = -40\left(\frac{1}{\sqrt{2}} + \frac{1}{\sqrt{2}}\right)$$

$$= -40 \times \frac{2}{\sqrt{2}}$$

$$= -40\sqrt{2} < 0 \;(-\text{ve})$$

$\Rightarrow p$ is maximum when $\theta = \dfrac{\pi}{4}$.

Therefore, p is maximum when

$$AB = 20\cos\frac{\pi}{4} = 20 \times \frac{1}{\sqrt{2}} = 10\sqrt{2}$$

$$AD = 20\sin\frac{\pi}{4} = 20 \times \frac{1}{\sqrt{2}} = 10\sqrt{2}$$

i.e., when adjacent sides are equal and each of them is $10\sqrt{2}$ cm.

That is, a rectangle which can be inscribed in a circle of radius 10 cm is a square of side $10\sqrt{2}$ cm. **Hence Proved.**

10. *Prove that the volume of the largest cone that can be inscribed in a sphere of radius 'R' is $\frac{8}{27}$ of the volume of the sphere.*

Sol. Let r and h be the radius and height of the cone

$$AC = h, AB = r$$

$$OA = AC - OC = h - R$$

C R h O R A r B

In Δ OAB

$$OB^2 = AB^2 + OA^2$$

$$R^2 = r^2 + (h - R)^2$$

$$\therefore \quad r^2 = R^2 - (h - R)^2$$

$$r^2 = R^2 - (h^2 + R^2 - 2hR)$$

$$r^2 = 2hR - h^2 \quad \ldots(i)$$

Volume of cone,

$$V = \frac{1}{3}\pi r^2 h$$

$$= \frac{1}{3}\pi(2hR - h^2)h$$

$$= \frac{\pi}{3}(2Rh^2 - h^3)$$

$$\frac{dV}{dh} = \frac{\pi}{3}(4Rh - 3h^2)$$

For maximum or minimum,

$$\frac{dV}{dh} = 0$$

$$\Rightarrow \quad \frac{\pi}{3}(4Rh - 3h^2) = 0$$

$$\Rightarrow \quad 4Rh = 3h^2$$

$$\Rightarrow \quad h = \frac{4}{3}R$$

$$\frac{d^2V}{dh^2} = \frac{\pi}{3}(4R - 6h)$$

$$\Rightarrow \quad \left.\frac{d^2V}{dh^2}\right|_{h=\frac{4}{3}R} = \frac{\pi}{3}\left(4R - 6 \times \frac{4}{3}R\right)$$

$$= \frac{\pi}{3}(4R - 8R) < 0$$

$\Rightarrow$ V is maximum at $h = \frac{4}{3}R$

$$\therefore \quad r^2 = 2.\frac{4R}{3}R - \left(\frac{4}{3}R\right)^2$$

$$= \frac{8}{3}.R^2 - \frac{16}{9}R^2$$

[From eq. (i),]

$$= \left(\frac{24-16}{9}\right)R^2$$

$$\Rightarrow \quad r^2 = \frac{8}{9}R^2$$

$\therefore$ Volume of largest cone,

$$V = \frac{1}{3}\pi r^2 h$$

$$= \frac{1}{3}\pi\left(\frac{8}{9}R^2\right)\cdot\left(\frac{4}{3}R\right)$$

$$= \frac{8}{27}\cdot\left(\frac{4}{3}\pi R^3\right)$$

$$= \frac{8}{27}\text{(volume of sphere)}$$

Hence Proved.

11. *Show that the right circular cone of least curved surface area and given volume has an altitude equal to $\sqrt{2}$ times the radius of the base.*

Sol. Let r, h, l be the radius, height and slant height of the cone respectively.

Volume of cone,

$$V = \frac{1}{3}\pi r^2 h$$

$$\Rightarrow \quad h = \frac{3V}{\pi r^2} \quad \ldots(i)$$

Curved surface area of cone,

$$A = \pi r l$$

$$\Rightarrow \quad A^2 = \pi^2 r^2 l^2$$

$$= \pi^2 r^2(r^2 + h^2) \quad [\because l^2 = r^2 + h^2]$$

$$= \pi^2 r^2\left(r^2 + \frac{9V^2}{\pi^2 r^4}\right) \quad \text{[Using (i)]}$$

$$= \pi^2 r^2\left(\frac{\pi^2 r^6 + 9V^2}{\pi^2 r^4}\right)$$

$$= \pi^2 r^4 + \frac{9V^2}{r^2}$$

Let $\quad S = A^2 = \pi^2 r^4 + \frac{9V^2}{r^2}$

$$\Rightarrow \quad \frac{dS}{dr} = 4\pi^2 r^3 + 9V^2\left(\frac{-2}{r^3}\right)$$

$$\Rightarrow \quad \frac{dS}{dr} = 4\pi^2 r^3 - \frac{18V^2}{r^3}$$

For max. or min, $\frac{dS}{dr} = 0$

$$\Rightarrow \quad 4\pi^2 r^3 - \frac{18V^2}{r^3} = 0$$

$$\Rightarrow \quad 4\pi^2 r^3 = \frac{18V^2}{r^3}$$

$$\Rightarrow \quad 4\pi^2 r^3 = \frac{18\left(\frac{1}{3}\pi r^2 h\right)^2}{r^3}$$

$$\Rightarrow \quad 4\pi^2 r^3 = 18\cdot\frac{1}{9}\frac{\pi^2 r^4 h^2}{r^3}$$

$$\Rightarrow \quad 2r^2 = h^2$$

$$\Rightarrow \quad h = \sqrt{2r^2} \text{ or } r = \frac{h}{\sqrt{2}}$$

$$\frac{d^2S}{dr^2} = 12\pi^2 r^2 + \frac{54V^2}{r^4}$$

$$\Rightarrow \quad \left.\frac{d^2S}{dr^2}\right|_{r=\frac{h}{\sqrt{2}}} = +\text{ve} > 0$$

⇒ Curved surface area of cone is minimum when $h = \sqrt{2}\, r$. **Hence Proved.**

12. *A square piece of tin of side 18 cm is to be made into a box without top, by cutting a square from each corner and folding up the flaps to form the box. What should be the side of the square to be cut off so that the volume of the box is the maximum possible ?*

Sol. Let x be the side of the square to be cut from each corner.

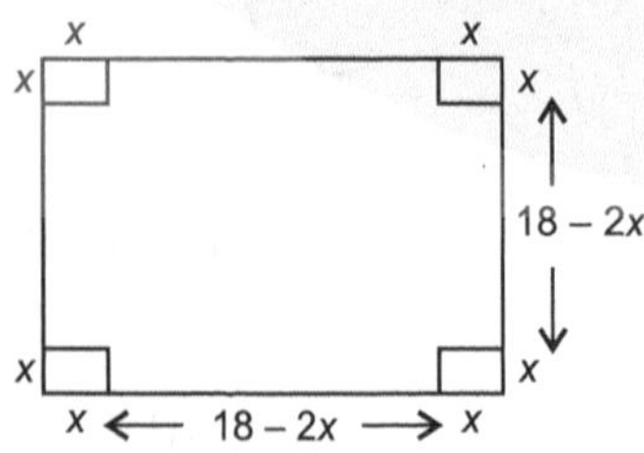

∴ For the box

$$l = 18 - 2x \text{ cm}$$

$$b = 18 - 2x \text{ cm}$$

$$h = x \text{ cm}$$

∴ Volume of the box

$$V = lbh = (18 - 2x)^2 . x$$

$$\frac{dV}{dx} = (18-2x)^2\, 1 + x .\, 2\,(18-2x)\,(-2)$$

$$= (18-2x)^2 - 4x\,(18-2x)$$

$$= (18-2x)\,(18-2x-4x)$$

$$= (18-2x)\,(18-6x)$$

$$= 2\,(9-x)\,2(9-3x)$$

$$\frac{dV}{dx} = 4\,(9-x)\,(9-3x) \quad \text{...(i}$$

For maxima or minima,

$$\frac{dV}{dx} = 0$$

$$\Rightarrow 4\,(9-x)\,(9-3x) = 0$$

$$\Rightarrow \quad 9 - x = 0 \text{ or } 9 - 3x = 0$$

$$\Rightarrow \quad x = 9 \text{ or } 3$$

but $x = 9$ is not possible [∵ then $l = b = 0$]

$$\therefore \quad x = 3$$

Differentiating, (i) w.r.t. x,

$$\frac{d^2V}{dx^2} = 4\,[(9-x)\,(-3) + (9-3x)\,(-1)]$$

$$= 4\,[-27 + 3x - 9 + 3x]$$

$$= 4\,[6x - 36]$$

$$= 24\,[x - 6]$$

$$\left.\frac{d^2V}{dx^2}\right|_{x=3} = 24\,(3-6)$$

$$= 24\,(-3) = -\text{ve} < 0$$

⇒ V is maximum when $x = 3$

Hence, volume of the box is maximum whe[n] the side of square to be cut from each corne[r] is 3 cm. **Ans**

13. *Show that the height of the cylinder of maximu[m] volume that can be inscribed in a sphere of radius R is $\frac{2R}{\sqrt{3}}$. Also find the maximum volume.*

Sol. Let r be radius and h be the height of th[e] cylinder

In Δ OMN

$$R^2 = r^2 + \left(\frac{h}{2}\right)^2 \quad [\text{ON = radius of circle = R}$$

$$\text{OM} = \text{OP} = \frac{h}{2}$$

$$\Rightarrow \quad r^2 = R^2 - \frac{h^2}{4}$$

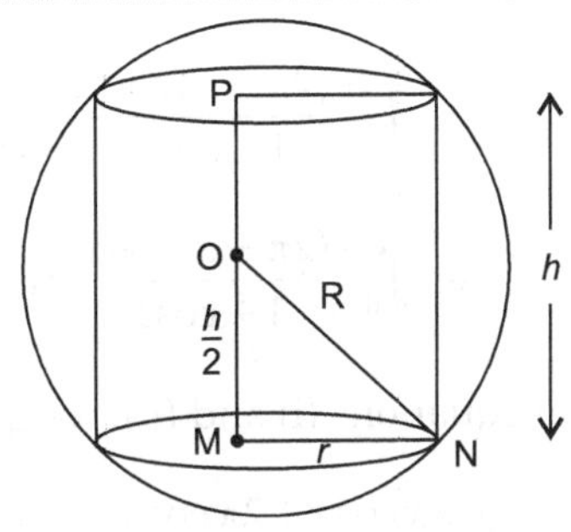

Volume of cylinder

$$V = \pi r^2 h$$

$$= \pi\left[R^2 - \frac{h^2}{4}\right]h \quad \text{...(i)}$$

$$\therefore \quad V = \pi\left[R^2 h - \frac{h^3}{4}\right]$$

$$\frac{dV}{dh} = \pi\left[R^2 - \frac{3h^2}{4}\right]$$

For max. or min. value

$$\frac{dV}{dh} = 0 \Rightarrow \pi\left[R^2 - \frac{3h^2}{4}\right] = 0$$

$$\Rightarrow \quad R^2 = \frac{3h^2}{4}$$

$$\Rightarrow \quad h = \frac{2R}{\sqrt{3}}$$

$$\frac{d^2V}{dh^2} = \pi\left(0 - \frac{6h}{4}\right) = -\frac{3}{2}\pi h$$

$$\Rightarrow \quad \left.\frac{d^2V}{dh^2}\right|_{h=\frac{2R}{\sqrt{3}}} = -\text{ve} < 0$$

$\Rightarrow$ V is maximum when

$$h = \frac{2R}{\sqrt{3}}$$

Hence Proved.

Maximum volume

$$= \pi.\left[R^2 - \frac{h^2}{4}\right]h$$

$$= \pi.\left[R^2 - \frac{4R^2}{3\times 4}\right]\frac{2R}{\sqrt{3}}$$

$$= \pi.\left[R^2 - \frac{R^2}{3}\right]\frac{2R}{\sqrt{3}}$$

$$= \pi \times \frac{2R^2}{3} \times \frac{2R}{\sqrt{3}}$$

$$= \frac{4\pi R^3}{3\sqrt{3}} \text{ cubic units.}$$

Ans.

Chapter 8. Integral Calculus

1. *Prove that :* $\int \frac{x - \sin x}{1 - \cos x} dx = -x\cot\left(\frac{x}{2}\right) + c$

where c is the constant of integration.

Sol. We have,

$$\text{L.H.S.} = \int \frac{x - \sin x}{1 - \cos x} dx$$

$$\text{L.H.S.} = \int \frac{x}{1 - \cos x} dx - \int \frac{\sin x}{1 - \cos x} dx$$

$$= \int \frac{x}{2\sin^2 x/2} dx - \int \frac{2\sin x/2 \cos x/2}{2\sin^2 x/2} dx$$

$$= \frac{1}{2}\int x \operatorname{cosec}^2 x/2 dx - \int \cot\frac{x}{2} dx$$

I II

$$= \frac{1}{2}\left\{x\left(-2\cot\frac{x}{2}\right) - \int 1\left(-2\cot\frac{x}{2}\right)dx\right\}$$

$$- \int \cot\frac{x}{2} dx + c$$

$$= -x\cot\frac{x}{2} + \int \cot\frac{x}{2} dx - \int \cot\frac{x}{2} dx + c$$

$$= -x\cot\left(\frac{x}{2}\right) + c$$

= R.H.S. **Hence Proved**

2. *Prove that :*

$$\int_1^2 \frac{\log x \, dx}{(1+x)^2} dx = \frac{5}{3}\log 2 - \log 3$$

Sol. We have,

$$\text{L.H.S.} = \int_1^2 \frac{\log x}{(1+x)^2} dx$$

Integrating by parts

$$= \int_1^2 \log x \frac{1}{(1+x)^2} dx$$

I II

On integrating by parts,

$$= \left[\log x\left(\frac{-1}{1+x}\right) - \int \frac{1}{x(1+x)} dx\right]_1^2$$

$$= \left[-\frac{\log x}{1+x} - \int \frac{(-1)}{x(1+x)} dx\right]_1^2$$

$$= \left[-\frac{\log x}{1+x} + \int \frac{1+x}{x(1+x)} dx - \int \frac{x}{x(1+x)} dx\right]_1^2$$

$$= \left[-\frac{\log x}{1+x} + \int \frac{1}{x} dx - \int \frac{1}{1+x} dx\right]_1^2$$

$$= \left[-\frac{\log x}{1+x} + \log x - \log(1+x)\right]_1^2$$

$$= \left(-\frac{\log 2}{3} + \log 2 - \log 3\right) - \left(-\frac{\log 1}{2} + \log 1 - \log 2\right)$$

$$= \frac{2}{3}\log 2 - \log 3 + \log 2 \qquad [\because \log 1 = 0]$$

$$= \frac{5}{3}\log 2 - \log 3$$

= R.H.S. **Hence Proved.**

3. *If f and g are continuous on [0, a] and satisfy f (x) = f (a – x) and g (x) + g (a – x) = 2. Show that*

$$\int_0^a f(x)\, g(x)\, dx = \int_0^a f(x)\, dx$$

Sol. Given $\int_0^a f(x)\, g(x)\, dx$

$$\Rightarrow \int_0^a f(x)\, g(x)\, dx = \int_0^a f(a-x)\, g(a-x)\, dx$$

$$\Rightarrow \int_0^a f(x)\, g(x)\, dx = \int_0^a f(a-x)[2 - g(x)]\, dx$$

$$\Rightarrow \int_0^a f(x)\, g(x)\, dx = 2\int_0^a f(x)\, dx - \int_0^a f(x)\, g(x)\, dx$$

$$\Rightarrow 2\int_0^a f(x)\, g(x)\, dx = 2\int_0^a f(x)\, dx$$

$$\Rightarrow \int_0^a f(x)\, g(x)\, dx = \int_0^a f(x)\, dx.$$ **Hence Proved.**

4. *Prove that* $\int_0^{2\pi} \frac{x\cos x}{1+\cos x} dx = 2\pi^2.$

Sol. Let

$$I = \int_0^{2\pi} \frac{x\cos x}{1+\cos x} dx \qquad \text{...(i)}$$

Using property, $\int_0^{2\pi} f(x)\, dx = \int_0^a f(a-x)\, dx$

$$I = \int_0^{2\pi} \frac{(2\pi - x)\cos(2\pi - x)}{1+\cos(2\pi - x)} dx$$

$$= \int_0^{2\pi} \frac{(2\pi - x)\cos x}{1+\cos x} dx \qquad \text{...(ii)}$$

Adding equations (i) and (ii), we get

$$I + I = \int_0^{2\pi} \frac{x\cos x + 2\pi\cos x - x\cos x}{1+\cos x} dx$$

$$\Rightarrow 2I = \int_0^{2\pi} \frac{2\pi\cos x}{(1+\cos x)} dx$$

$$= 2\pi \int_0^{2\pi} \frac{\cos x\,(1-\cos x)}{(1+\cos x)(1-\cos x)} dx$$

$$= 2\pi \int_0^{2\pi} \frac{\cos x - \cos^2 x}{\sin^2 x} dx$$

$$= 2\pi \int_0^{2\pi} \left[\frac{\cos x}{\sin^2 x} - \frac{\cos^2 x}{\sin^2 x}\right] dx$$

$$= 2\pi \int_0^{2\pi} (\cot x \operatorname{cosec} x - \cot^2 x)\, dx$$

$$= 2\pi \int_0^{2\pi} (\cot x \operatorname{cosec} x - \operatorname{cosec}^2 x + 1)\, dx$$

$$= 2\pi\,[-\operatorname{cosec} x + \cot x + x]_0^{2\pi}$$

$$= 2\pi\,[-\operatorname{cosec} 2\pi + \cot 2\pi + 2\pi - (-\operatorname{cosec} 0 + \cot 0 + 0)]$$

$$= 2\pi\,[2\pi]$$

$$\Rightarrow 2I = 4\pi^2$$

$$\Rightarrow I = 2\pi^2$$

$$\therefore \int_0^{2\pi} \frac{x\cos x}{1+\cos x} dx = 2\pi^2$$ **Hence Proved.**

5. *Prove that*

$$\int_0^{\pi/2} \frac{3\sin\theta + 4\cos\theta}{\sin\theta + \cos\theta} d\theta = \frac{7\pi}{4}.$$

Sol. Let

$$I = \int_0^{\pi/2} \frac{3\sin\theta + 4\cos\theta}{\sin\theta + \cos\theta} d\theta \qquad \text{... (i)}$$

Using property,

$$\int_0^a f(x)\, dx = \int_0^a f(a-x)\, dx$$

$$= \int_0^{\pi/2} \frac{3\sin(\pi/2 - \theta) + 4\cos(\pi/2 - \theta)}{\sin(\pi/2 - \theta) + \cos(\pi/2 - \theta)} d\theta$$

$$I = \int_0^{\pi/2} \frac{3\cos\theta + 4\sin\theta}{\cos\theta + \sin\theta} d\theta \quad \ldots \text{(ii)}$$

On adding (i) and (ii), we get

$$2I = \int_0^{\pi/2} \frac{7(\sin\theta + \cos\theta)}{\sin\theta + \cos\theta} d\theta$$

$$\Rightarrow \quad 2I = 7\int_0^{\pi/2} d\theta = 7[\theta]_0^{\pi/2}$$

$$\Rightarrow \quad 2I = \frac{7\pi}{2}$$

$$\Rightarrow \quad I = \frac{7\pi}{4}$$

$$\therefore \int_0^{\pi/3} \frac{3\sin\theta + 4\cos\theta}{\sin\theta + \cos\theta} d\theta = \frac{7\pi}{4}$$ **Hence Proved.**

6. *Prove that :* $\int_0^{\pi/2} \frac{\sin x}{\sin x + \cos x} dx = \frac{\pi}{4}$

Sol. Let $I = \int_0^{\pi/2} \frac{\sin x}{\sin x + \cos x} dx \quad \ldots$(i)

Then,

$$I = \int_0^{\pi/2} \frac{\sin\left(\frac{\pi}{2} - x\right)}{\sin\left(\frac{\pi}{2} - x\right) + \cos\left(\frac{\pi}{2} - x\right)} dx$$

$$\left[\text{Using} : \int_0^a f(x)\,dx = \int_0^a f(a-x)\,dx\right]$$

$$\Rightarrow \quad I = \int_0^{\pi/2} \sin 2x \log \tan x \, dx \quad \ldots \text{(ii)}$$

Adding equations (i) and (ii), we get

$$2I = \int_0^{\pi/2} \frac{\sin x}{\sin x + \cos x} dx + \int_0^{\pi/2} \frac{\cos x}{\sin x + \cos x} dx$$

$$\Rightarrow \quad 2I = \int_0^{\pi/2} \frac{\sin x + \cos x}{\sin x + \cos x} dx$$

$$= \int_0^{\pi/2} 1.dx = [x]_0^{\pi/2}$$

$$\Rightarrow \quad 2I = \frac{\pi}{2} - 0 = \frac{\pi}{2}$$

$$\Rightarrow \quad I = \frac{\pi}{4}.$$ **Hence Proved.**

7. *Prove that :*

$$\int_0^{\pi/2} \sin 2x \log \tan x \, dx = 0$$

Sol. Let

$$I = \int_0^{\pi/2} \sin 2x \log \cot x \, dx \quad \ldots \text{(i)}$$

$$\Rightarrow \quad I = \int_0^{\pi/2} \sin 2\left(\frac{\pi}{2} - x\right) \log \tan\left(\frac{\pi}{2} - x\right) dx$$

$$\left[\because \int_0^a f(x)\,dx = \int_0^a f(a-x)\,dx\right]$$

$$\Rightarrow \quad I = \int_0^{\pi/2} \sin 2x \log \cot x \, dx \quad \ldots \text{(ii)}$$

Adding equations (i) and (ii), we get

$$2I = \int_0^{\pi/2} \sin 2x \,\{\log \tan x + \log \cot x\}\, dx$$

$$= \int_0^{\pi/2} \sin 2x \log(\tan x \cot x)\, dx$$

$$\Rightarrow \quad 2I = \int_0^{\pi/2} (\sin 2x).(\log 1)\, dx = 0$$

$$[\because \log 1 = 0]$$

$$\Rightarrow \quad I = 0$$ **Hence Proved.**

8. Prove that : $\int_0^1 \log\left(\frac{1}{x} - 1\right) dx = 0$

Sol. $I = \int_0^1 \log\left(\frac{1}{x} - 1\right) dx$

$$= \int_0^1 \log\left(\frac{1-x}{x}\right) dx \quad \ldots \text{(i)}$$

$$= \int_0^1 \log\left\{\frac{1-(1-x)}{1-x}\right\} dx$$

$$\left[\text{Using} : \int_0^a f(x)\,dx = \int_0^a f(a-x)\,dx\right]$$

$$\Rightarrow \quad I = \int_0^1 \log\left(\frac{x}{1-x}\right) dx \quad \ldots \text{(ii)}$$

Adding equation (i) and (ii), we get

$$2I = \int_0^1 \left\{\log\left(\frac{1-x}{x}\right) + \log\left(\frac{x}{1-x}\right)\right\} dx$$

$$= \int_0^1 \log 1.\,dx$$

$$\Rightarrow \quad I = 0.$$ **Hence Proved.**

9. *Show that :*

$$\int_0^{\pi/2} f(\sin 2x) \sin x \, dx = \sqrt{2} \int_0^{\pi/4} f(\cos 2x) \cos x \, dx.$$

Sol. Let

$$I = \int_0^{\pi/2} f(\sin 2x) \sin x \, dx \quad \ldots \text{(i)}$$

Then, $I = \int_0^{\pi/2} f\left\{\sin 2\left(\frac{\pi}{2}-x\right)\right\} \sin\left(\frac{\pi}{2}-x\right) dx$

$\Rightarrow \quad I = \int_0^{\pi/2} f\{\sin(\pi-2x)\} \cos x \, dx$

$\Rightarrow \quad I = \int_0^{\pi/2} f(\sin 2x) \cos x \, dx \qquad \ldots(ii)$

Adding equations (i) and (ii), we get

$$2I = \int_0^{\pi/2} f(\sin 2x)(\sin x + \cos x) \, dx$$

$\Rightarrow \quad 2I = 2\int_0^{\pi/4} f(\sin 2x)(\sin x + \cos x) \, dx$

$\Rightarrow \quad 2I = 2\sqrt{2} \int_0^{\pi/4} f(\sin 2x) \left(\frac{1}{\sqrt{2}}\sin x + \frac{1}{\sqrt{2}}\cos x\right) dx$

$\Rightarrow \quad 2I = 2\sqrt{2} \int_0^{\pi/4} f(\sin 2x) \sin\left(x + \frac{\pi}{4}\right) dx$

$\Rightarrow \quad 2I = 2\sqrt{2} \int_0^{\pi/4} f\left\{\sin 2\left(\frac{\pi}{4}-x\right)\right\} \left\{\sin\left(\frac{\pi}{4}-x+\frac{\pi}{4}\right) dx\right\}$

$\Rightarrow \quad 2I = 2\sqrt{2} \int_0^{\pi/4} f\left\{\sin\left(\frac{\pi}{2}-2x\right)\right\} \left\{\sin\left(\frac{\pi}{2}-x\right) dx\right\}$

$\Rightarrow \quad 2I = 2\sqrt{2} \int_0^{\pi/4} f(\cos 2x) \cos x \, dx$

$\Rightarrow \quad I = \sqrt{2} \int_0^{\pi/4} f(\cos 2x) \cos x \, dx$

Hence,

$\int_0^{\pi/2} f(\sin 2x) \sin x \, dx$

$= \sqrt{2} \int_0^{\pi/4} f(\cos 2x) \cos x \, dx.$ **Hence Proved.**

10. *Prove that :*

$$\int_0^{2\pi} \frac{x \sin^{2n} x}{\sin^{2n} x + \cos^{2n} x} dx = \pi^2$$

Sol. Let

$$I = \int_0^{2\pi} \frac{x \sin^{2n} x}{\sin^{2n} x + \cos^{2n} x} dx \qquad \ldots(i)$$

Then,

$$I = \int_0^{2\pi} \frac{(2\pi - x)\sin^{2n}(2\pi - x)}{\sin^{2n}(2\pi - x) + \cos^{2n}(2\pi - x)} dx$$

$\Rightarrow \quad I = \int_0^{2\pi} \frac{(2\pi - x)\sin^{2n} x}{\sin^{2n} x + \cos^{2n} x} dx \qquad \ldots(ii)$

Adding equation (i) and (ii), we get

$$2I = \int_0^{2\pi} \frac{x \sin^{2n} x}{\sin^{2n} x + \cos^{2n} x} + \frac{(2\pi - x)\sin^{2n} x}{\sin^{2n} x + \cos^{2n} x} dx$$

$\Rightarrow \quad 2I = \int_0^{2\pi} \frac{2\pi \sin^{2n} x}{\sin^{2n} x + \cos^{2n} x} dx$

$\Rightarrow \quad I = \pi \int_0^{2\pi} \frac{\sin^{2n} x}{\sin^{2n} x + \cos^{2n} x} dx$

$\Rightarrow \quad I = 2\pi \int_0^{\pi} \frac{\sin^{2n} x}{\sin^{2n} x + \cos^{2n} x} dx$

$\Rightarrow \quad I = 4\pi \int_0^{\pi/2} \frac{\sin^{2n} x}{\sin^{2n} x + \cos^{2n} x} dx \qquad \ldots(iii)$

$\Rightarrow \quad I = 4\pi \int_0^{\pi/2} \frac{\sin^{2n}(\pi/2 - x)}{\sin^{2n}(\pi/2) - x + \cos^{2n}(\pi/2 - x)} dx$

$\Rightarrow \quad I = 4\pi \int_0^{\pi/2} \frac{\cos^{2n} x}{\cos^{2n} x + \sin^{2n} x} dx \qquad \ldots(iv)$

Adding equations (iii) and (iv), we get

$$2I = 4\pi \int_0^{\pi/2} \frac{\sin^{2n} x}{\sin^{2n} x + \cos^{2n} x} + \frac{\cos^{2n} x}{\sin^{2n} + \cos^{2n} x} dx$$

$$2I = 4\pi \int_0^{\pi/2} 1 . dx = 4\pi \times \frac{\pi}{2}$$

$\Rightarrow \quad I = \pi^2.$ **Hence Proved**

Chapter 9. Differential Equations

1. *Show that $y = Ax + \frac{B}{x}$, $x \neq 0$ is a solution of the differential equation*

$$x^2 \frac{d^2y}{dx^2} + x\frac{dy}{dx} - y = 0$$

Sol. We have,

$$y = Ax + \frac{B}{x}, x \neq 0 \qquad \ldots(i)$$

Differentiating both sides with respect to x, we get

Prove the Following Type Questions

$$\frac{dy}{dx} = A - \frac{B}{x^2}$$

Differentiating with respect to x, we get

$$\frac{d^2y}{dx^2} = \frac{2B}{x^3}$$

Substituting the values of y, $\frac{dy}{dx}$ and $\frac{d^2y}{dx^2}$ in

$x^2\frac{d^2y}{dx^2} + x\frac{dy}{dx} - y$, we get

$$\Rightarrow x^2\frac{d^2y}{dx^2} + x\frac{dy}{dx} - y$$

$$= x^2\left(\frac{2B}{x^3}\right) + x\left(A - \frac{B}{x^2}\right) - \left(Ax + \frac{B}{x}\right)$$

$$= \frac{2B}{x} + Ax - \frac{B}{x} - Ax - \frac{B}{x} = 0$$

Thus, the function $y = Ax + \frac{B}{x}$ satisfies the differential equation $x^2\frac{d^2y}{dx^2} + x\frac{dy}{dx} - y = 0$.

Hence, $y = Ax + \frac{B}{x}$ is a solution of the given differential equation. **Hence Proved.**

2. *Show that the function $y = (A + Bx)e^{3x}$ is a solution of the equation $\frac{d^2y}{dx^2} - 6\frac{dy}{dx} + 9y = 0$.*

Sol. We have,

$$y = (A + Bx)e^{3x} \quad \text{...(i)}$$

Differentiating (i) with respect to x, we get

$$\frac{dy}{dx} = Be^{3x} + 3e^{3x}(A + Bx) \quad \text{...(ii)}$$

Differentiating (ii) with respect to x, we get

$$\frac{d^2y}{dx^2} = 3Be^{3x} + 9e^{3x}(A + Bx) + 3Be^{3x}$$

$$\frac{d^2y}{dx^2} = 6Be^{3x} + 9e^{3x}(A + Bx) \quad \text{...(iii)}$$

Substituting the values,

$$\because \quad \frac{d^2y}{dx^2} - 6\frac{dy}{dx} + 9y = \{6Be^{3x} + 9e^{3x}(A + Bx)\}$$

$$- 6\{Be^{3x} + 3e^{3x}(A + Bx)\} + \{9(A + Bx)e^{3x}\} = 0$$

Thus, $y = (A + Bx)e^{3x}$ satisfies the given differential equation.

Hence, it is a solution of the given differential equation. **Hence Proved.**

3. *Show that the differential equation representing one parameter family of curves is $(x^2 - y^2) = c(x^2 + y^2)^2$ is $(x^3 - 3xy^2)\,dx = (y^3 - 3x^2y)\,dx$*

Sol. The given equation of one parameter family of curves is

$$x^2 - y^2 = c(x^2 + y^2)^2 \quad \text{...(i)}$$

Differentiating (i) with respect to x, we get

$$2x - 2y\frac{dy}{dx} = 2c(x^2 + y^2)\left(2x + 2y\frac{dy}{dx}\right)$$

$$\Rightarrow \left(x - y\frac{dy}{dx}\right) = 2c(x^2 + y^2)\left(x + y\frac{dy}{dx}\right) \quad \text{...(ii)}$$

On substituting the value of c obtained from (i) in (ii), we get

$$\left(x - y\frac{dy}{dx}\right) = \frac{2(x^2 - y^2)(x^2 + y^2)}{(x^2 + y^2)^2}\left(x + y\frac{dy}{dx}\right)$$

$$\Rightarrow (x^2 + y^2)\left(x - y\frac{dy}{dx}\right) = 2(x^2 - y^2)\left(x + y\frac{dy}{dx}\right)$$

$$\Rightarrow \{x(x^2 + y^2) - 2x(x^2 - y^2)\}$$

$$= \frac{dy}{dx}\{(2y(x^2 - y^2) + y(x^2 + y^2)\}$$

$$\Rightarrow \quad (3xy^2 - x^3) = \frac{dy}{dx}(3x^2y - y^3)$$

$\Rightarrow (x^3 - 3xy^2)\,dx = (y^3 - 3x^2y)\,dy$, which is the given differential equation. **Hence Proved.**

4. *Show that $y = a\cos(\log x) + b\sin(\log x)$ is a solution of the differential equation*

$$x^2\frac{d^2y}{dx^2} + x\frac{dy}{dx} + y = 0$$

Sol. We have,

$$y = a\cos(\log x) + b\sin(\log x)$$

Differentiating with respect to x, we get

$$\frac{dy}{dx} = -\frac{a\sin(\log x)}{x} + \frac{b\cos(\log x)}{x}$$

$$\Rightarrow \quad x\frac{dy}{dx} = -a\sin(\log x) + b\cos(\log x)$$

Again differentiating both sides with respect to x, we obtain

$$x\frac{d^2y}{dx^2} + \frac{dy}{dx} = -\frac{a\cos(\log x)}{x} - \frac{b\sin(\log x)}{x}$$

$$\Rightarrow x^2\frac{d^2y}{dx^2} + x\frac{dy}{dx} = -[a\cos(\log x) + b\sin(\log x)]$$

$$\Rightarrow x^2\frac{d^2y}{dx^2} + x\frac{dy}{dx} = -y$$

$$\Rightarrow x^2\frac{d^2y}{dx^2}+x\frac{dy}{dx}+y=0,$$

which is same as the given differential equation. Hence, $y=a\cos(\log x)+b\sin(\log x)$ is a solution of the given differential equation.

Hence Proved.

5. *Show that $y=cx+\frac{a}{c}$ is a solution of the differential equation $y=x\frac{dy}{dx}+\frac{a}{\frac{dy}{dx}}$.*

Sol. We have,

$$y = xc+\frac{a}{c} \quad ...(i)$$

Differentiating with respect to x, we get

$$\frac{dy}{dx} = c \quad ...(ii)$$

$$\therefore \quad x\frac{dy}{dx}+\frac{a}{\frac{dy}{dx}} = xc+\frac{a}{c} \quad \left[\text{Putting } \frac{dy}{dx}=c\right]$$

$$\Rightarrow \quad \frac{xdy}{dx}+\frac{a}{\frac{dy}{dx}} = y$$

Hence, $y = cx+\frac{a}{c}$ is a solution of the given differential equation. **Hence Proved.**

Chapter 11. Vectors

1. *If $\vec{a}=\hat{i}-2\hat{j}+3\hat{k}, \vec{b}=2\hat{i}+3\hat{j}-5\hat{k}$, prove that $\vec{a}$ and $\vec{a}\times\vec{b}$ are perpendicular.**

Sol. Given vectors are $\vec{a} = \hat{i}-2\hat{j}+3\hat{k}$

and $\vec{b} = 2\hat{i}+3\hat{j}-5\hat{k}$

$$\vec{a}\times\vec{b} = \begin{vmatrix}\hat{i} & \hat{j} & \hat{k}\\ 1 & -2 & 3\\ 2 & 3 & -5\end{vmatrix}$$

$$= \hat{i}(10-9)-\hat{j}(-5-6)+\hat{k}(3+4)$$

$$= \hat{i}+11\hat{j}+7\hat{k}$$

Now,

$$\vec{a}.\left(\vec{a}\times\vec{b}\right) = (\hat{i}-2\hat{j}+3\hat{k})\cdot(\hat{i}+11\hat{j}+7\hat{k})$$

$$= 1-22+21=0$$

$$\therefore \quad \vec{a}\cdot\left(\vec{a}\times\vec{b}\right) = 0$$

$\therefore\ \vec{a}$ and $\left(\vec{a}\times\vec{b}\right)$ are perpendicular.

Hence Proved

2. *If $\vec{a}+\vec{b}+\vec{c}=0$ then prove that*

$$\vec{a}\times\vec{b} = \vec{b}\times\vec{c} = \vec{c}\times\vec{a}$$

Sol. Given, $\vec{a}+\vec{b}+\vec{c}=0$

$$\Rightarrow \quad \vec{a} = -(\vec{b}+\vec{c}) \quad ...(i)$$

$$\therefore \quad \vec{a}\times\vec{b} = -(\vec{b}+\vec{c})\times\vec{b}$$

$$= -\vec{b}\times\vec{b}-\vec{c}\times\vec{b}$$

$$= \vec{0}-(-\vec{b}\times\vec{c})$$

$$= \vec{b}\times\vec{c} \quad ...(ii)$$

and

$$\vec{c}\times\vec{a} = \vec{c}\times(-\vec{b}-\vec{c})$$

$$= -\vec{c}\times\vec{b}-\vec{c}\times\vec{c}$$

$$= \vec{b}\times\vec{c}-0$$

$$= \vec{b}\times\vec{c} \quad ...(iii)$$

From equations (ii) and (iii), we get

$$\vec{a}\times\vec{b} = \vec{b}\times\vec{c} = \vec{c}\times\vec{a}$$ **Hence Proved.**

3. **(a)** *In a triangle ABC, using vectors, prove that $c^2=a^2+b^2-2ab\cos c$.*

(b) *Prove that $\vec{a}\cdot[(\vec{b}+\vec{c})\times(\vec{a}+2\vec{b}+3\vec{c})]=[\vec{a}\ \vec{b}\ \vec{c}]$*

Sol. (a) $\vec{a}+\vec{b}+\vec{c}=0$

$$\Rightarrow \quad \vec{a}+\vec{b} = -\vec{c}$$

Squaring on both sides, we get

$$(\vec{a}+\vec{b})(\vec{a}+\vec{b})=(-\vec{c})(-\vec{c})$$

$$\Rightarrow \quad b^2+a^2+2ab\cos(\pi-c)=c^2$$

$$\Rightarrow \quad b^2+a^2+2ab(-\cos c)=c^2$$

$$\Rightarrow \quad c^2=a^2+b^2-2ab\cos c$$ **Hence Proved.**

* Frequently asked previous years Board Exam Questions.

(b) L.H.S. $= \vec{a}.(\vec{b}+\vec{c})\times(\vec{a}+2\vec{b}+3\vec{c})$

$= \vec{a}.(\vec{b}\times\vec{a}+\vec{b}\times 2\vec{b}+\vec{b}\times 3\vec{c}+\vec{c}\times\vec{a}$
$+\vec{c}\times 2\vec{b}+\vec{c}\times 3\vec{c})$

$= \vec{a}.(\vec{b}\times\vec{a}+3\vec{b}\times\vec{c}+\vec{c}\times\vec{a}+2\vec{c}\times\vec{b})$

$= \vec{a}.(\vec{b}\times\vec{a})+\vec{a}.(3\vec{b}\times\vec{c})+\vec{a}.(\vec{c}\times\vec{a})+\vec{a}.(2\vec{c}\times\vec{b})$

$= 0+3[\vec{a}.(\vec{b}\times\vec{c})]+0+2[(\vec{a}.(\vec{c}\times\vec{b})]$

$= 3[\vec{a}\,\vec{b}\,\vec{c}]+2[\vec{a}\,\vec{c}\,\vec{b}]$

$= 3[\vec{a}\,\vec{b}\,\vec{c}]-2[\vec{a}\,\vec{b}\,\vec{c}]$

$= [\vec{a}\,\vec{b}\,\vec{c}]$ R.H.S. **Hence Proved.**

4. *Prove by vector method that angle bisectors of a triangle are concurrent.*

Sol. Let the position vector of A, B and C be a, b and c respectively. Let $|BC| = \alpha$, $|CA| = \beta$ and $|AB| = \gamma$ and let the angle bisectors AD and BE meet at P.

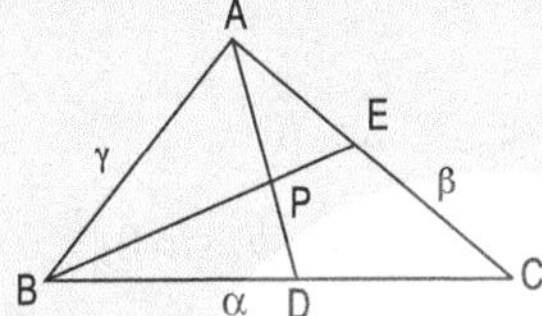

As AD is an angle bisector, we get,

$$\frac{AB}{AC}=\frac{BD}{DC}$$

$$\Rightarrow \frac{BD}{DC}=\frac{\gamma}{\beta}$$

$$\Rightarrow BD=\frac{\gamma\alpha}{\beta+\gamma}$$

Position vector of $D = \dfrac{\beta b+\gamma c}{\beta+\gamma}$

Since BP is an angle bisector,

$$\therefore \frac{BD}{BA}=\frac{DP}{PA}$$

$$\Rightarrow \left(\frac{\alpha\gamma}{\beta+\gamma}\right)\frac{1}{\gamma}=\frac{DP}{AP}$$

$DP : PA = \alpha : (\gamma+\beta)$

Position vector of $P = \dfrac{\alpha a+(\gamma+\beta)\left(\dfrac{\beta b+\gamma c}{\beta+\gamma}\right)}{\alpha+\beta+\gamma}$

$$= \frac{\alpha a+\beta b+\gamma c}{\alpha+\beta+\gamma}$$

Therefore, from consideration of symmetry, it can be said that the angle bisectors are concurrent. **Hence Proved.**

5. *Show that the line segment joining the mid-points of two sides of a triangle is parallel to third side and is half as long.*

Sol. Let $\vec{a}$ and $\vec{b}$ be the position vector of A and B respectively with respect to O, *i.e.* $OA = \vec{a}$ and $OB = \vec{b}$. Let D and E be the mid-points of OB and AB respectively. Then, the position vector of D is $\frac{1}{2}(\vec{b}+0) = \frac{1}{2}\vec{b}$, and the position vector of E is $\frac{1}{2}(\vec{a}+\vec{b})$.

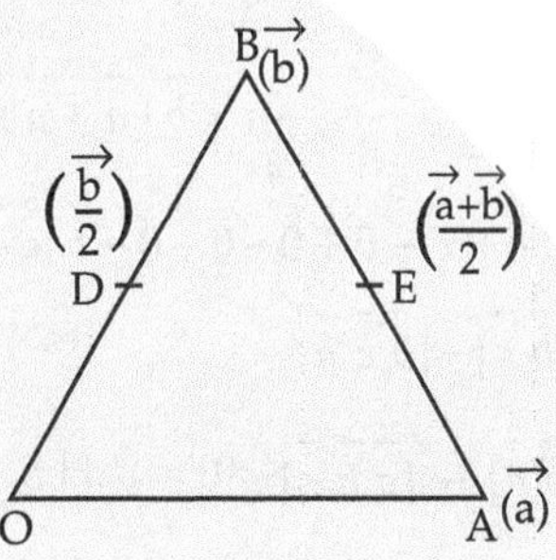

Thus, DE = position vector of E − position vector of D

$$= \frac{\vec{a}+\vec{b}}{2}-\frac{\vec{b}}{2}=\frac{\vec{a}}{2}=\frac{OA}{2}$$

Thus, DE is parallel to OA and $DE = \frac{1}{2}OA$.

Hence Proved.

6. *Prove that*

$$[\vec{a}+\vec{b}\;\;\vec{b}+\vec{c}\;\;\vec{c}+\vec{a}]=2[\vec{a}\,\vec{b}\,\vec{c}].$$

Sol. L. H. S. $= [\vec{a}+\vec{b}\;\;\vec{b}+\vec{c}\;\;\vec{c}+\vec{a}]$

$= (\vec{a}+\vec{b}).[(\vec{b}\times\vec{c})\times(\vec{c}+\vec{a})]$

$= (\vec{a}+\vec{b}).[\vec{b}\times\vec{c}+\vec{c}\times\vec{c}+\vec{b}\times\vec{a}+\vec{c}\times\vec{a}]$

$= [\vec{a}+\vec{b}].[\vec{b}\times\vec{c}+\vec{b}\times\vec{a}+\vec{c}\times\vec{a}]$ (as $\vec{c}\times\vec{c}=0$)

$= \vec{a}.(\vec{b}\times\vec{c})+\vec{b}.(\vec{b}\times\vec{c})+\vec{a}.(\vec{b}\times\vec{a})$
$+\vec{b}.(\vec{b}\times\vec{a})+\vec{a}.(\vec{c}\times\vec{a})+\vec{b}.(\vec{c}\times\vec{a})$

$= \vec{a}.(\vec{b}\times\vec{c})+\vec{b}.(\vec{c}\times\vec{a})$

[As rest of the terms are zero]

$= 2[\vec{a}\,\vec{b}\,\vec{c}]$ = R.H.S. **Hence Proved.**

7. *For any three vectors $\vec{a}, \vec{b}, \vec{c}$ prove*

$$\left[\vec{a}-\vec{b}\ \ \vec{b}-\vec{c}\ \ \vec{c}-\vec{a}\right]=0.$$

Sol. L.H.S. $= [\vec{a}-\vec{b}\ \ \vec{b}-\vec{c}\ \ \vec{c}-\vec{a}]\cdot$

$= [(\vec{a}-\vec{b})\times(\vec{b}-\vec{c})]\cdot(\vec{c}-\vec{a})$

$= (\vec{a}\times\vec{b}-\vec{a}\times\vec{c}-\vec{b}\times\vec{b}+\vec{b}\times\vec{c})\cdot(\vec{c}-\vec{a})$

$= (\vec{a}\times\vec{b}-\vec{a}\times\vec{c}+\vec{b}\times\vec{c})\cdot(\vec{c}-\vec{a})\quad (\because \vec{b}\times\vec{b}=0)$

$= \vec{a}\times\vec{b}\cdot\vec{c}-\vec{a}\times\vec{c}\cdot\vec{c}+\vec{b}\times\vec{c}\cdot\vec{c}$

$-\vec{a}\times\vec{b}\cdot\vec{a}+\vec{a}\times\vec{c}\cdot\vec{a}-\vec{b}\times\vec{c}\cdot\vec{a}$

$= \vec{a}\times\vec{b}\times\vec{c}-0+0-0+0-\vec{b}\times\vec{c}\times\vec{a}$

$= [\vec{a}\ \vec{b}\ \vec{c}]-[\vec{b}\ \vec{c}\ \vec{a}]$

$= [\vec{a}\ \vec{b}\ \vec{c}]-[\vec{a}\ \vec{b}\ \vec{c}] = 0 =$ R.H.S. **Hence Proved.**

8. *If D, E, F are mid-points of the sides of a triangle ABC, prove by vector method that :*

Area of Δ *DEF* $= \frac{1}{4}$ *(Area of* Δ *ABC).*

Sol. We consider A as origin then the position vectors of B and C may be taken as $\vec{b}$ and $\vec{c}$, respectively.

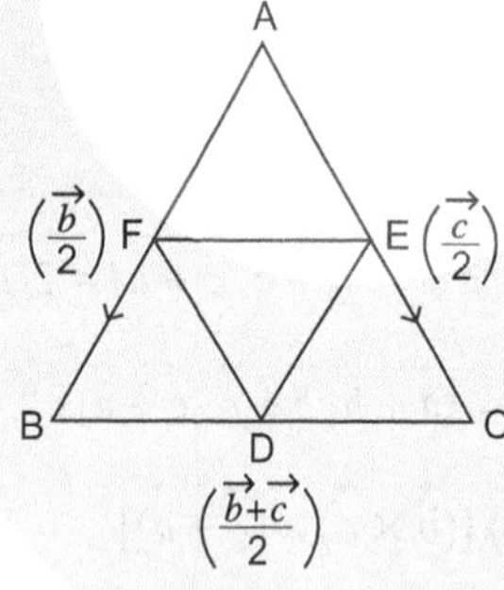

$\therefore$ Position vector of mid-point D $= \dfrac{\vec{b}+\vec{c}}{2}$

Position vector of E $= \dfrac{\vec{c}}{2}$

Position vector of F $= \dfrac{\vec{b}}{2}$

Hence, $\vec{DE}$ = Position Vector of E − Position Vector of D

$$= \frac{\vec{c}}{2}-\frac{\vec{b}+\vec{c}}{2} = -\frac{\vec{b}}{2}$$

And $\vec{DF}$ = Position vector of F − Position vector of D

$$= \frac{\vec{b}}{2}-\frac{\vec{b}+\vec{c}}{2} = -\frac{\vec{c}}{2}$$

Now, vector area of Δ DEF $= \frac{1}{2}(\vec{DE}\times\vec{DF})$

$$= \frac{1}{2}\left(\frac{-\vec{b}}{2}\right)\times\left(\frac{-\vec{c}}{2}\right) = \frac{1}{8}\vec{b}\times\vec{c}$$

$$= \frac{1}{4}\left(\frac{1}{2}\vec{b}\times\vec{c}\right)$$

$$= \frac{1}{4}\ [\text{Area of } \Delta \text{ ABC}]$$

$\therefore$ Area of Δ DEF $= \frac{1}{4}$ (Area of Δ ABC).

Hence Proved

9. *Prove that*

$$|\vec{a}\times\vec{b}|^2 = |\vec{a}|^2|\vec{b}|^2-(\vec{a}\,.\,\vec{b})^2.$$

Sol.

L. H. S. $= |\vec{a}\times\vec{b}|^2$

$= \big|\,|a|\,|b|\sin\theta\,\big|^2$

$= |a|^{-2}\,|b|^{-2}\sin^2\theta$

$= |a|^{-2}\,|b|^{-2}(1-\cos^2\theta)$

$= |a|^{-2}\,|b|^{-2}-|a|^{-2}\,|b|^{-2}\cos^2\theta.$

$= |a|^{-2}\,|b|^{-2}-(a\,.\,b)^2 =$ R. H. S.

Hence Proved

9. *Show that*

$$[\vec{a}\times\vec{b}\ \ \vec{b}\times\vec{c}\ \ \vec{c}\times\vec{a}] = [\vec{a}\ \vec{b}\ \vec{c}]^2.$$

Sol. L.H.S. $= [\vec{a}\times\vec{b}\ \ \vec{b}\times\vec{c}\ \ \vec{c}\times\vec{a}]$

$= (\vec{a}\times\vec{b}).[(\vec{b}\times\vec{c})\times(\vec{c}\times\vec{a})]$

Let $\vec{b}\times\vec{c}=\vec{d}$, then

$(\vec{b}\times\vec{c})\times(\vec{c}\times\vec{a}) = \vec{d}\times(\vec{c}\times\vec{a})$

$= (\vec{d}\,.\,\vec{a})\,\vec{c}-(\vec{d}\,.\,\vec{c})\,\vec{a}$

$= [(\vec{b}\times\vec{c}).\,\vec{a}]\,\vec{c}-[(\vec{b}\times\vec{c}).\,\vec{c}]\,\vec{a}$

$= [\vec{b}\ \vec{c}\ \vec{a}]\,\vec{c}-[\vec{b}\ \vec{c}\ \vec{c}]\,\vec{a}$

$= [\vec{b}\ \vec{c}\ \vec{a}]\,\vec{c}$ as $[\vec{b}\ \vec{c}\ \vec{c}] =$

Now,

$$[\vec{a}\times\vec{b}\ \ \vec{b}\times\vec{c}\ \ \vec{c}\times\vec{a}] = (\vec{a}\times\vec{b}).[\vec{b}\ \vec{c}\ \vec{a}]\vec{c}$$

$$= [\vec{b}\ \vec{c}\ \vec{a}][(\vec{a}\times\vec{b}).\vec{c}]$$

$$= [\vec{a}\ \vec{b}\ \vec{c}][\vec{a}\ \vec{b}\ \vec{c}]$$

$$= [a\ b\ c]^2 = \text{R.H.S.}$$

Hence Proved.

10. *Show that*

$$\hat{i}\times(\vec{a}\times\hat{i})+\hat{j}\times(\vec{a}\times\hat{j})+\hat{k}(\vec{a}\times\hat{k}) = 2a.$$

where $\vec{a} = a_1\hat{i}+a_2\hat{j}+a_3\hat{k}.$

Sol. L.H.S. $= \hat{i}\times(\vec{a}\times\hat{i})+\hat{j}\times(\vec{a}\times\hat{j})+\hat{k}(\vec{a}\times\hat{k})$

$$= [(\hat{i}.\hat{i})\vec{a}-(\hat{i}.\vec{a})\hat{i}]+[(\hat{j}.\hat{j})\vec{a}$$

$$-(\hat{j}.\vec{a})\hat{j}]+[(\hat{k}.\hat{k}).\vec{a}-(\hat{k}.\vec{a}).\hat{k}]$$

$$= \vec{a}-(\hat{i}.\vec{a})\hat{i}+\vec{a}-(\hat{j}.\vec{a})\hat{j}+\vec{a}-(\hat{k}.\vec{a})\hat{k}$$

$$= 3\vec{a}-[(\hat{i}.\vec{a})\hat{i}+(\hat{j}.\vec{a})\hat{j}+(\hat{k}.\vec{a}).\hat{k}]$$

$\because\ \hat{i}(\hat{i}.\vec{a}) = \hat{i}[\hat{i}(a_1\hat{i}+a_2\hat{j}+a_3\hat{k})] = a_1\hat{i}$

$\because\ \hat{j}(\hat{j}.\vec{a}) = \hat{j}[\hat{j}(a_1\hat{i}+a_2\hat{j}+a_3\hat{k})] = a_2\hat{j}$

$\because\ \hat{k}(\hat{k}.\vec{a}) = \hat{k}[\hat{k}(a_1\hat{i}+a_2\hat{j}+a_3\hat{k})] = a_3\hat{k}$

$$= 3\vec{a}-(a_1\hat{i}+a_2\hat{j}+a_3\hat{k})$$

$$= 3\vec{a}-\vec{a} = 2\vec{a}$$

= R.H.S. **Hence Proved.**

11. *Prove that**

$$\vec{a}.[(\vec{b}+\vec{c})\times(\vec{a}+3\vec{b}+4\vec{c})] = [\vec{a}\ \vec{b}\ \vec{c}]$$

Sol. L.H.S. $= \vec{a}.[(\vec{b}+\vec{c})\times(\vec{a}+3\vec{b}+4\vec{c})]$

$$= \vec{a}.[\vec{b}\times\vec{a}+3(\vec{b}\times\vec{b})+4(\vec{b}\times\vec{c})+\vec{c}\times\vec{a}$$

$$+3(\vec{c}\times\vec{b})+4(\vec{c}\times\vec{c})]$$

$$= \vec{a}.[\vec{b}\times\vec{a}+0+4(\vec{b}\times\vec{c})+\vec{c}\times\vec{a}$$

$$+3(\vec{c}\times\vec{b})+0]$$

$$= \vec{a}.(\vec{b}\times\vec{a})+4[\vec{a}.(\vec{b}\times\vec{c})]+\vec{a}.(\vec{c}\times\vec{a})$$

$$+3[\vec{a}.(\vec{c}\times\vec{b})]$$

$$= 0+4[\vec{a}\ \vec{b}\ \vec{c}]+0+3[\vec{a}\ \vec{c}\ \vec{b}]$$

$$= 4[\vec{a}\ \vec{b}\ \vec{c}]-3[\vec{a}\ \vec{b}\ \vec{c}]$$

$= [\vec{a}\ \vec{b}\ \vec{c}] =$ R.H.S. **Hence Proved.**

12. *Prove by vector method, that the parallelograms on the same base and between the same parallels are equal in area.*

Sol. *OACB* and *OAED* are the two parallelograms on the same base and between the same parallels.

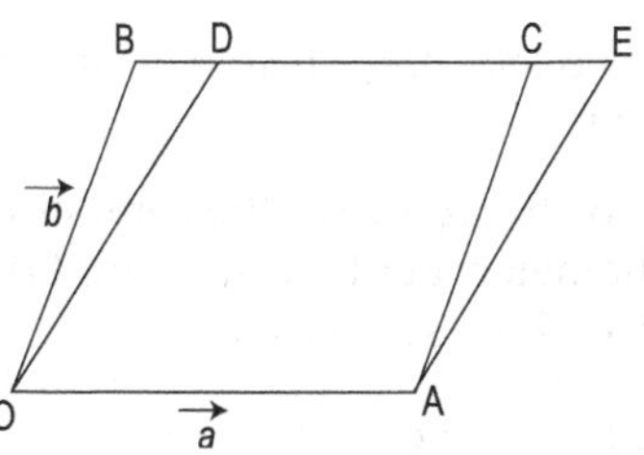

Taking *O* as origin, let $\vec{OA} = \vec{a}, \vec{OB} = \vec{b}$

Area of parallelogram *OACB* = $\vec{a}\times\vec{b}$

Since, *D* is any point on *BC*, we have

$$\vec{BD} = k.\vec{BC} = k.\vec{OA} = k.\vec{a}$$

where *k* is some scalar

$$\therefore\quad \vec{OD} = \vec{OB}+\vec{BD} = \vec{b}+k.\vec{a}$$

Area of parallelogram *OAED*

$$= \vec{OA}\times\vec{OD}$$

$$= \vec{a}\times(\vec{b}+k.\vec{a})$$

$$= \vec{a}\times\vec{b}+k\vec{a}\times\vec{a}$$

$$= \vec{a}\times\vec{b} \qquad (\text{as } \vec{a}\times\vec{a} = 0)$$

= Area of parallelogram *OACB*.

Hence Proved.

13. *Prove by a vector method that the diagonals of a rhombus are perpendicular.*

Sol. Let *ABCD* be a rhombus.

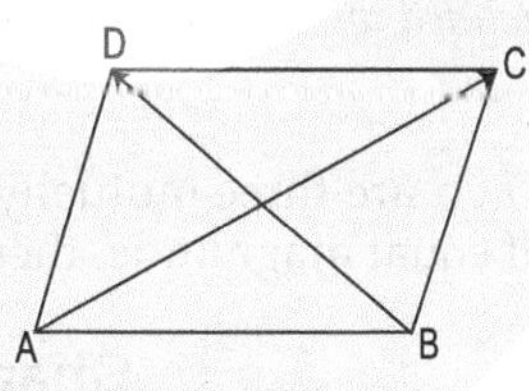

$$\therefore\quad \vec{AC} = \vec{AB}+\vec{BC};$$

And, $\vec{BD} = \vec{BC}+\vec{CD} = \vec{BC}-\vec{AB}$

$[\because \vec{CD} = -\vec{AB}]$

$$\vec{AC}.\vec{BD} = (\vec{BC}+\vec{AB}).(\vec{BC}-\vec{AB})$$

$$= \vec{BC}^2-\vec{AB}^2$$

$$= BC^2-AB^2$$

$$= 0 \quad (\because BC = AB)$$

$$\Rightarrow \vec{AC}.\vec{BD} = 0$$

* Frequently asked previous years Board Exam Questions.

Since, the dot product is zero, AC and BD are at right angles to each other. **Hence Proved.**

5. *Using vectors, prove that angle in a semi-circle is a right angle.*

Sol. Let O be the centre of the semi-circle and BA be the diameter. Let P be any point on the circumference of the semi-circle.

Let $\vec{OA} = \vec{a}$ then $\vec{OB} = -\vec{a}$

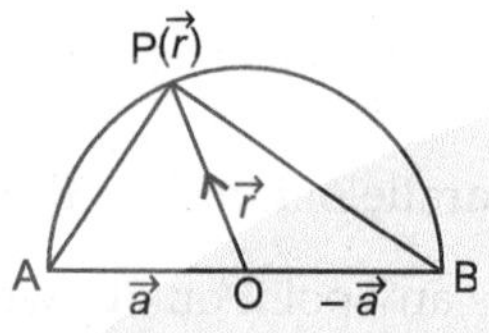

Let $\vec{OP} = \vec{r}$

$\therefore \quad \vec{AP} = \vec{OP} - \vec{OA} = \vec{r} - \vec{a}$

$\vec{BP} = \vec{OP} - \vec{OB} = \vec{r} - (-\vec{a}) = \vec{r} + \vec{a}$

$$\vec{AP} \cdot \vec{BP} = (\vec{r} - \vec{a}) \cdot (\vec{r} + \vec{a})$$

$$= \vec{r}^2 - \vec{a}^2$$

$$= \vec{a}^2 - \vec{a}^2 \qquad [\because r = a \text{ as, OP = OA}]$$

$$= 0$$

$\therefore \vec{AP}$ is perpendicular to $\vec{BP}$.

$\Rightarrow \quad \angle APB = 90°$. **Hence Proved.**

6. *If $\vec{a}, \vec{b}, \vec{c}$ are three mutually perpendicular vectors of equal magnitude, prove that $(\vec{a} + \vec{b} + \vec{c})$ is equally inclined with vectors $\vec{a}, \vec{b}$ and $\vec{c}$.*

Sol. Since, $\vec{a}, \vec{b}, \vec{c}$ are three mutually perpendicular vectors of equal magnitude, therefore

$$\vec{a}.\vec{b} = \vec{b}\cdot\vec{c} = \vec{c}\cdot\vec{a} = 0$$

and $|\vec{a}| = |\vec{b}| = |\vec{c}| = x$

Let $\quad \vec{d} = \vec{a} + \vec{b} + \vec{c}$

Then, $|\vec{d}^2| = |\vec{a} + \vec{b} + \vec{c}|^2$

$$= (\vec{a} + \vec{b} + \vec{c}) \cdot (\vec{a} + \vec{b} + \vec{c})$$

$$\Rightarrow \quad |\vec{d}^2| = \vec{a}\cdot\vec{a} + \vec{a}\cdot\vec{b} + \vec{a}\cdot\vec{c} + \vec{b}\cdot\vec{a} + \vec{b}\cdot\vec{b} + \vec{b}\cdot\vec{c} + \vec{c}\cdot\vec{a} + \vec{c}\cdot\vec{b} + \vec{c}\cdot\vec{c}$$

$$= 1 + 0 + 0 + 0 + 1 + 0 + 0 + 0 + 1 = 3$$

$$|\vec{d}| = \sqrt{3}$$

Let the angles between $\vec{d}$ and $\vec{a}, \vec{b}, \vec{c}$ be α, β, γ respectively. Then,

$$\cos\alpha = \frac{\vec{d}.\vec{a}}{|\vec{d}|\cdot|\vec{a}|}$$

$$= \frac{(\vec{a} + \vec{b} + \vec{c})\,\vec{a}}{\sqrt{3}.|\vec{a}|}$$

$$= \frac{\vec{a}\cdot\vec{a} + \vec{b}\cdot\vec{a} + \vec{c}\cdot\vec{a}}{\sqrt{3}\cdot|\vec{a}|}$$

$$= \frac{1+0+0}{\sqrt{3}\cdot|\vec{a}|} = \frac{1}{\sqrt{3}|\vec{a}|}$$

$$= \frac{1}{\sqrt{3}\,x}$$

Similarly, $\cos\beta = \frac{1}{\sqrt{3}\,x}$ and $\cos\gamma = \frac{1}{\sqrt{3}\,x}$

Thus, $\cos\alpha = \cos\beta = \cos\gamma$

i.e., $\quad \alpha = \beta = \gamma$ **Hence Proved.**

Chapter 13. Application of Integrals

1. *Show that the area included between the x-axis and the curve $a^2y = x^2(x + a)$ is $\frac{a^2}{12}$.*

Sol. Given curve is

$$y = \frac{x^2(x+a)}{a^2} \qquad \text{... (i)}$$

This curve cuts the x-axis at points where $y = 0$.

Putting, $y = 0$ in (i), we get

$$x = 0 \text{ or } x = -a$$

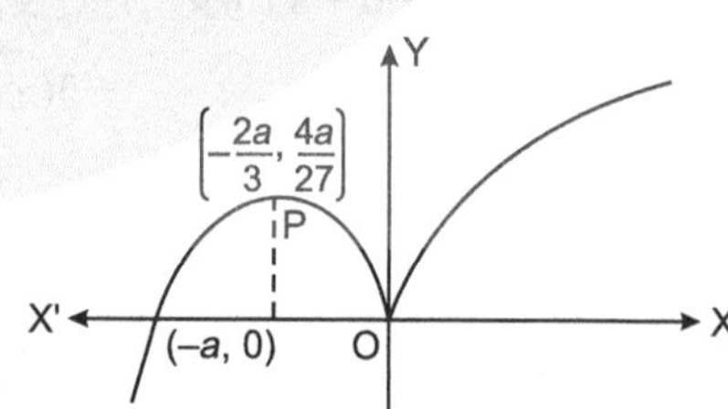

Hence, the required area included between the x-axis and the curve is given by

$$\int_{-a}^{0} \frac{x^2(x+a)}{a^2}\,dx = \int_{-a}^{0}\left(\frac{x^3}{a^2} + \frac{x^2}{a}\right)dx$$

$$= \left[\frac{x^4}{4a^2}\right]_{-a}^{0} + \left[\frac{x^3}{3a}\right]_{-a}^{0}$$

$$= \left[0 - \frac{a^4}{4a^2}\right] + \left[0 - \left(\frac{-a^3}{3a}\right)\right]$$

$$= -\frac{a^2}{4} + \frac{a^2}{3} = \frac{-3a^2 + 4a^2}{12}$$

$$= \frac{a^2}{12} \text{ sq. units}$$

Hence Proved.

2. *Prove that the curves $y^2 = 4x$ and $x^2 = 4y$ divide the area of the square bounded by $x = 0$, $x = 4$, $y = 4$ and $y = 0$ into three equal parts.*

Sol. The given curves

$$y^2 = 4x \quad \ldots(i)$$

and

$$x^2 = 4y \quad \ldots(ii)$$

are right hand parabola and upward parabola respectively.

To find point of intersection of (i) and (ii)

Put value of y from (ii) in equation (i)

$$\left(\frac{x^2}{4}\right)^2 = 4x$$

$$\Rightarrow \quad \frac{x^4}{16} = 4x$$

$$\Rightarrow \quad x^4 = 64x$$

$$\Rightarrow \quad x^4 - 64x = 0$$

$$\Rightarrow \quad x(x^3 - 64) = 0$$

$$\Rightarrow \quad x = 0, x^3 = 64 = 4^3$$

$$\Rightarrow \quad x = 0, x = 4$$

When $x = 0$, from (ii) $y = 0$

When $x = 4$, $y = \frac{x^2}{4} = \frac{16}{4} = 4$

$\therefore$ The two parabola meet in the points (0, 0) and (4, 4).

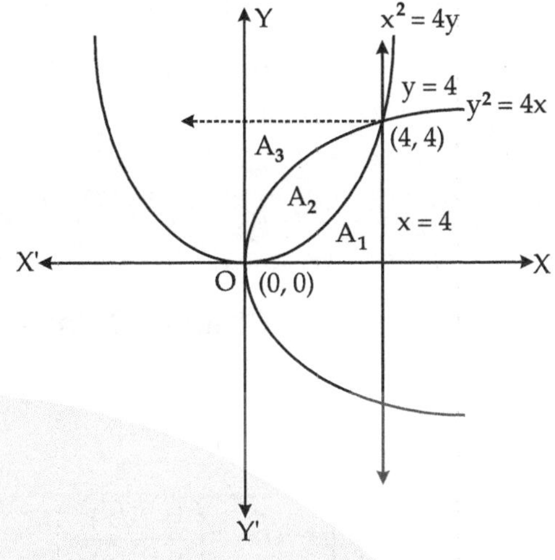

From the figure, we find that the square is divided into three parts

Let these parts be A_1, A_2 and A_3

We have to prove that

$$A_1 = A_2 = A_3$$

$$\text{Area of part III} = \int_0^4 y\,dx \text{ (parabola } x^2 = 4y)$$

$$= \int_0^4 \frac{x^2}{4}\,dx = \frac{1}{4}\left[\frac{x^3}{3}\right]_0^4$$

$$= \frac{1}{12}(4^3 - 0^3) = \frac{64}{12}$$

$$= \frac{16}{3} \text{ sq. units} \quad \ldots(i)$$

Area of part I = Area of square – Area of II and III

$$= 16 - \int_0^4 \sqrt{4x}\,dx$$

$$= 16 - \frac{2 \times 2}{3}\left[x^{3/2}\right]_0^4$$

$$= 16 - \frac{32}{3} = \frac{16}{3} \text{ sq. units} \quad \ldots(ii)$$

Area of part II = Area of square – Area of I – Area of III

$$= 16 - \frac{16}{3} - \frac{16}{3}$$

$$= \frac{16}{3} \text{ sq. units} \quad \ldots(iii)$$

From (i), (ii) and (iii) we get, area of 3 parts are equal.

Hence Proved.

□□

Computational Questions | Set 5 |

Chapter 3. Inverse Trigonometric Functions

1. *Simplify :* $\cos^{-1}\left(\frac{1-x}{1+x}\right), x \in [0, 1]$.

Sol. We have,

$$\cos^{-1}\left(\frac{1-x}{1+x}\right) = \cos^{-1}\left(\frac{1-(\sqrt{x})^2}{1+(\sqrt{x})^2}\right)$$

$$= 2\tan^{-1}(\sqrt{x})$$ **Ans.**

2. *Simplify :* $\tan^{-1}\frac{1-a}{1+a} - \tan^{-1}\frac{1-b}{1+b}$.

Sol. We have, $\tan^{-1}\frac{1-a}{1+a} - \tan^{-1}\frac{1-b}{1+b}$

$= [\tan^{-1}(1) - \tan^{-1}(a)] - [\tan^{-1}(1) - \tan^{-1}(b)]$

$= \tan^{-1} 1 - \tan^{-1} a - \tan^{-1} 1 + \tan^{-1} b$

$= \tan^{-1} b - \tan^{-1} a$

$= \tan^{-1}\left(\frac{b-a}{1+ab}\right)$ **Ans.**

3. *Simplify :* $\cos^{-1}\frac{3}{5} + \sin^{-1}\frac{5}{13}$.

Sol. We have, $\cos^{-1}\frac{3}{5} + \sin^{-1}\frac{5}{13}$.

$= \tan^{-1}\frac{4}{3} + \tan^{-1}\frac{5}{12}$

$$= \tan^{-1}\left(\frac{\frac{4}{3}+\frac{5}{12}}{1-\frac{4}{3}\times\frac{5}{12}}\right) = \tan^{-1}\left(\frac{\frac{16+5}{12}}{\frac{9-5}{9}}\right)$$

$$= \tan^{-1}\left(\frac{21}{12}\times\frac{9}{4}\right)$$

$$= \tan^{-1}\left(\frac{63}{16}\right)$$ **Ans.**

4. *Simplify :*

$$\tan^{-1}\frac{1}{3}+\tan^{-1}\frac{1}{5}+\tan^{-1}\frac{1}{7}+\tan^{-1}\frac{1}{8}$$

Sol. We have,

$$\tan^{-1}\frac{1}{3}+\tan^{-1}\frac{1}{5}+\tan^{-1}\frac{1}{7}+\tan^{-1}\frac{1}{8}$$

$$= \left(\tan^{-1}\frac{1}{3}+\tan^{-1}\frac{1}{8}\right)+\left(\tan^{-1}\frac{1}{5}+\tan^{-1}\frac{1}{7}\right)$$

$$= \tan^{-1}\left(\frac{\frac{1}{3}+\frac{1}{8}}{1-\frac{1}{3}\times\frac{1}{8}}\right)+\tan^{-1}\left(\frac{\frac{1}{5}+\frac{1}{7}}{1-\frac{1}{5}\times\frac{1}{7}}\right)$$

$$= \tan^{-1}\left(\frac{\frac{8+3}{24}}{\frac{24-1}{24}}\right)+\tan^{-1}\left(\frac{\frac{7+5}{35}}{\frac{35-1}{35}}\right)$$

$$= \tan^{-1}\left(\frac{11}{23}\right)+\tan^{-1}\left(\frac{12}{34}\right)$$

$$= \tan^{-1}\left(\frac{11}{23}\right)+\tan^{-1}\left(\frac{6}{17}\right)$$

$$= \tan^{-1}\left(\frac{\frac{11}{23}+\frac{6}{17}}{1-\frac{11}{23}\times\frac{6}{17}}\right)$$

$$= \tan^{-1}\left(\frac{\frac{187+138}{391}}{\frac{391-66}{391}}\right)$$

$$= \tan^{-1}\left(\frac{325}{325}\right)$$

$$= \tan^{-1} 1 = \frac{\pi}{4}$$ **Ans.**

5. *Simplify :* $\tan(2\cot^{-1} x)$.

Sol. Let, $\cot^{-1} x = a$

$x = \cot a$

Now, $\tan(2\cot^{-1} x) = \tan(2a)$ $[\because \cot^{-1} x = a]$

$$= \frac{2\tan a}{1-\tan^2 a}$$

$$= \frac{\frac{2}{x}}{1-\left(\frac{1}{x}\right)^2} \quad \left[\because \cot a = x,\ \tan a = \frac{1}{x}\right]$$

$$= \frac{2x}{x^2-1}$$ **Ans.**

Chapter 4. Determinants

1. *Find the roots of the equation*

$$\begin{vmatrix} 2-x & 3 & 3 \\ 3 & 4-x & 5 \\ 3 & 5 & 4-x \end{vmatrix} = 0.$$

Sol.

Given, $\begin{vmatrix} 2-x & 3 & 3 \\ 3 & 4-x & 5 \\ 3 & 5 & 4-x \end{vmatrix} = 0$

Applying $R_2 \to R_2 - R_3$, we obtain

$$\Rightarrow \begin{vmatrix} 2-x & 3 & 3 \\ 0 & -x-1 & 1+x \\ 3 & 5 & 4-x \end{vmatrix} = 0$$

Now applying $C_2 \to C_2 + C_3$, we obtain

$$\Rightarrow \begin{vmatrix} 2-x & 6 & 3 \\ 0 & 0 & 1+x \\ 3 & 9-x & 4-x \end{vmatrix} = 0$$

Expand along R_2, we get

$\Rightarrow \quad -(1+x)\{(2-x)(9-x) - 6 \times 3\} = 0$

$\Rightarrow \quad -(1+x)\{18 - 9x - 2x + x^2 - 18\} = 0$

$\Rightarrow \quad -(1+x)(x^2 - 11x) = 0$

$\Rightarrow \quad -x(1+x)(x-11) = 0$

$\Rightarrow \quad x = 0, -1, 11$ **Ans.**

2. *Suppose f(x) is a function satisfying.*

(a) $f(0) = 2, f(1) = 1.$

(b) *f has a minimum value at* $x = \frac{5}{2}$ *and*

(c) for all x,

$$f'(x) = \begin{vmatrix} 2ax & 2ax-1 & 2ax+b+1 \\ b & b+1 & -1 \\ 2(ax+b) & 2ax+2b+1 & 2ax+b \end{vmatrix}$$

Where a, b are some constants. Determine the constants a, b, and f(x).

Sol. Given

$$f'(x) = \begin{vmatrix} 2ax & 2ax-1 & 2ax+b+1 \\ b & b+1 & -1 \\ 2(ax+b) & 2ax+2b+1 & 2ax+b \end{vmatrix}$$

Applying $R_3 \to R_3 - R_1 - 2R_2$, we get,

$$f'(x) = \begin{vmatrix} 2ax & 2ax-1 & 2ax+b+1 \\ b & b+1 & -1 \\ 0 & 0 & 1 \end{vmatrix}$$

Expanding along R_3, we get

$$= \begin{vmatrix} 2ax & 2ax-1 \\ b & b+1 \end{vmatrix}$$

Applying $C_2 \to C_2 - C_1$, we obtain,

$$= \begin{vmatrix} 2ax & -1 \\ b & 1 \end{vmatrix}$$

$\therefore f'(x) = 2ax + b$

On integrating, we get

$$f(x) = ax^2 + bx + c$$

Now, $f'\left(\frac{5}{2}\right) = 0$

$\Rightarrow \quad 5a + b = 0$ ...(i)

and $f(0) = 2$

$\Rightarrow \quad c = 2$ ...(ii)

and $f(1) = 1$

$\Rightarrow \quad a + b + c = 1$ ...(iii)

Solving (i), (ii) and (iii), we get,

$$a = \frac{1}{4}, b = \frac{-5}{4}, c = 2$$

$\therefore \quad f(x) = \frac{1}{4}x^2 - \frac{5}{4}x + 2$ **Ans.**

Chapter 5. Matrices

1. *If* $\begin{bmatrix} x+y & y-z \\ z-2x & y-x \end{bmatrix} = \begin{bmatrix} 3 & -1 \\ 1 & 1 \end{bmatrix}$, *find x, y, z.*

Sol. Given,

$$\begin{bmatrix} x+y & y-z \\ z-2x & y-x \end{bmatrix} = \begin{bmatrix} 3 & -1 \\ 1 & 1 \end{bmatrix}$$

Equating corresponding elements of equal matrices, we get,

$x + y = 3$ ... (i)

$z - 2x = 1$ ...(ii)

$y - z = -1$ ...(iii)

$y - x = 1$ ...(iv)

Adding equations (i) and (iv), we get,

$$\begin{aligned} x + y &= 3 \\ -x + y &= 1 \\ \hline 2y &= 4 \end{aligned}$$

$\Rightarrow \quad y = 2$

Substituting $y = 2$ in equation (i), we get,

$$x + 2 = 3$$

$\Rightarrow \qquad x = 3 - 2 = 1$

Substituting $y = 2$ in equation (iii), we get,

$$2 - z = -1$$

$$\Rightarrow \qquad z = 2 + 1 = 3$$

$\therefore \; x = 1, y = 2, z = 3$ **Ans.**

2. *If* $A = \begin{bmatrix} 3 & 1 \\ 7 & 5 \end{bmatrix}$, *find the values of x and y such that* $A^2 + xI_2 = yA$.

Sol.

Given, $\quad A = \begin{bmatrix} 3 & 1 \\ 7 & 5 \end{bmatrix}$

$\therefore \quad A^2 = A \cdot A$

$$= \begin{bmatrix} 3 & 1 \\ 7 & 5 \end{bmatrix}\begin{bmatrix} 3 & 1 \\ 7 & 5 \end{bmatrix}$$

$$= \begin{bmatrix} 9+7 & 3+5 \\ 21+35 & 7+25 \end{bmatrix}$$

$$= \begin{bmatrix} 16 & 8 \\ 56 & 32 \end{bmatrix}$$

Also, $\quad A^2 + xI_2 = y \cdot A$

$$\Rightarrow \begin{bmatrix} 16 & 8 \\ 56 & 32 \end{bmatrix} + x\begin{bmatrix} 1 & 0 \\ 0 & 1 \end{bmatrix} = y\begin{bmatrix} 3 & 1 \\ 7 & 5 \end{bmatrix}$$

$$\Rightarrow \begin{bmatrix} 16 & 8 \\ 56 & 32 \end{bmatrix} + \begin{bmatrix} x & 0 \\ 0 & x \end{bmatrix} = \begin{bmatrix} 3y & y \\ 7y & 5y \end{bmatrix}$$

$$\Rightarrow \begin{bmatrix} 16+x & 8 \\ 56 & 32+x \end{bmatrix} = \begin{bmatrix} 3y & y \\ 7y & 5y \end{bmatrix}$$

Equating corresponding elements of equal matrices, we get

$$y = 8$$

and $\quad 16 + x = 3y$

$$\therefore \quad x = 3 \times 8 - 16 = 24 - 16 = 8$$

Thus, the required value of x is 8 and y is 8.

Ans.

3. *If* $A = \begin{bmatrix} 3 & -2 \\ 4 & -2 \end{bmatrix}$, *find x such that* $A^2 = xA - 2I$. *Hence, find* A^{-1}.

Sol.

Given, $\quad A = \begin{bmatrix} 3 & -2 \\ 4 & -2 \end{bmatrix}$

$$\therefore \quad A^2 = \begin{bmatrix} 3 & -2 \\ 4 & -2 \end{bmatrix}\begin{bmatrix} 3 & -2 \\ 4 & -2 \end{bmatrix}$$

$$= \begin{bmatrix} 3\times3+(-2)\times4 & 3\times(-2)+(-2)\times(-2) \\ 4\times3+(-2)\times4 & 4\times(-2)+(-2)\times(-2) \end{bmatrix}$$

$$= \begin{bmatrix} 9-8 & -6+4 \\ 12-8 & -8+4 \end{bmatrix}$$

$$= \begin{bmatrix} 1 & -2 \\ 4 & -4 \end{bmatrix}$$

Now, $\quad A^2 = xA - 2I$

$$\Rightarrow \begin{bmatrix} 1 & -2 \\ 4 & -4 \end{bmatrix} = x\begin{bmatrix} 3 & -2 \\ 4 & -2 \end{bmatrix} - 2\begin{bmatrix} 1 & 0 \\ 0 & 1 \end{bmatrix}$$

$$\Rightarrow \begin{bmatrix} 1 & -2 \\ 4 & -4 \end{bmatrix} = \begin{bmatrix} 3x & -2x \\ 4x & -2x \end{bmatrix} - \begin{bmatrix} 2 & 0 \\ 0 & 2 \end{bmatrix}$$

$$\Rightarrow \begin{bmatrix} 1 & -2 \\ 4 & -4 \end{bmatrix} = \begin{bmatrix} 3x-2 & -2x \\ 4x & -2x-2 \end{bmatrix}$$

Equating corresponding elements of equal matrices, we get

$$3x - 2 = 1$$

$$\Rightarrow \quad 3x = 3 \quad \Rightarrow \quad x = 1$$

Now, $\quad A^2 = xA - 2I$

$$\Rightarrow \quad A^2 = 1 \cdot A - 2I$$

Multiplying both sides by A^{-1}, we get

$$\Rightarrow \quad A^2 \cdot A^{-1} = AA^{-1} - 2I.A^{-1}$$

$$\Rightarrow \quad A(AA^{-1}) = I - 2A^{-1}$$

$$\Rightarrow \quad AI = I - 2A^{-1} \qquad [\because AA^{-1} = I]$$

$$\Rightarrow \quad 2A^{-1} = I - AI$$

$$\Rightarrow \quad 2A^{-1} = \begin{bmatrix} 1 & 0 \\ 0 & 1 \end{bmatrix} - \begin{bmatrix} 3 & -2 \\ 4 & -2 \end{bmatrix}\begin{bmatrix} 1 & 0 \\ 0 & 1 \end{bmatrix}$$

$$\Rightarrow \quad 2A^{-1} = \begin{bmatrix} 1 & 0 \\ 0 & 1 \end{bmatrix} - \begin{bmatrix} 3 & -2 \\ 4 & -2 \end{bmatrix}$$

$$\Rightarrow \quad 2A^{-1} = \begin{bmatrix} 1-3 & 0+2 \\ 0-4 & 1+2 \end{bmatrix}$$

$$= \begin{bmatrix} -2 & 2 \\ -4 & 3 \end{bmatrix}$$

$$\therefore \quad A^{-1} = \frac{1}{2}\begin{bmatrix} -2 & 2 \\ -4 & 3 \end{bmatrix}$$

$$\Rightarrow \quad A^{-1} = \begin{bmatrix} -1 & 1 \\ -2 & \frac{3}{2} \end{bmatrix}$$ **Ans.**

4. *For the matrix* $A = \begin{bmatrix} 3 & 1 \\ 7 & 5 \end{bmatrix}$, *find x and y such that* $A^2 + xI = yA$. *Hence, find* A^{-1}.

Sol.

Given, $\quad A = \begin{bmatrix} 3 & 1 \\ 7 & 5 \end{bmatrix}$

Computational Questions

$$\therefore \quad A^2 = \begin{bmatrix} 3 & 1 \\ 7 & 5 \end{bmatrix}\begin{bmatrix} 3 & 1 \\ 7 & 5 \end{bmatrix}$$

$$= \begin{bmatrix} 9+7 & 3+5 \\ 21+35 & 7+25 \end{bmatrix} = \begin{bmatrix} 16 & 8 \\ 56 & 32 \end{bmatrix}$$

Since $A^2 + xI = yA$,

$$\therefore \begin{bmatrix} 16 & 8 \\ 56 & 32 \end{bmatrix} + x\begin{bmatrix} 1 & 0 \\ 0 & 1 \end{bmatrix} = y\begin{bmatrix} 3 & 1 \\ 7 & 5 \end{bmatrix}$$

$$\Rightarrow \begin{bmatrix} 16+x & 8 \\ 56 & 32+x \end{bmatrix} = \begin{bmatrix} 3y & y \\ 7y & 5y \end{bmatrix}$$

Equating corresponding elements of equal matrices, we get,

$$16 + x = 3y, \qquad 8 = y,$$
$$56 = 7y \quad \text{and} \quad 32 + x = 5y$$

On solving, we get $y = 8$ and $x = 8$.

Substituting these values in $A^2 + xI = yA$, we get

$$A^2 + 8I = 8A$$

Multiplying both sides by A^{-1}, we get,

$$\Rightarrow \quad A^{-1}(A^2 + 8I) = 8AA^{-1} = 8I$$

$$\Rightarrow [A^{-1}.A].A + 8A^{-1}I = 8I$$

$$\Rightarrow \quad IA + 8A^{-1} = 8I$$

$$\Rightarrow \quad 8A^{-1} = 8I - IA$$

$$= 8\begin{bmatrix} 1 & 0 \\ 0 & 1 \end{bmatrix} - \begin{bmatrix} 3 & 1 \\ 7 & 5 \end{bmatrix}$$

$$= \begin{bmatrix} 5 & -1 \\ -7 & 3 \end{bmatrix}$$

Hence, $A^{-1} = \frac{1}{8}\begin{bmatrix} 5 & -1 \\ -7 & 3 \end{bmatrix}$ **Ans.**

5. *Find the product of the matrices A and B where* $A = \begin{bmatrix} -5 & 1 & 3 \\ 7 & 1 & -5 \\ 1 & -1 & 1 \end{bmatrix}$, $B = \begin{bmatrix} 1 & 1 & 2 \\ 3 & 2 & 1 \\ 2 & 1 & 3 \end{bmatrix}$. *Hence, solve the following equations by matrix method*

$$x + y + 2z = 1$$
$$3x + 2y + z = 7$$
$$2x + y + 3z = 2$$

Sol. Given, $A = \begin{bmatrix} -5 & 1 & 3 \\ 7 & 1 & -5 \\ 1 & -1 & 1 \end{bmatrix}$, $B = \begin{bmatrix} 1 & 1 & 2 \\ 3 & 2 & 1 \\ 2 & 1 & 3 \end{bmatrix}$

$$\therefore \quad AB = \begin{bmatrix} -5+3+6 & -5+2+3 & -10+1+9 \\ 7+3-10 & 7+2-5 & 14+1-15 \\ 1-3+2 & 1-2+1 & 2-1+3 \end{bmatrix}$$

$$\Rightarrow \quad AB = \begin{bmatrix} 4 & 0 & 0 \\ 0 & 4 & 0 \\ 0 & 0 & 4 \end{bmatrix}$$

$$\Rightarrow \quad AB = 4\begin{bmatrix} 1 & 0 & 0 \\ 0 & 1 & 0 \\ 0 & 0 & 1 \end{bmatrix}$$

$$\Rightarrow \quad AB = 4I_3$$ **Ans.**

$$\Rightarrow \quad B\left(\frac{1}{4}A\right) = I_3$$

Multiplying both sides by B^{-1}

$$B^{-1}B\left(\frac{1}{4}A\right) = B^{-1}I_3$$

$$\Rightarrow \quad B^{-1} = \frac{1}{4}A$$

$$\Rightarrow \quad B^{-1} = \frac{1}{4}\begin{bmatrix} -5 & 1 & 3 \\ 7 & 1 & -5 \\ 1 & -1 & 1 \end{bmatrix}$$

The given system of equation is

$$x + y + 2z = 1$$
$$3x + 2y + z = 7$$
$$2x + y + 3z = 2$$

This system can be written as

$$\begin{bmatrix} 1 & 1 & 2 \\ 3 & 2 & 1 \\ 2 & 1 & 3 \end{bmatrix}\begin{bmatrix} x \\ y \\ z \end{bmatrix} = \begin{bmatrix} 1 \\ 7 \\ 2 \end{bmatrix} \quad \text{or} \quad BX = C$$

where, $X = \begin{bmatrix} x \\ y \\ z \end{bmatrix}$, $C = \begin{bmatrix} 1 \\ 7 \\ 2 \end{bmatrix}$

As B^{-1} exists, the given system has a unique solution

$$X = B^{-1}C$$

$$\Rightarrow \quad X = \frac{1}{4}\begin{bmatrix} -5 & 1 & 3 \\ 7 & 1 & -5 \\ 1 & -1 & 1 \end{bmatrix}\begin{bmatrix} 1 \\ 7 \\ 2 \end{bmatrix}$$

$$\Rightarrow \quad \begin{bmatrix} x \\ y \\ z \end{bmatrix} = \frac{1}{4}\begin{bmatrix} -5+7+6 \\ 7+7-10 \\ 1-7+2 \end{bmatrix}$$

$$\Rightarrow \quad \begin{bmatrix} x \\ y \\ z \end{bmatrix} - \begin{bmatrix} 2 \\ 1 \\ -1 \end{bmatrix}$$

Equating corresponding elements of equal matrices, we get

$$\Rightarrow \quad x = 2, y = 1, z = -1$$

Hence, the solution of given system of equations is $x = 2, y = 1, z = -1$. **Ans.**

6. *Find the adjoint of the matrix*

$$A = \begin{bmatrix} 1 & 0 & -1 \\ 3 & 4 & 5 \\ 0 & -6 & -7 \end{bmatrix}$$ *and hence, find the matrix* A^{-1}.

Sol.

Given, $$A = \begin{bmatrix} 1 & 0 & -1 \\ 3 & 4 & 5 \\ 0 & -6 & -7 \end{bmatrix}$$

$$|A| = \begin{vmatrix} 1 & 0 & -1 \\ 3 & 4 & 5 \\ 0 & -6 & -7 \end{vmatrix}$$

$$= 1\{4 \times -7 - (-6) \times 5\} - 1\{3 \times -6 - 0\}$$

$$= \{-28 + 30\} - \{-18\}$$

$$= -28 + 30 + 18 = 20$$

$$|A| \neq 0$$

$\therefore$ A^{-1} exists

Co–factors, $$A_{11} = \begin{vmatrix} 4 & 5 \\ -6 & -7 \end{vmatrix} = -28 + 30 = 2$$

$$A_{12} = -\begin{vmatrix} 3 & 5 \\ 0 & -7 \end{vmatrix} = -(-21) = 21$$

$$A_{13} = \begin{vmatrix} 3 & 4 \\ 0 & -6 \end{vmatrix} = -18,$$

$$A_{21} = -\begin{vmatrix} 0 & -1 \\ -6 & -7 \end{vmatrix} = 6,$$

$$A_{22} = \begin{vmatrix} 1 & -1 \\ 0 & -7 \end{vmatrix} = -7,$$

$$A_{23} = -\begin{vmatrix} 1 & 0 \\ 0 & -6 \end{vmatrix} = 6,$$

$$A_{31} = \begin{vmatrix} 0 & -1 \\ 4 & 5 \end{vmatrix} = 4,$$

$$A_{32} = -\begin{vmatrix} 1 & -1 \\ 3 & 5 \end{vmatrix} = -8,$$

$$A_{33} = \begin{vmatrix} 1 & 0 \\ 3 & 4 \end{vmatrix} = 4$$

$$\text{adj } A = \begin{bmatrix} 2 & 21 & -18 \\ 6 & -7 & 6 \\ 4 & -8 & 4 \end{bmatrix}^T$$

$$= \begin{bmatrix} 2 & 6 & 4 \\ 21 & -7 & -8 \\ -18 & 6 & 4 \end{bmatrix}$$

Hence, $$A^{-1} = \frac{1}{|A|} \text{adj } A$$

$$= \frac{1}{20}\begin{bmatrix} 2 & 6 & 4 \\ 21 & -7 & -8 \\ -18 & 6 & 4 \end{bmatrix}$$ **Ans**

7. *Find* A^{-1}, *where* $A = \begin{bmatrix} 4 & 2 & 3 \\ 1 & 1 & 1 \\ 3 & 1 & -2 \end{bmatrix}$ *Hence, solve th*

following system of linear equations :

$$4x + 2y + 3z = 2$$
$$x + y + z = 1$$
$$3x + y - 2z = 5$$

Sol.

Here, $$|A| = \begin{vmatrix} 4 & 2 & 3 \\ 1 & 1 & 1 \\ 3 & 1 & -2 \end{vmatrix}$$

$$= 4(-2-1) - 2(-2-3) + 3(1-3$$

$$= -12 + 10 - 6 = -8 \neq 0$$

$\Rightarrow A^{-1}$ exists and

$$A^{-1} = \frac{1}{|A|} \text{adj}(A)$$

Now, adjoint of matrix A

$$\therefore \quad A_{11} = \begin{vmatrix} 1 & 1 \\ 1 & -2 \end{vmatrix} = -3,$$

$$A_{12} = -\begin{vmatrix} 1 & 1 \\ 3 & -2 \end{vmatrix} = 5,$$

$$A_{13} = \begin{vmatrix} 1 & 1 \\ 3 & 1 \end{vmatrix} = -2$$

$$A_{21} = -\begin{vmatrix} 2 & 3 \\ 1 & -2 \end{vmatrix} = 7,$$

$$A_{22} = \begin{vmatrix} 4 & 3 \\ 3 & -2 \end{vmatrix} = -17,$$

$$A_{23} = -\begin{vmatrix} 4 & 2 \\ 3 & 1 \end{vmatrix} = 2$$

$$A_{31} = \begin{vmatrix} 2 & 3 \\ 1 & 1 \end{vmatrix} = -1,$$

$$A_{32} = -\begin{vmatrix} 4 & 3 \\ 1 & 1 \end{vmatrix} = -1,$$

$$A_{33} = \begin{vmatrix} 4 & 2 \\ 1 & 1 \end{vmatrix} = 2$$

$$\therefore \quad \text{adj}(A) = \begin{bmatrix} -3 & 5 & -2 \\ 7 & -17 & 2 \\ -1 & -1 & 2 \end{bmatrix}^T$$

$$= \begin{bmatrix} -3 & 7 & -1 \\ 5 & -17 & -1 \\ -2 & 2 & 2 \end{bmatrix}$$

$$\therefore \quad A^{-1} = \frac{1}{|A|}\text{adj}(A)$$

$$= -\frac{1}{8}\begin{bmatrix} -3 & 7 & -1 \\ 5 & -17 & -1 \\ -2 & 2 & 2 \end{bmatrix}$$

$$\Rightarrow \quad A^{-1} = \frac{1}{8}\begin{bmatrix} 3 & -7 & 1 \\ -5 & 17 & 1 \\ 2 & -2 & -2 \end{bmatrix}$$ **Ans.**

The given system of equations is

$$4x + 2y + 3z = 2$$
$$x + y + z = 1$$
$$3x + y - 2z = 5$$

This system can be written as $AX = B$

where $A = \begin{bmatrix} 4 & 2 & 3 \\ 1 & 1 & 1 \\ 3 & 1 & -2 \end{bmatrix}$, $X = \begin{bmatrix} x \\ y \\ z \end{bmatrix}$ and $B = \begin{bmatrix} 2 \\ 1 \\ 5 \end{bmatrix}$

As $|A| \neq 0$, the given system has a unique solution $X = A^{-1}B$

$$\Rightarrow \quad X = \frac{1}{8}\begin{bmatrix} 3 & -7 & 1 \\ -5 & 17 & 1 \\ 2 & -2 & -2 \end{bmatrix}\begin{bmatrix} 2 \\ 1 \\ 5 \end{bmatrix}$$

$$= \frac{1}{8}\begin{bmatrix} 6-7+5 \\ -10+17+5 \\ 4-2-10 \end{bmatrix} = \frac{1}{8}\begin{bmatrix} 4 \\ 12 \\ -8 \end{bmatrix}$$

$$\Rightarrow \quad \begin{bmatrix} x \\ y \\ z \end{bmatrix} = \begin{bmatrix} 1/2 \\ 3/2 \\ -1 \end{bmatrix}$$

Equating corresponding elements of equal matrices, we get

$$\Rightarrow \quad x = \frac{1}{2}, y = \frac{3}{2}, z = -1$$

Hence, the solution of the given system of equations is

$$x = \frac{1}{2}, y = \frac{3}{2}, z = -1.$$ **Ans.**

8. *If* $A = \begin{bmatrix} 1 & 2 & -3 \\ 2 & 3 & 2 \\ 3 & -3 & -4 \end{bmatrix}$, *find* A^{-1} *and hence, the following system of linear equations :*

$$x + 2y - 3z = -4$$
$$2x + 3y + 2z = 2$$
$$3x - 3y - 4z = 11$$

Sol. Given,

$$A = \begin{bmatrix} 1 & 2 & -3 \\ 2 & 3 & 2 \\ 3 & -3 & -4 \end{bmatrix}$$

$$\therefore |A| = 1(-12+6) - 2(-8-6) - 3(-6-9)$$
$$= -6 + 28 + 45$$
$$= 67 \neq 0$$

$\Rightarrow A^{-1}$ exists.

Now cofactors of A,

$$A_{11} = \begin{vmatrix} 3 & 2 \\ -3 & -4 \end{vmatrix} = -6$$

$$A_{12} = -\begin{vmatrix} 2 & 2 \\ 3 & -4 \end{vmatrix} = 14$$

$$A_{13} = \begin{vmatrix} 2 & 3 \\ 3 & -3 \end{vmatrix} = -15$$

$$A_{21} = -\begin{vmatrix} 2 & -3 \\ -3 & -4 \end{vmatrix} = 17$$

$$A_{22} = \begin{vmatrix} 1 & -3 \\ 3 & -4 \end{vmatrix} = 5$$

$$A_{23} = -\begin{vmatrix} 1 & 2 \\ 3 & -3 \end{vmatrix} = 9$$

$$A_{31} = \begin{vmatrix} 2 & -3 \\ 3 & 2 \end{vmatrix} = 13$$

$$A_{32} = -\begin{vmatrix} 1 & -3 \\ 2 & 2 \end{vmatrix} = -8$$

$$A_{33} = \begin{vmatrix} 1 & 2 \\ 2 & 3 \end{vmatrix} = -1$$

$$\text{Adj}(A) = \begin{bmatrix} -6 & 17 & 13 \\ 14 & 5 & -8 \\ -15 & 9 & -1 \end{bmatrix}$$

$$\therefore \quad A^{-1} = \frac{1}{|A|}\text{Adj}(A)$$

$$\Rightarrow \quad A^{-1} = \frac{1}{67}\begin{bmatrix} -6 & 17 & 13 \\ 14 & 5 & -8 \\ -15 & 9 & -1 \end{bmatrix}$$ **Ans.**

The given system of equation can be written as

$$AX = B$$

$$\begin{bmatrix} 1 & 2 & -3 \\ 2 & 3 & 2 \\ 3 & -3 & -4 \end{bmatrix}\begin{bmatrix} x \\ y \\ z \end{bmatrix} = \begin{bmatrix} -4 \\ 2 \\ 11 \end{bmatrix}$$

As $|A| \neq 0$, thus the system has a unique solution

$$\therefore \quad X = A^{-1}B$$

$$= \frac{1}{67}\begin{bmatrix} -6 & 17 & 13 \\ 14 & 5 & -8 \\ -15 & 9 & -1 \end{bmatrix}\begin{bmatrix} -4 \\ 2 \\ 11 \end{bmatrix}$$

$$= \frac{1}{67}\begin{bmatrix} 24 + 34 + 143 \\ -56 + 10 - 88 \\ 60 + 18 - 11 \end{bmatrix}$$

$$= \frac{1}{67}\begin{bmatrix} 201 \\ -134 \\ 67 \end{bmatrix} = \begin{bmatrix} 3 \\ -2 \\ 1 \end{bmatrix}$$

Equating corresponding elements of equal matrices, we get, $x = 3$, $y = -2$, $z = 1$. **Ans.**

Chapter 6. Differential Calculus

1. *Find the values of a and b such that the following function f (x) is a continuous function.*

$$\begin{cases} 5, & x \leq 2 \\ ax + b, & 2 < x < 10 \\ 21, & x \geq 10 \end{cases}$$

Sol. Given function is continuous. For the continuity at $x = 2$ and $x = 10$.

Continuity at $x = 2$

$$\text{L.H.L.} = \lim_{x \to 2^-} f(x)$$

$$= \lim_{x \to 2^-} 5 = 5$$

$$\text{R.H.L.} = \lim_{x \to 2^+} f(x)$$

$$= \lim_{x \to 2^+} (ax + b)$$

$$= \lim_{h \to 0} \{a(2 + h) + b\} \quad [\because x = 2 + h]$$

$$= 2a + b$$

Since, $f(x)$ is continuous at $x = 2$

So, $\quad$ L.H.L. = R.H.L. = $f(2)$

$$\therefore \quad 2a + b = 5 \quad \ldots(i)$$

Continuity at $x = 10$

$$\text{L.H.L.} = \lim_{x \to 10^-} f(x)$$

$$= \lim_{x \to 10^-} (ax + b)$$

$$= \lim_{h \to 0} [a(10 - h) + b] \quad [\because x = 10 - h]$$

$$= 10a + b$$

$$\text{R.H.L.} = \lim_{x \to 10^+} f(x)$$

$$= \lim_{x \to 10^+} 21$$

$$= 21$$

Since, $f(x)$ is continuous at $x = 10$

So, $\quad$ L.H.L. = R.H.L. = $f(10)$

$$10a + b = 21 \quad \ldots(ii)$$

Solving (i) and (ii), we get,

$$a = 2 \text{ and } b = 1.$$ **Ans.**

2. *If the function f (x) defined by*

$$f(x) = \begin{cases} 2a + 5x & , \text{ if } x > 1 \\ 11 & , \text{ if } x = 1 \\ 6ax - 7b & , \text{ if } x < 1 \end{cases}$$

is continuous at x = 1. Find the values of a and b.

Sol. It is given that the function $f(x)$ is continuous at $x = 1$.

L.H.L. = R.H.L. = $f(1)$

$$\lim_{x \to 1^-} f(x) = \lim_{x \to 1^+} f(x) = f(1)$$

$$\lim_{x \to 1} (6ax - 7b) = \lim_{x \to 1} (2a + 5x) = 11$$

$$6a - 7b = 2a + 5 = 11$$

which gives,

$$6a - 7b = 11 \quad \ldots(i)$$

$$2a + 5 = 11 \quad \ldots(ii)$$

Solving (ii) gives,

$$2a = 11 - 5$$

$$a = \frac{6}{2} = 3 \quad \ldots(iii)$$

Solving (i) and (iii), we get,

$$a = 3 \text{ and } b = 1.$$ **Ans.**

3. *For what choice of a and b is the function*

$$f(x) = \begin{cases} 1+x^2, & x \le 2 \\ ax+b, & x > 2 \end{cases}$$

is differentiable at $x = 2$.

Sol. Since, $f(x)$ is differentiable at $x = 2$ and we know that every differentiable function is continuous.

$\therefore f(x)$ is continuous at $x = 2$ whens

i.e., L.H.L. = R.H.L. = $f(2)$

$$\lim_{x\to 2^-} f(x) = \lim_{x\to 2^+} f(x) = f(2)$$

$$\lim_{x\to 2} 1+x^2 = \lim_{x\to 2} (ax+b) = 5$$

$$5 = 2a + b \quad \text{...(i)}$$

Also $f(x)$ is differentiable at $x = 2$ When,

(L.H.D. at $x = 2$) = (R.H.D. at $x = 2$)

$$\lim_{x\to 2^-} \frac{f(x)-f(2)}{x-2} = \lim_{x\to 2^+} \frac{f(x)-f(2)}{x-2}$$

$$\lim_{x\to 2} \frac{(1+x^2)-(1+2^2)}{x-2} = \lim_{x\to 2} \frac{(ax+b)-(1+2^2)}{x-2}$$

$$\lim_{x\to 2} \frac{x^2-2^2}{x-2} = \lim_{x\to 2} \frac{ax+b-5}{x-2}$$

$$\lim_{x\to 2} (x+2) = \lim_{x\to 2} \frac{ax+b-(2a+b)}{x-2}$$

$$[\because 5 = 2a+b]$$

$$\lim_{x\to 2} (x+2) = \lim_{x\to 2} \frac{a(x-2)}{x-2}$$

$$4 = a \quad \text{...(ii)}$$

Solving (i) and (ii), we get,

$a = 4$ and $b = -3$ **Ans.**

4. *For what choice of a and b is the function*

$$f(x) = \begin{cases} a+6x+x^2, & \text{if } x \le 1 \\ bx+5, & \text{if } x > 1 \end{cases}$$

is differentiable at $x = 1$.

Sol. Since, $f(x)$ is differentiable at $x = 1$ and we know that every differentiable function is continuous.

$\therefore f(x)$ is continuous at $x = 1$

i.e., L.H.L. = R.H.L. = $f(1)$

$$\lim_{x\to 1^-} f(x) = \lim_{x\to 1^+} f(x) = f(1)$$

$$\lim_{x\to 1} a+6x+x^2 = \lim_{x\to 1} bx+5$$

$$\Rightarrow a+7 = b+5$$

$$\Rightarrow a-b = -2 \quad \text{...(i)}$$

Also, $f(x)$ is differentiable at $x = 1$ when,

(L.H.D. at $x = 1$) = (R.H.D. at $x = 1$)

$$\lim_{x\to 1^-} \frac{f(x)-f(1)}{x-1} = \lim_{x\to 1^+} \frac{f(x)-f(1)}{x-1}$$

$$\lim_{x\to 1} \frac{a+6x+x^2-(a+7)}{x-1} = \lim_{x\to 1} \frac{bx+5-(b+5)}{x-1}$$

$$\lim_{x\to 1} \frac{x^2+6x-7}{x-1} = \lim_{x\to 1} \frac{bx+5-(b+5)}{x-1}$$

$$[\because a+7 = b+5]$$

$$\lim_{x\to 1} \frac{(x+7)(x-1)}{x-1} = \lim_{x\to 1} \frac{bx-b}{x-1}$$

$$\lim_{x\to 1} (x+7) = \lim_{x\to 1} \frac{b(x-1)}{x-1}$$

$$1+7 = b$$

$$\Rightarrow b = 8 \quad \text{...(ii)}$$

Solving (i) and (ii), we get,

$a = 6$ and $b = 8$. **Ans.**

5. *It is given that Rolle's theorem holds good for the function $f(x) = x^3 + ax^2 + bx$, $x \in [1, 2]$ at the point $x = 4/3$. Find the values of a and b.*

Sol. Given,

$$f(x) = x^3 + ax^2 + bx$$

Now, $f(x)$ is a polynomial.

So, it is continuous on R *i.e.,* on [1, 2]

$$\therefore \quad f'(x) = 3x^2 + 2ax + b$$

which exists in the interval (1, 2)

$\therefore f(x)$ is derivable on (1, 2)

$$\therefore \quad f(1) = (1)^3 + a(1)^2 + b(1) = 1 + a + b$$

and

$$f(2) = (2)^3 + a(2)^2 + b(2) = 8 + 4a + 2b$$

It is given that Rolle's theorem holds for $f(x)$ on [1, 2]

then, $f(1) = f(2)$

$$1 + a + b = 8 + 4a + 2b$$

$$3a + b + 7 = 0 \quad \text{...(i)}$$

Since, Rolle's theorem holds at point $x = \frac{4}{3}$

$$\Rightarrow f'(c) = 0$$

$$3\left(\frac{4}{3}\right)^2 + 2a\left(\frac{4}{3}\right) + b - 0$$

$$\frac{16}{3} + \frac{8a}{3} + b = 0$$

$$16 + 8a + 3b = 0 \quad \text{...(ii)}$$

Solving (i) and (ii), we get,

$a = -5$ and $b = 8$ **Ans.**

Chapter 7. Applications of Derivatives

1. *The slope of the curve $ax + by - 2 + xy = 0$ at (1, 1) is 2. Find the values of a and b.*

Sol. Given, the equation of curve,

$$ax + by - 2 + xy = 0 \quad \text{...(i)}$$

On differentiating w.r.t. x, we get,

$$a + b\frac{dy}{dx} + x\frac{dy}{dx} + y = 0$$

$$b\frac{dy}{dx} + x\frac{dy}{dx} = -(a + y)$$

$$\frac{dy}{dx}(b + x) = -(a + y)$$

$$\frac{dy}{dx} = \frac{-(a+y)}{b+x}$$

when $x = 1$ and $y = 1$, slope of the curve is 2

i.e., $$\left(\frac{dy}{dx}\right)_{(1,1)} = \frac{-(a+1)}{b+1} = 2$$

$$-(a + 1) = 2(b + 1)$$

$$-a - 1 = 2b + 2$$

$$a + 2b + 3 = 0 \quad \text{...(ii)}$$

Since, the point (1, 1) lies on (i)

$$\therefore \quad a(1) + b(1) - 2 + (1)(1) = 0$$

$$a + b - 2 + 1 = 0$$

$$a + b - 1 = 0 \quad \text{...(iii)}$$

Solving (ii) and (iii), we get,

$$a = 5 \text{ and } b = -4.$$ **Ans.**

2. *Find the equations of the tangent and the normal to the curve $y = (x^2 + x + 1)(x^2 - 3x + 2)$ at the points, where it cuts x-axis.*

Sol. Given, the equation of curve,

$$y = (x^2 + x + 1)(x^2 - 3x + 2) \quad \text{...(i)}$$

At x-axis, $y = 0$

So, putting $y = 0$ in eq. (i), we get

$$0 = (x^2 + x + 1)(x^2 - 3x + 2)$$

$$0 = (x^2 + x + 1)(x - 1)(x - 2)$$

$$(x^2 + x + 1) \neq 0$$

$$\Rightarrow \quad x - 1 = 0 \text{ and } x - 2 = 0$$

$$\Rightarrow \quad x = 1, x = 2$$

$\therefore$ The points of intersection of curve with x-axis are (1, 0) and (2, 0).

From (i), $y = (x^2 + x + 1)(x^2 - 3x + 2)$

On differentiating both sides w.r.t. x, we get,

$$\frac{dy}{dx} = (2x + 1)(x^2 - 3x + 2) + (x^2 + x + 1)(2x - 3)$$

when $x = 1$ and $y = 0$

$$\left(\frac{dy}{dx}\right)_{(1,0)} = (2(1) + 1)((1)^2 - 3(1) + 2) + ((1)^2 + (1) + 1)(2(1) - 3)$$

$$= 3.(0) + 3.(-1)$$

$$\left(\frac{dy}{dx}\right)_{(1,0)} = -3 \quad \text{...(i)}$$

when $x = 2$ and $y = 0$

$$\left(\frac{dy}{dx}\right)_{(2,0)} = (2(2) + 1)((2)^2 - 3(2) + 2) + ((2)^2 + 2 + 1)(2(2) - 3)$$

$$= 5\cdot(0) + 7(1) = 7$$

Now, the equations of the tangents at (1, 0) and (2, 0) are :

At (1, 0),

$$y - 0 = \left(\frac{dy}{dx}\right)_{(1,0)}(x - 1)$$

$$y = -3(x - 1) \quad \text{[From (i)]}$$

$$y + 3x - 3 = 0$$

At (2, 0),

$$y - 0 = \left(\frac{dy}{dx}\right)_{(2,0)}(x - 2)$$

$$y = 7(x - 2)$$

$$y = 7x - 14$$

$$y - 7x + 14 = 0$$

i.e., $y + 3x - 3 = 0$ and $y - 7x + 14 = 0$ **Ans**

And the equations of the normals at (1, 0) and (2, 0) are

At (1, 0),

$$y - 0 = \frac{1}{\left(\frac{dy}{dx}\right)_{(1,0)}}(x - 1)$$

$$y = \frac{1}{-3}(x - 1)$$

$$-3y - x + 1 = 0$$

$$\text{or, } 3y + x - 1 = 0$$

At (2, 0),

$$y - 0 = \frac{1}{\left(\frac{dy}{dx}\right)_{(2,0)}}(x - 2)$$

$$y = \frac{1}{7}(x-2)$$

$$7y - x + 2 = 0$$

i.e., $3y + x - 1 = 0$ and $7y - x + 2 = 0$ **Ans.**

3. *For the curve $y = 4x^3 - 2x^5$, find all points at which the tangent passes through the origin.*

Sol. Let (x_1, y_1) be the point at which the tangent passes through origin

then, $y_1 = 4(x_1)^3 - 2(x_1)^5$ ...(i)

Given, $y = 4x^3 - 2x^5$

On differentiating w.r.t. x, we get,

$$\frac{dy}{dx} = 12x^2 - 10x^4$$

At $x = x_1$ and $y = y_1$

$$\left(\frac{dy}{dx}\right)_{(x_1, y_1)} = 12(x_1)^2 - 10(x_1)^4$$

$$= 12x_1^2 - 10x_1^4$$

Now, the equation of tangent at (x_1, y_1) is

$$y - y_1 = \left(\frac{dy}{dx}\right)_{(x_1, y_1)} (x - x_1)$$

$$y - y_1 = (12x_1^2 - 10x_1^4)(x - x_1)$$

It is given that the tangent passes through origin

So, $0 - y_1 = (12x_1^2 - 10x_1^4)(0 - x_1)$

$y_1 = 12x_1^3 - 10x_1^5$ ...(ii)

Subtracting (ii) from (i), we get

$$0 = -8x_1^3 + 8x_1^5$$

$$-8x_1^3(1 - x_1^2) = 0$$

$\Rightarrow$ $x_1 = 0$ or $x_1 = \pm 1$

Now, when $x_1 = 0$

$\Rightarrow$ $y_1 = 0$ [Using (ii)]

when $x_1 = 1$

$\Rightarrow$ $y_1 = 2$ [Using (ii)]

and when $x_1 = -1$

$\Rightarrow$ $y_1 = -2$ [Using (ii)]

Hence, the points are (0, 0), (1, 2) and (– 1, – 2).

Ans.

4. *It is given that at $x = 1$ and $x = 2$, the function $y = ax^2 - b \log x + x$ attains its maximum value. Find the values of a and b.*

Sol. We have,

$$y = ax^2 - b\log x + x$$

On differentiating w.r.t. x, we get

$$\frac{dy}{dx} = 2ax - \frac{b}{x} + 1$$

Since, y attains its maximum value at $x = 1$ and $x = 2$.

So, $\frac{dy}{dx} = 0$ at $x = 1$ and $x = 2$

i.e., at $x = 1$,

$$\left(\frac{dy}{dx}\right)_{(x=1)} = 2a - b + 1 = 0 \quad ...(i)$$

and at $x = 2$,

$$\left(\frac{dy}{dx}\right)_{(x=2)} = 4a - \frac{b}{2} + 1 = 0 \quad ...(ii)$$

Solving (i) and (ii), we get

$$a = -\frac{1}{6} \text{ and } b = -\frac{2}{3}$$

Now, $$\frac{d^2y}{dx^2} = 2a + \frac{b}{x^2}$$

when, $a = -\frac{1}{6}$ and $b = -\frac{2}{3}$

$$\frac{d^2y}{dx^2} = 2\left(\frac{-1}{6}\right) - \frac{2}{3x^2}$$

$$\frac{d^2y}{dx^2} = -\frac{1}{3} - \frac{2}{3x^2}$$

$$\therefore \left(\frac{d^2y}{dx^2}\right)_{x=1} = \frac{-1}{3} - \frac{2}{3} = \frac{-3}{3} = -1 < 0$$

and $$\left(\frac{d^2y}{dx^2}\right)_{x=2} = \frac{-1}{3} - \frac{2}{3 \times 4}$$

$$= \frac{-3}{6} = \frac{-1}{2} < 0$$

Hence, y is maximum at $x = 1$ and $x = 2$, when $a = \frac{-1}{6}$ and $b = \frac{-2}{3}$. **Ans.**

5. *A ladder 5 m long is leaning against a wall. The bottom of the ladder is pulled along the ground away from the wall at the rate of 2 m/sec. How fast its height on the wall decreasing when the foot of the ladder is 4 m away from the wall ?*

Sol. Let AB = 5 cm be the length of ladder.

Also, the position of ladder at any time t.

Let OA = y and OB = x

Then, $(AB)^2 = (OA)^2 + (OB)^2$

$(5)^2 = y^2 + x^2$ ...(i)

Since, the bottom of the ladder is pulled along the ground away from the wall at the rate $\frac{dx}{dt}$.

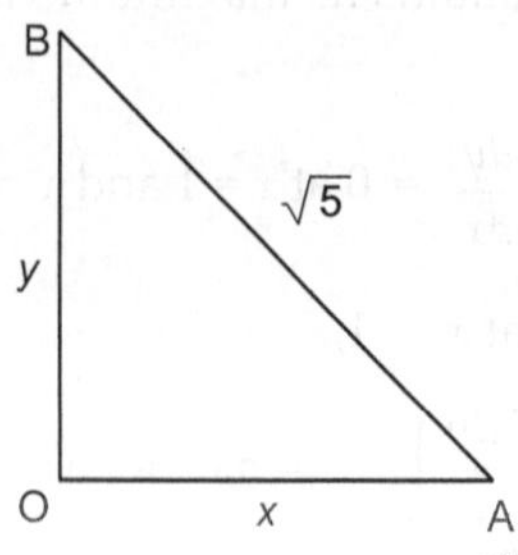

i.e., $\frac{dx}{dt} = 2$ m/sec

From (i) $x^2 + y^2 = 25$

On differentiating both sides w.r.t. t, we get

$$2x\frac{dx}{dt} + 2y\frac{dy}{dt} = 0$$

$$2x(2) + 2y\cdot\frac{dy}{dt} = 0 \qquad \left[\because \frac{dx}{dt} = 2\right]$$

$$= -4x$$

$$\frac{dy}{dt} = \frac{-2x}{y} \qquad \text{...(ii)}$$

when $x = 4$,

$$y = \sqrt{25-16} = \sqrt{9} = 3$$

and $\left(\frac{dy}{dt}\right)_{(4,3)} = \frac{-2(4)}{3} = \frac{-8}{3}$ m/sec

Hence, the rate of decrease in the height of the ladder on the wall is $\frac{8}{3}$ m/sec. **Ans.**

Chapter 9. Differential Equations

1. *Find the general solution of the differential equation $\frac{dy}{dx} = \frac{1-\cos x}{1+\cos x}$.*

Sol. Given,

$$\frac{dy}{dx} = \frac{1-\cos x}{1+\cos x}$$

$$dy = \frac{1-\cos x}{1+\cos x}dx$$

$$dy = \frac{2\sin^2 x/2}{2\cos^2 x/2}dx$$

$$dy = \tan^2 x/2\, dx$$

$$dy = \left(\sec^2\frac{x}{2} - 1\right)dx$$

On integrating both sides, we get,

$$y = \int\left(\sec^2\frac{x}{2} - 1\right)dx + c$$

$$y = \frac{\tan x/2}{1/2} - x + c$$

$$y = 2\tan\frac{x}{2} - x + c. \qquad \textbf{Ans.}$$

2. *Find the general solution of the differential equation*

$$\frac{dy}{dx} + \frac{\cos x}{1+\sin x}y = \frac{x}{1+\sin x}.$$

Sol. Given differential equation is

$$\frac{dy}{dx} + \frac{\cos x}{1+\sin x}y = \frac{x}{1+\sin x} \qquad \text{...(i)}$$

which is a linear differential equation of form

$$\frac{dy}{dx} + P.y = Q$$

Here, $P = \frac{\cos x}{1+\sin x}$, $Q = \frac{x}{1+\sin x}$

$\therefore$ Integrating factor (I.F.)

$$= e^{\int P dx}$$

$$= e^{\int \frac{\cos x}{1+\sin x}dx}$$

$$= e^{\log(1+\sin x)}$$

$$= 1 + \sin x$$

Multiplying both sides of (i) by I.F., we get

$$(1+\sin x)\frac{dy}{dx} + y\cos x = x$$

Integrating w.r.t. x we get

$$\int\frac{d}{dx}[y(1+\sin x)]\cdot dx = \int x\cdot dx$$

$$y(1+\sin x) = \int x\, dx + c$$

$$y(1+\sin x) = \frac{x^2}{2} + c$$

$$\therefore \quad y = \frac{2c+x^2}{2(1+\sin x)}. \qquad \textbf{Ans.}$$

3. *Find the general solution of the differential equation*

$$= \frac{dy}{dx} = \frac{\sqrt{1-y^2}}{\sin^{-1} y - x}.$$

Computational Questions

Sol. We have,

$$\frac{dy}{dx} = \frac{\sqrt{1-y^2}}{\sin^{-1} y - x}$$

$$\Rightarrow \quad \frac{dx}{dy} = \frac{\sin^{-1} y}{\sqrt{1-y^2}} - \frac{x}{\sqrt{1-y^2}}$$

$$\Rightarrow \quad \frac{dx}{dy} + \frac{x}{\sqrt{1-y^2}} = \frac{\sin^{-1} y}{\sqrt{1-y^2}} \quad \ldots(i)$$

which is a linear differential equation of form

$$\frac{dx}{dy} + \text{P}x = \text{Q}$$

$$\therefore \quad \text{I.F.} = e^{\int P\, dx}$$

$$= e^{\int \frac{1}{\sqrt{1-y^2}} dx} \quad \left[\because P = \frac{1}{\sqrt{1-y^2}}\right]$$

$$= e^{\sin^{-1} y}$$

Multiplying both sides of (i) by I.F., we get

$$e^{\sin^{-1} y} \frac{dy}{dx} + e^{\sin^{-1} y} \cdot \frac{x}{\sqrt{1-y^2}} = \frac{\sin^{-1} y \cdot e^{\sin^{-1} y}}{\sqrt{1-y^2}}$$

Integrating both sides w.r.t. y, we get,

$$\Rightarrow \quad x\, e^{\sin^{-1} y} = \int e^{\sin^{-1} y} \cdot \frac{\sin^{-1} y}{\sqrt{1-y^2}} dy + c$$

$$\Rightarrow \quad x\, e^{\sin^{-1} y} = \int e^t \cdot t\, dt + c$$

$$\left[\begin{array}{l} \because \text{Putting } \sin^{-1} y = t \\ \dfrac{1}{\sqrt{1-y^2}} dy = dt \end{array}\right]$$

$$\Rightarrow \quad x\, e^{\sin^{-1} y} = e^t \cdot t - e^t + c$$

$$\Rightarrow \quad x\, e^{\sin^{-1} y} = e^{\sin^{-1} y} (\sin^{-1} y - 1) + c. \quad \textbf{Ans.}$$

4. *Find the general solution of the differential equation* $x \log x \cdot \dfrac{dy}{dx} + y = \dfrac{2}{x} \log x$.

Sol. Given, $x \log x \cdot \dfrac{dy}{dx} + y = \dfrac{2}{x} \log x \quad \ldots(i)$

$$\Rightarrow \quad \frac{dy}{dx} + \frac{y}{x \log x} = \frac{2}{x^2} \quad \ldots(ii)$$

which is a linear differential equation of the form

$$\frac{dy}{dx} + Py = \text{Q}$$

Here, $\quad \text{P} = \dfrac{1}{x \log x}, \ \text{Q} = \dfrac{2}{x^2}$

$$\therefore \quad \text{I.F.} = e^{\int P dx}$$

$$= e^{\int \frac{1}{x \log x} dx}$$

$$= e \log (\log x)$$

$$= \log x$$

Multiplying equation (ii) by I.F., we get,

$$\log x \cdot \frac{dy}{dx} + \frac{y}{x} = \frac{2 \log x}{x^2}$$

Integrating both sides, we get,

$$y \log x = \int \frac{2}{x^2} \log x\, dx + c$$

$$y \log x = 2 \int \underset{\text{I}}{\log x} \cdot \underset{\text{II}}{x^{-2}}\, dx + c$$

On integrating by parts,

$$\Rightarrow \quad y \log x = 2\left[\frac{\log x \cdot x^{-1}}{-1} - \int \frac{1}{x} \cdot \frac{x^{-1}}{-1} dx\right] + c$$

$$\Rightarrow \quad y \log x = 2\left[-\frac{\log x}{x} + \int x^{-2} dx\right] + c$$

$$\Rightarrow \quad y \log x = 2\left[-\frac{\log x}{x} + \frac{x^{-1}}{-1}\right] + c$$

$$\Rightarrow \quad y \log x = \frac{2}{x}(-\log x - 1) + c$$

$$\Rightarrow \quad y \log x = -\frac{2}{x}(\log x + 1) + c. \quad \textbf{Ans.}$$

5. *Find the particular solution of the differential equation :* $\dfrac{dy}{dx} - 2y = \cos 3x$, *given that* $y = 1$ *when* $x = 0$.

Sol. Given,

$$\frac{dy}{dx} - 2y = \cos 3x \quad \ldots(i)$$

which is a linear differential equation of the form $\dfrac{dy}{dx} + Py = \text{Q}$.

Here, $\quad \text{P} = -2$ and $\text{Q} = \cos 3x$

$$\therefore \quad \text{I.F.} = e^{\int P dx}$$

$$= e^{\int -2\, dx} = e^{-2x}$$

Multiplying both sides of (i) by I.F.

$$e^{-2x} \frac{dy}{dx} - 2y e^{-2x} = \cos 3x \cdot e^{-2x}$$

Integrating both sides w.r.t. x, we get

$$y e^{-2x} = \int \underset{I}{e^{-2x}} . \underset{II}{\cos 3x}\, dx + c \quad ...(ii)$$

On integrating by parts,

$$y e^{-2x} = \frac{1}{3} e^{-2x} \sin 3x - \int \frac{-2}{3} \cdot e^{-2x} \sin 3x\, dx + c$$

$$\Rightarrow y e^{-2x} = \frac{1}{3} e^{-2x} \sin 3x + \frac{2}{3} \int \underset{I}{e^{-2x}} . \underset{II}{\sin 3x}\, dx + c$$

$$\Rightarrow y e^{-2x} = \frac{1}{3} e^{-2x} \sin 3x + \frac{2}{3}\left[\frac{-1}{3} e^{-2x} \cos 3x - \int -2 \cdot e^{-2x} \left(\frac{-\cos 3x}{3}\right) dx\right] + c$$

$$y e^{-2x} = \frac{1}{3} e^{-2x} \sin 3x + \frac{2}{3}\left[\frac{-1}{3} e^{-2x} \cos 3x - \frac{2}{3} \int e^{-2x} \cos 3x\, dx\right] + c$$

$$\Rightarrow y e^{-2x} = \frac{1}{3} e^{-2x} \sin 3x - \frac{2}{9} e^{-2x} \cos 3x - \frac{4}{9} \int e^{-2x} \cos 3x\, dx + c$$

$$\Rightarrow y e^{-2x} = \frac{e^{-2x}}{9} (3 \sin 3x - 2 \cos 3x) - \frac{4}{9} y e^{-2x} + c$$

[Using (ii)]

$$\left(1 + \frac{4}{9}\right) y e^{-2x} = \frac{e^{-2x}}{9} (3 \sin 3x - 2 \cos 3x) + c$$

$$\Rightarrow \frac{13}{9} \cdot y e^{-2x} = \frac{e^{-2x}}{9} (3 \sin 3x - 2 \cos 3x) + c$$

$$\Rightarrow y e^{-2x} = \frac{e^{-2x}}{13} (3 \sin 3x - 2 \cos 3x) + 9c$$

$$\Rightarrow y e^{-2x} = \frac{e^{-2x}}{13} (3 \sin 3x - 2 \cos 3x) + K \quad ...(iii)$$

$[\because K = 9c]$

It is given that $y = 1$ when $x = 0$. Putting these values in (iii), we get

$$(1)\, e^{-0} = \frac{e^{-0}}{13} (3 \sin 0 - 2 \cos 0) + K$$

$$1 = \frac{1}{13} (0 - 2) + K$$

$$\Rightarrow 1 + \frac{2}{13} = K$$

$$\Rightarrow K = \frac{15}{13}$$

Putting this value of K in (iii), we get

$$y e^{-2x} = \frac{e^{-2x}}{13} (3 \sin 3x - 2 \cos 3x) + \frac{15}{13}.$$

Ans

Chapter 11. Vectors

1. *Find the unit vector perpendicular to the plane of vectors $\vec{a}$ and $\vec{b}$ where $\vec{a} = \hat{i} - \hat{j} + \hat{k}$, $\vec{b} = \hat{i} + 2\hat{j}$.*

Sol. Consider, $\vec{a} \times \vec{b} = \begin{vmatrix} \hat{i} & \hat{j} & \hat{k} \\ 1 & -1 & 1 \\ 1 & 2 & 0 \end{vmatrix}$

$$= \hat{i}(0-2) - \hat{j}(0-1) + \hat{k}(2+1)$$

$$= -2\hat{i} + \hat{j} + 3\hat{k}$$

$$\Rightarrow |\vec{a} \times \vec{b}| = \sqrt{(-2)^2 + (1)^2 + (3)^2}$$

$$= \sqrt{4+1+9} = \sqrt{14}$$

$\therefore$ Unit vector perpendicular to $\vec{a}$ and $\vec{b}$ is given as

$$\hat{n} = \frac{\vec{a} \times \vec{b}}{|\vec{a} \times \vec{b}|}$$

$$= \frac{-2\hat{i} + \hat{j} + 3\hat{k}}{\sqrt{14}}$$

Ans.

2. *Find the unit vector perpendicular to the two vectors $\hat{i} + 2\hat{j} - \hat{k}$ and $2\hat{i} + 3\hat{j} + \hat{k}$.*

Sol.

Let $\vec{A} = \hat{i} + 2\hat{j} - \hat{k}$, $\vec{B} = 2\hat{i} + 3\hat{j} + \hat{k}$

$\because \quad \vec{A} \times \vec{B} = (AB \sin\theta)\, \hat{n}$

$$\therefore \quad \hat{n} = \frac{\vec{A} \times \vec{B}}{(AB \sin\theta)} = \frac{\vec{A} \times \vec{B}}{|\vec{A} \times \vec{B}|}$$

$$\Rightarrow \quad \vec{A} \times \vec{B} = \begin{vmatrix} \hat{i} & \hat{j} & \hat{k} \\ 1 & 2 & -1 \\ 2 & 3 & 1 \end{vmatrix}$$

$$= \hat{i}(2+3) - \hat{j}(1+2) + \hat{k}(3-4)$$

$$= 5\hat{i} - 3\hat{j} - \hat{k}$$

Computational Questions

$\therefore \quad \hat{n} = \frac{\vec{A} \times \vec{B}}{|\vec{A} \times \vec{B}|} = \frac{5\hat{i} - 3\hat{j} - \hat{k}}{\sqrt{25+9+1}}$

$\hat{n} = \frac{5\hat{i} - 3\hat{j} - \hat{k}}{\sqrt{35}}$ **Ans.**

Chapter 12. Three Dimensional Geometry

1. (a) *Find the shortest distance between the lines :*

$\frac{x-8}{3} = \frac{y+9}{-16} = \frac{z-10}{7}$ *and* $\frac{x-15}{3} = \frac{y-29}{8} = \frac{5-z}{5}$

(b) *Find the equation of the plane passing through the line of intersection of the planes $x + 2y + 3z - 5 = 0$ and $3x - 2y - z + 1 = 0$ and cutting off equal intercepts on the x and z axes.*

Sol. (a) Given lines are :

$$\frac{x-8}{3} = \frac{y+9}{-16} = \frac{z-10}{7} = t \quad \ldots(i)$$

and $$\frac{x-15}{3} = \frac{y-29}{8} = \frac{z-5}{-5} = s \quad \ldots(ii)$$

Any point on (i) is M $(8 + 3 - t, - 9 - 16t, 10 + 7t)$ and any point on (ii) is

N $(15 + 3s, 29 + 8s, 5 - 5s)$

$\therefore$ Direction numbers of the line MN are

$(15 + 3s - 8 - 3t, 29 + 8s + 9 + 16t, 5 - 5s - 10 - 7t)$

$= (7 + 3s - 3t, 38 + 8s + 16t, - 5 - 5s - 7t)$

Now, MN will be the shortest distance between (i) and (ii) if MN is perpendicular to both.

i.e., if $3 (7 + 3s - 3t) + (- 16) (38 + 8s + 16t) + 7 (- 5 - 5s - 7t) = 0$

and $3 (7 + 3s - 3t) + 8 (38 + 8s + 16t) + (- 5) (- 5 - 5s - 7t) = 0$

i.e., if $- 154s - 314t - 622 = 0$

$\Rightarrow \quad 77s + 157t + 311 = 0 \quad \ldots(iii)$

and $\quad 98s + 154t + 350 = 0$

$\Rightarrow \quad 49s + 77t + 175 = 0 \quad \ldots(iv)$

On solving (iii) and (iv) simultaneously, we get

$t = - 1$ and $s = - 2$

$t = - 1$ gives M (5, 7, 3) and

$s = - 2$ gives N (9, 13, 15)

$\therefore$ The shortest distance between the given lines

$$MN = \sqrt{(9-5)^2 + (13-7)^2 + (15-3)^2}$$

$$= \sqrt{196} = 14 \text{ units.}$$ **Ans.**

(b) The equation of any plane passing through the lines of intersection of the planes

$x + 2y + 3z - 5 = 0$ and $3x - 2y - z + 1 = 0$ is

$(x + 2y + 3z - 5) + k (3x - 2y - z + 1) = 0 \quad \ldots(i)$

For the intercept on x – axis, on putting $y = 0$ and $y = 0$, we get

$$x + 3kx - 5 + k = 0$$

$$\Rightarrow \quad x = \frac{5-k}{3k+1}$$

For the intercept on z-axis, on putting $x = 0$, $y = 0$, we get

$$3z - kz - 5 + k = 0$$

$$\Rightarrow \quad z = \frac{5-k}{3-k}$$

$\therefore$ Intercepts on x-axis and z-axis made by the plane (i) are

$\frac{5-k}{3k+1}$ and $\frac{5-k}{3-k}$ respectively.

$$\Rightarrow \quad \frac{5-k}{3k+1} = \frac{5-k}{3-k}$$

$$\Rightarrow \quad - 8k + k^2 + 15 = - 3k^2 + 14k + 5$$

$$\Rightarrow \quad - 4k^2 + 22k - 10 = 0$$

$$\Rightarrow \quad 4k^2 - 22k + 10 = 0$$

$$\Rightarrow \quad (4k - 2) (k - 5) = 0$$

$$\Rightarrow \quad k = \frac{1}{2}, 5$$

On putting $k = 5$ in (i), we notice that the plane passes through origin and hence, it cannot make intercepts on axes. Therefore, $k = 1/2$ is the only admissible value.

Substituting $k = 1/2$ in equation (i), the equation of the required plane is

$$x + 2y + 3z - 5 + 1/2 (3x - 2y + z + 1) = 0$$

$$\Rightarrow 2x + 4y + 6z - 10 + 3x - 2y + z + 1 = 0$$

$$\Rightarrow \quad 5x + 2y + 7z - 9 = 0$$ **Ans.**

2. (a) *Find the shortest distance between the lines whose vector equations are*

$$\vec{r} = (4\hat{i} - \hat{j} + 2\hat{k}) + \lambda (\hat{i} + 2\hat{j} - 3\hat{k})$$

and $$\vec{r} = (2\hat{i} + \hat{j} - 2\hat{k}) + \mu (3\hat{i} + 2\hat{j} - 4\hat{k}).$$

(b) *Find the equation of the plane passing through the line of intersection of the planes $x + 2y + 3z - 4 = 0$ and $3z - y = 0$ and perpendicular to the plane $3x + 4y - 2z + 6 = 0$.*

Sol. (a) The vector equations of the lines are,

$$\vec{r} = (4\hat{i} - \hat{j} + 2\hat{k}) + \lambda(\hat{i} + 2\hat{j} - 3\hat{k})$$

and $\vec{r} = (2\hat{i}+\hat{j}-\hat{k})+\mu(3\hat{i}+2\hat{j}-4\hat{k})$

Then, the shortest distance between them is

$$d=\left|\frac{(\vec{a_2}-\vec{a_1})\,(\vec{b_1}\times\vec{b_2})}{|\vec{b_1}\times\vec{b_2}|}\right|$$

Now, $\vec{a_2}-\vec{a_1} = (2\hat{i}+\hat{j}-\hat{k})-(4\hat{i}-\hat{j}+2\hat{k})$

$= -2\hat{i}+2\hat{j}-3\hat{k}$

Also $\vec{b_1}\times\vec{b_2} = \begin{vmatrix}\hat{i} & \hat{j} & \hat{k}\\ 1 & 2 & -3\\ 3 & 2 & -4\end{vmatrix}$

$= \hat{i}(-8+6)-\hat{j}(-4+9)+\hat{k}(2-6)$

$= -2\hat{i}-5\hat{j}-4\hat{k}$

$\Rightarrow |\vec{b_1}\times\vec{b_2}| = \sqrt{(-2)^2+(-5)^2+(-4)^2}$

$= \sqrt{4+25+16}$

$= \sqrt{45} = 3\sqrt{5}$

and $(\vec{a_2}-\vec{a_1})\,(\vec{b_1}\times\vec{b_2})$

$= (-2\hat{i}+2\hat{j}-3\hat{k})\,(-2\hat{i}-5\hat{j}-4\hat{k})$

$= 4-10+12 = 6$

$\therefore$ Shortest distance

$= \left(\frac{6}{3\sqrt{5}}\right) = \frac{2}{\sqrt{5}}$ units. **Ans.**

(b) Equation of any plane passing through the intersection of given planes is given by

$(x+2y+3z-4)+\lambda(3z-y) = 0$

$\Rightarrow x+(2-\lambda)y+(3+3\lambda)z-4 = 0$ ...(i)

But this is perpendicular to the plane $3x+4y-2z+6=0$

$\therefore l_1 l_2+m_1 m_2+n_1 n_2 = 0$

$\Rightarrow 3+4(2-\lambda)-2(3+3\lambda) = 0$

$\Rightarrow 3+8-4\lambda-6-6\lambda = 0$

$\Rightarrow -10\lambda = -5$

$\Rightarrow \lambda = \frac{1}{2}$

Hence, the required equation of plane is given by

$x+\left(2-\frac{1}{2}\right)y+\left(3+\frac{3}{2}\right)z-4 = 0$

$\Rightarrow 2x+3y+9z-8 = 0$

$\Rightarrow 2x+3y+9z = 8$ **Ans.**

3. (a) *Find the equation of a line passing through the point (– 1, 3, – 2) and perpendicular to the lines :*

$$\frac{x}{1}=\frac{y}{2}=\frac{z}{3} \text{ and } \frac{x+2}{-3}=\frac{y-1}{2}=\frac{z+1}{5}.$$

(b) *Find the equation of planes parallel to the plane 2x – 4y + 4z = 7 and which are at a distance of five units from the point (3, – 1, 2).*

Sol. (a) Any line through the point (– 1, 3, – 2) is

$$\frac{x-(-1)}{a}=\frac{y-3}{b}=\frac{z-(-2)}{c} \quad ...(i)$$

It is perpendicular to the lines

$$\frac{x}{1}=\frac{y}{2}=\frac{z}{3} \quad ...(ii)$$

and $$\frac{x+2}{-3}=\frac{y-1}{2}=\frac{z+1}{5} \quad ...(iii)$$

Then $1.a+2.b+3.c = 0$

$\Rightarrow a+2b+3c = 0$ ...(iv)

and $-3.a+2.b+5.c = 0$

$\Rightarrow -3a+2b+5c = 0$ ...(v)

Subtracting (v) from (iv), we get,

$4a-2c = 0$

$\Rightarrow c = 2a$ ...(vi)

From (iv), we get,

$a+2b+3.2a = 0$

$\Rightarrow 2b = -7a$

$\Rightarrow b = \frac{-7a}{2}$ ...(vii)

From (vi) and (vii), we get,

$a:b:c = a:-\frac{7a}{2}:2a$

$\Rightarrow a:b:c = 2:-7:4$

Putting these values in (i), the equation of the required lines are

$$\frac{x+1}{2}=\frac{y-3}{-7}=\frac{z+2}{4}$$ **Ans.**

(b) The given plane is $2x-4y+4z=7$ ...(i)

Equation of any plane parallel to (i) is given as,

$$2x - 4y + 4z + k = 0 \quad \ldots(ii)$$

Now (ii) is at a distance of 5 units from the point (3, – 1, 2) if

$$\frac{|2\times 3 - 4.(-1) + 4\times 2 + k|}{\sqrt{(2)^2+(-4)^2+(4)^2}} = 5$$

$$\Rightarrow \quad \frac{|18+k|}{6} = 5$$

$$\Rightarrow \quad |18+k| = 30$$

$$\Rightarrow \quad 18+k = 30$$

or $\quad 18 + k = -30$

$$\Rightarrow \quad k = 12$$

or $\quad k = -48$

Substituting these values of k in (ii) the equation of the required planes are

$$2x - 4y + 4z + 12 = 0$$

and $\quad 2x - 4y + 4z - 48 = 0$ **Ans.**

4. **(a)** *Find the vector equation of the line passing through the point (–1, 2, 1) and parallel to the line* $\vec{r} = 2\hat{i} + 3\hat{j} - \hat{k} + \lambda(\hat{i} - 2\hat{j} + \hat{k})$. *Also, find the distance between these lines.*

(b) *Find the equation of the plane passing through the points A (2, 1, –3), B (–3, –2, 1) and C (2, 4, – 1).*

Sol. **(a)** The given line is

$$\vec{r} = 2\hat{i} + 3\hat{j} - \hat{k} + \lambda(\hat{i} - 2\hat{j} + \hat{k}) \quad \ldots(i)$$

It passes through the point A_1 with position vector $\vec{a_1} = 2\hat{i} + 3\hat{j} - \hat{k}$ and is parallel to the vector

$$\vec{b} = \hat{i} - 2\hat{j} - \hat{k}$$

A₂(–1, 2, 1)

b

N

A₁(2, 3, –1)

The equation of the line passing through the point

A_2 (– 1, 2, 1) with position vector

$$\vec{a_2} = -\hat{i} + 2\hat{j} + \hat{k}$$

and parallel to the line (i) is

$$\vec{r} = -\hat{i} + 2\hat{j} + \hat{k} + \mu(\hat{i} - 2\hat{j} + \hat{k}) \quad \ldots(ii)$$

Also $\vec{a_2} - \vec{a_1} = -3\hat{i} - \hat{j} + 2\hat{k}$

$$|\vec{b}| = \sqrt{(1)^2+(2)^2+(1)^2} = \sqrt{6}$$

and $|\vec{b} \times (\vec{a_2} - \vec{a_1})|$

$$= \begin{vmatrix} \hat{i} & \hat{j} & \hat{k} \\ 1 & -2 & 1 \\ -3 & -1 & 2 \end{vmatrix}$$

$$= \hat{i}(-4+1) - \hat{j}(2+3) + \hat{k}(-1-6)$$

$$= -3\hat{i} - 5\hat{j} - 7\hat{k}$$

$$\therefore \quad |\vec{b} \times (\vec{a_2} - \vec{a_1})| = \sqrt{(-3)^2+(-5)^2+(-7)^2}$$

$$= \sqrt{9+25+49} = \sqrt{83}$$

∴ The distance between the parallel lines (i) and (ii) is

$$= \frac{|\vec{b} \times (\vec{a_2} - \vec{a_1})|}{|b|}$$

$$= \frac{\sqrt{83}}{\sqrt{6}} = \sqrt{\frac{83}{6}} \text{ units}$$ **Ans.**

(b) Equation of any plane passing through A (2, 1, – 3) is

$$a(x-2) + b(y-1) + c(z+3) = 0 \quad \ldots(i)$$

Since, it passes through B (– 3, – 2, 1) and C (2, 4, – 1)

we get

$$a(-3-2) + b(-2-1) + c(1+3) = 0$$

$$\Rightarrow \quad -5a - 3b + 4c = 0$$

$$\Rightarrow \quad 5a + 3b - 4c = 0 \quad \ldots(ii)$$

and $\quad a(2-2) + b(4-1) + c(-1+3) = 0$

$$\Rightarrow \quad 3b + 2c = 0 \quad \ldots(iii)$$

Solving (ii) and (iii), we get

$$a : b : c = 18 : -10 : 15$$

Replacing in (i) required equation is given as,

$$18(x-2) - 10(y-1) + 15(z+3) = 0$$

$$18x - 36 - 10y + 10 + 15z + 45 = 0$$

$$18x - 10y + 15z + 19 = 0$$ **Ans.**

5. *Find the Cartesian equation of the plane passing through the line of intersection of the planes.*

$$\vec{r}\cdot(2\hat{i} + 3\hat{j} - 4\hat{k}) + 5 = 0$$

and $\vec{r}\cdot(\hat{i} - 5\hat{j} + 7\hat{k}) + 2 = 0$

and intersecting y-axis at (0, 3).

Sol. Any plane passing through the intersection of the given planes is

$[\vec{r}\cdot(2\hat{i}+3\hat{j}-4\hat{k})+5]+\lambda[\vec{r}\cdot(\hat{i}-5\hat{j}+7\hat{k})+2]=0$

$\vec{r}\,[(2+\lambda)\hat{i}+(3-5\lambda)\hat{j}+(-4+7\lambda)\hat{k}] = -5-2\lambda$ ...(i)

Given equation (i) intersects the y-axis at (0, 3) so at y-axis

coordinate of $x = 0$
coordinate of $y = 3$
coordinate of $z = 0$

As at y-axis coordinate of x and z will be 0.

$\therefore \quad \vec{r} = 0\hat{i}+3\hat{j}+0\hat{k}$

$$(0\hat{i}+3\hat{j}+0\hat{k})\cdot[(2+\lambda)\hat{i}+(3-5\lambda)\hat{j}+(-4+7\lambda)\hat{k}] = -5-2\lambda$$

$$0\,(2+\lambda)+3\,(3-5\lambda)+0\,(-4+7\lambda)+5+2\lambda = 0$$

$$\lambda = \frac{14}{13}$$

$\therefore$ The required plane is

$$\left(2+\frac{14}{13}\right)x+\left(3-5\times\frac{14}{13}\right)y+\left(-4+7\times\frac{14}{13}\right)z + 5+2\times\left(\frac{14}{13}\right) = 0$$

$$\frac{(26+14)}{13}x+\left(\frac{(39-70)}{13}\right)y+\frac{(-52+98)}{13}z + \frac{65+28}{13} = 0$$

or $40x - 31y + 46z + 93 = 0$ which is the required equation. **Ans.**

6. *A plane meets the co-ordinate axis in A, B, C such that the centroid of the triangle ABC is the point (p, q, r) then prove that the equation of the plane is*

$$\frac{x}{p}+\frac{y}{q}+\frac{z}{r} = 3.$$

Sol. Let the equation of the plane be

$$\frac{x}{a}+\frac{y}{b}+\frac{z}{c} = 1 \quad ...(i)$$

The plane meets the coordinate axis in $A\,(a, 0, 0)$, $B\,(0, b, 0)$, $C\,(0, 0, c)$ respectively.

The centroid of the triangle ABC is $\left(\frac{a}{3},\frac{b}{3},\frac{c}{3}\right)$.

$$\therefore \quad p = \frac{a}{3},\; q = \frac{b}{3},\; r = \frac{c}{3}$$

$\Rightarrow a = 3p,\; b = 3q,\; c = 3r$

Substituting these values in (i), we get,

$$\frac{x}{3p}+\frac{y}{3q}+\frac{z}{3r} = 1$$

$$\Rightarrow \quad \frac{x}{p}+\frac{y}{q}+\frac{z}{r} = 3$$

which is the required equation of plane.

Hence Proved

7. *Find the equation of the plane passing through the point (2, 3, 4) and making equal intercepts on axes.*

Sol. Let the equation of the plane is

$$\frac{x}{a}+\frac{y}{b}+\frac{z}{c} = 1$$

As it makes equal intercepts on axes *i.e.,*

$$a = b = c = a \text{ (let)}$$

$$\Rightarrow \quad \frac{x}{a}+\frac{y}{a}+\frac{z}{a} = 1$$

$$\Rightarrow \quad x+y+z = a \quad ...(i)$$

As it passes through (2, 3, 4),

$$\therefore \quad 2+3+4 = a = 9$$

Substituting this values in equation (i), we get,

$$x+y+z = 9$$ **Ans.**

8. *Find the equations of the two planes passing through the points (0, 4, –3) and (6, – 4, 3), if the sum of their intercepts on the three axes is zero.*

Sol. Let equation of the plane be

$$\frac{x}{a}+\frac{y}{b}+\frac{z}{c} = 1$$

Since, (0, 4, – 3) and (6, – 4, 3) lie on it,

$$\therefore \quad \frac{4}{b}-\frac{3}{c} = 1 \quad ...(i)$$

$$\text{and} \quad \frac{6}{a}-\frac{4}{b}+\frac{3}{c} = 1 \quad ...(ii)$$

From (i) and (ii), we have

$$\frac{6}{a} = 2$$

$$\Rightarrow \quad a = 3$$

$$\text{and} \quad a+b+c = 0$$

$$\Rightarrow \quad b+c = -3$$

Putting in (i),

$$\frac{4}{b}+\frac{3}{3+b} = 1$$

$$\Rightarrow \quad b^2-4b-12 = 0$$

$$\Rightarrow \quad (b-6)(b+2) = 0$$

$$\Rightarrow \quad b = 6 \text{ or } b = -2$$

$$\text{Then,} \quad c = -9 \text{ or } c = -1$$

$\therefore$ Equations of plane are given as,

$$\frac{x}{3}+\frac{y}{6}-\frac{z}{9} = 1 \text{ and } \frac{x}{3}-\frac{y}{2}-\frac{z}{1} = 1$$ **Ans.**

9. *A plane is passing through the point (2, – 3, 1) and perpendicular to the straight line joining the points*

(3, 4, – 1) and (2, – 1, 5). Find the equation of the plane.

Sol. The straight line joining (3, 4, –1) and (2, –1, 5) is a normal to the plane. Therefore, their direction ratios of the normal are (1, 5, –6).

It passes through (2, – 3, 1).

∴ Equation of the plane is

$$(x - 2) + 5(y + 3) - 6(z - 1) = 0$$

$$\Rightarrow \quad x - 2 + 5y + 15 - 6z + 6 = 0$$

$$\Rightarrow \quad x + 5y - 6z + 19 = 0$$ **Ans.**

10. *Find the equation of the plane which is parallel to x-axis and passes through the points (2, 3, 1), (4, – 5, 3).*

Sol. The equation of plane parallel to x-axis is

$$by + cz + d = 0 \quad \text{...(i)}$$

As it passes through (2, 3, 1),

$$3b + c + d = 0 \quad \text{...(ii)}$$

As it passes through (4, – 5, 3) also,

$$-5b + 3c + d = 0. \quad \text{...(iii)}$$

From equations (ii) and (iii) by cross-multiplication method we get,

$$\frac{b}{1-3} = \frac{-c}{3+5} = \frac{d}{9+5} = k \text{ (let)}$$

$$\Rightarrow \quad b = -2k, c = -8k, d = 14k$$

Hence, equation (i) becomes

$$-2ky - 8kz + 14k = 0$$

$$\Rightarrow \quad y + 4z - 7 = 0$$ **Ans.**

11. *Show that the four points (0, 4, 3), (–1, –5, –3), (–2, –2, 1) and (1, 1, –1) are coplanar and find the equation of the common plane.*

Sol. Equation of plane passing through (0, 4, 3) is

$$a(x - 0) + b(y - 4) + c(z - 3) = 0 \quad \text{...(i)}$$

As it passes through (– 1, – 5, – 3)

$$\Rightarrow \quad -a - 9b - 6c = 0$$

$$\Rightarrow \quad a + 9b + 6c = 0 \quad \text{...(ii)}$$

The plane passes through (– 2, – 2, 1)

$$\Rightarrow \quad -2a - 6b - 2c = 0$$

$$\Rightarrow \quad a + 3b + c = 0 \quad \text{...(iii)}$$

From equations (ii) and (iii), by cross-multiplication method, we get

$$\frac{a}{9-18} = \frac{-b}{1-6} = \frac{c}{3-9} = k$$

$$\Rightarrow \quad a = -9k, b = 5k, c = -6k$$

Substituting these values in equation (i), we get

$$-9k(x) + 5k(y - 4) - 6k(z - 3) = 0$$

$$\Rightarrow \quad -9x + 5y - 20 - 6z + 18 = 0$$

$$\Rightarrow \quad 9x - 5y + 6z + 2 = 0$$

Let the plane $9x - 5y + 6z + 2 = 0$ passes through the point (1, 1, – 1), to prove this

$$\therefore \quad \text{L.H.S.} = 9 \times 1 - 5 \times 1 + 6(-1) + 2$$
$$= 0 = \text{R.H.S.}$$

∴ Hence, (0, 4, 3), (– 1, – 5, – 3), (– 2, – 2, 1) and (1, 1, – 1) are coplanar. **Hence Proved.**

12. *Find the equation of the plane through A (–1, 1, 1) and B (4, 2, –1) and perpendicular to the plane x – 2y + 2z = 3.*

Sol. Any plane through (– 1, 1, 1) is given by

$$a(x + 1) + b(y - 1) + c(z - 1) = 0 \quad \text{... (i)}$$

Equation (i) passes through (4, 2, – 1)

Substituting in (i), we get,

$$5a + b - 2c = 0 \quad \text{... (ii)}$$

Equation (i) is perpendicular to $x - 2y + 2z = 3$

$$\therefore \quad 1.a - 2.b + 2.c = 0$$

i.e., $$a - 2b + 2c = 0 \quad \text{...(iii)}$$

Solving (ii) and (iii) by method of cross-multiplication, we get,

$$\frac{a}{2-4} = \frac{b}{-2-10} = \frac{c}{-10-1} = k$$

$$a = -2k, b = -12k, c = -11k$$

Substituting these values in (i), we get,

$$-2k(x + 1) - 12k(y - 1) - 11k(z - 1) = 0$$

∴ $2x + 12y + 11z - 21 = 0$ is the required plane. **Ans.**

13. *Find the equation of a plane through the point (–1, –1, 2) and perpendicular to the planes 3x + 2y – 3z = 1 and 5x – 4y + z = 5.*

Sol. Equation of plane is $a(x + 1) + b(y + 1) + c(z - 2) = 0$.

Where a, b, c are direction ratio of normal of the plane, whose equation is to find out.

$$3a + 2b - 3c = 0 \quad \text{...(i)}$$

$$5a - 4b + c = 0 \quad \text{...(ii)}$$

When two planes are perpendicular to each other, their normals will also be perpendicular to each other.

By solving (i) and (ii), by cross-multiplication method, we get

$$\frac{a}{-10} = \frac{b}{-18} = \frac{c}{-22} = k$$

$$\therefore a = -10k, b = -18k, c = -22k$$

By substituting values of a, b, c in equation of plane, we get

$$-10k(x+1) - 18k(y+1) - 22k(z-2) = 0$$
$$\Rightarrow 10(x+1) + 18(y+1) + 22(z-2) = 0$$
$$\Rightarrow 5(x+1) + 9(y+1) + 11(z-2) = 0$$
$$\Rightarrow 5x + 9y + 11z - 8 = 0$$ **Ans.**

14. *Find the equation of a plane per- pendicular to the plane $2x - 3y + 6z + 8 = 0$ and passing through the intersection of $x + 2y + 3z - 4 = 0$ and $2x - y - z + 5 = 0$.*

Sol. Equation of any plane passing through the intersection of given planes is given by

$$(x + 2y + 3z - 4) + \lambda(2x - y - z + 5) = 0$$
$$\Rightarrow (1 + 2\lambda)x + (2 - \lambda)y + (3 - \lambda)z + (5\lambda - 4) = 0 \quad \dots \text{(i)}$$

But this is perpendicular to the plane $2x - 3y + 6z + 8 = 0$ if normal to them are perpendicular to each other

i.e., $$l_1l_2 + m_1m_2 + n_1n_2 = 0$$
$$\Rightarrow 2(1 + 2\lambda) - 3(2 - \lambda) + 6(3 - \lambda) = 0$$
$$\Rightarrow 2 + 4\lambda - 6 + 3\lambda + 18 - 6\lambda = 0$$
$$\Rightarrow \lambda = -14$$

Substituting this value in equation (i), we get,

$$(1 - 28)x + (2 + 14)y + (3 + 14)z + (-70 - 4) = 0$$
$$\Rightarrow -27x + 16y + 17z - 74 = 0$$
$$\Rightarrow 27x - 16y - 17z + 74 = 0$$ **Ans.**

15. *Find the equation of the plane passing through (1, 2, 3) and prependicular to the straight line $\frac{x}{-2} = \frac{y}{4} = \frac{z}{3}$.*

Sol. Direction ratios of the given line are $-2, 4, 3$. Since, the given line is perpendicular to the required plane, therefore, direction ratios of a normal to the plane are proportional to $-2, 4, 3$.

Hence, the equation of the plane passing through the point (1, 2, 3) and having the given line as normal is

$$-2(x - 1) + 4(y - 2) + 3(z - 3) = 0$$
$$\Rightarrow 2x - 4y - 3z + 15 = 0$$ **Ans.**

16. *Find the equation of the plane which contains the line $\frac{x-1}{2} = \frac{y+1}{-1} = \frac{z-3}{4}$ and is perpendicular to the plane $x + 2y + z = 12$.*

Sol. The equation of plane passing through $(1, -1, 3)$ is $A(x-1) + B(y+1) + C(z-3) = 0$ which is perpendicular to the given palne.

$\therefore \quad 2A - B + 4C = 0$ and $A + 2B + C = 0$

Eliminating A, B, C, from the above equations, we get

$$\begin{vmatrix} x-1 & y+1 & z-3 \\ 2 & -1 & 4 \\ 1 & 2 & 1 \end{vmatrix} = 0$$

$$\Rightarrow (x-1)(-1-8) - (y+1)(2-4) + (z-3)(4+1) = 0$$
$$\Rightarrow (-9)(x-1) - (-2)(y+1) + 5(z-3) = 0$$
$$\Rightarrow -9x + 9 + 2y + 2 + 5z - 15 = 0$$
$$\Rightarrow -9x + 2y + 5z - 4 = 0$$
$$\Rightarrow 9x - 2y - 5z + 4 = 0$$ **Ans.**

17. *Find the angle between the line*

$$\frac{x-6}{3} = \frac{y-7}{2} = \frac{z-7}{-2}$$

and the plane $x + y + 2z = 0$.

Sol. Given line

$$\frac{x-6}{3} = \frac{y-7}{2} = \frac{z-7}{-2}$$

and the plane $x + y + 2z = 0$.

Let l_1, m_1 and n_1 be the direction cosines of normal to the plane, hence

$$\therefore \quad l_1 = \frac{1}{\sqrt{6}}, \ m^1 = \frac{1}{\sqrt{6}}, \ n^1 = \frac{2}{\sqrt{6}}.$$

Let l_2, m_2 and n_2 be the direction cosines of the line given

$$\therefore \quad l^2 = \frac{3}{\sqrt{17}}, \ m^2 = \frac{2}{\sqrt{17}}, \ n^2 = \frac{-2}{\sqrt{17}}.$$

Let 'θ' be the angle between normal of the plane and line, then

$$\cos\theta = l_1l_2 + m_1m_2 + n_1n_2$$
$$\Rightarrow \cos(90° - \theta)$$
$$= \frac{1}{\sqrt{6}} \times \frac{3}{\sqrt{17}} + \frac{1}{\sqrt{6}} \times \frac{2}{\sqrt{17}} + \frac{2}{\sqrt{6}} \times \left(\frac{-2}{\sqrt{17}}\right)$$
$$= \frac{3+2-4}{\sqrt{6} \times \sqrt{17}} = \frac{1}{\sqrt{17}} \times \frac{1}{\sqrt{6}}$$

∴ Angle between line and plane

$$\theta = 90° - \cos^{-1}\frac{1}{\sqrt{6}.\sqrt{17}}$$ **Ans.**

18. *A plane passes through the point (4, 2, 4) and is perpendicular to the planes $2x + 5y + 4z + 1 = 0$ and $4x + 7y + 6z + 2 = 0$. Find the equation of the plane.*

Sol. Let the required plane be

$$ax + by + cz + d = 0 \quad \dots \text{(i)}$$

Since, it passes through the point (4, 2, 4)

$$\therefore \quad 4a + 2b + 4c + d = 0 \quad \dots \text{(ii)}$$

Since, the plane is perpendicular to the planes

$$2x + 5y + 4z + 1 = 0$$

and $$4x + 7y + 6z + 2 = 0$$

Therefore, the normal to the plane (i) is perpendicular to the normal of these two planes.

$$2a + 5b + 4c + 1 = 0 \quad \text{...(iii)}$$

$$4a + 7b + 6c + 2 = 0 \quad \text{...(iv)}$$

From (iii) and (iv), we have,

$$\frac{a}{2} = \frac{b}{4} = \frac{c}{-6} = k$$

$\Rightarrow$ $a = 2k, b = 4k, c = -6k$

$\therefore$ We get, $8k + 8k - 24k + d = 0$ [From (ii)]

$\Rightarrow$ $d = 8k$

Substituting these value in (i), we get,

$$2kx + 4ky - 6kz + 8k = 0$$

$\Rightarrow$ $x + 2y - 3z + 4 = 0$ **Ans.**

19. *Find the equation of the plane through the intersection of the planes.*

$\vec{r}\cdot(\hat{i} + 3\hat{j} - \hat{k}) = 9$ and $\vec{r}\cdot(2\hat{i} - \hat{j} + \hat{k}) = 3$ passing through the origin.

Sol. Equation of the line passing through the intersection of given plane is,

$$\vec{r}\cdot(\hat{i} + 3\hat{j} - \hat{k}) - 9 + \lambda[\vec{r}\cdot(2\hat{i} - \hat{j} + \hat{k}) - 3] = 0$$

$$\vec{r}\cdot[(1+2\lambda)\hat{i} + \hat{j}(3-\lambda) + \hat{k}(-1+\lambda)] - 9 - 3\lambda = 0$$

$\because$ Plane passes throught origin (0, 0, 0), then

vector equation = $0\cdot\hat{i} + 0\cdot\hat{j} + 0\cdot\hat{k}$

It satisfies the equation of plane,

$\therefore$ $-9 - 3\lambda = 0$

$\Rightarrow$ $3\lambda = -9$

$\Rightarrow$ $\lambda = -3$

So, equation of required plane

$$\vec{r}(-5\hat{i} + 6\hat{j} - 4\hat{k}) = 0$$ **Ans.**

Chapter 13. Application of Integrals

1. *Draw a rough sketch of the curve $y = x^2 - 5x + 6$ and find the area bounded by the curve and the x-axis.*

Sol. Given curve is $y = x^2 - 5x + 6$

$\Rightarrow$ $y = (x - 3)(x - 2)$

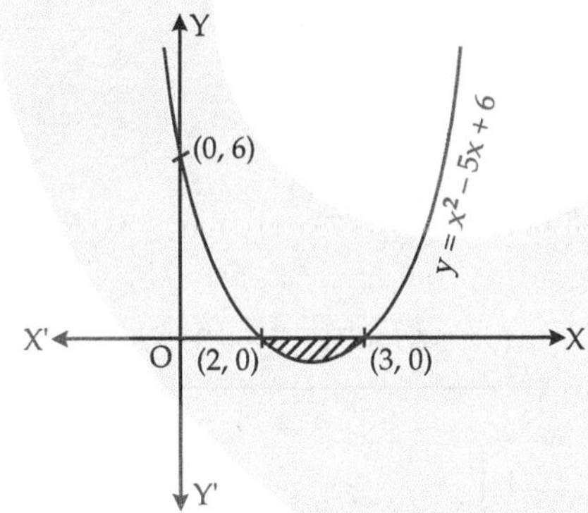

(i) Curve does not pass through the origin.

(ii) The curve cuts the x-axis at (2, 0) and (3, 0)

(iii) The curve cuts the y-axis at (0, 6).

The area bounded by the curve and the x-axis is given by

$$\int_2^3 y\,dx = \int_2^3 (x^2 - 5x + 6)\,dx$$

$$= \left[\frac{x^3}{3} - \frac{5x^2}{2} + 6x\right]_2^3$$

$$= \left[\frac{27}{3} - 5\times\frac{9}{2} + 6\times 3\right] - \left[\frac{8}{3} - 5\times\frac{4}{2} + 6\times 2\right]$$

$$= 9 - \frac{45}{2} + 18 - \frac{8}{3} + 10 - 12$$

$$= 25 - \frac{151}{6} = -\frac{1}{6} \text{ sq. units.}$$

But area cannot be negative.

$\therefore$ Area = 1 sq. units. **Ans.**

2. *Draw a rough sketch of the curve $x^2 + y = 9$ and find the area enclosed by the curve, the x-axis and the lines $x + 1 = 0$ and $x - 2 = 0$.*

Sol. The given equation of the curve is

$$x^2 + y = 9$$

$\Rightarrow$ $$x^2 = -(y - 9)$$

The graphical representation of the curve and the lines is given as,

Since, the curve intersects the y-axis at the point (0, 9)

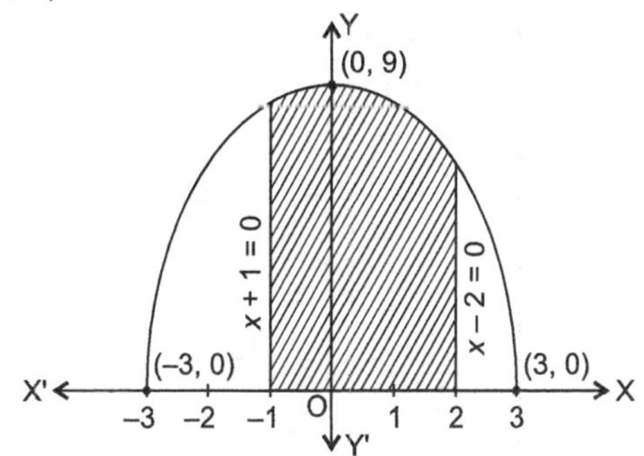

$\therefore$ Required Area $= \int_{-1}^{2} (9 - x^2)\, dx$

$= \left[9x - \frac{x^3}{3}\right]_{-1}^{2}$

$= \left[\left(18 - \frac{8}{3}\right) - \left(-9 + \frac{1}{3}\right)\right]$

$= 27 - 3$

$= 24$ sq. units. **Ans.**

3. *Draw a rough sketch of the curve $y^2 + 1 = x$, $x \le 2$. Find the area enclosed by the curve and the line $x = 2$.*

Sol. The equation of the given curve is

$$y^2 + 1 = x, \quad x \le 2$$

$$\Rightarrow \quad y^2 = x - 1, x \le 2$$

The graphical respresentation of the given curve and line is given as,

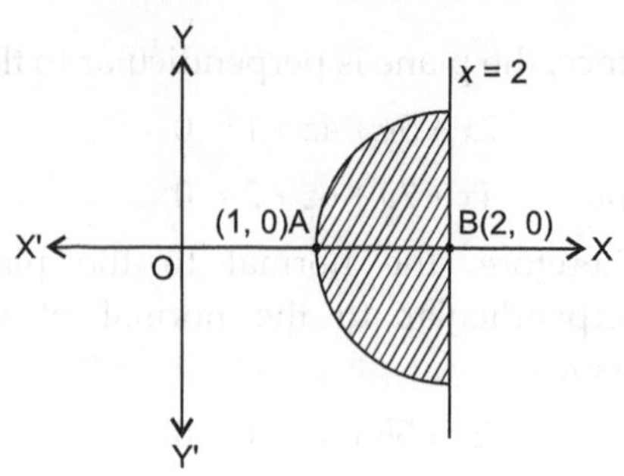

The Required Area $= 2\int_{1}^{2} \sqrt{x-1}\, dx$

$= \frac{[2(x-1)^{3/2}]_1^2}{3/2}$

$= 2 \times \frac{2}{3} \times [1^{3/2}]$

$= \frac{4}{3}$ sq. units. **Ans**

Chapter 14. Application of Calculus

1. *Given that the total cost function for x units of a commodity is*

$$C(x) = \frac{x^3}{3} + 3x^2 - 7x + 16$$

(i) Find the Marginal Cost (MC)

(ii) Find the Average Cost (AC)

(iii) Prove that

Marginal Average Cost, (MAC) $= \frac{x(MC) - C(x)}{x^2}$.

Sol. Given,

$$C(x) = \frac{x^3}{3} + 3x^2 - 7x + 16$$

(i) Marginal cost, (M.C.)

$= \frac{d}{dx}\left[\frac{x^3}{3} + 3x^2 - 7x + 16\right]$

$= x^2 + 6x - 7$ **Ans.**

(ii) Average cost,

(AC) $= \frac{\text{Total cost}}{\text{Output}}$

$= \frac{\frac{x^3}{3} + 3x^2 - 7x + 16}{x}$

$= \frac{x^2}{3} + 3x - 7 + \frac{16}{x}$ **Ans.**

(iii) Marginal Average cost, (MAC)

$= \frac{d}{dx}$ [A.C.]

$= \frac{d}{dx}\left[\frac{x^2}{3} + 3x - 7 + \frac{16}{x}\right]$

$= \frac{2x}{3} + 3 - \frac{16}{x^2}$

Again, $\frac{x(\text{MC}) - \text{C}(x)}{x^2}$

$= \frac{x(x^2 + 6x - 7) - \frac{x^3}{3} - 3x^2 + 7x - 16}{x^2}$

$= \frac{1}{x^2}\left[x^3 + 6x^2 - 7x - \frac{x^3}{3} - 3x^2 + 7x - 16\right]$

$= \frac{1}{x^2}\left[x^3 - \frac{x^3}{3} + 3x^2 - 16\right]$

$= \frac{2x^3}{3x^2} + 3 - \frac{16}{x^2}$

$= \frac{2}{3}x + 3 - \frac{16}{x^2}$

$\Rightarrow$ Marginal average cost (MAC)

$= \frac{x(MC) - C(x)}{x^2}$. **Hence Proved.**

Computational Questions

2. *The cost function C (x) of a firm is given by C (x) = $3x^2 - 6x + 5$. Find*

(i) the average cost and

(ii) the marginal cost when x = 2.

Sol. Given, $C = C(x) = 3x^2 - 6x + 5$

(i) Average Cost (*AC*)

$$= \frac{C}{x}$$

$$= \frac{1}{x}(3x^2 - 6x + 5)$$

$$= 3x - 6 + \frac{5}{x}$$

$\therefore$ when $x = 2$,

$$AC = 3 \times 2 - 6 + \frac{5}{2}$$

$$= 6 - 6 + \frac{5}{2}$$

$$= \frac{5}{2} = 2{\cdot}5$$

(ii) Marginal Cost (*MC*)

$$= \frac{dC}{dx}$$

$$= \frac{d}{dx}(3x^2 - 6x + 5)$$

$$= 6x - 6$$

$\therefore$ when $x = 2, MC = 6 \times 2 - 6$

$$= 12 - 6 = ₹\,6$$ **Ans.**

3. *The total cost C (x) of a firm is given by C (x) = $0{\cdot}005x^3 - 0{\cdot}2x^2 - 30x + 2000$, where x is the output. Determine*

(i) the average cost

(ii) the marginal average cost

(iii) the marginal cost.

Sol. (i) Given, total cost is

$$C(x) = 0{\cdot}005x^3 - 0{\cdot}2x^2 - 30x + 2000$$

$$\Rightarrow AC = 0{\cdot}005x^2 - 0{\cdot}2x - 30 + \frac{2000}{x}$$

(ii) The marginal average cost is

$$MAC = \frac{d}{dx}(AC)$$

$$= \frac{d}{dx}\left[0{\cdot}005x^2 - 0{\cdot}2x - 30 + \frac{2000}{x}\right]$$

$$= 2 \times 0{\cdot}005x - 0{\cdot}2 + 2000\left(-\frac{1}{x^2}\right)$$

$$= 0{\cdot}01x - 0{\cdot}2 - \frac{2000}{x^2}$$

(iii) The marginal cost is,

$$MC = \frac{dC}{dx}$$

$$= \frac{d}{dx}[0{\cdot}005x^3 - 0{\cdot}2x^2 - 30x + 2000]$$

$$= 3 \times 0{\cdot}005x^2 - 2 \times 0{\cdot}2x - 30$$

$$= 0{\cdot}015x^2 - 0{\cdot}4x - 30$$ **Ans.**

4. *The average cost function AC for a commodity is given by AC = $x + 5 + \frac{36}{x}$ in terms of output x. Find the*

(i) Total cost and the marginal cost as the functions of x.

(ii) Output for which AC increases.

Sol. (i) Let C be the total cost function. Then,

$$\text{Average Cost } (AC) = \frac{C}{x}$$

$$\Rightarrow \quad C = AC.x$$

$$\Rightarrow \quad C = \left(x + 5 + \frac{36}{x}\right)x$$

$$\Rightarrow \quad C = x^2 + 5x + 36$$

Let *MC* be the marginal cost function, then

$$MC = \frac{dC}{dx}$$

$$= \frac{d}{dx}(x^2 + 5x + 36)$$

$$MC = 2x + 5$$ **Ans.**

(ii) Differentiating *AC* w.r.t. *x*, we get

$$\text{Now,} \quad \frac{d}{dx}(AC) = \frac{d}{dx}\left(x + 5 + \frac{36}{x}\right)$$

$$= 1 - \frac{36}{x^2}$$

For *AC* to be increasing $\frac{d}{dx}(AC) > 0$

$$\Rightarrow \quad 1 - \frac{36}{x^2} > 0$$

$$\Rightarrow \quad x^2 - 36 > 0$$

$$\Rightarrow \quad (x - 6)(x + 6) > 0$$

$$\Rightarrow \quad x < -6 \text{ or } x > 6$$

$$\Rightarrow \quad x > 6 \qquad [\because x > 0]$$

Hence, the average cost increases, if the output $x > 6$. **Ans.**

5. *A company produces a commodity with ₹ 24,000 fixed cost. The variable cost is estimated to be 25% of the total revenue recovered on selling the product at a rate of ₹ 8 per unit. Find the following :*

(i) Cost function

(ii) Revenue function

(iii) Breakeven point.

Sol. Suppose that x number of units are produced and sold.

(i) As each units variable cost is 25% of revenue.

$\therefore$ The variable cost of x units

$$= 25\% \text{ of } ₹\, 8x$$
$$= ₹\, 2x$$

$\therefore$ Total cost of producing x units

$$C(x) = \text{Total fixed cost} + \text{Total variable cost}$$
$$= ₹\,(24{,}000 + 2x) \qquad \textbf{Ans.}$$

(ii) Price of one unit = ₹ 8

$\therefore$ Total revenue on selling x units

$$= R(x) = ₹\, 8x \qquad \textbf{Ans.}$$

(iii) At breakeven values

$$C(x) = R(x)$$
$$24{,}000 + 2x = 8x$$
$$24{,}000 = 6x$$
$$x = 4{,}000$$

Hence, the breakeven point is 4,000 units.

Ans.

6. *A firm has the cost function* $C = \frac{x^3}{3} - 7x^2 + 111x + 50$ *and demand function* $x = 100 - p$.

(i) *Write the total revenue function in terms of x.*

(ii) *Formulate the total profit function P in terms of x.*

(iii) *Find the profit maximising level of output x.*

Sol. Cost function is

$$C(x) = \frac{x^3}{3} - 7x^2 + 111x + 50$$

Demand function is

$$x = 100 - p$$
$$\Rightarrow \quad p = 100 - x$$

(i) Revenue function,

$$R(x) = p.x = x(100 - x)$$
$$= 100x - x^2 \qquad \ldots\text{(i)}$$

(ii) Profit function,

$$P(x) = \text{Revenue} - \text{Cost}$$
$$= R(x) - C(x)$$
$$= 100x - x^2 - \frac{x^3}{3} + 7x^2 - 111x - 50$$
$$= -\frac{x^3}{3} + 6x^2 - 11x - 50 \qquad \ldots\text{(ii)}$$

(iii) Differentiating equation (ii) w.r.t. x, we get

$$\frac{dP}{dx} = -x^2 + 12x - 11 \qquad \ldots\text{(iii)}$$

Now, $\frac{dP}{dx} = 0$

$$\Rightarrow \quad -x^2 + 12x - 11 = 0$$
$$\Rightarrow \quad x^2 - 12x + 11 = 0$$
$$\Rightarrow \quad x^2 - 11x - x + 11 = 0$$
$$\Rightarrow \quad x(x - 11) - 1(x - 11) = 0$$
$$\Rightarrow \quad (x - 1)(x - 11) = 0$$
$$\Rightarrow \quad x = 1, 11$$

Again differentiating, we get

$$\frac{d^2P}{dx^2} = 12 - 2x \qquad \ldots\text{(iv)}$$

At $x = 1$, $\frac{d^2P}{dx^2} = 10$

$$\Rightarrow \quad \frac{d^2P}{dx^2} > 0 \text{ (Minimum value)}$$

At $x = 11$, $\frac{d^2P}{dx^2} = -10$

$$\Rightarrow \quad \frac{d^2P}{dx^2} < 0 \qquad \text{(Maximum value)}$$

Hence, the profit maximising level of output is $x = 11$. **Ans.**

7. *A company is selling a certain product. The demand function of the product is linear. The company can sell 2000 units when the price is ₹ 8 per unit and when the price is ₹ 4 per unit, it can sell 3000 units. Determine*

(i) *The demand function.*

(ii) *The total revenue function.*

Sol. Given, the demand function is linear

$$\because \quad P(x) = ax + b$$

In I[st] case, $x = 8$, $P(x) = 2000$

$$\therefore \quad 2000 = 8a + b \qquad \ldots\text{(i)}$$

In IInd case $x = 4$, $P(x) = 3000$

$$3000 = 4a + b \qquad \ldots\text{(ii)}$$

Solving equation (i) and equation (ii), we get,

$$b = 4000, a = -250$$

$\therefore$ **(i)** Demand function = $4000 - 250x$ **Ans.**

(ii) Total revenue = $x(4000 - 250x)$

$$= 4000x - 250x^2 \qquad \textbf{Ans.}$$

8. *The demand function for a commodity is given by* $P = ae^{-x/300}$, *where P is the price per unit. Given that the price is ₹ 7 per unit when 600 units of product are produced. Find the total average and marginal revenue function. Also, find the price unit when the marginal revenue is zero.*

Sol. Given,

$$P = ae^{-x/300} \text{ (demand function)} \quad ...(i)$$

When $P = 7, x = 600$

$\therefore$ $7 = ae^{\frac{600}{-300}}$

$\Rightarrow$ $7 = ae^{-2}$

$\Rightarrow$ $a = 7e^2$

Substituting in (i), we get

$$P = 7e^2 . e^{-x/300}$$

$$\Rightarrow P = 7e^{\frac{(600-x)}{300}} \quad ...(ii)$$

Now, total revenue function

$$TR = Px = 7\,x\,e^{(600-x)/300}$$ **Ans.**

Marginal revenue function

$$MR = \frac{d}{dx}(TR) = \frac{d}{dx}\left(7.x.e^{\frac{600-x}{300}}\right)$$

$$= 7\left[e^{\frac{600-x}{300}} + xe^{\frac{600-x}{300}} \times \frac{-1}{300}\right]$$

$$= 7e^{\frac{600-x}{300}}\left(1 - \frac{x}{300}\right) = \frac{7}{300}(300-x)\,e^{\frac{600-x}{300}}$$ **Ans.**

when, $MR = 0$, we have $300 - x = 0$

$\Rightarrow$ $x = 300$

Substituting in (ii), we get, $P = 7e^{(600-300)\,300} = 7e$.

Hence, the price per unit = ₹ $7e$. **Ans.**

9. *The demand function of a monopolist is given by $P = 1500 - 2x - x^2$. Find the marginal revenue for any level of output x. Also find marginal revenue (MR), when $x = 10$.*

Sol. Let R be the revenue function, then

$$R = Px$$
$$= (1500 - 2x - x^2)x$$
$$= 1500x - 2x^2 - x^3$$

$$\therefore \quad MR = \frac{dR}{dx}$$

$$\Rightarrow \quad MR = \frac{d}{dx}(1500x - 2x^2 - x^3)$$
$$= 1500 - 4x - 3x^2$$

$$\therefore \quad (MR)_{x=10} = 1500 - 4 \times 10 - 3 \times 10^2$$
$$= 1500 - 40 - 300$$
$$= ₹1160$$ **Ans.**

10. *The demand for a certain product is represented by the equation $p = 500 + 25x - \frac{x^2}{3}$ in rupees where x is the number of units and p is the price per unit. Find :*

(i) Marginal revenue function.

(ii) The marginal revenue when 10 units are sold.

Sol. We have,

$$p = 500 + 25\,x - \frac{x^2}{3}$$

(i) Now, Total Revenue function (R) = px

$$R = 500\,x + 25x^2 - \frac{x^2}{3}$$

$\therefore$ Marginal Revenue function (MR) = $\frac{dR}{dx}$

$$= 500 + 50\,x - \frac{2x}{3}$$ **Ans.**

(ii) Marginal Revenue when 10 units are sold *i.e.*, when $x = 10$,

$$(MR)_{10} = 500 + 50\,(10) - \frac{2 \times 10}{3}$$
$$= 500 + 500 - \frac{20}{3}$$
$$= 1000 - \frac{20}{3}$$
$$= \frac{3000 - 20}{3} = \frac{2080}{3}$$
$$= ₹\,693{\cdot}34$$ **Ans.**

11. *In a factory, it is found that the number of units (x) produced in a day depends upon the number of workers (n) and is obtained by the relation, $x = \frac{5n}{\sqrt{n+5}}$. The demand function of the product is $P = \frac{2}{x} + x$.*

Determine the marginal revenue, when $n = 20$.

Sol. Given, $P = \frac{2}{x} + x$

$\therefore$ Total revenue (R) = Px

$$= \left(\frac{2}{x} + x\right)x$$
$$= 2 + x^2$$

$\therefore$ Marginal Revenue (MR)

$$- \frac{dR}{dx} - \frac{d}{dx}(2 + x^2)$$
$$= 2x \quad ...(i)$$

Also, given that $x = \frac{5n}{\sqrt{n+5}}$

when $n = 20$, $x = \frac{5 \times 20}{\sqrt{20+5}}$

$$= \frac{5 \times 20}{5} = 20$$

Hence, from (i), when $n = 20$

$$MR = 2 \times 20 = ₹40$$ **Ans.**

12. *$C(x) = 5x + 350$ and $R(x) = 50x - x^2$, are respectively the total cost and the total revenue functions for a company that produces and sells x units of a particular product.*

*Find **(i)** the breakeven values, **(ii)** the values of x that produce a profit, **(iii)** the values of x that result in a loss.*

Sol. (i) For breakeven value,

$$C(x) = R(x)$$

$\therefore \quad 5x + 350 = 50x - x^2$

$\Rightarrow \quad x^2 - 45x + 350 = 0$

Solving,

$x^2 - 35x - 10x + 350 = 0$

$\Rightarrow x(x - 35) - 10(x - 35) = 0$

$\Rightarrow \quad x = 10, 35$

$\therefore$ For breakeven the company can produce and sell 35 units or 10 units.

(ii) The values of x that will produce profit are given by $R(x) > C(x)$

$\Rightarrow \quad 50x - x^2 > 5x + 350$

$\Rightarrow \quad 45x - x^2 - 350 > 0$

$\Rightarrow \quad x^2 - 45x + 350 < 0$

$\Rightarrow \quad (x - 10)(x - 35) < 0$

By the method of intervals, we find that, this inequality is satisfied only when x lies between 10 and 35.

i.e., $10 < x < 35$. Hence for profit, the number of units produced should lie between 10 and 35.

(iii) The values of x that will result in loss given by $C(x) > R(x)$

i.e., $5x + 350 > 50x - x^2$

$\Rightarrow \quad x^2 - 45x + 350 > 0$

$\Rightarrow \quad (x - 10)(x - 35) > 0$

This inequality is satisfied when

$x < 10$ or $x > 35$. **Ans.**

13. *Find the relationship between the slopes of marginal revenue curve and the average revenue curve, for the demand function, $x = \frac{b - P}{a}$ where x denotes the numbers of units sold at the price P per unit.*

Sol. Given, $$x = \frac{b - P}{a}$$

$\Rightarrow \quad P = b - ax$

Total Revenue, $R = Px = (b - ax)x$

$= bx - ax^2$

Average Revenue, $(AR) = \frac{R}{x} = \frac{bx - ax^2}{x}$

$= b - ax$

$\therefore$ Slope of (AR) Curve $= \frac{d}{dx}(AR)$

$= \frac{d}{dx}(b - ax)$

$= -a$ ...(i)

Now, Marginal Revenue (MR)

$= \frac{dR}{dx} = \frac{d}{dx}(bx - ax^2)$

$= b - 2ax$

$\therefore$ Slope of (MR) Curve $= \frac{d}{dx}(b - 2ax)$

$= -2a$ ...(ii)

From (i) and (ii), we find that slope of (MR) curve is twice the slope of (AR) curve. **Ans.**

14. *The demand function for a manufacturers, product is, $P = \frac{180 - x}{4}$ where x is the number of units and P is the price per unit. At what value of x will there be maximum revenue ? What is the maximum revenue ?*

Sol. Given, $$P = \frac{180 - x}{4}$$

Revenue, $R = Px$

$$= \frac{x(180 - x)}{4}$$

$$= \frac{180x - x^2}{4} \quad ...(i)$$

Differentiating w.r.t. x, we get

$$\therefore \quad \frac{dR}{dx} = \frac{1}{4}(180 - 2x) \quad ...(ii)$$

For maximum revenue,

$$\frac{dR}{dx} = 0$$

$\therefore \quad \frac{1}{4}(180 - 2x) = 0$

$\Rightarrow \quad 180 - 2x = 0$

$\Rightarrow \quad 2x = 180$

$\Rightarrow \quad x = 90$

Differentiating (ii) w.r.t. x,

$$\therefore \quad \frac{d^2R}{dx^2} = \frac{1}{4} \times (-2) = -\frac{1}{2} < 0.$$

$\therefore$ R is maximum for $x = 90$.

Substituting in (i), we get,

$$R = \frac{180 \times 90 - 90^2}{4}$$

$$= \frac{16200 - 8100}{4}$$

$$= \frac{8100}{4} = ₹\,2025$$

Hence, the maximum revenue is ₹ 2025. **Ans.**

15. *The demand function is* $x = \dfrac{24-2p}{3}$ *where x is the number of units demanded and p is the price per unit. Find :*

(i) The revenue function R in terms of p.

(ii) The price and the number of units demanded for which the revenue is maximum.

Sol. Given, $x = \dfrac{24-2p}{3}$

(i) Revenue function, $R(p) = px$

$$= p\left(\frac{24-2p}{3}\right)$$

$$= 8p - \frac{2}{3}p^2$$

(ii) For maximum revenue, $\dfrac{dR(p)}{dp}$

$$= 8 - \frac{4}{3}p = 0$$

$$\Rightarrow \quad p = 6$$

and $$\frac{d^2R(p)}{dp^2} = -\frac{4}{3} \ (< 0)$$

$\Rightarrow$ R (p) is maximum at $p = 6$

For maximum revenue, price per unit is 6

$$\therefore \quad x = \frac{24 - 2 \times 6}{3}$$

$$= \frac{24-12}{3} = \frac{12}{3}$$

$$= 4$$

$\therefore$ Price is 6 and number of units demanded is 4 for which the revenue is maximum. **Ans.**

Chapter 15. Linear Regression

1. *The coefficient of correlation between the values denoted by X and Y is 0·5. The mean of X is 3 and that of Y is 5. Their standard deviations are 5 and 4 respectively. Find :*

(i) the two lines of regression.

(ii) the expected value of Y, when X is given 14.

(iii) the expected value of X, when Y is given 9.

Sol. Given, $\bar{x} = 3$, $\bar{y} = 5$, $\sigma_x = 5$, $\sigma_y = 4$, $r = 0·5$

$$\therefore \quad b_{yx} = r \cdot \frac{\sigma_y}{\sigma_x} = 0·5 \times \frac{4}{5} = 0·4,$$

$$b_{xy} = r \cdot \frac{\sigma_x}{\sigma_y} = 0·5 \times \frac{5}{4} = 0·625$$

(i) Regression equation of y on x is

$$y - \bar{y} = b_{yx}(x - \bar{x})$$

$$\Rightarrow \quad y - 5 = (0·4)(x-3)$$

$$\Rightarrow \quad y - 5 = 0·4x - 1·2$$

$$\Rightarrow \quad y = 0·4x + 3·8$$

and regression equation of x on y is

$$x - \bar{x} = b_{xy}(y - \bar{y})$$

$$\Rightarrow \quad x - 3 = (0·625)(y-5)$$

$$\Rightarrow \quad x - 3 = 0·625y - 3·125$$

$$\Rightarrow \quad x = 0·625\,y - 0·125$$ **Ans.**

(ii) Putting, $x = 14$ in line of regression of y on x

$$y = 0·4 \times 14 + 3·8$$

$$= 5·6 + 3·8 = 9·4$$ **Ans.**

(iii) Putting, $y = 9$ in line of regression of x on y

$$x = 0·625 \times 9 - 0·125$$

$$= 5·625 - 0·125 = 5·5$$ **Ans.**

2. *For the given lines of regression,* $3x - 2y = 5$ *and* $x - 4y = 7$, *find :*

(a) regression coefficients b_{yx} *and* b_{xy}.

(b) coefficient of correlation $r(x, y)$.

Sol.

(a) Given, the lines of regression are

$$3x - 2y = 5 \quad \ldots(i)$$

$$x - 4y = 7 \quad \ldots(ii)$$

Let equation (i) be the regression line of y on x. Then, line (ii) is the regression line of x on y. Let us check whether our assumption is correct or not.

Rewriting equations (i) and (ii)

$$y = \frac{3x}{2} - \frac{5}{2}$$

and $x = 4y + 7$

which gives

$$b_{yx} = \frac{3}{2} \text{ and } b_{xy} = 4$$

$$\therefore b_{yx} \cdot b_{xy} = \frac{3}{2} \times 4 = 6 > 1 \quad \text{(not possible)}$$

So, our assumption is wrong and equation (i) is the regression line of x on y and (ii) is the regression line of y on x.

Now, writing equation (i) as

$$x = \frac{2}{3}y + \frac{5}{3}, \text{ we get}$$

$$b_{xy} = \frac{2}{3}$$

and writing (ii) as $y = \frac{x}{4} - \frac{7}{4}$, we get

$$b_{yx} = \frac{1}{4}$$

(b) Coefficient of correlation (r)

$$= \pm\sqrt{b_{xy} \cdot b_{yx}}$$

$$= \pm\sqrt{\frac{1}{4} \times \frac{2}{3}}$$

$$= \pm\sqrt{\frac{1}{6}} = \pm\frac{\sqrt{6}}{6}$$

But b_{xy} and b_{yx} being both ± ve.

$\therefore$ r is also + ve.

$$\therefore \quad r = \frac{\sqrt{6}}{6}$$ **Ans.**

3. *From the equations of the two regression lines, $4x + 3y + 7 = 0$ and $3x + 4y + 8 = 0$, find :*

(a) Mean of x and y.

(b) Regression coefficients.

(c) Coefficient of correlation.

Sol. (a) The equation of the two lines of regression are

$$4x + 3y + 7 = 0 \quad \ldots\text{(i)}$$

$$3x + 4y + 8 = 0 \quad \ldots\text{(ii)}$$

Solving (i) and (ii) simultaneously, we get

$$x = -4/7 \text{ and } y = -11/7$$

We know that the regression lines intersect at $(\bar{x}, \bar{y})$, then

$$\Rightarrow \quad \bar{x} = -4/7 \text{ and } \bar{y} = -11/7$$ **Ans.**

(b) Let equation (i) be the regression equation of y on x and (ii) the regression equation of x on y then from (i)

$$y = -\frac{4}{3}x - \frac{7}{3}$$

$$\Rightarrow \quad b_{yx} = -\frac{4}{3}$$

and from (ii) $$x = -\frac{4}{3}y - \frac{8}{3}$$

$$\Rightarrow \quad b_{xy} = -\frac{4}{3}$$

But $r^2 = b_{yx} \cdot b_{xy} = \left(-\frac{4}{3}\right) \times \left(-\frac{4}{3}\right) > 1$

which is impossible. Hence, equation (i) is the regression line of x on y and equation (ii) is the regression equation of y on x. From (i)

$$x = -\frac{3}{4}y - \frac{7}{4}$$

$$\therefore \quad b_{xy} = -\frac{3}{4}$$

From (ii) $$y = -\frac{3}{4}x - 2$$

$$\therefore \quad b_{yx} = -\frac{3}{4}$$ **Ans.**

(c) Coefficient of correlation,

$$|r| = \sqrt{b_{yx} \cdot b_{xy}} = \sqrt{-\frac{3}{4} \times -\frac{3}{4}}$$

$$\Rightarrow \quad r = \pm\frac{3}{4}$$

But r has the same sign as regression coefficients

$$\therefore \quad r = -\frac{3}{4}$$ **Ans.**

4. *The following results were obtained with respect to two variable x and y : $\Sigma x = 30$, $\Sigma y = 42$, $\Sigma xy = 199$, $\Sigma x^2 = 184$, $\Sigma y^2 = 318$, $n = 6$. Find the following :*

(i) *The regression coefficients.*

(ii) *Correlation coefficient between x and y.*

(iii) *Regression equation of y on x.*

(iv) *The likely value of y when x = 10.*

Sol. (i) Here $n = 6$,

$$\bar{x} = \frac{\Sigma x}{6}$$

$$= \frac{30}{6} = 5$$

Computational Questions

$$\bar{y} = \frac{\Sigma y}{6} = \frac{42}{6} = 7$$

The regression coefficients are given as,

$$b_{yx} = \frac{\Sigma xy - \frac{1}{n}\Sigma x \cdot \Sigma y}{\Sigma x^2 - \frac{1}{n}(\Sigma x)^2}$$

$$= \frac{199 - \frac{1}{6}(30)(42)}{184 - \frac{1}{6}(30)^2}$$

$$= \frac{199 - 210}{184 - 150}$$

$$= \frac{-11}{34} = -0{\cdot}323$$

and $$b_{xy} = \frac{\Sigma xy - \frac{1}{n}\Sigma x\, \Sigma y}{\Sigma y^2 - \frac{1}{n}(\Sigma y)^2}$$

$$= \frac{199 - \frac{1}{6}(30)(42)}{318 - \frac{1}{6}(42)^2}$$

$$= \frac{199 - 210}{318 - 294}$$

$$= \frac{-11}{24}$$

$$= -0{\cdot}458$$ **Ans.**

(ii) Correlation coefficient between x and y,

$$r^2 = b_{yx} \cdot b_{xy}$$

$$= (-0{\cdot}323) \times (-0{\cdot}458)$$

$$= 0{\cdot}1479$$

$$r = \sqrt{0{\cdot}1479}$$

Since b_{xy}, b_{yx} and r have the same sign.

$$\therefore \quad r = -0{\cdot}384$$ **Ans.**

(iii) Now, regression line of y on x is given as

$$y - \bar{y} = b_{yx}(x - \bar{x})$$

$$\Rightarrow \quad y - 7 = \frac{-11}{34}(x - 5)$$

$$\Rightarrow \quad 34y - 238 = -11x + 55$$

$$\Rightarrow \quad 11x + 34y - 293 = 0$$ **Ans.**

(iv) Putting, $x = 10$ in regression equation of y on x.

$$\therefore \quad 11 \times 10 + 34y - 293 = 0$$

$$\Rightarrow \quad 34y = 183$$

$$\Rightarrow \quad y = \frac{183}{34}$$

$$\Rightarrow \quad y = 5{\cdot}382$$ **Ans.**

5. *Two regression lines are represented by $4x + 10y = 9$ and $6x + 3y = 4$. Find the line of regression of y on x.*

Sol. Assuming $4x + 10y = 9$ to be regression line of y on x.

$$10y = -4x + 9$$

$$\Rightarrow \quad y = -\frac{4}{10}x + \frac{9}{10}$$

$$\Rightarrow \quad b_{yx} = \frac{-2}{5}$$

Then, $6x + 3y = 4$ will be the regression line of x on y.

$$\therefore \quad 6x = -3y + 4$$

$$\Rightarrow \quad x = -\frac{3}{6}y + \frac{4}{6}$$

$$\Rightarrow \quad b_{xy} = -\frac{1}{2}$$

Now, $$b_{yx} \times b_{xy} = -\frac{2}{5} \times -\frac{1}{2} = \frac{1}{5} < 1$$

which is true. Hence, our assumption is correct, *i.e.,* the regression line of y on x is $4x + 10y = 9$.

Ans.

6. *Two regression lines are represented by $2x + 3y - 10 = 0$ and $4x + y - 5 = 0$.*

Find the line of regression of y on x.

Sol. Let the line of regression of y on x is

$$2x + 3y - 10 = 0$$

$$\Rightarrow \quad 3y = -2x + 10$$

or $$y = \frac{-2}{3}x + \frac{10}{3}$$

$$\Rightarrow \quad b_{yx} = -2/3$$

Then, the line of regression of x on y is

$$4x + y - 5 = 0$$

$$\Rightarrow 4x = -y + 5 \text{ or } x = -\frac{1}{4}y + \frac{5}{4}$$

$$\Rightarrow \quad b_{xy} = -\frac{1}{4}$$

Now, for such case, $b_{yx} \cdot b_{xy} < 1$

$$\Rightarrow \quad b_{yx} \cdot b_{xy} = -\frac{2}{3}\left(-\frac{1}{4}\right) - \frac{1}{6} < 1$$

which is true. Hence, our assumption is correct

$\therefore$ Line of regression of y on x is

$$2x + 3y - 10 = 0$$ **Ans.**

□□

Evaluate

Set 6

Chapter 3. Inverse Trigonometric Functions

1. *Evaluate :* $\cos\left(\operatorname{cosec}^{-1}\frac{13}{5}\right)$.

Sol. We have, $\cos\left(\operatorname{cosec}^{-1}\frac{13}{5}\right)$

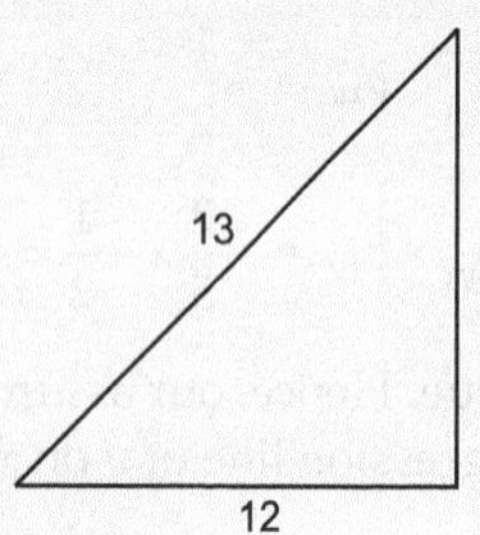

$= \cos\left(\cos^{-1}\frac{12}{13}\right) \qquad \left[\because \operatorname{cosec}^{-1}\frac{13}{5} = \cos^{-1}\frac{12}{13}\right]$

$= \frac{12}{13}$ **Ans.**

2. *Evaluate :* $\cos\left(\cot^{-1}\frac{3}{4}\right)$.

Sol. We have, $\cos\left(\cot^{-1}\frac{3}{4}\right)$

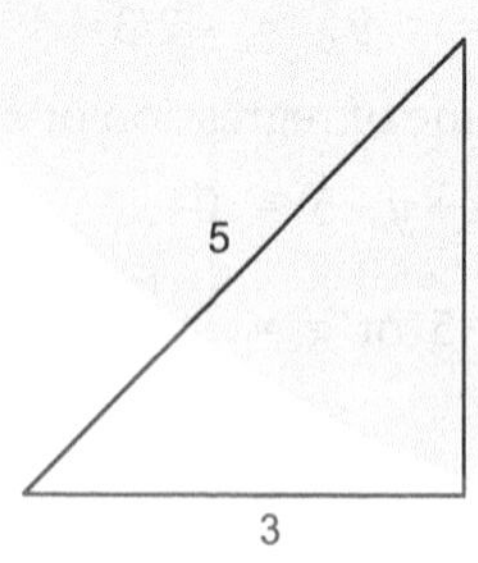

$= \cos\left(\cos^{-1}\frac{3}{5}\right) \qquad \left[\because \cot^{-1}\frac{3}{4} = \cos^{-1}\frac{3}{5}\right]$

$= \frac{3}{5}$ **Ans.**

3. *Evaluate :*

$\sec^2 (\tan^{-1} 2) + \operatorname{cosec}^2 (\cot^{-1} 3)$.

Sol. $\sec^2 (\tan^{-1} 2) + \operatorname{cosec}^2 (\cot^{-1} 3)$

$= \{\sec (\tan^{-1} 2)\}^2 + \{\operatorname{cosec} (\cot^{-1} 3)\}^2$

$= \left\{\sec\left(\tan^{-1}\frac{2}{1}\right)\right\}^2 + \left\{\operatorname{cosec}\left(\cot^{-1}\frac{3}{1}\right)\right\}^2$

$= \left[\sec(\sec^{-1}\sqrt{5})\right\}^2 + \left\{\operatorname{cosec}(\operatorname{cosec}^{-1}\sqrt{10})\right\}^2$

$= (\sqrt{5})^2 + (\sqrt{10})^2$

$= 5 + 10$

$= 15.$ **Ans**

4. *Solve for x : $\cos (\sin^{-1} x) = \frac{1}{9}$.*

Sol. Given,

$$\cos (\sin^{-1} x) = \frac{1}{9}$$

$\Rightarrow \quad \cos\left[\cos^{-1}\sqrt{1-x^2}\right] = \frac{1}{9}$

$\Rightarrow \quad \sqrt{1-x^2} = \frac{1}{9}$

On squaring both sides, we get

$\Rightarrow \quad 1 - x^2 = \frac{1}{81}$

$\Rightarrow \quad 81 - 81x^2 = 1$

$\Rightarrow \quad 81x^2 = 80$

$\Rightarrow \quad x^2 = \frac{80}{81}$

$\Rightarrow \quad x = \pm\sqrt{\frac{80}{81}}$

$\therefore \quad x = \pm\frac{4}{9}\sqrt{5}$ **Ans.**

5. *Evaluate*

$\cos [2 \cos^{-1} x + \sin^{-1} x]$ at $x = \frac{1}{5}$.

Sol. Let $\cos^{-1} x = \theta$

then, $x = \cos\theta;\ 0 \le \theta \le \pi$

$\Rightarrow \quad \sin^{-1} x = \sin^{-1} [\cos\theta]$

$\Rightarrow \quad \sin^{-1} x = \sin^{-1}\left[\sin\left(\frac{\pi}{2}-\theta\right)\right]$

$= \frac{\pi}{2} - \theta$

Thus, $\cos(2\cos^{-1}x + \sin^{-1}x)$

$$= \cos\left(2\theta + \frac{\pi}{2} - \theta\right) = \cos\left(\theta + \frac{\pi}{2}\right)$$

$$= -\sin\theta = -\sqrt{1-\cos^2\theta}$$

$$= -\sqrt{1-x^2} = -\sqrt{1-\frac{1}{25}} \quad \left[\because x = \frac{1}{5}\right]$$

$$= \frac{-2\sqrt{6}}{5}$$ **Ans.**

6. *Evaluate :* $\tan^{-1}\left\{\frac{\sqrt{1+x^2}+\sqrt{1-x^2}}{\sqrt{1+x^2}-\sqrt{1-x^2}}\right\}$.

Sol. We have, $\tan^{-1}\left\{\frac{\sqrt{1+x^2}+\sqrt{1-x^2}}{\sqrt{1+x^2}-\sqrt{1-x^2}}\right\}$

Let $x^2 = \cos 2\theta$, we get

$$= \tan^{-1}\left\{\frac{\sqrt{1+\cos 2\theta}+\sqrt{1-\cos 2\theta}}{\sqrt{1+\cos 2\theta}-\sqrt{1-\cos 2\theta}}\right\}$$

$$= \tan^{-1}\left\{\frac{\sqrt{2\cos^2\theta}+\sqrt{2\sin^2\theta}}{\sqrt{2\cos^2\theta}-\sqrt{2\sin^2\theta}}\right\}$$

$$= \tan^{-1}\left\{\frac{\cos\theta+\sin\theta}{\cos\theta-\sin\theta}\right\} = \tan^{-1}\left\{\frac{1+\tan\theta}{1-\tan\theta}\right\}$$

$$= \tan^{-1}\left\{\tan\left(\frac{\pi}{4}+\theta\right)\right\}$$

$$= \frac{\pi}{4}+\theta \quad \left[\because -1 < x < 1 \Rightarrow 0 < x^2 < 1 \Rightarrow 0 < 2\theta < \frac{\pi}{2} \Rightarrow 0 < \theta < \frac{\pi}{4}\right]$$

$$= \frac{\pi}{4} + \frac{1}{2}\cos^{-1}x^2 \quad [\because x^2 = \cos 2\theta]$$ **Ans.**

7. *Evaluate :* $\sin^{-1}\frac{3}{5} - \sin^{-1}\frac{8}{17}$.

Sol. We have,

$$\sin^{-1}\frac{3}{5} - \sin^{-1}\frac{8}{17}$$

$$= \cos^{-1}\frac{4}{5} - \cos^{-1}\frac{15}{17}$$

$$\left[\because \sin^{-1}\frac{3}{5} = \cos^{-1}\frac{4}{5} \text{ and } \sin^{-1}\frac{8}{17} = \cos^{-1}\frac{15}{17}\right]$$

$$= \cos^{-1}\left\{\frac{4}{5}\times\frac{15}{17} + \sqrt{1-\left(\frac{4}{5}\right)^2}\times\sqrt{1-\left(\frac{15}{17}\right)^2}\right\}$$

$$\left[\therefore \cos^{-1}x - \cos^{-1}y = \cos^{-1}xy + \sqrt{1-x^2}\sqrt{1-y^2}\right]$$

$$= \cos^{-1}\left\{\frac{4}{5}\times\frac{15}{17} + \sqrt{1-\frac{16}{25}}\times\sqrt{1-\frac{225}{289}}\right\}$$

$$= \cos^{-1}\left\{\frac{12}{17} + \frac{3}{5}\times\frac{8}{17}\right\} = \cos^{-1}\left\{\frac{12}{17} + \frac{24}{85}\right\}$$

$$= \cos^{-1}\left\{\frac{60+24}{85}\right\} = \cos^{-1}\left\{\frac{84}{85}\right\}$$ **Ans.**

8. *Evaluate :* $\tan^{-1} 2 + \tan^{-1} 3$

Sol. We have, $\tan^{-1} 2 + \tan^{-1} 3$ ...(i)

Since, $\tan^{-1} x + \tan^{-1} y$

$$= \pi + \tan^{-1}\left(\frac{x+y}{1-xy}\right), \text{ if } xy > 1$$

Then, (i) becomes

$$= \pi + \tan^{-1}\left\{\frac{2+3}{1-2\times 3}\right\}$$

$$= \pi + \tan^{-1}\left(\frac{5}{1-6}\right)$$

$$= \pi + \tan^{-1}(-1)$$

$$= \pi - \frac{\pi}{4} = \frac{3\pi}{4}$$ **Ans.**

9. *Evaluate :* $\tan\left(2\tan^{-1}\frac{1}{5}\right)$.

Sol. We have, $\tan\left(2\tan^{-1}\frac{1}{5}\right)$.

$$= \tan\left\{\tan^{-1}\left(\frac{2\times\frac{1}{5}}{1-\frac{1}{25}}\right)\right\}$$

$$\left[\because 2\tan^{-1}x = \tan^{-1}\left(\frac{2x}{1-x^2}\right) \text{if } -1 < x < 1\right]$$

$$= \tan\left(\tan^{-1}\frac{5}{12}\right) = \frac{5}{12}$$ **Ans.**

10. *Evaluate tan* $\left[2\tan^{-1}\frac{1}{5} - \frac{\pi}{4}\right]$.

Sol. We know that,

$$2\tan^{-1}x = \tan^{-1}\left[\frac{2x}{1-x^2}\right]$$

$$\therefore \tan\left[2\tan^{-1}\frac{1}{5} - \frac{\pi}{4}\right]$$

$$= \tan\left[\tan^{-1}\left[\frac{2\times\frac{1}{5}}{1-\left(\frac{1}{5}\right)^2}\right] - \tan^{-1}(1)\right]$$

Evaluate

$$= \tan\left[\tan^{-1}\left[\frac{2/5}{24/25}\right] - \tan^{-1}(1)\right]$$

$$= \tan\left[\tan^{-1}\left(\frac{5}{12}\right) - \tan^{-1}(1)\right]$$

$$= \tan\left[\tan^{-1}\left[\frac{\frac{5}{12}-1}{1+\frac{5}{12}\times 1}\right]\right]$$

$$= \tan\left[\tan^{-1}\left(\frac{-7}{17}\right)\right] = \frac{-7}{17}$$ **Ans.**

11. *Evaluate :*

$$\tan\left[2\tan^{-1}\left(\frac{1}{2}\right) - \cot^{-1}(3)\right]$$

Sol. Consider,

$$\tan\left[2\tan^{-1}\left(\frac{1}{2}\right) - \cot^{-1}(3)\right]$$

$$= \tan\left[\tan^{-1}\left(\frac{2\times\frac{1}{2}}{1-\left(\frac{1}{2}\right)^2}\right) - \cot^{-1}(3)\right]$$

$$= \tan\left[\tan^{-1}\left(\frac{1}{1-\left(\frac{1}{4}\right)}\right) - \tan^{-1}\frac{1}{3}\right]$$

$$= \tan\left[\tan^{-1}\left(\frac{4}{3}\right) - \tan^{-1}\left(\frac{1}{3}\right)\right]$$

$$= \tan\left[\tan^{-1}\left(\frac{\frac{4}{3}-\frac{1}{3}}{1+\frac{4}{9}}\right)\right]$$

$$= \tan\left[\tan^{-1}\left(\frac{\frac{4-1}{3}}{\frac{9+4}{9}}\right)\right]$$

$$= \tan\left[\tan^{-1}\left(\frac{1}{13/9}\right)\right]$$

$$= \tan\left[\tan^{-1}\left(\frac{9}{13}\right)\right]$$

$$= \frac{9}{13}$$ **Ans.**

Chapter 4. Determinants

1. *Evaluate* $\begin{vmatrix} (b+c)^2 & a^2 & a^2 \\ b^2 & (c+a)^2 & b^2 \\ c^2 & c^2 & (a+b)^2 \end{vmatrix}$

Sol. Consider,

$$\begin{vmatrix} (b+c)^2 & a^2 & a^2 \\ b^2 & (c+a)^2 & b^2 \\ c^2 & c^2 & (a+b)^2 \end{vmatrix}$$

Operating $C_1 \to C_1 - C_3$ and $C_2 \to C_2 - C_3$, we obtain

$$= \begin{vmatrix} (b+c)^2 - a^2 & 0 & a^2 \\ 0 & (c+a)^2 - b^2 & b^2 \\ c^2 - (a+b)^2 & c^2 - (a+b)^2 & (a+b)^2 \end{vmatrix}$$

$$= \begin{vmatrix} (b+c+a)(b+c-a) & 0 & a^2 \\ 0 & (c+a+b)(c+a-b) & b^2 \\ (c-a-b)(c+a+b) & (c-a-b)(c+a+b) & (a+b)^2 \end{vmatrix}$$

Taking $(a + b + c)$ common from C_1 and C_2, we get

$$= (a+b+c)^2 \begin{vmatrix} b+c-a & 0 & a^2 \\ 0 & c+a-b & b^2 \\ c-a-b & c-a-b & (a+b)^2 \end{vmatrix}$$

Applying $R_3 \to R_3 - R_1 - R_2$, we obtain

$$= (a+b+c)^2 \begin{vmatrix} b+c-a & 0 & a^2 \\ 0 & c+a-b & b^2 \\ -2b & -2a & 2ab \end{vmatrix}$$

Multiplying C_1 and C_2 by a and b, respectively

$$= (a+b+c)^2 \frac{1}{ab} \begin{vmatrix} (b+c-a)a & 0 & a^2 \\ 0 & (c+a-b)b & b^2 \\ -2ab & -2ab & 2ab \end{vmatrix}$$

Applying $C_1 \to C_1 + C_3$ and $C_2 \to C_2 + C_3$, we obtain

Evaluate

$$= (a+b+c)^2 \frac{1}{ab} \begin{vmatrix} ab+ac & a^2 & a^2 \\ b^2 & bc+ab & b^2 \\ 0 & 0 & 2ab \end{vmatrix}$$

Expanding along R_3, we get

$$= (a+b+c)^2 \frac{1}{ab}.2ab \begin{vmatrix} ab+ac & a^2 \\ b^2 & bc+ac \end{vmatrix}$$

Taking a common from R_1 and b common from R_2 we get

$$= 2ab(a+b+c)^2.\frac{1}{ab}.ab \begin{vmatrix} b+c & a \\ b & c+a \end{vmatrix}$$

$= 2ab\ (a + b + c)^2\ [(b + c)\ (c + a) - ab]$

$= 2ab\ (a + b + c)^2\ c\ (a + b + c)$

$= 2abc\ (a + b + c)^3.$ **Ans.**

2. *Evaluate the value of Δ where*

$$\Delta = \begin{vmatrix} 2 & 0 & 0 & 1 \\ -6 & -4 & 0 & 2 \\ -5 & -1 & -4 & 2 \\ -3 & 2 & 6 & -3 \end{vmatrix}$$

Sol. Applying $C_1 \to C_1 - 2C_4$, we obtain

$$\Rightarrow \quad \Delta = \begin{vmatrix} 0 & 0 & 0 & 1 \\ -10 & -4 & 0 & 2 \\ -9 & -1 & -4 & 2 \\ 3 & 2 & 6 & -3 \end{vmatrix}$$

Expanding along R_1, we get

$$\Delta = (-1)^{1+4} \begin{vmatrix} -10 & -4 & 0 \\ -9 & -1 & -4 \\ 3 & 2 & 6 \end{vmatrix}$$

$$\Delta = - \begin{vmatrix} -10 & -4 & 0 \\ -9 & -1 & -4 \\ 3 & 2 & 6 \end{vmatrix}$$

Applying $C_1 \to C_1 - C_2$ and $C_3 \to C_3 - 2C_1$, we obtain

$$\Rightarrow \quad \Delta = - \begin{vmatrix} -10+4 & -4 & 20 \\ -9+1 & -1 & -4+18 \\ 3-2 & 2 & 6-6 \end{vmatrix}$$

$$= - \begin{vmatrix} -6 & -4 & 20 \\ -8 & -1 & 14 \\ 1 & 2 & 0 \end{vmatrix}$$

Now applying $C_2 \to C_2 - 2C_1$, we obtain

$$\Rightarrow \quad \Delta = - \begin{vmatrix} -6 & -4+12 & 20 \\ -8 & -1+16 & 14 \\ 1 & 2-2 & 0 \end{vmatrix}$$

$$= - \begin{vmatrix} -6 & 8 & 20 \\ -8 & 15 & 14 \\ 1 & 0 & 0 \end{vmatrix}$$

Expanding along R_3, we get

$\Delta = -1\ (8 \times 14 - 15 \times 20)$

$= -1\ (112 - 300) = -1\ (-188)$

$= 188$ **Ans.**

Chapter 6. Differential Calculus

1. *Using L'Hospital's Rule, evaluate :*

$$\lim_{x\to 0}(1+\sin x)^{\cot x}.$$

Sol. Consider, $\lim_{x\to 0}(1+\sin x)^{\cot x}$

Let $\quad y = \lim(1+\sin x)^{\cot x}$

Take log on both sides,

$$\log y = \log \lim_{x\to 0}(1+\sin x)^{\cot x}$$

$$= \lim_{x\to 0} \log(1+\sin x)^{\cot x}$$

$$= \lim_{x\to 0}[\cot x \log(1+\sin x)]$$

$$= \lim_{x\to 0}\left[\frac{\log(1+\sin x)}{\tan x}\right]$$

$$\left(\frac{0}{0}\text{ form}\right)$$

$$= \lim_{x\to 0} \frac{\frac{1}{(1+\sin x)} \times \cos x}{\sec^2 x}$$

[using L'Hospital's Rule]

$$= \lim_{x\to 0} \frac{\cos x}{\sec^2 x(1+\sin x)}$$

$$= \frac{1}{1(1+0)} = 1$$

$\Rightarrow \quad \log y = 1$

$\Rightarrow \quad \log y = \log_e e$

$\Rightarrow \quad y = e$ **Ans.**

Evaluate

2. *Evaluate :*

$$\lim_{x\to\pi/2}\left[x\tan x-\frac{\pi}{2}\sec x\right]$$

Sol. Consider,

$$\lim_{x\to\frac{\pi}{2}}\left[x\tan x-\frac{\pi}{2}\sec x\right]$$

$$=\lim_{x\to\frac{\pi}{2}}\left[\frac{x\sin x}{\cos x}-\frac{\pi}{2}\times\frac{1}{\cos x}\right]$$

$$=\lim_{x\to\frac{\pi}{2}}\left[\frac{x\sin x-\frac{\pi}{2}}{\cos x}\right]\qquad\left(\frac{0}{0}\text{ form}\right)$$

By L'Hospital's Rule,

$$=\lim_{x\to\frac{\pi}{2}}\left[\frac{x\cos x+\sin x}{-\sin x}\right]$$

$$=-1$$ **Ans.**

3. *Using L' Hospital's Rule, evaluate :*

$$\lim_{x\to 0}\left(\frac{e^x-e^{-x}-2x}{x-\sin x}\right)$$

Sol. Consider,

$$\lim_{x\to 0}\left(\frac{e^x-e^{-x}-2x}{x-\sin x}\right)\qquad\left(\frac{0}{0}\text{ form}\right)$$

Using L' Hospital's Rule,

$$=\lim_{x\to 0}\frac{e^x+e^{-x}-2}{1-\cos x}\qquad\left(\frac{0}{0}\text{ form}\right)$$

Again using L' Hospital's Rule,

$$=\lim_{x\to 0}\frac{e^x-e^{-x}}{\sin x}\qquad\left(\frac{0}{0}\text{ form}\right)$$

Again using L' Hospital's Rule,

$$=\lim_{x\to 0}\frac{e^x-e^{-x}}{\cos x}$$

$$=\frac{1+1}{2}=2$$ **Ans.**

4. *Evaluate :* $\lim_{x\to\frac{\pi}{2}}(\cos x\cdot\log\tan x)$.

Sol. Consider,

$$\lim_{x\to\pi/2}(\cos x.\log\tan x)\qquad(0.\ \infty\text{ form})$$

$$=\lim_{x\to\pi/2}\frac{\log\tan x}{\sec x}\qquad\left(\frac{\infty}{\infty}\text{ form}\right)$$

Using L'Hospital's Rule,

$$=\lim_{x\to\pi/2}\frac{\frac{1}{\tan x}\times\sec^2 x}{\sec x.\tan x}$$

$$=\lim_{x\to\pi/2}\frac{\sec^2 x}{\tan^2 x\cdot\sec x}$$

$$=\lim_{x\to\pi/2}\frac{\sec x}{\tan^2 x}$$

$$=\lim_{x\to\pi/2}\frac{1}{\cos x}\times\frac{\cos^2 x}{\sin^2 x}$$

$$=\lim_{x\to\pi/2}\frac{\cos x}{\sin^2 x}$$

$$=\frac{\cos\pi/2}{\sin^2\pi/2}=\frac{0}{1}=0$$ **Ans**

5. *Evaluate*

$$\lim_{x\to 0}\frac{\sin x-x+\frac{1}{6}x^3}{x^3}.$$

Sol. Consider,

$$\lim_{x\to 0}\frac{\sin x-x+\frac{1}{6}x^3}{x^3}\qquad\left(\frac{0}{0}\text{ form}\right)$$

Using L'Hospital's Rule,

$$=\lim_{x\to 0}\frac{\cos x-1+\frac{1}{6}\times 3x^2}{3x^2}\qquad\left(\frac{0}{0}\text{ form}\right)$$

Again using L'Hospital's Rule

$$=\lim_{x\to 0}\frac{-\sin x-0+\frac{1}{2}\times 2x}{6x}\qquad\left(\frac{0}{0}\text{ form}\right)$$

Again using L'Hospital's Rule

$$=\lim_{x\to 0}\frac{-\cos x+1}{6}=\frac{0}{6}=0$$ **Ans.**

6. *Evaluate :* $\lim_{y\to 0}\frac{y-\tan^{-1}y}{y-\sin y}$

Sol. Consider, $\lim_{y\to 0}\frac{y-\tan^{-1}y}{y-\sin y}\qquad\left(\frac{0}{0}\text{ form}\right)$

$$=\lim_{y\to 0}\frac{1-\frac{1}{1+y^2}}{1-\cos y}\qquad\left(\frac{0}{0}\text{ form}\right)$$

Using L'Hospital's Rule, we get

$$= \lim_{y\to 0}\frac{0-(-1)(1+y^2)^{-2}.2y}{\sin y}$$

$$= \lim_{y\to 0}\frac{2y}{(1+y^2)^2.\sin y} = \lim_{y\to 0}\frac{2}{(1+y^2)^2}.\left(\frac{y}{\sin y}\right)$$

$$= \frac{2}{(1+0)^2}.1 = 2\left[\because \lim_{y\to 0}\frac{\sin y}{y}=1\right]$$ **Ans.**

7. *Using L'Hospital's rule, evaluate :*

$$\lim_{x\to 0}\frac{x-\sin x}{x^2\sin x}.$$

Sol. Given, $\lim_{x\to 0}\frac{x-\sin x}{x^2\sin x}$

$$= \lim_{x\to 0}\frac{x-\sin x}{x^3}.\frac{x}{\sin x}$$

$$= \lim_{x\to 0}\frac{x-\sin x}{x^3}.1 \qquad \left(\because \lim_{x\to 0}\frac{\sin x}{x}=1\right)$$

Applying L'Hospital's Rule, we have

$$= \lim_{x\to 0}\frac{x-\sin x}{x^3} \qquad \left(\frac{0}{0}\text{form}\right)$$

$$= \lim_{x\to 0}\frac{1-\cos x}{3x^2} \qquad \left(\frac{0}{0}\text{form}\right)$$

$$= \lim_{x\to 0}\frac{\sin x}{6x} \qquad \left(\frac{0}{0}\text{form}\right)$$

$$= \frac{1}{6}\lim_{x\to 0}\frac{\sin x}{x} = \frac{1}{6}.$$ **Ans.**

8. *Using L' Hospital's rule, evaluate :*

$$\lim_{x\to 0}\left(\frac{1}{x^2}-\frac{\cot x}{x}\right)$$

Sol. $\lim_{x\to 0}\left(\frac{1}{x^2}-\frac{\cot x}{x}\right)$

$$= \lim_{x\to 0}\frac{\sin x - x\cos x}{x^2\sin x}$$

$$= \lim_{x\to 0}\frac{\sin x - x\cos x}{x^3}.\lim_{x\to 0}\frac{x}{\sin x}$$

$$= \lim_{x\to 0}\frac{\sin x - x\cos x}{x^3}.1\left(\frac{0}{0}\text{form}\right)\left[\because \lim_{x\to 0}\frac{\sin x}{x}=1\right]$$

Applying L' Hospital's rule, we get

$$= \lim_{x\to 0}\frac{\cos x-(\cos x - x\sin x)}{3x^2}$$

$$= \lim_{x\to 0}\frac{x\sin x}{3x^2}$$

$$= \lim_{x\to 0}\frac{\sin x}{3x} \qquad \left(\frac{0}{0}\text{form}\right)$$

Applying L' Hospital's rule, we get

$$= \frac{1}{3}\lim_{x\to 0}\frac{\sin x}{x} = \frac{1}{3}$$ **Ans.**

Chapter 7. Applications of Derivatives

1. *Evaluate the angle of intersection of $xy = 6$ and $x^2y = 12$.*

Sol. Given curves are

$$xy = 6 \qquad \ldots(i)$$

and $$x^2y = 12 \qquad \ldots(ii)$$

From (i), $y = \frac{6}{x}$

Putting this in (ii)

$$x^2\left(\frac{6}{x}\right) = 12$$

$$6x = 12$$

$$\Rightarrow \quad x = 2$$

Putting $x = 2$ in (i), we get

$$y = \frac{6}{2} = 3$$

Thus, the curves intersect at point (2, 3)

Now, differentiating (i) w.r.t. x, we get

$$x\frac{dy}{dx}+y = 0$$

$$\frac{dy}{dx} = \frac{-y}{x}$$

$$m_1 = \left(\frac{dy}{dx}\right)_{(2,3)}$$

$$= \frac{-3}{2}$$

Differentiating (ii), w.r.t. x, we get

$$x^2\frac{dy}{dx}+2xy = 0$$

$$\frac{dy}{dx} = -\frac{2y}{x}$$

Evaluate

$$m_2 = \left(\frac{dy}{dx}\right)_{(2,3)}$$

$$= \frac{-2(3)}{2} = -3$$

Let θ be the angle of intersection of curves (i) and (ii) then,

$$\tan\theta = \frac{m_1 - m_2}{1 + m_1 m_2} = \frac{-(3/2)+3}{1+(-3/2)(-3)}$$

$$= \frac{3/2}{11/2} = \frac{3}{11}$$

$$\theta = \tan^{-1}\left(\frac{3}{11}\right)$$ **Ans.**

2. *Evaluate the approximate value of $\log_e(4{\cdot}01)$ given that $\log_e 4 = 1{\cdot}3863$.*

Sol. Let $y = \log_e x$

then, $\frac{dy}{dx} = \frac{1}{x}$

Let $x = 4$ and $x + \delta x = 4{\cdot}01$

⇒ $\delta x = 0{\cdot}01$

∴ $y + \delta y = \log_e(x + \delta x)$

$= \log_e(4{\cdot}01)$

$\delta y = \log_e(4{\cdot}01) - y$

$\delta y = \log_e(4{\cdot}01) - \log_e x$

$\delta y = \log_e(4{\cdot}01) - \log_e 4$

$\log_e(4{\cdot}01) = \delta y + \log_e 4$

$= \delta y + 1{\cdot}3863$

Now, $\delta y = \frac{dy}{dx}.\delta x$

$= \frac{1}{x}(0.01) = \frac{0.01}{4} = 0.0025$

∴ $\log e\,(4{\cdot}01) = 0{\cdot}0025 + 1{\cdot}3863$

$= 1{\cdot}3888.$ **Ans.**

3. *Evaluate the approximate value of* $\sqrt{0{\cdot}037}$

Sol. Let $y = \sqrt{x}$

then, $\frac{dy}{dx} = \frac{1}{2\sqrt{x}}$

Let $x = 0{\cdot}040$ and $x + \delta x = 0{\cdot}037$

⇒ $\delta x = -0{\cdot}003.$

$y + \delta y = \sqrt{x+\delta x} = \sqrt{0{\cdot}037}$

$\delta y = \sqrt{0{\cdot}037} - y$

$\delta y = \sqrt{0{\cdot}037} - \sqrt{0{\cdot}040}$

$\sqrt{0{\cdot}037} = \delta y + \sqrt{0{\cdot}040}$

Now, $\delta y = \frac{dy}{dx}.\delta x = \frac{1}{2\sqrt{x}}(-0.003)$

$= \frac{(-0.003)}{2\sqrt{0.040}} = \frac{-3}{400}$

∴ $\sqrt{0{\cdot}037} = \frac{-3}{400} + 0.2$

$= 0{\cdot}2 - 0{\cdot}0075$

$= 0{\cdot}1925$ **Ans.**

4. *Evaluate the approximate value of $f(2{\cdot}01)$, where $f(x) = 4x^2 + 5x + 2$.*

Sol. Let $y = f(x)$

Let $x = 2$ and $x + \delta x = 2{\cdot}01$

⇒ $\delta x = 0{\cdot}01$

For $x = 2, y = f(2)$

$y = 4(2)^2 + 5(2) + 2$

$y = 16 + 10 + 2 = 28$

Now, $y = f(x)$

$y = 4x^2 + 5x + 2$

$\frac{dy}{dx} = 8x + 5$

$\left(\frac{dy}{dx}\right)_{x=2} = 8(2) + 5$

$= 16 + 5 = 21$

Now, $\delta y = \frac{dy}{dx}.\delta x = 21 \times 0{\cdot}01 = 0{\cdot}21$

∴ $f(2{\cdot}01) = y + \delta y$

$= 28 + 0{\cdot}21$

$= 28{\cdot}21.$ **Ans.**

5. *Evaluate the approximate value of $f(5{\cdot}001)$, where $f(x) = x^3 - 7x^2 + 15$.*

Sol. Let $y = f(x)$

Let $x = 5$ and $x + \delta x = 5{\cdot}001$

⇒ $\delta x = 0{\cdot}001$

For $x = 5$, $y = f(5)$

$y = (5)^3 - 7(5)^2 + 15$

$= 125 - 175 + 15$

$= -35$

Now, $y = f(x)$

$y = x^3 - 7x^2 + 15$

$\frac{dy}{dx} = 3x^2 - 14x$

Evaluate

$$\left(\frac{dy}{dx}\right)_{x=5} = 3(5)^2 - 14(5)$$
$$= 75 - 70$$
$$= 5$$

Now, $\delta y = \frac{dy}{dx}.\delta x$
$$= 5 \times 0{\cdot}001 = 0{\cdot}005$$

$\therefore$ $f(5{\cdot}001) = y + \delta y$
$$= -35 + 0{\cdot}005$$
$$= -34{\cdot}995.$$ **Ans.**

6. *Evaluate the values of k for which $f(x) = kx^3 - 9kx^2 + 9x + 3$ is increasing on R.*

Sol. Since, the function $f(x)$ is increasing on R, then
$$f'(x) > 0 \ \forall\ x \in R$$
$$3kx^2 - 18kx + 9 > 0 \text{ for all } x \in R$$
$$kx^2 - 6kx + 3 > 0 \text{ for all } x \in R$$
$[\because ax^2 + bx + c > 0 \text{ for all } x \in R \Rightarrow a > 0 \text{ and } b^2 - 4ac < 0]$

So, $k > 0$ and $36k^2 - 12k < 0$
$$k > 0 \text{ and } 12k(3k-1) < 0$$
$$k > 0 \text{ and } k(3k-1) < 0$$
$$3k - 1 < 0 \qquad [\because k > 0]$$
$$k < \frac{1}{3}$$
$$k \in (0, 1/3)$$

Hence, $f(x)$ is increasing on R for k $(0, 1/3)$. **Ans.**

Chapter 8. Integral Calculus

1. *Integrate* $\int \frac{1+\tan^2 x}{\sqrt{1-\tan^2 x}} dx.$

Sol. Let $\tan x = t$

$\therefore \sec^2 x.dx = dt$
$$\int \frac{1+\tan^2 x\, dx}{\sqrt{1-\tan^2 x}} = \int \frac{\sec^2 x}{\sqrt{1-\tan^2 x}} dx$$
$$= \int \frac{1}{\sqrt{1-t^2}} dt$$
$$= \sin^{-1} t = \sin^{-1}[\tan x] + c,$$
where 'c' is constant of integration. **Ans.**

2. *Evaluate* $\int x^2 (e^{x^3}) \cos(2e^{x^3})\, dx.$

Sol. Given, $\int x^2 (e^{x^3}) \cos(2e^{x^3})\, dx$

Putting $2e^{x^3} = t$

Differentiating w.r.t. x, we get
$$2e^{x^3} \times 3x^2 = \frac{dt}{dx}$$
$$\therefore \quad dx = \frac{dt}{2e^{x^3} \times 3x^2}$$

Now, $\int x^2 (e^{x^3}) \cos t \frac{dt}{2e^{x^3} \times 3x^2}$
$$= \frac{1}{6}\int \cos t\, dt$$
$$= \frac{1}{6}\sin t + c$$
$$= \frac{1}{6}\sin(2e^{x^3}) + c$$ **Ans.**

3. *Evaluate* $\int \frac{1}{x+\sqrt{x}} dx.$

Sol. $\int \frac{1}{x+\sqrt{x}} dx$

Put $\sqrt{x} = t$
$$\Rightarrow \quad x = t^2$$
$$\Rightarrow \quad dx = 2t\, dt$$
$$\therefore \quad \int \frac{dx}{x+\sqrt{x}} = \int \frac{2t dt}{t^2 + t} = 2\int \frac{1}{t+1} dt$$
$$= 2 \log |t+1| + c$$
$$= 2\log(\sqrt{x}+1) + c.$$ **Ans.**

4. *Evaluate :* $\int \frac{\text{cosec } x}{\log \tan\left(\frac{x}{2}\right)} dx.$

Sol. Consider, $\int \frac{\text{cosec } x}{\log \tan\left(\frac{x}{2}\right)} dx$

Put $\log \tan\left(\frac{x}{2}\right) = t$ ...(i)

On differentiating w.r.t. x, we get
$$\Rightarrow \quad \frac{1}{\tan\frac{x}{2}} \cdot \frac{1}{2}\sec^2\frac{x}{2} = \frac{dt}{dx}$$
$$\Rightarrow \quad \frac{1}{2} \cdot \frac{\cos x/2}{\sin x/2 \cos^2 x/2} dx = dt$$
$$\Rightarrow \quad \frac{1}{2\sin\frac{x}{2}\cos\frac{x}{2}} dx = dt$$

$\Rightarrow \quad \frac{1}{\sin x} dx = dt$

$\Rightarrow \quad \operatorname{cosec} x\, dx = dt \quad \ldots\text{(ii)}$

Now, on dividing both sides by 't' and then integrating

$$\therefore \quad \int \frac{\operatorname{cosec} x\, dx}{\log \tan\left(\frac{x}{2}\right)} = \int \frac{dt}{t}$$

$$\int \frac{\operatorname{cosec} x\, dx}{\log \tan\left(\frac{x}{2}\right)} = \log t + c$$

Put the value of 't'

$$\int \frac{\operatorname{cosec} x\, dx}{\log \tan\left(\frac{x}{2}\right)} = \log\left[\log \tan \frac{x}{2}\right] + c.$$

Ans.

5. *Evaluate* $\int \frac{\cos x}{\sin x + \sqrt{\sin x}} dx.$

Sol. Consider,

$$I = \int \frac{\cos x}{\sin x + \sqrt{\sin x}} dx.$$

Let $\sin x = t^2$

$\Rightarrow \cos x\, dx = 2t\, dt$

$$\therefore \quad I = 2\int \frac{t\, dt}{t^2 + t} = 2\int \frac{dt}{1+t}$$

$$= 2 \log_e |t + 1| + c$$

$$= 2 \log_e |\sqrt{\sin x} + 1| + c. \quad \textbf{Ans.}$$

6. *Evaluate :* $\int \frac{1}{x^2} \sin^2\left(\frac{1}{x}\right) dx$

Sol.

We have, $\int \frac{1}{x^2} \sin^2\left(\frac{1}{x}\right) dx$

Put $\quad \frac{1}{x} = t$

$$-\frac{1}{x^2} dx = dt$$

$$\Rightarrow \int \frac{1}{x^2} \sin^2\left(\frac{1}{x}\right) dx = -\int \sin^2 t\, dt$$

$$= -\int \frac{(1 - \cos 2t)\, dt}{2}$$

$$= \frac{1}{2}\int \cos 2t\, dt - \frac{1}{2}\int dt$$

$$= \frac{1}{2}\left(\frac{\sin 2t}{2}\right) - \frac{t}{2} + c$$

(where c is constant of integration)

$$= \frac{1}{4} \sin\left(\frac{2}{x}\right) - \frac{1}{2x} + c \quad \textbf{Ans.}$$

7. *Evaluate :* $\int \tan^3 x\, dx$

Sol. Here, $\int \tan^3 x\, dx$

$$= \int \tan x \cdot \tan^2 x\, dx$$

$$= \int \tan x (\sec^2 x - 1)\, dx$$

$$= \int \tan x \cdot \sec^2 x\, dx - \int \tan x\, dx$$

In the first integral, put $u = \tan x$ so that,

$$\frac{du}{dx} = \sec^2 x$$

$$= \int u \cdot du - \log |\sec x| + c$$

$$= \frac{u^2}{2} - \log |\sec x| + c$$

$$\therefore \quad \int \tan^3 x\, dx = \frac{\tan^2 x}{2} - \log |\sec x| + c -$$

Ans

8. *Evaluate*

$$\int \frac{\log(\log x)}{x} dx.$$

Sol. Let $I = \int \frac{\log(\log x)}{x} dx$

Substituting $\log x = t$, we get

$$\Rightarrow \quad \frac{1}{x} dx = dt$$

$$\Rightarrow \quad I = \int \log t\, dt = \int \log t \times 1\, dt$$

Integration by parts,

$$= \log t \int 1.\, dt - \int \frac{1}{t} - .\, t.dt$$

$$= t \log t - t + c$$

$$= \log x \log(\log x) - \log x + c \quad \textbf{Ans.}$$

9. *Evaluate the integral*

$$\int \frac{\sqrt{\tan x}}{\sin 2x} dx.$$

Sol. Consider, $\int \frac{\sqrt{\tan x}}{\sin 2x} dx$

$$= \int \frac{\sqrt{\sin x}}{\sqrt{\cos x}} \frac{1}{2 \sin x \cos x} dx$$

$$= \int \frac{1}{2} \cdot \frac{1}{\sqrt{\sin x}} \frac{1}{\sqrt{\cos x}} \frac{dx}{\cos x}$$

Multiplying and dividing by $\sqrt{(\cos x)}$, we get

$$= \int \frac{1}{2} \cdot \frac{\sqrt{\cos x}}{\sqrt{\sin x} \cos^2 x} dx$$

$$= \frac{1}{2}\int \frac{\sec^2 x}{\sqrt{\tan x}} dx$$

Evaluate

Let $\tan x = t$, $\sec^2 x\, dx = dt$

$$= \frac{1}{2}\int \frac{dt}{\sqrt{t}} = \frac{1}{2} \times 2\sqrt{t}$$

$$= \sqrt{t} = \sqrt{\tan x} + c$$ **Ans.**

10. Evaluate : $\int \frac{1}{\sqrt{1+\sin x}} dx.$

Sol. Let $I = \int \frac{1}{\sqrt{1+\sin x}} dx$

$$= \int \frac{1}{\sqrt{\sin^2\frac{x}{2} + \cos^2\frac{x}{2} + 2\sin\frac{x}{2}\cos\frac{x}{2}}} dx$$

$$= \int \frac{1}{\sqrt{\left(\sin\frac{x}{2} + \cos\frac{x}{2}\right)^2}} dx$$

$$= \int \frac{1}{\sin\frac{x}{2} + \cos\frac{x}{2}} dx$$

$$= \frac{1}{\sqrt{2}}\int \frac{1}{\sin\frac{x}{2}\cdot\frac{1}{\sqrt{2}} + \cos\frac{x}{2}\cdot\frac{1}{\sqrt{2}}} dx.$$

$$= \frac{1}{\sqrt{2}}\int \frac{1}{\sin\frac{x}{2}\cos\frac{\pi}{4} + \cos\frac{x}{2}\sin\frac{\pi}{4}} dx$$

$$= \frac{1}{\sqrt{2}}\int \frac{1}{\sin\left(\frac{x}{2} + \frac{\pi}{4}\right)} dx$$

$$= \frac{1}{\sqrt{2}}\int \operatorname{cosec}\left(\frac{x}{2} + \frac{\pi}{4}\right) dx$$

$$= \frac{1}{\sqrt{2}} 2\log\left|\operatorname{cosec}\left(\frac{\pi}{4} + \frac{x}{2}\right) - \cot\left(\frac{x}{2} + \frac{\pi}{4}\right)\right| + c$$

$$= \sqrt{2}\log\left|\operatorname{cosec}\left(\frac{\pi}{4} + \frac{x}{2}\right) - \cot\left(\frac{\pi}{4} + \frac{x}{2}\right)\right| + c$$ **Ans.**

11. *Evaluate* $\int \frac{x+\sin x}{1+\cos x} dx.$

Sol. Consider, $\int \frac{x+\sin x}{1+\cos x} dx$

$$= \int \frac{x+\sin x}{2\cos^2 x/2} dx$$

$$= \int \frac{x}{2\cos^2 x/2} dx + \int \frac{2\sin x/2 \cos x/2}{2\cos^2 x/2} dx$$

$$= \frac{1}{2}\int x \sec^2 x/2\, dx + \int \tan x/2\, dx$$

Integrating by parts,

$$= \frac{1}{2}\left[x.\int \sec^2 x/2 dx - \int\left(\frac{d}{dx}(x).\int \sec^2 x/2 dx\right)dx\right] + \int \tan x/2\, dx$$

$$= \frac{1}{2}\left[x.\frac{\tan x/2}{1/2} - \int \frac{\tan x/2 dx}{1/2}\right] + \int \tan\frac{x}{2} dx + c$$

$$= x\tan x/2 - \int \tan x/2\, dx + \int \tan x/2\, dx + c$$

$$= x\tan x/2 + c$$ **Ans.**

12. *Evaluate* $\int e^x (\tan x + \log \sec x)\, dx.$

Sol. Consider, $\int e^x (\tan x + \log \sec x)\, dx$

$$= \int e^x \tan x\, dx + \int e^x \log \sec x\, dx$$

Integrating by parts,

$$= \int e^x \tan x\, dx + \log(\sec x)\int e^x dx - \int\left\{\frac{d}{dx}(\log \sec x).\int(e^x dx)\right\}dx$$

$$= \int e^x \tan x\, dx + \log \sec x\, e^x - \int e^x \tan x\, dx + c$$

$$= e^x \log \sec x + c$$ **Ans.**

13. *Evaluate* $\int e^x \frac{(2+\sin 2x)}{\cos^2 x} dx.$

Sol. Consider, $\int e^x\left(\frac{2}{\cos^2 x} + \frac{2\sin x\cos x}{\cos^2 x}\right)dx$

$$= \int e^x (2\sec^2 x + 2\tan x)\, dx$$

$$= 2\int e^x (\sec^2 x + \tan x)\, dx$$

$$= 2\left[\int e^x \sec^2 x\, dx + \int e^x \tan x\, dx\right]$$

$$= 2\left[\int e^x \sec^2 x\, dx + \tan x\int e^x dx - \int\frac{d\tan x}{dx}\int e^x dx\right]$$

$$= 2\left[\int e^x \sec^2 x\, dx + e^x \tan x\, dx - \int e^x \sec^2 x\, dx\right]$$

$$= 2e^x \tan x + c$$ **Ans.**

14. *Evaluate* $\int e^{-2x} \sin x\ dx.$

Sol. Let $I = \int e^{-2x} \sin x\ dx$

Integrating by parts,

$$I = \sin x\int e^{-2x} dx - \int \cos x\,\frac{(e^{-2x})}{-2}.\ dx$$

$$I = \sin x\,.\,\frac{e^{-2x}}{-2} + \frac{1}{2}\int \cos x.e^{-2x}.\ dx$$

$$I = \frac{-1}{2}e^{-2x}.\sin x + \frac{1}{2}\left[\cos x\int e^{-2x}dx - \int(-\sin x)\frac{e^{-2x}}{-2}dx\right.$$

$$I = \frac{-1}{2}e^{-2x}\sin x + \frac{1}{2}\left[\cos x\frac{e^{-2x}}{-2} - \frac{1}{2}\int \sin x\, e^{-2x}dx\right]$$

$$I = \frac{-1}{2}e^{-2x}\sin x - \frac{\cos x\, e^{-2x}}{4} - \frac{1}{4}I$$

$$\frac{5}{4}I = \frac{-1}{2}e^{-2x}\sin x - \frac{\cos x\, e^{-2x}}{4} + c$$

where c is constant of integration.

$$I = \frac{-2}{5}e^{-2x}\sin x - \frac{1}{5}e^{-2x}\cos x + c$$ **Ans.**

15. *Evaluate*

$$\int e^x\left[\frac{1+\sin x}{1+\cos x}\right]dx.$$

Sol. Consider, $\int e^x\left[\frac{1+\sin x}{1+\cos x}\right]dx$

$$= \int e^x \frac{(1+\sin x)(1-\cos x)}{(1+\cos x)(1-\cos x)}dx$$

$$= \int \frac{e^x\cdot(1-\cos x+\sin x-\sin x\cos x)}{1-\cos^2 x}dx$$

$$= \int \frac{e^x}{\sin^2 x}[1-\cos x+\sin x-\sin x\cos x]\,dx$$

$$= \int e^x . \text{cosec}^2 x\, dx - \int e^x . \text{cosec}\, x . \cot x . dx$$

$$+ \int e^x . \text{cosec}\, x\, dx - \int e^x . \cot x . dx$$

On integrating by parts

$$= e^x . \int \text{cosec}^2 x . dx - \int e^x . (-\cot x) . dx$$

$$+ e^x . \int -\text{cosec}\, x . \cot x . dx$$

$$- \int e^x . \text{cosec}\, x . dx + \int e^x \text{cosec}\, x . dx$$

$$- \int e^x \cot x . dx$$

$$= e^x . \int \text{cosec}^2 x . dx + \int e^x \cot x\, dx$$

$$+ e^x . \text{cosec}\, x - \int e^x \text{cosec}\, x . dx$$

$$+ \int e^x \text{cosec}\, x . dx - \int e^x \cot x . dx$$

$$= e^x . (-\cot x) + e^x \text{cosec}\, x + c$$

where c is constant of integration.

$$= e^x[\text{cosec}\, x - \cot x] + c$$ **Ans.**

16. *Evaluate* $\int x^2 \sin^{-1} x\, dx.$

Sol. Let $I = \int x^2 \sin^{-1} x\, dx$

Integrating by parts, taking x^2 as second function.

$$I = \sin^{-1} x \int x^2 dx - \int \frac{d}{dx}(\sin^{-1} x)\int x^2 dx$$

$$I = \frac{x^3}{3}\sin^{-1} x - \int \frac{x^3}{3}\frac{dx}{\sqrt{1-x^2}}$$

Let $\int \frac{x^3}{3}\frac{dx}{\sqrt{1-x^2}} = I_1$

$$I_1 = \frac{1}{3}\int \frac{dt}{2}\frac{(1-t)}{\sqrt{t}} \qquad \left(\begin{array}{l}\text{Put } 1-x^2 = t\\ \therefore -2x\, dx = dt\end{array}\right)$$

$$= \frac{1}{3}\left(\frac{1}{2}\int \frac{dt}{\sqrt{t}} - \frac{1}{2}\int \sqrt{t}\, dt\right)$$

$$= \frac{1}{3}\left(\frac{1}{2}2\sqrt{t} - \frac{1}{2}\times\frac{2}{3}\times t^{3/2}\right)$$

$$= \frac{1}{3}\left(\sqrt{t} - \frac{1}{3}t^{3/2}\right)$$

$$= \frac{1}{3}\left[\sqrt{1-x^2} - \frac{1}{3}(1-x^2)^{3/2}\right]$$

$$\therefore I = \frac{x^3}{3}\sin^{-1} x + \frac{1}{3}\sqrt{1-x^2} - \frac{1}{9}(1-x^2)^{3/2} + c$$

Ans.

17. *Evaluate :* $\int \tan^{-1}\sqrt{\frac{(1-x)}{1+x}}\, dx.$

Sol. Let $I = \int \tan^{-1}\sqrt{\frac{(1-x)}{1+x}}\, dx$

Putting $x = \cos\theta$ and $dx = -\sin\theta\, d\theta$, we get

$$= \int \tan^{-1}\sqrt{\frac{1-\cos\theta}{1+\cos\theta}}\, dx$$

$$= \int \tan^{-1}\sqrt{\frac{2\sin^2\theta/2}{2\cos^2\theta/2}}\, dx$$

$$= \int \tan^{-1}\left(\tan\frac{\theta}{2}\right)(-\sin\theta)\, d\theta$$

$$= -\frac{1}{2}\int \theta\sin\theta\, d\theta$$

On integrating by parts

$$= -\frac{1}{2}\left[\theta(-\cos\theta) - \int 1(-\cos\theta)\, d\theta\right]$$

$$= \frac{1}{2}\theta\cos\theta - \frac{1}{2}\int\cos\theta\, d\theta$$

$$= \frac{1}{2}\theta\cos\theta - \frac{1}{2}\sin\theta$$

$$= \frac{1}{2}\theta\cos\theta - \frac{1}{2}\sqrt{1-\cos^2\theta}$$

$$= \frac{1}{2}(x - \cos^{-1} x - \sqrt{1-x^2}) + c.$$ **Ans.**

18. *Integrate :* $\int \frac{x+3}{\sqrt{x^2+4x+5}}dx.$

Sol. Let $t = x^2 + 4x + 5$

Then, $dt = (2x + 4)\, dx$

$$\Rightarrow \int \frac{x+3}{\sqrt{x^2+4x+5}}dx = \frac{1}{2}\int \frac{(2x+4)+2}{\sqrt{x^2+4x+5}}dx$$

$$\Rightarrow \frac{1}{2}\int \frac{2x+4}{\sqrt{x^2+4x+5}}dx + \frac{1}{2}\int \frac{2}{\sqrt{x^2+4x+5}}dx$$

$$= \frac{1}{2}\int \frac{dt}{\sqrt{t}} + \int \frac{2}{\sqrt{(x+2)^2+1^2}}dx$$

$$= \frac{1}{2}[2\sqrt{t} + 2\log|\sqrt{1+(x+2)^2}+(x+2)|] + c$$

$$= \sqrt{x^2+4x+5} + \log|\sqrt{(1+(x+2)^2)} + (x+2)| + c.$$

Ans.

19. *Evaluate* $\int \frac{x}{(x+1)^2}dx.$

Sol. Consider, $\int \frac{x}{(x+1)^2}dx$

$$= \int \left[\frac{1}{x+1} - \frac{1}{(x+1)^2}\right]dx$$

$$= \int \frac{1}{x+1}dx - \int \frac{1}{(x+1)^2}dx$$

$$= \log(x+1) - \frac{(x+1)^{-1}}{(-1)} + c$$

$$= \log(x+1) + \frac{1}{(x+1)} + c.$$ **Ans.**

20. *Evaluate* $\int \frac{2\sin 2\theta - \cos\theta}{6-\cos^2\theta - 4\sin\theta}d\theta$

Sol. Consider,

$$I = \int \frac{2\sin 2\theta - \cos\theta}{6-\cos^2\theta-4\sin\theta}d\theta$$

$$\Rightarrow I = \int \frac{4\sin\theta\cos\theta - \cos\theta}{6-1+\sin^2\theta-4\sin\theta}d\theta$$

$$= \int \frac{\cos\theta\,(4\sin\theta-1)}{\sin^2\theta-4\sin\theta+5}d\theta$$

Let $\sin\theta = t$

$\cos\theta\, d\theta = dt$

$$\therefore I = \int \frac{4t-1}{t^2-4t+5}dt = \int \frac{4t-8+7}{t^2-4t+5}dt$$

$$\Rightarrow = 2\int \frac{2t-4}{t^2-4t+5} + 7\int \frac{dt}{t^2-4t+5}$$

$$= 2\int \frac{2t-4}{t^2-4t+5}dt + 7\int \frac{dt}{(t^2-2)^2+1}$$

$= 2\log(t^2 - 4t + 5) + 7\tan^{-1}(t-2) + c$

$= 2\log(\sin^2\theta - 4\sin\theta + 5)$

$+ 7\tan^{-1}(\sin\theta - 2) + c$ **Ans.**

21. *Evaluate*

$$\int \frac{(x-1)}{(x-3)(x-2)^2}dx.$$

Sol. Let $\frac{x-1}{(x-3)(x-2)^2}$

$$= \frac{A}{(x-3)} + \frac{B}{(x-2)} + \frac{C}{(x-2)^2}$$

$$= \frac{A(x-2)^2 + B(x-3)(x-2) + C(x-3)}{(x-3)(x-2)^2}$$

$\Rightarrow (x-1) = A(x^2 + 4 - 4x)$

$+ B(x^2 - 5x + 6) + C(x-3)$

$= x^2(A + B) + x(-4A - 5B + C)$

$+ (4A + 6B - 3C)$

Comparing coefficients on both sides, we get

$\Rightarrow A + B = 0 \Rightarrow A = -B$

and $-4A - 5B + C = 1$

$\Rightarrow C = 4A + 5B + 1$

$= -4B + 5B + 1$

$= B + 1$

and $4A + 6B - 3C = -1$

$\Rightarrow -4B + 6B - 3B - 3 = -1$

$-B = -1 + 3$

$\Rightarrow -B = 2$ or $B = -2$

$\Rightarrow A = 2$ and $C = -1$

So, $\int \frac{x-1}{(x-3)(x-2)}dx$

$$= \int \frac{2}{(x-3)}dx - \int \frac{2}{x-2}dx - \int \frac{1}{(x-2)^2}dx$$

$$= 2\log(x-3) - 2\log(x-2) + \frac{1}{(x-2)} + c$$ **Ans.**

22. *Evaluate :*

$$\int \frac{dx}{x\{6(\log x)^2 + 7\log x + 2\}}$$

Evaluate

Sol. Let $\log x = t$

$$\Rightarrow \quad \frac{1}{x}dx = dt$$

$$\therefore \int \frac{dx}{x\{6(\log x)^2 + 7\log x + 2\}}$$

$$= \int \frac{dt}{6t^2 + 7t + 2}$$

$$= \int \frac{dt}{(3t+2)(2t+1)}$$

By using partial fraction

$$\text{Let } \frac{1}{(3t+2)(2t+1)} = \frac{A}{3t+2} + \frac{B}{2t+1}$$

$$\therefore \quad 1 = A(2t+1) + B(3t+2)$$

On putting $t = -\frac{2}{3}$ and $t = -\frac{1}{2}$ respectively, we get

$$\therefore \quad 1 = A\left(2\left(-\frac{2}{3}\right)+1\right)$$

$$\Rightarrow \quad A = -3$$

$$1 = B\left(3\left(-\frac{1}{2}\right)+2\right)$$

$$\Rightarrow \quad B = 2$$

$$\therefore \int \frac{dt}{(3t+2)(2t+1)}$$

$$= \int \frac{-3}{3t+2}dt + \int \frac{2}{2t+1}dt$$

$$= -3\log\frac{|3t+2|}{3} + 2\log\frac{|2t+1|}{2} + c$$

$$= \log\left|\frac{2t+1}{3t+2}\right| + c$$

$$= \log\left|\frac{2\log x + 1}{3\log x + 2}\right| + c. \quad \textbf{Ans.}$$

23. *Evaluate :* $\int \frac{x^2 - 5x - 1}{x^4 + x^2 + 1}dx.$

Sol. Let $I = \int \frac{x^2 - 5x - 1}{x^4 + x^2 + 1}dx = \frac{Ax+B}{x^2+x+1} + \frac{Cx+D}{x^2-x+1}$

$[\because x^4 + x^2 + 1 = (x^2 + x + 1)(x^2 - x + 1)]$

$\Rightarrow x^2 - 5x - 1 = Ax^3 - Ax^2 + Ax + Bx^2 - Bx + B$

$+ Cx^3 + Dx^2 + Cx^2 + Dx + Cx + D.$

On comparing the coefficients, we get

$$\Rightarrow \quad A + C = 0 \quad \ldots(i)$$

$$\Rightarrow -A + B + D + C = 1 \quad \ldots(ii)$$

$$\Rightarrow \quad A - B + C + D = -5 \quad \ldots(iii)$$

$$\text{and} \quad B + D = 1 \quad \ldots(iv)$$

Solving (i), (ii), (iii) and (iv) simultaneously, we get

$$A = 0, B = 3, C = 0, D = -2$$

Putting values of A, B, C and D in (I), we get

$$\int \frac{x^2 - 5x - 1}{x^4 + x^2 + 1}dx$$

$$= \int \frac{3}{x^2 + x + 1}dx - \int \frac{2\,dx}{x^2 - x + 1}$$

$$= 3\int \frac{dx}{\left(x+\frac{1}{2}\right)^2 + \left(\frac{\sqrt{3}}{2}\right)^2} - 2\int \frac{dx}{\left(x-\frac{1}{2}\right)^2 + \left(\frac{\sqrt{3}}{2}\right)^2}$$

$$= 3\frac{1}{\sqrt{3}}\tan^{-1}\frac{\left(x+\frac{1}{2}\right)}{\sqrt{\frac{3}{2}}} - 2\frac{1}{\sqrt{3}}\tan^{-1}\frac{\left(x-\frac{1}{2}\right)}{\frac{\sqrt{3}}{2}} + c$$

$$= \frac{3\times 2}{\sqrt{3}}\tan^{-1}\left(\frac{2x+1}{\sqrt{3}}\right) - \frac{4}{\sqrt{3}}\tan^{-1}\left(\frac{2x+1}{\sqrt{3}}\right) + c$$

$$= 2\sqrt{3}\tan^{-1}\left(\frac{2x+1}{\sqrt{3}}\right) - \frac{4}{\sqrt{3}}\tan^{-1}\left(\frac{2x+1}{\sqrt{3}}\right) + c.$$

Ans

24. *Evaluate :* $\int \frac{\sec x}{1 + \operatorname{cosec} x}dx$

Sol.

$$\int \frac{\sec x}{1 + \operatorname{cosec} x}dx$$

$$= \int \frac{1}{\cos x\left(1 + \frac{1}{\sin x}\right)}dx$$

$$= \int \frac{\sin x}{\cos x(\sin x + 1)}dx$$

$$= \int \frac{\sin x \cos x}{\cos^2 x(1 + \sin x)}dx$$

$$= \int \frac{\sin x \cos x}{(1 - \sin^2 x)(1 + \sin x)}dx$$

$$= \int \frac{\sin x \cos x}{(1 - \sin x)(1 + \sin x)(1 + \sin x)}dx$$

$$= \int \frac{\sin x \cos x}{(1-\sin x)(1+\sin x)^2} dx$$

Now, let $\sin x = t$, on differentiating, we get

$$\cos x\, dx = dt$$

$$\therefore \quad \int \frac{\sec x}{1+\operatorname{cosec} x} dx = \int \frac{t}{(1-t)(1+t)^2} dt$$

Let $\quad \frac{t}{(1-t)(1+t)^2} = \frac{A}{1-t} + \frac{B}{1+t} + \frac{C}{(1+t)^2}$

$$\Rightarrow \quad t = A(1+t)^2 + B(1-t)(1+t) + C(1-t) \quad \ldots(i)$$

On putting $t = 1$ and $t = -1$, we get

$$1 = A \cdot 4 \quad \text{and} \quad -1 = C \cdot 2$$

$$\Rightarrow \quad A = \frac{1}{4} \quad \text{and} \quad C = -\frac{1}{2}$$

On equating coefficient of t^2 on both sides of (i), we get

$$0 = A - B$$

$$\Rightarrow \quad B = A = \frac{1}{4}$$

$$\therefore \int \frac{\sec x}{1+\operatorname{cosec} x} dx$$

$$= \frac{1}{4}\int \frac{1}{1-t} dt + \frac{1}{4}\int \frac{dt}{1+t} - \frac{1}{2}\int \frac{dt}{(1+t)^2}$$

$$= \frac{1}{4}\frac{\log|1-t|}{-1} + \frac{1}{4}\log|1+t| - \frac{1}{2}\frac{(1-t)^{-1}}{-1} + c$$

$$= \frac{1}{4}\log\left|\frac{1+t}{1-t}\right| + \frac{1}{2(1-t)} + c$$

$$= \frac{1}{4}\log\left|\frac{1+\sin x}{1-\sin x}\right| + \frac{1}{2(1+\sin x)} + c. \quad \textbf{Ans.}$$

25. *Evaluate :* $\int \frac{\sin 2x}{(1+\sin x)(2+\sin x)} dx$

Sol.

We have, $\quad I = \int \frac{\sin 2x}{(1+\sin x)(2+\sin x)} dx$

$$I = \int \frac{2 \cdot \sin x \cos x}{(1+\sin x)(2+\sin x)} dx$$

$$I = \int \frac{2t\, dt}{(1+t)(2+t)}$$

[Put $\sin x = t \Rightarrow \cos x\, dx = dt$]

By partial fractions,

Let $\quad \frac{2t}{(1+t)(2+t)} = \frac{A}{(1+t)} + \frac{B}{(2+t)}$

$$2t = A(2+t) + B(1+t)$$

Putting $t = -1$ and $t = -2$, we get

$$-2 = A(2-1)$$

$$-2 = A$$

and $\quad -4 = B(1-2)$

$$-4 = -B$$

$$\Rightarrow \quad A = -2 \text{ and } B = 4$$

Now, $I = -2\int \frac{dt}{1+t} + 4\int \frac{dt}{2+t}$

$$I = -2\log|1+t| + 4\log|2+t| + c$$

$$I = -2\log|1+\sin x| + 4\log|2+\sin x| + c$$

$$= \log|1+\sin x|^{-2} + \log|2+\sin x|^4 + c$$

$$= \log\left|\frac{(2+\sin x)^4}{(1+\sin x)^2}\right| + c \quad \textbf{Ans.}$$

26. *Evaluate :* $\int \frac{2y^2}{y^2+4} dy.$

Sol. Given, $\int \frac{2y^2}{y^2+4} dy$

$$= 2\int \frac{y^2+4-4}{y^2+4} dy$$

$$= 2\int dy - 8\int \frac{dy}{y^2+4}$$

$$= 2y - 8 \cdot \frac{1}{2}\tan^{-1}\left(\frac{y}{2}\right) + c$$

$$\left[\because \int \frac{dx}{x^2+a^2} = \frac{1}{a}\tan^{-1}\left(\frac{x}{a}\right)\right]$$

$$= 2y - 4\tan^{-1}\left(\frac{y}{2}\right) + c. \quad \textbf{Ans.}$$

27. *Evaluate :* $\int \frac{\sin x + \cos x}{\sqrt{9+16\sin 2x}} dx$

Sol. Let $I = \int \frac{\sin x + \cos x}{\sqrt{9+16\sin 2x}} dx$

$$= \int \frac{\sin x + \cos x}{\sqrt{9+16\sin 2x + 16 - 16}} dx$$

$$= \int \frac{\sin x + \cos x}{\sqrt{25 - 16(1-\sin 2x)}} dx$$

$$= \int \frac{\sin x + \cos x}{\sqrt{25 - 16(\sin^2 x + \cos^2 - x\sin 2x)}} dx$$

Evaluate

$$\Rightarrow I = \int \frac{\sin x + \cos x}{\sqrt{25 - 16 (\sin x - \cos x)^2}} dx$$

Let $(\sin x - \cos x) = t$

$(\cos x + \sin x)\, dx = dt$

$$\Rightarrow \quad I = \int \frac{dt}{\sqrt{(5)^2 - (4t)^2}} = \frac{1}{4} \sin^{-1}\left(\frac{4t}{5}\right) + c$$

$$\left[\because \int \frac{1}{\sqrt{a^2 - x^2}} dx = \sin^{-1}\left(\frac{x}{a}\right) + c\right]$$

Putting the value of t, we get

$$I = \frac{1}{4} \sin^{-1}\left[\frac{4(\sin x - \cos x)}{5}\right] + c$$ **Ans.**

28. *Using definite integral as the limit of a sum, evaluate*

$$\lim_{n \to \infty}\left[\frac{1}{1+n^3} + \frac{4}{8+n^3} + \frac{9}{27+n^3} + \text{.....to } n \text{ terms}\right].$$

Sol. Let

$$L = \lim_{n \to \infty}\left[\frac{1}{1+n^3} + \frac{4}{8+n^3} + \frac{9}{27+n^3} + \text{... to } n \text{ terms}\right]$$

We see, $$T_r = \frac{r^2}{r^3 + n^3} = \frac{\frac{r^2}{n^2} \cdot \frac{1}{n}}{\frac{r^3}{n^3} + 1}$$

$$\therefore \quad L = \lim_{n \to \infty}\left[\sum_{r=1}^{n} T_r\right]$$

$$= \lim_{n \to \infty} \int_1^n \frac{\frac{r^2}{n^2} \cdot \frac{1}{n}}{\frac{r^3}{n^3} + 1} dr$$

Let, $\frac{r}{n} = x$

$\Rightarrow \quad \frac{dr}{n} = dx$

When $r = 1$

$\Rightarrow \quad x = \frac{1}{n}$

$r = n$

$\Rightarrow \quad x = 1$

$$\Rightarrow \quad L = \lim_{n \to \infty} \int_{1/n}^{1} \frac{x^2}{x^3 + 1} dx$$

$$\Rightarrow \quad L = \frac{1}{3} \int_0^1 \frac{3x^2 dx}{x^3 + 1}$$

$$= \frac{1}{3}\left[\log(x^3 + 1)\right]_0^1$$

$$= \frac{1}{3}[\log 2 - \log 1]$$

$$= \frac{1}{3} \log 2.$$ **Ans.**

29. *Evaluate* $\int_{-1}^{1} e^x dx$ *as limit of sums.*

Sol. We have,

$$\int_a^b f(x)\, dx = \lim_{h \to 0} h\, [f(a) + f(a + h) + f(a + 2h) + \ldots + f(a + (n - 1)h)]$$

where, $h = \frac{b - a}{n}$

Here, $a = -1, b = 1,$

$$f(x) = e^x \text{ and } h = \frac{1 - (-1)}{n} = \frac{2}{n}$$

$$\therefore \quad I = \int_{-1}^{1} e^x dx$$

$$I = h\, [f(-1) + f(-1 + h) + f(-1 + 2h) + \ldots + f(-1 + (n - 1)\, h)]$$

$$I = \lim_{h \to 0} h\, [e^{-1} + e^{-1+h} + e^{-1+2h} + \ldots + e^{-1+(n-1)h}]$$

$$I = \lim_{h \to 0} he^{-1}\, [1 + e^h + e^{2h} + \ldots + e^{(n-1)h}]$$

$$I = \lim_{h \to 0} he^{-1}\left[\left\{\frac{(e^h)^n - 1}{e^h - 1}\right\}\right]$$

$$I = \lim_{h \to 0} e^{-1}\left\{\frac{e^2 - 1}{\left(\frac{e^h - 1}{h}\right)}\right\} \qquad \left[\because h = \frac{2}{n}\right]$$

$$I = e^{-1}\left(\frac{e^2 - 1}{1}\right) = e - e^{-1}$$ **Ans.**

30. *Evaluate* $\int_0^1 e^{2-3x} dx$ *as limit of sums.*

Sol. We have,

$$\int_a^b f(x)\, dx = \lim_{h \to 0} h\, [f(a) + f(a + h) + f(a + 2h) + \ldots + f(a + (n - 1)\, h)], \text{ where } h = \frac{b - a}{n}$$

Here, $a = 0, b = 1$ and $f(x) = e^{2-3x}$

$\therefore \quad h = \frac{1}{n}$

$\Rightarrow \quad nh = 1$

Evaluate

$$\therefore \quad I = \int_0^1 e^{2-3x}$$

$$I = \lim_{h\to 0} [f(0) + f(h) + f(2h) + \dots + f((n-1)h)]$$

$$I = \lim_{h\to 0} h\,[e^2 + e^{2-3h} + e^{2-3(2h)} + \dots + e^{2-3(n-1)h}]$$

$$I = \lim_{h\to 0} he^2\,[1 + e^{-3h} + e^{-3(2h)} + \dots + e^{-3(n-1)h}]$$

$$I = \lim_{h\to 0} he^2 \left\{\frac{(e^{-3h})^n - 1}{e^{-3h} - 1}\right\}$$

$$I = \lim_{h\to 0} he^2 \left\{\frac{(e^{-3hn} - 1}{e^{-3h} - 1}\right\} \qquad [\because nh = 1]$$

$$I = e^2 \lim_{h\to 0} \frac{e^{-3} - 1}{\left(\frac{e^{-3h} - 1}{-3h}\right)} \times \left(\frac{-1}{3}\right)$$

$$I = e^2 (e^{-3} - 1) \times \left(\frac{-1}{3}\right)$$

$$I = -\frac{1}{3}(e^{-1} - e^2)$$

$$I = \frac{1}{3}(e^2 - e^{-1})$$ **Ans.**

31. *Evaluate* $\int_a^b x^2\,dx$ *as limit of sums.*

Sol. We have,

$$\int_a^b x^2\,dx = \lim_{h\to 0} h\,[f(a) + f(a+h) + f(a+2h) + \dots + f(a + (n-1)h)] \text{ where } h = \frac{b-a}{n}$$

$$\therefore I = \int_a^b x^2\,dx$$

$$I = \lim_{h\to 0} h\,[a^2 + (a+h)^2 + (a+2h)^2 + \dots + (a + (n-1)h)^2]$$

$$I = \lim_{h\to 0} h\,[na^2 + 2ah\,(1 + 2 + 3 + \dots + (n-1)) + h^2\{1^2 + 2^2 + \dots + (n-1)^2\}]$$

$$I = \lim_{h\to 0} h\left\{nu^2 + 2ah \times \frac{n(n-1)}{2} + h^2 \times \frac{n(n-1)(2n-1)}{6}\right]$$

$$I = \lim_{h\to 0} \left\{(nh)\,a^2 + a(nh)(nh-h) + \frac{1}{6}(nh)(nh-h)(2nh-h)\right]$$

$$I = \lim_{h\to 0} \left\{(b-a)a^2 + a(b-a)(b-a-h) + \frac{1}{6}(b-a)(b-a-h)(2(b-a)-h)\right\}$$

$$\left[\because h = \frac{b-a}{n} \Rightarrow nh = b-a\right]$$

$$I = \left\{(b-a)a^2 + a(b-a)^2 + \frac{2}{6}(b-a)^3\right\}$$

$$I = \frac{(b-a)}{3}[3a^2 + 3a(b-a) + (b-a)^2]$$

$$I = \frac{(b-a)}{3}(3a^2 + 3ab - 3a^2 + b^2 - 2ab + a^2)$$

$$I = \frac{1}{3}(b-a)(a^2 + ab + b^2)$$

$$I = \frac{1}{3}(b^3 - a^3)$$ **Ans.**

32. *Evaluate* $\int_{-1}^{1}\{x - [x]\}\,dx$, *where* $[x]$ *is greatest integer function.*

Sol. Consider,

$$\int_{-1}^{1}\{x-[x]\}\,dx = \int_{-1}^{1} x\,dx - \int_{-1}^{1}[x]\,dx$$

$$= 0 - \int_{-1}^{1}[x]\,dx$$

$\because x$ is an odd function

$$= -\int_{-1}^{0}(-1)\,dx + \int_0^1 0\cdot dx$$

$$= [x]_{-1}^{0} = 1.$$ **Ans.**

33. *Evaluate*

$$\int_0^{\pi/4} \frac{2\cos 2x}{1+\sin 2x}\,dx.$$

Sol. Let $I = \int_0^{\pi/4} \frac{2\cos 2x}{1+\sin 2x}\,dx$

Let $1 + \sin 2x = t$

Differentiating both sides w.r.t. x, we get

$$\cos 2x \,.\, 2 = \frac{dt}{dx}$$

$$\Rightarrow \quad 2\cos 2x \,.\, dx = dt$$

At $x = 0$, $t = 1 + \sin 2 \times 0 = 1 + 0 = 1$

At $x = \frac{\pi}{4}$, $t = 1 + \sin 2 \times \frac{\pi}{4} = 1 + 1 = 2$

$$\Rightarrow \quad I = \int_1^2 \frac{dt}{t}$$

$$= [\log_e t]_1^2$$

$$= \log 2 - \log 1$$

$$= \log 2$$ **Ans.**

34. *Evaluate :*

$$\int_0^{\pi/2} \frac{\sin x \cos x}{\cos^2 x + 3\cos x + 2}.$$

Sol. Consider,

$$\int_0^{\pi/2} \frac{\sin x \cos x}{\cos^2 x + 3\cos x + 2} dx$$

Let, $\cos x = t$

$\therefore$ $-\sin x\, dx = dt$

When $x = 0, t = 1$

and $x = \dfrac{\pi}{2}, t = 0$

$$\therefore \quad I = \int_1^0 \frac{-t.dt}{t^2+3t+2}$$

$$= \int_0^1 \frac{t\,dt}{t^2+3t+2}$$

Now, $\dfrac{t}{t^2+3t+2} = \dfrac{t}{(t+2)(t+1)}$

Taking partial fractions, we get

$$\frac{t}{t^2+3t+2} = \frac{A}{t+2} + \frac{B}{t+1}$$

$$t = At + A + Bt + 2B$$

Equating coefficients of like terms, we get

$A + B = 1, A + 2B = 0$

$\Rightarrow$ $B = -1$ and $A = 2$

$$\therefore \quad I = \int_0^1 \left(\frac{2}{t+2} - \frac{1}{t+1}\right) dt$$

$$= [2\log(t+2) - \log(t+1)]_0^1$$

$$= 2\log 3 - \log 2 - (2\log 2 - \log 1)$$

$$= 2\log 3 - 3\log 2$$

$$= \log 9 - \log 8$$

$$= \log \frac{9}{8}.$$

Ans.

35. *Evaluate :*

$$\int_0^1 x \tan^{-1} x\, dx.$$

Sol. Let $I = \int_0^1 x \tan^{-1} x \,.\, dx$

On integrating by parts

$$= \tan^{-1} x \cdot \int_0^1 x.\,dx - \int_0^1 \frac{x^2}{2}.\frac{1}{1+x^2} dx$$

$$= \left[\frac{x^2}{2}.\tan^{-1} x\right]_0^1 - \frac{1}{2}\left[\int_0^1 \frac{x^2+1-1}{x^2+1}.\,dx\right]$$

$$= \left[\frac{x^2}{2}.\tan^{-1} x\right]_0^1 - \frac{1}{2}\left[\int_0^1 \left(1 - \frac{1}{1+x^2}\right) dx\right]$$

$$= \frac{1}{2}.\frac{\pi}{4} - 0 - \frac{1}{2}\left[x - \tan^{-1} x\right]_0^1$$

$$= \frac{\pi}{8} - \frac{1}{2}\left[1 - 0 - \frac{\pi}{4} - 0\right]$$

$$= \frac{\pi}{8} + \frac{\pi}{8} - \frac{1}{2}$$

$$= \frac{\pi}{4} - \frac{1}{2}.$$

Ans

36. *Evaluate :*

$$\int_{-5}^{5} |x+2|\, dx.$$

Sol. Let $I = \int_{-5}^{5} |x+2|\, dx$

$$= \int_{-5}^{-2} |x+2|\, dx + \int_{-2}^{5} |x+2|\, dx$$

$$= \int_{-5}^{-2} (x+2)\, dx + \int_{-2}^{5} (x+2)\, dx$$

$$= -\left[\frac{x^2}{2} + 2x\right]_{-5}^{-2} + \left[\frac{x^2}{2} + 2x\right]_{-2}^{5}$$

$$= -\frac{1}{2}(4-25) - 2(-2+5)$$

$$+ \frac{1}{2}(25-4) + 2(5+2$$

$$= \frac{21}{2} - 6 + \frac{21}{2} + 14 = 29$$

Ans

37. *Evaluate :* $\displaystyle\int_0^1 \frac{x\,e^x}{(1+x)^2} dx.$

Sol. Consider, $\displaystyle\int_0^1 \frac{x\,e^x}{(1+x)^2} dx.$

$$= \int_0^1 \frac{(x+1)-1}{(1+x)^2}.\,e^x\, dx$$

$$= \int_0^1 \frac{(x+1)e^x}{(x+1)^2} dx - \int_0^1 \frac{e^x}{(x+1)^2} dx$$

$$= \int_0^1 \frac{e^x}{x+1} dx - \int_0^1 \frac{e^x}{(x+1)^2} dx$$

Integrating by parts, we get

$$= \left[e^x \frac{1}{x+1}\right]_0^1 - \int_0^1 e^x \left[-\frac{1}{(x+1)^2}\right] dx$$

$$- \int_0^1 \frac{e^x}{(x+1)^2} dx$$

$$= \left[\frac{1}{2}e^1 - 1.e^0\right] + \int_0^1 \frac{1}{(x+1)^2} . e^x dx$$

$$- \int_0^1 \frac{1}{(x+1)^2} . e^x \, dx$$

$$= \frac{1}{2}e - 1 = \frac{e}{2} - 1$$ **Ans.**

38. *Evaluate* $\int_{-3}^{3} |x + 2| dx.$

Sol. Let $|x + 1| = f(x)$ We have

$$f(x) = x + 2, \text{ if } x \geq -2$$
$$= -(x + 2) \text{ if } x < -2$$

$$\therefore \int_{-3}^{3} |x+2| \, dx = \int_{-3}^{-2} -(x+2) \, dx + \int_{-2}^{3} (x+2) \, dx$$

$$= -\left[\frac{x^2}{2} + 2x\right]_{-3}^{-2} + \left[\frac{x^2}{2} + 2x\right]_{-2}^{3}$$

$$= -\left[2 - 4 - \frac{9}{2} - 2(-3)\right] + \left[\frac{9}{2} + 6 - 2 + 4\right]$$

$$= -\left[4 - \frac{9}{2}\right] + \left[\frac{9}{2} + 8\right]$$

$$= \frac{1}{2} + \frac{9}{2} + 8$$

$= 5 + 8 = 13$ **Ans.**

39. *Evaluate* $\int_a^b \frac{\log x}{x} dx$.

Sol. Let $I = \int_a^b \frac{\log x}{x} dx$

On integrating by parts

$$= \log x \int_a^b \frac{1}{x} \cdot dx - \int_a^b \frac{1}{x} \cdot \log x . \, dx$$

$$= \log x \int_a^b \frac{1}{x} - I$$

$$2I = \left[\log x . \log x\right]_a^b$$

$$\Rightarrow I = \frac{1}{2}[(\log x)^2]_a^b$$

$$I = \frac{1}{2}\{(\log b)^2 - (\log a)^2\}$$

$$= \frac{1}{2}\{(\log b + \log a)(\log b - \log a)\}$$

$$= \frac{1}{2}\log ab . \log\frac{b}{a}$$ **Ans.**

40. *Evaluate*

$$\int_0^{1/2} \frac{\sin^{-1} x}{(1-x^2)^{3/2}} dx.$$

Sol. Given, $\int_0^{1/2} \frac{\sin^{-1} x}{(1-x^2)^{3/2}} dx$

Let $\sin^{-1} x = t$

$\Rightarrow x = \sin t$

$\therefore dx = \cos t \, dt$

$$\therefore \int_0^{1/2} \frac{\sin^{-1} x}{(1-x^2)^{3/2}} dx = \int_0^{\pi/6} \frac{t \cos t \, dt}{\cos^3 t}$$

$\left[\text{When } x = 0, t = 0; \quad x = \frac{1}{2}, t = \frac{\pi}{6}\right]$

$$= \int_0^{\pi/6} \frac{t}{\cos^2 t} dt$$

$$= \int_0^{\pi/6} t \sec^2 t \, dt$$

Integrating by parts, we get

$$= \left[t \int \sec^2 t - \int \left\{\frac{d}{dt}(t) . \int \sec^2 t\right\} dt\right]_0^{\pi/6}$$

$$= [t . \tan t - \int \tan t]_0^{\pi/6} = [t . \tan t - \log \sec t]_0^{\pi/6}$$

$$= \left[\frac{\pi}{6}\tan\frac{\pi}{6} - \log\sec\frac{\pi}{6}\right] - [0 - \log\sec 0]$$

$$= \left[\frac{\pi}{6}\left(\frac{1}{\sqrt{3}}\right) - \log\frac{2}{\sqrt{3}}\right] - [0 - 0]$$

$$= \frac{\pi}{6\sqrt{3}} - \log\frac{2}{\sqrt{3}}.$$ **Ans.**

41. *Evaluate :* $\int_0^3 f(x) \, dx$ *, where*

$$f(x) = \begin{cases} \cos 2x, 0 & \leq x \leq \frac{\pi}{2} \\ 3 & \frac{\pi}{2} \leq x \leq 3. \end{cases}$$

Sol. $$\int_0^3 f(x) \, dx = \int_0^{\pi/2} f(x) \, dx + \int_{\pi/2}^3 f(x) \, dx$$

$$= \int_0^{\pi/2} \cos 2x \, dx + \int_{\pi/2}^3 3 \, dx$$

$$= \left[\frac{\sin 2x}{2}\right]_0^{\pi/2} + [3x]_{\pi/2}^3$$

$$= \left[\frac{\sin 2 \times \pi/2}{2} - \frac{\sin 2 \times 0}{2}\right] + [3 \times 3 - 3\pi/2]$$

Evaluate

$$= \frac{\sin\pi}{2} - 0 + 9 - \frac{3\pi}{2} \quad [\because \sin\pi = 0]$$

$$= 9 - \frac{3\pi}{2} = 3\left(3 - \frac{\pi}{2}\right)$$ **Ans.**

42. *Evaluate :* $\int_1^2 \frac{\sqrt{x}}{\sqrt{3-x}+\sqrt{x}}dx.$

Sol. Let $I = \int_1^2 \frac{\sqrt{x}}{\sqrt{3-x}+\sqrt{x}}dx$ ...(i)

Using property, $\int_a^b f(x)dx = \int_a^b f(a+b-x)\,dx$

$$I = \int_1^2 \frac{\sqrt{3-x}}{\sqrt{3-(3-x)}+\sqrt{3-x}}dx \quad ...(ii)$$

Adding (i) and (ii), we get

$$2I = \int_1^2 \frac{\sqrt{x}+\sqrt{3-x}}{\sqrt{x}+\sqrt{3-x}}dx = \int_1^2 1\,dx$$

$$= [x]_1^2 = 2-1 = 1$$

$$\therefore \quad I = \frac{1}{2}$$ **Ans.**

43. *Using properties of definite integrals, evaluate :*

$$\int_0^{\pi/2} \frac{\sqrt{\sin x}}{\sqrt{\sin x}+\sqrt{\cos x}}dx.$$

Sol. Let, $I = \int_0^{\pi/2} \frac{\sqrt{\sin x}}{\sqrt{\sin x}+\sqrt{\cos x}}dx$ ...(i)

Using property, $\int_0^a f(x)\,dx = \int_0^a f(a-x)\,dx$

$$= \int_0^{\pi/2} \frac{\sqrt{\sin\left(\frac{\pi}{2}-x\right)}}{\sqrt{\sin\left(\frac{\pi}{2}-x\right)}+\sqrt{\cos\left(\frac{\pi}{2}-x\right)}}dx$$

(Using property)

$$I = \int_0^{\pi/2} \frac{\sqrt{\cos x}}{\sqrt{\cos x}+\sqrt{\sin x}} \quad ...(ii)$$

Adding equation (i) and (ii),

$$2I = \int_0^{\pi/2} \frac{\sqrt{\sin x}+\sqrt{\cos x}}{\sqrt{\sin x}+\sqrt{\cos x}}dx$$

$$2I = \int_0^{\pi/2} 1\,dx = [x]_0^{\pi/2}$$

$$I = \frac{1}{2}\left[\frac{\pi}{2}-0\right]$$

$$I = \frac{\pi}{4}$$ **Ans.**

44. *Evaluate :* $\int_0^1 \log\left(\frac{1}{x}-1\right)dx.$

Sol. Consider, $\int_0^1 \log\left(\frac{1}{x}-1\right)dx = I$ (say)

Using property, $\int_0^a f(x)\,dx = \int_0^a f(a-x)\,dx$

$$I = \int_0^1 \log\left(\frac{1}{(1-x)}-1\right)dx$$

$$= \int_0^1 \log\left(\frac{1-(1-x)}{1-x}\right)dx$$

$$= \int_0^1 \log\left(\frac{x}{1-x}\right)dx$$

$$= \int_0^1 \log\left(\frac{1-x}{x}\right)^{-1} dx$$

$$= \int_0^1 (-1)\log\left(\frac{1-x}{x}\right)dx$$

$$I = -\int_0^1 \log\left(\frac{1-x}{x}\right)dx$$

$$I = -I$$

$$2I = 0$$

$$\therefore \quad I = 0$$

$$\therefore \int_0^1 \log\left(\frac{1}{x}-1\right)dx = 0.$$ **Ans.**

45. *Evaluate :*

$$\int_0^{\pi/2} \sin 2x \ \log(\tan x)\,dx \cdot$$

Sol. Let $I = \int_0^{\pi/2} \sin 2x\,.\log(\tan x)\,dx$ ...(i)

Also, $I = \int_0^{\pi/2} \sin 2\left(\frac{\pi}{2}-x\right).\log\left[\tan\left(\frac{\pi}{2}-x\right)\right].dx$

$$\left\{\because \int_0^a f(x).dx = \int_0^a f(x-a)\,.\,dx\right\}$$

$$= \int_0^{\pi/2} \sin(\pi-2x).\log\cot x.\,dx$$

$$= \int_0^{\pi/2} \sin 2x.\log\cot x.\,dx \quad ...(ii)$$

Adding equations (i) and (ii), we get

$$2I = \int_0^{\pi/2} \sin 2x\,(\log\tan x + \log\cot x)\,dx$$

$$= \int_0^{\pi/2} \sin 2x \, \log(\tan x \, . \cot x) \, dx$$

$$= \int_0^{\pi/2} \sin 2x \, . \log 1 . \, dx \qquad [\because \log 1 = 0]$$

$$= 0.$$

$$\Rightarrow \quad \int_0^{\pi/2} \sin 2x \, \log(\tan x) \, dx = 0 \qquad \textbf{Ans.}$$

46. *Evaluate :*

$$\int_0^{\pi/2} \frac{\sqrt{\sec x}}{\sqrt{\sec x} + \sqrt{\operatorname{cosec} x}} dx.$$

Sol. Let,

$$I = \int_0^{\pi/2} \frac{\sqrt{\sec x}}{\sqrt{\sec x} + \sqrt{\operatorname{cosec} x}} dx \qquad \ldots(i)$$

Using property, $\int_0^a f(x)\, dx = \int_0^a f(a-x)\, dx$

$$I = \int_0^{\pi/2} \frac{\sqrt{\sec(\pi/2 - x)}}{\sqrt{\sec(\pi/2 - x)} + \sqrt{\operatorname{cosec}(\pi/2 - x)}}$$

$$= \int_0^{\pi/2} \frac{\sqrt{\operatorname{cosec} x}}{\sqrt{\operatorname{cosec} x} + \sqrt{\sec x}} dx \qquad \ldots(ii)$$

By adding equations (i) and (ii), we get

$$\therefore \quad 2I = \int_0^{\pi/2} \frac{\sqrt{\sec x}}{\sqrt{\operatorname{cosec} x} + \sqrt{\sec x}} \cdot dx + \int_0^{\pi/2} \frac{\sqrt{\operatorname{cosec} x}}{\sqrt{\sec x} + \sqrt{\operatorname{cosec} x}} dx$$

$$\therefore \quad 2I = \int_0^{\pi/2} \frac{(\sqrt{\sec x} + \sqrt{\operatorname{cosec} x})}{(\sqrt{\sec x} + \sqrt{\operatorname{cosec} x})} dx$$

$$\therefore \quad 2I = \int_0^{\pi/2} 1 \, dx$$

$$\therefore \quad 2I = [x]_0^{\pi/2} = \frac{\pi}{2}$$

$$\therefore \quad I = \frac{\pi}{4}$$

Hence, $\int_0^{\pi/2} \frac{\sqrt{\sec x}}{\sqrt{\operatorname{cosec} x} + \sqrt{\sec x}} dx = \frac{\pi}{4}.$ **Ans.**

47. *Evaluate*

$$\int_0^{\pi/2} \log \sin dx.$$

Sol. Let $I = \int_0^{\pi/2} \log \sin dx$ ...(i)

Using property, $\int_0^a f(x)\, dx = \int_0^a f(a-x)\, dx$

$$I = \int_0^{\pi/2} \log \sin\left(\frac{\pi}{2} - x\right) dx$$

$$= \int_0^{\pi/2} \log \cos x \, dx \qquad \ldots(ii)$$

Adding equations (i) and (ii), we get

$$2I = \int_0^{\pi/2} \log(\sin x) + \log(\cos x) \, dx$$

$$= \int_0^{\pi/2} \log \sin x \cos x \, dx$$

$$= \int_0^{\pi/2} \log \frac{\sin 2x}{2} dx$$

$$= \int_0^{\pi/2} (\log \sin 2x - \log 2) \, dx$$

$$= \int_0^{\pi/2} \log \sin 2x \, dx - \int_0^{\pi/2} \log 2 \, dx$$

Putting, $t = 2x,$
$dt = 2dx$

$$= \left[\frac{1}{2} \int_0^{\pi} \log \sin t \, dt\right] - \log 2 [x]_0^{\pi/2}$$

$[\because \sin(\pi - t) = \sin t]$

$$= \left[\frac{1}{2} . 2 \int_0^{\pi/2} \log \sin t \, dt\right] - \frac{\pi}{2} \log 2$$

$$2I = I - \frac{\pi}{2} \log 2$$

$$I = -\frac{\pi}{2} \log 2. \qquad \textbf{Ans.}$$

48. *Evaluate*

$$\int_0^{\pi/2} \frac{x \sin x \cos x}{\cos^4 x + \sin^4 x} dx.$$

Sol. Let $I = \int_0^{\pi/2} \frac{x \sin x \cos x}{\cos^4 x + \sin^4 x} dx$ ...(i)

Using property, $\int_0^a f(x)\, dx = \int_0^a f(a-x)\, dx$

$$I = \int_0^{\pi/2} \frac{\left(\frac{\pi}{2} - x\right) \sin\left(\frac{\pi}{2} - x\right) \cos\left(\frac{\pi}{2} - x\right)}{\cos^4\left(\frac{\pi}{2} - x\right) + \sin^4\left(\frac{\pi}{2} - x\right)} dx$$

$$I = \int_0^{\pi/2} \frac{\left(\frac{\pi}{2} - x\right) \sin x \cos x}{\cos^4 + \sin^4 x} \qquad \ldots(ii)$$

Adding equations (i) and (ii), we get

$$2I = \int_0^{\pi/2} \frac{\pi}{2} \frac{\cos x \sin x}{2\sin^4 x + \cos^4 x} dx$$

Dividing numerator and denominator by $\cos^4 x$, we get

$$2I = \frac{\pi}{2}\int_0^{\pi/2} \frac{\tan x \sec^2 x}{\tan^4 x + 1} dx$$

Put $\tan^2 x = t$

$2 \tan x \sec^2 x \, dx = dt$

When $x = 0$, $t = 0$ and

When $x = \frac{x}{2}$, $t = \infty$

$$2I = \frac{\pi}{2}\int_0^{\infty} \frac{1dt}{2(1+t^2)} = \frac{\pi}{8}\int_0^{\infty} \frac{dt}{(1+t^2)}$$

$$= \frac{\pi}{8}\left[\tan^{-1} t\right]_0^{\infty} = \frac{\pi}{8}\left(\frac{\pi}{2} - 0\right) = \frac{\pi^2}{16}$$ **Ans.**

49. *Evaluate :* $\int_0^{\pi/2} \frac{1}{1+\tan x} dx.$

Sol. $I = \int_0^{\pi/2} \frac{1}{1+\tan x} dx$

$$\therefore \quad I = \int_0^{\pi/2} \frac{\cos x \, dx}{\cos x + \sin x} \quad \text{...(i)}$$

Using property, $\int_0^a f(x)\,dx = \int_0^a f(a-x)\,dx$

$$= \int_0^{\pi/2} \frac{\left(\cos\frac{\pi}{2} - x\right)dx}{\cos\left(\frac{\pi}{2} - x\right) + \sin\left(\frac{\pi}{2} - x\right)}$$

$$\therefore \quad I = \int_0^{\pi/2} \frac{\sin x \, dx}{\sin x + \cos x} \quad \text{...(ii)}$$

Adding equations (i) and (ii), we get

$$\therefore \quad 2I = \int_0^{\pi/2} \frac{\sin x + \cos x}{\sin x + \cos x} dx$$

$$= \int_0^{\pi/2} 1.dx = [x]_0^{\pi/2} = \frac{\pi}{2}$$

Hence, $\int_0^{\pi/2} \frac{1}{1+\tan x} dx = \frac{\pi}{4}.$ **Ans**

50. *Using properties of definite integrals, evaluate :*

$$\int_0^{\pi/2} \frac{\sin x - \cos x}{1 + \sin x \cos x} dx$$

Sol. Let $I = \int_0^{\pi/2} \frac{\sin x - \cos x}{1+\sin x \cos x} dx$

$$I = \int_0^{\pi/2} \frac{\sin x}{1+\sin x \cos x} dx - \int_0^{\pi/2} \frac{\cos x}{1+\sin x \cos x} dx$$

Here, $a + b - x = \frac{\pi}{2} - x$, so using the property $\int_a^b f(x)\,dx = \int_a^b f(a+b-x)\,dx$ in first integral, we have

$$I = \int_0^{\pi/2} \frac{\sin\left(\frac{\pi}{2} - x\right)}{1+\sin\left(\frac{\pi}{2} - x\right)\cos\left(\frac{\pi}{2} - x\right)} dx - \int_0^{\pi/2} \frac{\cos x}{1+\sin x \cos x} dx$$

$$I = \int_0^{\pi/2} \frac{\cos x}{1+\cos x \sin x} dx - \int_0^{\pi/2} \frac{\cos x}{1+\sin x \cos x} dx$$

$$I = 0$$ **Ans.**

Chapter 11. Vectors

1. *Evaluate :*

$$\vec{a} \times (\vec{b} + \vec{c}) + \vec{b} \times (\vec{c} + \vec{a}) + \vec{c} \times (\vec{a} + \vec{b}).$$

Sol. We have $\vec{a} \times (\vec{b} + \vec{c}) + \vec{b} \times (\vec{c} + \vec{a}) + \vec{c} \times (\vec{a} + \vec{b})$

$$= (\vec{a} \times \vec{b}) + (\vec{a} \times \vec{c}) + (\vec{b} \times \vec{c}) + (\vec{b} \times \vec{a}) + (\vec{c} \times \vec{a}) + (\vec{c} \times \vec{b})$$

$$= (\vec{a} \times \vec{b}) + (\vec{a} \times \vec{c}) + (\vec{b} \times \vec{c}) - (\vec{a} \times \vec{b}) - (\vec{a} \times \vec{c}) - (\vec{b} \times \vec{c})$$

$= 0.$ $\quad [\because \vec{a} \times \vec{b} = -(\vec{b} \times \vec{a})]$

Ans.

2. *If* $\vec{a} = \hat{i} + \hat{j} + 2\hat{k}$ *and* $\vec{b} = -2\hat{j} + 4\hat{k}$. *Evaluate* $\vec{a}.\vec{b}$.

Sol. $\vec{a}.\vec{b}=(\hat{i}+\hat{j}+2\hat{k})\cdot(0\hat{i}-2\hat{j}+4\hat{k})$

$= (1)(0) + (1)(-2) + (2)(4)$

$= 0 - 2 + 8 = 6$ **Ans.**

3. *If* $\vec{a}=-\hat{i}+3\hat{j}+2\hat{k}$ *and* $\vec{b}=\hat{i}-2\hat{j}+5\hat{k}$. *Evaluate* $\vec{a}\times\vec{b}$.

Sol. $\vec{a}\times\vec{b}=\begin{vmatrix}\hat{i} & \hat{j} & \hat{k}\\ -1 & 3 & 2\\ 1 & -2 & 5\end{vmatrix}$

$= (15-(-4))\hat{i} - (-5-2)\hat{j} + (2-3)\hat{k}$

$= 19\hat{i}+7\hat{j}-\hat{k}$ **Ans.**

4. *If* $|\vec{a}\times\vec{b}|=\vec{a}.\vec{b}$ *then, evaluate the angle between the two vectors* $\vec{a}$ *and* $\vec{b}$.

Sol. Given,

$|\vec{a}\times\vec{b}|=\vec{a}.\vec{b}$

Now, $|\vec{a}|\ |\vec{b}|\sin\theta=|\vec{a}|\ |\vec{b}|\cos\theta$

$(\because |\vec{a}\times\vec{b}|=|\vec{a}||\vec{b}|\sin\theta$

and $\vec{a}.\vec{b}=|\vec{a}||\vec{b}|\cos\theta)$

$\Rightarrow \sin\theta = \cos\theta$

Dividing both sides by cos θ, we get

$\frac{\sin\theta}{\cos\theta} = 1$

$\tan\theta = 1$

$\theta = \tan^{-1}(1) = \frac{\pi}{4}.$

∴ Angle between the two vectors $\vec{a}$ and $\vec{b}$ is $\frac{\pi}{4}$. **Ans.**

5. *Evaluate the area of the triangle with vertices A (2, 4, 1), B (3, 2, 4) and C (4, 1, 3).*

Sol. Let $\vec{a},\vec{b}$ and $\vec{c}$ are the position vectors of the vertices A, B and C of ΔABC respectively.
Now, we have

$\vec{a}=2\hat{i}+4\hat{j}+\hat{k},$

$\vec{b}=3\hat{i}+2\hat{j}+4\hat{k}$

and $\vec{c}=4\hat{i}+\hat{j}+3\hat{k}$

Then, vector area of Δ ABC

$=\frac{1}{2}[(\vec{a}\times\vec{b})+(\vec{b}\times\vec{c})+(\vec{c}\times\vec{a})]$

$=\frac{1}{2}\left(\begin{vmatrix}\hat{i} & \hat{j} & \hat{k}\\ 2 & 4 & 1\\ 3 & 2 & 4\end{vmatrix}+\begin{vmatrix}\hat{i} & \hat{j} & \hat{k}\\ 3 & 2 & 4\\ 4 & 1 & 3\end{vmatrix}+\begin{vmatrix}\hat{i} & \hat{j} & \hat{k}\\ 4 & 1 & 3\\ 2 & 4 & 1\end{vmatrix}\right)$

$=\frac{1}{2}(14\hat{i}-5\hat{j}-8\hat{k}+2\hat{i}+7\hat{j}-5\hat{k}-11\hat{i}+2\hat{j}+14\hat{k})$

$=\frac{1}{2}(5\hat{i}+4\hat{j}+\hat{k})$

∴ Area of ΔABC $=\left|\frac{1}{2}(5\hat{i}+4\hat{j}+\hat{k})\right|$

$=\frac{1}{2}\sqrt{(5)^2+(4)^2+(1)^2}$

$=\frac{1}{2}\sqrt{25+16+1}$

$=\frac{\sqrt{42}}{4}$ sq. units. **Ans.**

6. *Evaluate the area of a parallelogram having diagonals* $\hat{i}-3\hat{j}+4\hat{k}$ *and* $2\hat{i}+\hat{j}-2\hat{k}$.

Sol. We have

$\vec{d_1}=\hat{i}-3\hat{j}+4\hat{k}$

and $\vec{d_2}=2\hat{i}+\hat{j}-2\hat{k}$

Then, $\vec{d_1}\times\vec{d_2}=\begin{vmatrix}\hat{i} & \hat{j} & \hat{k}\\ 1 & -3 & 4\\ 2 & 1 & -2\end{vmatrix}$

$= (6-4)\hat{i}-(-2-8)\hat{j}+(1+6)\hat{k}$

$= 2\hat{i}+10\hat{j}+7\hat{k}$

Now, $\vec{d_1}\times\vec{d_2} = \sqrt{(2)^2+(10)^2+(7)^2}$

$=\sqrt{4+100+49}$

$=\sqrt{153}=3\sqrt{17}$

Hence, area of parallelogram

$=\frac{1}{2}|d_1\times d_2|$

$=\frac{1}{2}(3\sqrt{17})$ sq. units

$=\frac{3}{2}\sqrt{17}$ sq. units. **Ans.**

7. *Evaluate :* $\hat{i}\cdot(\hat{j}\times\hat{k})+(\hat{i}\times\hat{k})\cdot\hat{j}$.

Sol. We have

$$\hat{i}\cdot(\hat{j}\times\hat{k})+(\hat{i}\times\hat{k})\cdot\hat{j} = \hat{i}\cdot\hat{i}+(-\hat{j})\cdot\hat{j}$$

$$(\because \hat{j}\times\hat{k}=\hat{i} \text{ and } \hat{i}\times\hat{k}=-\hat{j})$$

$$= |\hat{i}|^2-|\hat{j}|^2$$

$$= 1-1=0.$$ **Ans.**

8. *Evaluate :*

$$(\vec{a}-\vec{b})\,[(\vec{b}-\vec{c})\times(\vec{c}-\vec{a})]$$

Sol. We have,

$$(\vec{a}-\vec{b})\,[(\vec{b}-\vec{c})\times(\vec{c}-\vec{a})]$$

Now, the vectors are coplanar.

as $(\vec{a}-\vec{b})+(\vec{b}-\vec{c})+(\vec{c}-\vec{a})=0$

Then, $[(\vec{a}-\vec{b}),(\vec{b}-\vec{c}),(\vec{c}-\vec{a})]=0$

(Volume of paralleopiped)

i.e., $[(\vec{a}-\vec{b})[(\vec{b}-\vec{c})\times(\vec{c}-\vec{a})]=0$ **Ans.**

9. *Evaluate :*

$$\vec{p}\cdot(\vec{q}+\vec{r})\times(\vec{p}+2\vec{q}+3\vec{r})$$

Sol. Given,

$$\vec{p}\cdot(\vec{q}+\vec{r})\times(\vec{p}+2\vec{q}+3\vec{r})$$

$$=\vec{p}(\vec{q}\times\vec{p}+\vec{q}\times 2\vec{q}+\vec{q}\times 3\vec{r}+\vec{r}\times\vec{p}$$
$$\vec{r}\times 2\vec{q}+\vec{r}\times 3\vec{r})$$

$$=\vec{p}(\vec{q}\times\vec{p}+0+3\vec{q}\times\vec{r}+\vec{r}\times\vec{p}+2\vec{r}\times\vec{q}+0)$$

$$=\vec{p}(\vec{q}\times\vec{p})+\vec{p}(3\vec{q}\times\vec{r})+\vec{p}(\vec{r}\times\vec{p})+\vec{p}(2\vec{r}\times\vec{q})$$

$$=[\vec{p}\,\vec{q}\,\vec{p}]+3[\vec{p}\,\vec{q}\,\vec{r}]+[\vec{p}\,\vec{r}\,\vec{p}]+2[\vec{p}\,\vec{r}\,\vec{q}]$$

$$=0+3[\vec{p}\,\vec{q}\,\vec{r}]+0+2[\vec{p}\,\vec{r}\,\vec{q}]$$

$$=3[\vec{p}\,\vec{q}\,\vec{r}]-2[\vec{p}\,\vec{q}\,\vec{r}] \qquad (\because [\vec{p}\,\vec{q}\,\vec{r}]=-[\vec{p}\,\vec{q}\,\vec{r}])$$

$$=[\vec{p}\,\vec{q}\,\vec{r}]$$ **Ans.**

Chapter 13. Application of Integrals

1. *Sketch the graph* $y=|x+1|$. *Evaluate* $\int_{-3}^{1}|x+1|\,dx$. *What does this value represent on the graph ?*

Sol. We have,

$$y = |x+1|$$

$$y = \begin{cases} x+1 & \text{if, } x+1\ge 0 \\ -(x+1), & \text{if } x+1<0 \end{cases}$$

i.e.,
$$y = |x+1|$$
$$= \begin{cases} x+1 & \text{if, } x\ge -1 \\ -x-1, & \text{if } x<-1 \end{cases}$$

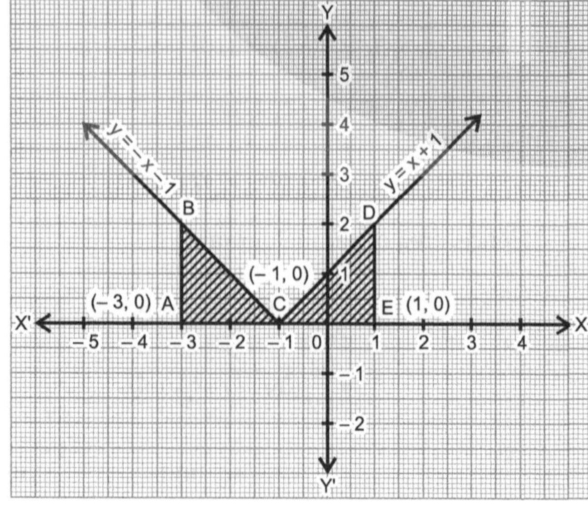

Now, $y=x+1$ and $y=-x-1$ are straight lines, plotting these lines represent the rough sketch of $y=|x+1|$

$$\int_{-3}^{1}|x+1|\,dx=\int_{-3}^{-1}|x+1|\,dx+\int_{-1}^{1}|x+1|\,dx$$

$$=\int_{-3}^{-1}-(x+1)dx+\int_{-1}^{1}(x+1)dx$$

$$=-\left[\frac{(x+1)^2}{2}\right]_{-3}^{-1}+\left[\frac{(x+1)^2}{2}\right]_{-1}^{1}$$

$$=-\left[0-\frac{4}{2}\right]+\left[\frac{4}{2}-0\right]$$

$$= 4 \text{ sq. units.}$$

This value represents the area of the shaded region, *i.e.*, 4 sq. units. **Ans.**

2. *Sketch the graph* $y=|x-5|$. *Evaluate* $\int_{0}^{1}|x-5|\,dx$. *What does this value of the integral represent on the graph ?*

Sol. We have,

$$y = |x-5|$$
$$= |x-5|$$
$$= \begin{cases} x-5, & \text{if } x-5\ge 0 \\ -(x-5), & \text{if } x-5<0 \end{cases}$$

i.e., $y = |x-5|$

$$= \begin{cases} x-5, & \text{if } x \geq 5 \\ -(x-5), & \text{if } x < 5 \end{cases}$$

Now, $y = x - 5$ and $y = -x + 5$ are straight lines, plotting these lines represent the rough sketch of $y = |x-5|$

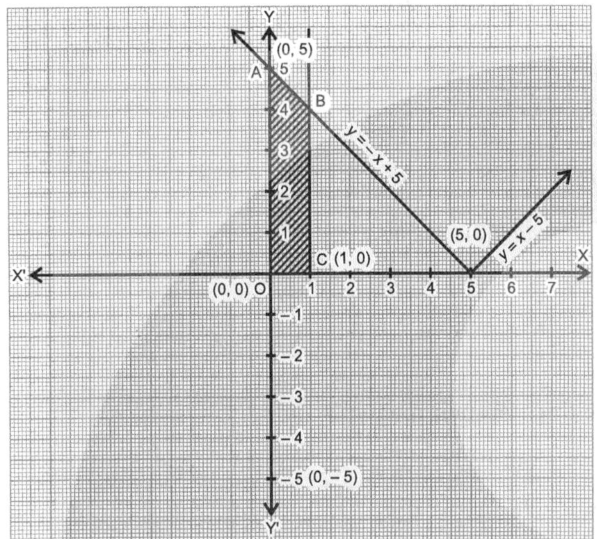

$$\int_0^1 |x-5|\, dx = \int_0^1 -(x-5)dx = \int_0^1 (5-x)\, dx$$

$$= \left[5x - \frac{x^2}{2}\right]_0^1$$

$$= \left[5 - \frac{1}{2}\right]$$

$$= \frac{9}{2}$$

This value of the integral represents the area of the shaded region, *i.e.,* $\frac{9}{2}$ sq. units. **Ans.**

3. *Sketch the graph* $y = |x+3|$. *Evaluate* $\int_{-6}^{0} |x+3|$ *dx. What does this integral represent on the graph ?*

Sol. We have,

$$y = |x+3|$$

$$= \begin{cases} x+3, & \text{if, } x+3 \geq 0 \\ -(x+3), & \text{if } x+3 < 0 \end{cases}$$

i.e., $y = |x+3|$

$$= \begin{cases} x+3, & \text{if, } x \geq -3 \\ -x-3, & \text{if } x < -3 \end{cases}$$

Now, $y = x + 3$ and $y = -x - 3$ are straight lines, plotting these lines represent the rough sketch of $y = |x+3|$

$$\int_{-6}^{0} |x+3|\, dx = \int_{-6}^{-3} |x+3|\, dx + \int_{-3}^{0} |x+3|\, dx$$

$$= \int_{-6}^{-3} -(x+3)\, dx + \int_{-3}^{0} (x+3)\, dx$$

$$= -\left[\frac{(x+3)^2}{2}\right]_{-6}^{-3} + \left[\frac{(x+3)^2}{2}\right]_{-3}^{0}$$

$$= -\left[\frac{(-3+3)^2}{2} - \frac{(-6+3)^2}{2}\right]$$

$$+ \left[\frac{(0+3)^2}{2} - \frac{(-3+3)^2}{2}\right]$$

$$= -\left(0 - \frac{9}{2}\right) + \left(\frac{9}{2} - 0\right) = 9 \text{ sq. units.}$$

The value of this integral represents the area of the shaded region, *i.e.,* 9 sq. units. **Ans.**

Data Based Questions

Set 7

Chapter 8. Linear Regression

1. *The marks obtained by 10 students in English and Hindi are given below :*

English	10	25	13	25	22	11	12	25	21	20
Hindi	12	22	16	15	18	18	17	23	24	17

Find the line of regression in which Hindi is taken as independent variables.

Sol. Let x denote the marks in English and y in Hindi.

Here $\bar{x} = \dfrac{184}{10} = 18{\cdot}4$, and $\bar{y} = \dfrac{182}{10} = 18{\cdot}2$

Since the values of $\bar{x}$ and $\bar{y}$ are fractional, we take the assumed means.

Let the assumed mean for x-series be 20, and that for y-series be 18.

x	y	$d_x = x - 20$	$d_y = y - 18$	d_xd_y	d_x^2	d_y^2
10	12	− 10	− 6	60	100	36
25	22	5	4	20	25	16
13	16	− 7	− 2	14	49	4
25	15	5	− 3	− 15	25	9
22	18	2	0	0	4	0
11	18	− 9	0	0	81	0
12	17	− 8	− 1	8	64	1
25	23	5	5	25	25	25
21	24	1	6	6	1	36
20	17	0	− 1	0	0	1
$\Sigma x = 184$	$\Sigma y = 182$	$\Sigma d_x = -16$	$\Sigma d_y = 2$	$\Sigma d_xd_y = 118$	$\Sigma d_x^2 = 374$	$\Sigma d_y^2 = 128$

Now, $$b_{xy} = \frac{\Sigma d_xd_y - \dfrac{\Sigma d_x \cdot \Sigma d_y}{n}}{\Sigma d_y^2 - \dfrac{(\Sigma d_y)^2}{n}} = \frac{118 - \dfrac{(-16)(2)}{10}}{128 - \dfrac{4}{10}}$$

$$= \frac{118 + 3{\cdot}2}{\dfrac{1276}{10}} = \frac{121{\cdot}2 \times 10}{1276}$$

$$= \frac{1212}{1276} = 0{\cdot}95$$

∴ The line of regression is given as

$$x - \bar{x} = b_{xy}(y - \bar{y}) \quad \Rightarrow x - 18{\cdot}4 = 0{\cdot}95\,(y - 18{\cdot}2)$$

$$\Rightarrow \quad x - 18{\cdot}4 = 0{\cdot}95y - 17{\cdot}29 \Rightarrow x = 0{\cdot}95y + 1{\cdot}11$$ **Ans.**

2. *Find the equations of the two lines of regression for the following observations :*

(3, 6), (4, 5), (5, 4), (6, 3), (7, 2).

Find an estimate of y for x = 2·5.

Sol.

We have

$$\Sigma x_i = 3+4+5+6+7 = 25$$

$$\Sigma y_i = 6+5+4+3+2 = 20$$

$$\Sigma x_i^2 = 9+16+25+36+49 = 135$$

$$\Sigma y_i^2 = 36+25+16+9+4 = 90$$

$$\Sigma x_i y_i = 18+20+20+18+14 = 90$$

$$n = 5,\ \bar{x} = \frac{\Sigma x_i}{n} = \frac{25}{5} = 5,\ \bar{y} = \frac{20}{5} \Rightarrow \bar{y} = 4$$

$$\therefore \quad b_{yx} = \frac{\Sigma x_i y_i - \frac{(\Sigma x_i)(\Sigma y_i)}{n}}{\Sigma x_i^2 - \frac{(\Sigma x_i)^2}{n}} = \frac{90 - \frac{25\times 20}{5}}{135 - \frac{625}{5}} = \frac{90-100}{10} = -1$$

$$b_{xy} = \frac{\Sigma x_i y_i - \frac{(\Sigma x_i)(\Sigma y_i)}{n}}{\Sigma y_i^2 - \frac{(\Sigma y_i)^2}{n}} = \frac{90 - \frac{25\times 20}{5}}{90 - \frac{400}{5}} = \frac{90-100}{90-80} = -1.$$

The regression lines are given as,

$$y - \bar{y} = b_{yx}(x - \bar{x}) \quad \text{and} \quad x - \bar{x} = b_{xy}(y - \bar{y})$$

$$\Rightarrow \quad y - 4 = -x + 5 \qquad x - 5 = -y + 4$$

$$\Rightarrow \quad x + y = 9 \qquad x + y = 9$$

$$\therefore \quad \text{At } x = 2{\cdot}5 \qquad y = 9 - 2{\cdot}5 = 6{\cdot}5.$$

Ans.

3. (a) *Find the equation of the regression line of y on x, if the observations (x, y) are as follows :*
(1, 4), (2, 8), (3, 2), (4, 12), (5, 10), (6, 14), (7, 16), (8, 6), (9, 18)
*Also, find the estimated value of y when x = 14.**[1]

(b) *Find the regression coefficients b_{yx}, b_{xy} and correlation coefficient 'r' for the following data :*
*(2, 8), (6, 8), (4, 5), (7, 6), (5, 2).**

Sol. (a)

x	y	xy	x^2
1	4	4	1
2	8	16	4
3	2	6	9
4	12	48	16
5	10	50	25
6	14	84	36
7	16	112	49
8	6	48	64
9	18	162	81
$\Sigma x = 45$	$\Sigma y = 90$	$\Sigma xy = 530$	$\Sigma x^2 = 285$

Here, $n = 9$

* Frequently asked previous years Board Exam Questions.

$$b_{yx} = \frac{\Sigma xy - \frac{1}{n}\Sigma x\Sigma y}{\Sigma x^2 - \frac{1}{n}(\Sigma x)^2} \qquad b_{yx} = \frac{530 - \frac{1}{9}\times 45\times 90}{285 - \frac{1}{9}\times 45\times 45}$$

$$b_{yx} = \frac{530-450}{285-225} \qquad b_{yx} = \frac{80}{60} = \frac{4}{3}$$

$$\bar{x} = \frac{\Sigma x}{n} = \frac{45}{9} = 5$$

$$\bar{y} = \frac{\Sigma y}{n} = \frac{90}{9} = 10$$

$\therefore$ Regression equation of y on x is

$$y - \bar{y} = b_{yx}(x - \bar{x})$$

$$\Rightarrow \quad y - 10 = \frac{4}{3}(x-5) \qquad \Rightarrow \quad 3y - 30 = 4x - 20$$

$$\Rightarrow \quad 4x - 3y + 10 = 0$$ **Ans.**

Now, when $x = 14$

$$4\times 14 - 3y + 10 = 0$$

$$\Rightarrow \quad 3y = 56 + 10$$

$$\Rightarrow \quad 3y = 66$$

$$\Rightarrow \quad y = 22$$ **Ans.**

(b)

x	x^2	y	y^2	xy
2	4	8	64	16
6	36	8	64	48
4	16	5	25	20
7	49	6	36	42
5	25	2	4	10
$\Sigma x = 24$	$\Sigma x^2 = 130$	$\Sigma y = 29$	$\Sigma y^2 = 193$	$\Sigma xy = 136$

$$b_{yx} = \frac{\Sigma xy - \frac{1}{n}\Sigma x\Sigma y}{\Sigma x^2 - \frac{1}{n}(\Sigma x)^2} = \frac{136 - \frac{1}{5}(24)(29)}{130 - \frac{1}{5}(24)^2}$$

$$= \frac{\frac{136\times 5 - 696}{5}}{\frac{130\times 5 - 576}{5}} = \frac{680-696}{650-576}$$

$$= \frac{-16}{74} = -0.22 \qquad b_{xy} = \frac{\Sigma xy - \frac{1}{n}\Sigma x\Sigma y}{\Sigma y^2 - \frac{1}{n}(\Sigma y)^2}$$

$$= \frac{136 - \frac{1}{5}(24)(29)}{193 - \frac{1}{5}(29)^2} = \frac{136\times 5 - 696}{193\times 5 - 841}$$

Data Based Questions

$$= \frac{680-696}{965-841} \qquad = \frac{-16}{124} = -0.13$$

$$r = \pm\sqrt{b_{yx}\,b_{xy}} = \pm\sqrt{(-0.22)(-0.13)} \qquad = \pm\sqrt{0.0286} = \pm 0.17$$

Since, r has same sign as regression coefficients

$\therefore \quad r(x, y) = -0.17 \qquad \because \quad b_{yx} = -0.22, b_{xy} = -0.13$

and $\quad r = -0.17$ **Ans.**

4. *The Personnel Manager of a factory wants to find a measure which he can use to fix the monthly income of persons applying for a job in production department. As an experiment project, he collected data on 7 persons from that department referring to years of service and their monthly income.*

Years of service (*x*)	*11*	*7*	*9*	*5*	*8*	*6*	*10*
Income in ₹ (*y*)	*10*	*8*	*6*	*5*	*9*	*7*	*11*

(i) Find the regression equation (y) on years of service (x).

(ii) Using it what initial start would you recommen for a person applying for a job after having served in similar capacity in another factory for 13 years.

Sol. (i) Consider the given data,

x	y	$d_x = x - 8$	$d_y = y - 8$	d_x^2	d_xd_y
11	10	3	2	9	6
7	8	– 1	0	1	0
9	6	1	– 2	1	– 2
5	5	– 3	– 3	9	9
8	9	0	1	0	0
6	7	– 2	– 1	4	2
10	11	2	3	4	6
$\Sigma x = 56$	$\Sigma y = 56$			$\Sigma d_x^2 = 28$	$\Sigma d_xd_y = 21$

$$\therefore \quad \bar{x} = \frac{\Sigma x}{7} = \frac{56}{7} = 8; \; \bar{y} = \frac{\Sigma y}{7} = \frac{56}{7} = 8$$

The regression equation of y on x is $y - \bar{y} = \dfrac{\Sigma d_xd_y}{\Sigma d_x^2}(x - \bar{x})$

$$\Rightarrow \quad y - 8 = \frac{21}{28}(x-8) \qquad \Rightarrow \quad y - 8 = \frac{3}{4}(x-8)$$

$$\Rightarrow \quad 4y - 32 = 3x - 24 \qquad \Rightarrow \quad 4y = 3x + 8$$ **Ans.**

(ii) When $x = 13$, $4y = 39 + 8 \qquad \Rightarrow \quad y = \dfrac{47}{4} = 11{\cdot}75$

Hence, the initial start should be ₹ 11·75. **Ans.**

5. *Given that the observations are :*

(9, – 4), (10, – 3), (11, – 1), (12, 0), (13, 1), (14, 3), (15, 5), (16, 8).

Find the two lines of regression and estimate the value of y when x = 13·5.

Sol.

x	x^2	y	y^2	xy
9	81	– 4	16	– 36
10	100	– 3	9	– 30
11	121	– 1	1	– 11
12	144	0	0	0
13	169	1	1	13
14	196	3	9	42
15	225	5	25	75
16	256	8	64	128
$\Sigma x = 100$	$\Sigma x^2 = 1292$	$\Sigma y = 9$	$\Sigma y^2 = 125$	$\Sigma xy = 181$

$$\bar{x} = \frac{\Sigma x}{N} = \frac{100}{8} = 12{\cdot}5 \qquad (\because N = 8)$$

$$\bar{y} = \frac{\Sigma y}{N} = \frac{9}{8} = 1{\cdot}125$$

$$b_{yx} = \frac{\Sigma xy - \frac{1}{N}\Sigma x \,\Sigma y}{\Sigma x^2 - \frac{1}{N}(\Sigma x)^2} = \frac{181 - \frac{9\times 100}{8}}{1292 - \frac{10000}{8}} = \frac{181 - 112{\cdot}5}{1292 - 1250} = \frac{181 - 112{\cdot}5}{42} = \frac{68{\cdot}5}{42} = 1{\cdot}63$$

$$b_{xy} = \frac{\Sigma xy - \frac{1}{N}\Sigma x\,\Sigma y}{\Sigma y^2 - \frac{1}{N}(\Sigma y)^2} = \frac{181 - \frac{900}{8}}{125 - \frac{81}{8}} = \frac{181 - 112{\cdot}5}{125 - 10{\cdot}125} = \frac{68{\cdot}5}{114{\cdot}875} = 0{\cdot}596$$

Regression line of y on x, $\quad y - \bar{y} = b_{xy}\,(x - \bar{x})$

$$y - 1{\cdot}125 = 1{\cdot}63\,(x - 12{\cdot}5)$$

$\Rightarrow \quad y - 1{\cdot}125 = 1{\cdot}63x - 20{\cdot}375$

$\Rightarrow \quad y = 1{\cdot}63x - 19{\cdot}25$

Regression line of x on y $\quad x - \bar{x} = b_{xy}\,(y - \bar{y})$

$$x - 12{\cdot}5 = 0{\cdot}596\,(y - 1{\cdot}125)$$

$$x - 12{\cdot}5 = 0{\cdot}596y - 0{\cdot}67$$

$$x = 0{\cdot}596y + 11{\cdot}83$$

Now, $\quad y = 1{\cdot}63x - 19{\cdot}25$

For $x = 13{\cdot}5$,

The estimated value of $\quad y = 1{\cdot}63 \times 13{\cdot}5 - 19{\cdot}25$

$$= 22{\cdot}005 - 19{\cdot}25 = 2{\cdot}755 \approx 2{\cdot}8$$ **Ans.**

6. *The marks obtained by 10 candidates in English and Mathematics are given below :*

Marks in English	20	13	18	21	11	12	17	14	19	15
Marks in Mathematics	17	12	23	25	14	8	19	21	22	19

Estimate the probable score for Mathematics if the marks obtained in English are 24.

Sol. Let x denote the marks in English and y in Mathematics.

x	y	$d_x = x - 16$	$d_y = y - 18$	$d_x \cdot d_y$	$(d_x)^2$	$(d_y)^2$
20	17	4	– 1	– 4	16	1
13	12	– 3	– 6	18	9	36
18	23	2	5	10	4	25
21	25	5	7	35	25	49
11	14	– 5	– 4	20	25	16
12	8	– 4	– 10	40	16	100
17	19	1	1	1	1	1
14	21	– 2	3	– 6	4	9
19	22	3	4	12	9	16
15	19	– 1	1	– 1	1	1
$\Sigma x = 160$	$\Sigma y = 180$			$\Sigma d_x d_y = 125$	$\Sigma(d_x)^2 = 110$	$\Sigma(d_y)^2 = 254$

$$\bar{x} = \frac{160}{10} = 16 \text{ and } \bar{y} = \frac{180}{10} = 18.$$

$\therefore$ A line of best fit is $y - \bar{y} = \dfrac{\Sigma d_x d_y}{\Sigma (d_x)^2}(x - \bar{x})$

$$y - 18 = \frac{125}{110}(x - 16)$$

when $x = 24$,

then
$$y - 18 = \frac{125}{110}(24 - 16)$$

$$y - 18 = \frac{25}{22} \times 8$$

$$22y - 396 = 200$$

$$22y = 396 + 200$$

$$y = \frac{596}{22} = 27{\cdot}09 \text{ or } 27 \text{ marks (nearly)}$$ **Ans.**

7. *The following table shows the mean and standard deviation of the marks of Mathematics and Physics scored by the students in a school :*

	Mathematics	*Physics*
Mean	*84*	*81*
Standard Deviation	*7*	*4*

The correlation co-efficient between the given marks is 0·86. Estimate the likely marks in Physics if the marks in Mathematics are 92.

Sol. Given, $\bar{x} = 84$, $\bar{y} = 81$

$\sigma_x = 7$, $\sigma_y = 4$

$r = 0{\cdot}86$

$$b_{yx} = r \cdot \frac{\sigma_y}{\sigma_x}$$

$$= 0{\cdot}86 \times \frac{4}{7} = 0{\cdot}49$$

∴ Regression equation of y on x is given as,

$$y - \bar{y} = b_{yx}(x - \bar{x})$$

$$\Rightarrow \quad y - 81 = 0{\cdot}49\,(x - 84)$$

$$\Rightarrow \quad y - 81 = 0{\cdot}49x - 41{\cdot}16$$

$$\Rightarrow \quad y = 0{\cdot}49x + 39{\cdot}84$$

Putting $x = 92$, we get

$$\Rightarrow \quad y = 45{\cdot}08 + 39{\cdot}84 = 84{\cdot}92$$

Hence, the marks in Physics are likely to be 84·92. **Ans.**

8. *The following observations are given :*

(1, 4), (2, 8), (3, 2), (4, 12), (5, 10), (6, 14), (7, 16), (8, 6), (9, 18)

Estimate the value of y when the value of x is 10 and also estimate the value of x when the value of y = 5.

Sol. Here, we have

x	x^2	y	y^2	xy
1	1	4	16	4
2	4	8	64	16
3	9	2	4	6
4	16	12	144	48
5	25	10	100	50
6	36	14	196	84
7	49	16	256	112
8	64	6	36	48
9	81	18	324	162
$\Sigma x = 45$	$\Sigma x^2 = 285$	$\Sigma y = 90$	$\Sigma y^2 = 1140$	$\Sigma xy = 530$

$$\bar{x} = \frac{45}{9} = 5, \quad \bar{y} = \frac{90}{9} = 10$$

$$\therefore \quad b_{yx} = \frac{\Sigma xy - \frac{1}{n}(\Sigma x\,\Sigma y)}{\Sigma x^2 - \frac{1}{n}(\Sigma x)^2} = \frac{530 - \frac{1}{9}\times 45\times 90}{285 - \frac{1}{9}\times 45\times 45} = \frac{80}{60} = \frac{4}{3}$$

$$b_{xy} = \frac{\Sigma xy - \frac{1}{n}(\Sigma x\,\Sigma y)}{\Sigma y^2 - \frac{1}{n}(\Sigma y)^2} = \frac{530 - \frac{1}{9}\times 45\times 90}{1140 - \frac{1}{9}\times 90\times 90} = \frac{80}{240} = \frac{1}{3}$$

∴ Regression line of y on x is given as,

$$y - \bar{y} = b_{yx}(x - \bar{x})$$

$$\Rightarrow \quad y - 10 = \frac{4}{3}(x - 5) \quad \Rightarrow \quad 3y - 30 = 4x - 20 \quad \Rightarrow \quad 3y = 4x + 10$$

Given $x = 10$, we get

$$3y = 4 \times 10 + 10 = 50$$

$$\Rightarrow \quad y = \frac{50}{3} = 16{\cdot}67$$

Now, regression line of x on y is given as,

$$x - \bar{x} = b_{xy}(y - \bar{y})$$

$$\Rightarrow \quad x - 5 = \frac{1}{3}(y - 10)$$

$$\Rightarrow \quad 3x - 15 = y - 10$$

$$3x = y + 5$$

Given $y = 5$, we get $3x = 5 + 5$

$$\Rightarrow \quad x = \frac{10}{3} = 3{\cdot}33$$ **Ans.**

9. *The following table shows the sales and advertisement expenditure of a firm :*

	Sales (₹ in crores)	***Advertisement Expenditure (₹ in crores)***
Mean	*40*	*6*
Standard deviation	*10*	*1·5*

Coefficient of correlation = $r = 0{\cdot}9$
Estimate the likely sales for a proposed advertisement expenditure of ₹ 10 crores.

Sol. Given,

	Sales (₹ in crores)	**Expenditure (₹ in crores)**
	x	y
Mean	40	6
Standard deviation	10	1·5

Co-efficient of correlation $= r = 0{\cdot}9$

Regression coefficient y on x $= b_{yx} = r\frac{\sigma_y}{\sigma_n}$

Regression coefficient x on y $= b_{xy} = r\frac{\sigma_x}{\sigma_y} = 0{\cdot}9 \times \frac{10}{1{\cdot}5} = \frac{90}{15} = 6$

$\therefore$ The required regression equation of x on y is

$$(x - \bar{x}) = b_{xy}(y - \bar{y})$$

$$\Rightarrow \quad x - 40 = \frac{90}{15}(y - 6)$$

$$\Rightarrow \quad x - 40 = 6y - 36$$

$$\Rightarrow \quad x = 6y + 4$$

Given, when $y = 10$

$$x = 6 \times 10 + 4$$

$$\therefore \quad x = ₹\ 64 \text{ crores}$$ **Ans.**

Chapter 16. Linear Programming

1. *(Transportation Problem) A brick manufacturer has two depots A and B, with stocks of 30,000 and 20,000 bricks respectively. He receives orders from three builders P, Q and R for 15,000, 20,000 and 15,000 bricks respectively. The cost (in ₹) of transporting 1,000 bricks to the builders from the depots are given below :*

From \ To	Transportation cost per 1000 bricks (in ₹)		
	P	*Q*	*R*
A	40	20	20
B	20	60	40

The manufacture wishes to find how to fulfil the order so that transportation cost is minimum. Formulate the linear programming problem.

Sol. To simplify, assume that 1 unit = 1,000 bricks.

Suppose that depot A supplies x units to P and y units to Q, so that depot A supplies $(30 - x - y)$ bricks to builder R.

Now, as P requires a total of 15,000 bricks, it requires $(15 - x)$ units from depot B. Similarly, Q requires $(20 - y)$ units from B and R requires $15 - (30 - x - y) = (x + y - 15)$ units from B.

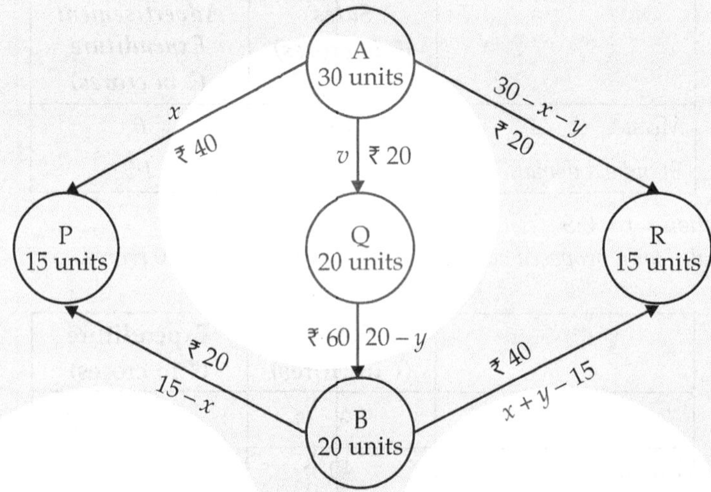

Using the transportation cost given in table, total transportion cost (in rupees) is

$$Z = 40x + 20y + 20\,(30 - x - y) + 20\,(15 - x) + 60\,(20 - y) + 40\,(x + y - 15)$$
$$= 40x - 20y + 1{,}500.$$

Obviously the constraints are that all quantities of bricks, supplied from A and B to P, Q, R are non-negative, *i.e.*, $x \geq 0, y \geq 0, 30 - x - y \geq 0, 15 - x \geq 0, 20 - y \geq 0, x + y - 15 \geq 0$. Instead of minimizing $Z = 40x - 20y + 1{,}500$. It is easier to minimize $Z = 40x - 20y$.

Hence, mathematical formulation of the linear programming problem is

Minimize $Z = 40x - 20y$

Subject to the constraints

$$x + y \geq 15,\ x + y \leq 30,$$
$$x \leq 15,\ y \leq 20$$
$$x \geq 0,\quad y \geq 0$$

Ans.

Solve the Following Type Questions

Set 8

Chapter 3. Inverse Trigonometric Functions

1. *Solve : $\sin(2\tan^{-1}x) = 1$*[*]

Sol. Given, $\sin(2\tan^{-1}x) = 1$

$$\Rightarrow \quad 2\tan^{-1}x = \sin^{-1}1$$

$$\Rightarrow \quad 2\tan^{-1}x = \frac{\pi}{2}$$

$$\Rightarrow \quad \tan^{-1}x = \frac{\pi}{4}$$

$$\Rightarrow \quad x = \tan\frac{\pi}{4}$$

$$\Rightarrow \quad x = 1$$

Ans.

2. *Solve for x :*[*]

$$\tan^{-1}\left(\frac{x-1}{x-2}\right) + \tan^{-1}\left(\frac{x+1}{x+2}\right) = \frac{\pi}{4}$$

Sol. Given, $\tan^{-1}\left(\frac{x-1}{x-2}\right) + \tan^{-1}\left(\frac{x+1}{x+2}\right) = \frac{\pi}{4}$

$$\Rightarrow \quad \tan^{-1}\left[\frac{\left(\frac{x-1}{x-2}\right)+\left(\frac{x+1}{x+2}\right)}{1-\left(\frac{x-1}{x-2}\right)\left(\frac{x+1}{x+2}\right)}\right] = \frac{\pi}{4}$$

$$\Rightarrow \quad \tan^{-1}\left[\frac{(x-1)(x+2)+(x+1)(x-2)}{(x-2)(x+2)-(x-1)(x+1)}\right] = \frac{\pi}{4}$$

$$\Rightarrow \quad \tan^{-1}\left[\frac{(x^2+x-1)+(x^2-x-2)}{(x^2-4)-(x^2-1)}\right] = \frac{\pi}{4}$$

$$\Rightarrow \quad \left[\frac{(x^2+x-2)+(x^2-x-2)}{(x^2-4)-(x^2-1)}\right] = \tan\frac{\pi}{4}$$

$$\Rightarrow \quad \frac{x^2+x-2+x^2-x-2}{x^2-4-x^2+1} = 1$$

$$\Rightarrow \quad \frac{2x^2-4}{-3} = 1$$

$$\Rightarrow \quad 2x^2 - 4 = -3$$

$$\Rightarrow \quad 2x^2 = -3+4$$

$$\Rightarrow \quad 2x^2 = 1$$

$$\Rightarrow \quad x^2 = \frac{1}{2}$$

$$\Rightarrow \quad x = \pm\frac{1}{\sqrt{2}}$$

Ans.

3. *Solve for x if $\tan(\cos^{-1}x) = \frac{2}{\sqrt{5}}$*

Sol. We have $\tan(\cos^{-1}x) = \frac{2}{\sqrt{5}}$

$$\tan\left(\tan^{-1}\frac{\sqrt{1-x^2}}{x}\right) = \frac{2}{\sqrt{5}}$$

$$\left(\because \cos^{-1}x = \tan^{-1}\left(\frac{\sqrt{1-x^2}}{x}\right)\right)$$

$$\Rightarrow \quad \frac{\sqrt{1-x^2}}{2} = \frac{2}{\sqrt{5}}$$

On squaring both sides, we get

$$\frac{1-x^2}{x^2} = \frac{4}{5}$$

$$\Rightarrow \quad 5-5x^2 = 4x^2$$

$$\Rightarrow \quad 9x^2 = 5$$

$$\Rightarrow \quad x = \sqrt{\frac{5}{9}} = \frac{\sqrt{5}}{3}$$

Ans.

4. *Solve : $\cos^{-1}(\sin\cos^{-1}x) = \frac{\pi}{6}$.*

Sol. Given, $\cos^{-1}(\sin\cos^{-1}x) = \frac{\pi}{6}$

$$\Rightarrow \quad \sin\cos^{-1}x = \cos\frac{\pi}{6}$$

$$\Rightarrow \quad \sin\cos^{-1}x = \sin\frac{\pi}{3}$$

$$\left[\because \cos\frac{\pi}{6} = \cos\left(\frac{\pi}{2}-\frac{\pi}{3}\right) = \sin\frac{\pi}{3}\right]$$

$$\Rightarrow \quad \cos^{-1}x = \frac{\pi}{3}$$

$$\Rightarrow \quad x = \cos\frac{\pi}{3}$$

$$(\because \cos(\theta) = \cos(-\theta))$$

Hence, $x = \pm\frac{1}{2}$.

Ans.

---

* Frequently asked previous years Board Exam Questions.

5. *Find the value of*

$$[\sin \cot^{-1} \cos (\tan^{-1} x)].$$

Sol. $[\sin \cot^{-1} \cos (\tan^{-1} x)]$

Consider, $\cos (\tan^{-1} x) = \cos\left(\cos^{-1}\dfrac{1}{\sqrt{1+x^2}}\right)$

$$= \frac{1}{\sqrt{1+x^2}} \quad \ldots\text{(i)}$$

Now, $\cot^{-1}\left(\dfrac{1}{\sqrt{1+x^2}}\right) = \tan^{-1}(\sqrt{1+x^2})$

$$= \sin^{-1}\left(\frac{\sqrt{1+x^2}}{\sqrt{1+1+x^2}}\right)$$

$$= \sin^{-1}\left(\frac{\sqrt{1+x^2}}{\sqrt{2+x^2}}\right) \quad \ldots\text{(ii)}$$

Now, $\sin [\cot^{-1} (\cos (\tan^{-1} x))]$

$$= \sin\left(\sin^{-1}\left(\frac{\sqrt{1+x^2}}{\sqrt{2+x^2}}\right)\right)$$

[From equations (i), (ii)]

$$= \sqrt{\frac{1+x^2}{2+x^2}}$$ **Ans.**

6. *Solve the equation for x :*

$$\sin^{-1}\frac{5}{x} + \sin^{-1}\frac{12}{x} = \frac{\pi}{2}, \; x \neq 0.$$

Sol.

The given equation is

$$\sin^{-1}\frac{5}{x} + \sin^{-1}\frac{12}{x} = \frac{\pi}{2}$$

$$\Rightarrow \quad \sin^{-1}\frac{12}{x} = \frac{\pi}{2} - \sin^{-1}\frac{5}{x}$$

$$\Rightarrow \quad \frac{12}{x} = \sin\left(\frac{\pi}{2} - \sin^{-1}\frac{5}{x}\right)$$

$$\Rightarrow \quad \frac{12}{x} = \cos\left(\sin^{-1}\frac{5}{x}\right)$$

$$\Rightarrow \quad \frac{12}{x} = \sqrt{1-\left(\frac{5}{x}\right)^2}$$

$[\because \cos (\sin^{-1} x) = \sqrt{1-x^2}\,]$

$$\Rightarrow \quad \left(\frac{12}{x}\right)^2 = 1 - \frac{25}{x^2}$$

$$\Rightarrow \quad \frac{144}{x^2} + \frac{25}{x^2} = 1$$

$$\Rightarrow \quad \frac{169}{x^2} = 1$$

$$\Rightarrow \quad x^2 = 169$$

$$\Rightarrow \quad x = \pm\sqrt{169}$$

$$\Rightarrow \quad x = 13, -13$$

But $x = -13$ doesn't satisfy the given equation.

Hence the solution of the given equation is $x = 13$. **Ans.**

7. *If* $\sin^{-1}\dfrac{2a}{1+a^2} + \cos^{-1}\dfrac{1-b^2}{1+b^2} = 2\tan^{-1} x$. *Solve for x.*

Sol. We have

$$\sin^{-1}\frac{2a}{1+a^2} + \cos^{-1}\frac{1-b^2}{1+b^2} = 2\tan^{-1} x$$

$$\Rightarrow \quad 2\tan^{-1} a + 2\tan^{-1} b = 2\tan^{-1} x$$

$$\Rightarrow \quad \tan^{-1} a + \tan^{-1} b = \tan^{-1} x$$

$$\Rightarrow \quad \tan^{-1}\left(\frac{a+b}{1-ab}\right) = \tan^{-1} x$$

$$\Rightarrow \quad x = \frac{a+b}{1-ab}.$$ **Ans.**

8. *Solve the equation for x :*

$$\sin^{-1}x + \sin^{-1}(1-x) = \cos^{-1}x, \; x \neq 0$$

Sol. We have,

$$\sin^{-1}x + \sin^{-1}(1-x) = \cos^{-1}x, \; x \neq 0$$

$$\sin^{-1}\left[x\sqrt{1-(1-x)^2} + (1-x)\sqrt{1-x^2}\right] = \cos^{-1}x$$

$$\Rightarrow \sin^{-1}\left[x\sqrt{1-1-x^2+2x} + (1-x)\sqrt{1-x^2}\right] = \sin^{-1}(\sqrt{1-x^2})$$

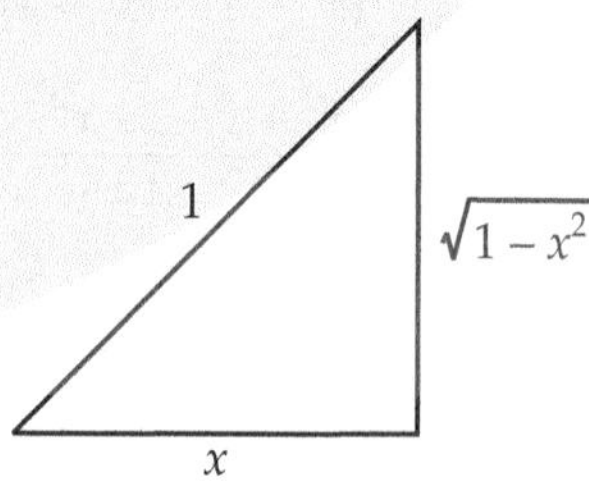

$$\Rightarrow \quad x\sqrt{2x-x^2} + (1-x)\sqrt{1-x^2} = \sqrt{1-x^2}$$

$$\Rightarrow \quad x\sqrt{2x-x^2} = \sqrt{1-x^2} - (1-x)\sqrt{1-x^2}$$

$$\Rightarrow \quad x\sqrt{2x-x^2} = \sqrt{1-x^2}\,(1-1+x)$$

$$\Rightarrow \quad x\sqrt{2x-x^2} = \sqrt{1-x^2}\,.x$$

$$\Rightarrow x\left(\sqrt{2x-x^2}-\sqrt{1-x^2}\right)=0$$

$\Rightarrow$ Either $x = 0$

or $\sqrt{2x-x^2}=\sqrt{1-x^2}$

On squaring both sides, we get

$$2x - x^2 = 1 - x^2$$
$$2x = 1$$
$$x = \frac{1}{2}$$
$$\Rightarrow x = 0, \frac{1}{2}$$ **Ans.**

Chapter 4. Determinants

1. *Solve for x by properties of determinants*

$$\begin{vmatrix} x+a & b & c \\ c & x+b & a \\ a & b & x+c \end{vmatrix} = 0.$$

Sol. Given, $\begin{vmatrix} x+a & b & c \\ c & x+b & a \\ a & b & x+c \end{vmatrix} = 0$

Applying $R_1 \to R_1 - R_3$, we obtain

$$\Rightarrow \begin{vmatrix} x & 0 & -x \\ c & x+b & a \\ a & b & x+c \end{vmatrix} = 0$$

Applying $C_1 \to C_1 + C_3$, we obtain

$$\Rightarrow \begin{vmatrix} 0 & 0 & -x \\ c+a & x+b & a \\ a+x+c & b & x+c \end{vmatrix} = 0$$

Now expanding along R_1, we get

$\Rightarrow \quad -x\{b(c+a) - (x+b)(a+x+c)\} = 0$

$\Rightarrow -x\{bc + ab - xa - x^2 - xc - ab - bx - bc\} = 0$

$\Rightarrow \quad -x(-xa - x^2 - xc - xb) = 0$

$\Rightarrow \quad -x\{-x(a+x+c+b)\} = 0$

$\Rightarrow \quad x^2\{x+a+b+c\} = 0$

$\Rightarrow$ Either $x = 0$ or $x = -(a+b+c)$ **Ans.**

2. *Use properties of determinants to solve for x :*

$$\begin{vmatrix} x-3 & 1 & 1 \\ 1 & x-3 & 1 \\ 1 & 1 & x-3 \end{vmatrix} = 0.$$

Sol.

Given, $\begin{vmatrix} x-3 & 1 & 1 \\ 1 & x-3 & 1 \\ 1 & 1 & x-3 \end{vmatrix} = 0.$

Applying $C_1 \to C_1 + C_2 + C_3$, we get

$$\begin{vmatrix} x-1 & 1 & 1 \\ x-1 & x-3 & 1 \\ x-1 & 1 & x-3 \end{vmatrix} = 0.$$

Taking $(x-1)$ as common

$$(x-1)\begin{vmatrix} 1 & 1 & 1 \\ 1 & x-3 & 1 \\ 1 & 1 & x-3 \end{vmatrix} = 0.$$

Applying $R_2 \to R_2 - R_1$ and $R_3 \to R_3 - R_1$, we get

$$(x-1)\begin{vmatrix} 1 & 1 & 1 \\ 0 & x-4 & 0 \\ 0 & 0 & x-4 \end{vmatrix} = 0.$$

Now expanding along C_1

$(x-1)[1((x-4)^2 - 0)] = 0$

$\Rightarrow \quad (x-1)(x-4)^2 = 0$

$\Rightarrow x = 1, 4, 4$

Either $x = 1$ or $x = 4$. **Ans.**

Chapter 5. Matrices

1. *Solve for x and y if*

$$\begin{bmatrix} x^2 \\ y^2 \end{bmatrix} + 2\begin{bmatrix} 2x \\ 3y \end{bmatrix} = 3\begin{bmatrix} 7 \\ -3 \end{bmatrix}.$$

Sol.

Given, $\begin{bmatrix} x^2 \\ y^2 \end{bmatrix} + 2\begin{bmatrix} 2x \\ 3y \end{bmatrix} = 3\begin{bmatrix} 7 \\ -3 \end{bmatrix}$

$$\Rightarrow \begin{bmatrix} x^2 \\ y^2 \end{bmatrix} + \begin{bmatrix} 4x \\ 6y \end{bmatrix} = \begin{bmatrix} 21 \\ -9 \end{bmatrix}$$

$$\Rightarrow \begin{bmatrix} x^2+4x \\ y^2+6y \end{bmatrix} = \begin{bmatrix} 21 \\ -9 \end{bmatrix}$$

Equating corresponding elements of equal matrices, we get

$$x^2 + 4x = 21$$
$$\Rightarrow x^2 + 4x - 21 = 0$$
$$\Rightarrow x^2 + 7x - 3x - 21 = 0$$
$$\Rightarrow x(x+7) - 3(x+7) = 0$$
$$\Rightarrow (x+7)(x-3) = 0$$
$$\Rightarrow x = -7, 3$$ **Ans.**

and $y^2 + 6y = -9$

$\Rightarrow y^2 + 6y + 9 = 0$

$\Rightarrow (y+3)^2 = 0$

$\Rightarrow y + 3 = 0$

$\Rightarrow y = -3$ **Ans.**

2. *Solve the matrix equation*

$$\begin{bmatrix} x^2 \\ y^2 \end{bmatrix} - 3\begin{bmatrix} x \\ 2y \end{bmatrix} = \begin{bmatrix} -2 \\ 9 \end{bmatrix}$$

Sol. Given,

$$\begin{bmatrix} x^2 \\ y^2 \end{bmatrix} - 3\begin{bmatrix} x \\ 2y \end{bmatrix} = \begin{bmatrix} -2 \\ 9 \end{bmatrix}$$

$$\Rightarrow \begin{bmatrix} x^2 \\ y^2 \end{bmatrix} + \begin{bmatrix} -3x \\ -6y \end{bmatrix} = \begin{bmatrix} -2 \\ 9 \end{bmatrix}$$

$$\Rightarrow \begin{bmatrix} x^2 - 3x \\ y^2 - 6y \end{bmatrix} = \begin{bmatrix} -2 \\ 9 \end{bmatrix}$$

Equating corresponding elements of equal matrices, we get

$x^2 - 3x = -2$

$\Rightarrow x^2 - 3x + 2 = 0$

$\Rightarrow (x-1)(x-2) = 0$

$\Rightarrow x = 1, 2$

and $y^2 - 6y = 9$

$\Rightarrow y^2 - 6y - 9 = 0$

$\Rightarrow y = 3 \pm 3\sqrt{2}$ **Ans.**

3. *Without solving comment on the nature of equations in the following system.*

$2x + 3y - 4z = 1$

$3x - y + 2z = -2$

$5x - 9y + 14z = 3.$

Sol. The given system of equation can be written as

$$\begin{bmatrix} 2 & 3 & -4 \\ 3 & -1 & 2 \\ 5 & -9 & 14 \end{bmatrix}\begin{bmatrix} x \\ y \\ z \end{bmatrix} = \begin{bmatrix} 1 \\ -2 \\ 3 \end{bmatrix}$$

or $AX = B$

where $A = \begin{bmatrix} 2 & 3 & -4 \\ 3 & -1 & 2 \\ 5 & -9 & 14 \end{bmatrix}, B = \begin{bmatrix} 1 \\ -2 \\ 3 \end{bmatrix}, X = \begin{bmatrix} x \\ y \\ z \end{bmatrix}$

$$|A| = \begin{vmatrix} 2 & 3 & -4 \\ 3 & -1 & 2 \\ 5 & -9 & 14 \end{vmatrix}$$

$= 2\{(-1)\,14 - 2(-9)\} - 3\{3 \times 14 - 2 \times 5\} + (-4)\{3(-9) - 5(-1)\}$

$= 2(-14 + 18) - 3(42 - 10) - 4(-27 + 5)$

$= 8 - 96 + 88 = 0$

$\Rightarrow |A| = 0$

$\therefore A^{-1}$ will not exist.

$\therefore$ The given system of equation is inconsistent *i.e.*, it has no solution. **Ans.**

4. *Given two matrices A and B*

$$A = \begin{bmatrix} 1 & -2 & 3 \\ 1 & 4 & 1 \\ 1 & -3 & 2 \end{bmatrix} \text{ and } B = \begin{bmatrix} 11 & -5 & -14 \\ -1 & -1 & 2 \\ -7 & 1 & 6 \end{bmatrix}$$

find AB and use this result to solve the following system of equations :

$x - 2y + 3z = 6, \quad x + 4y + z = 12, \quad x - 3y + 2z = 1$

Sol. Given,

$$A = \begin{bmatrix} 1 & -2 & 3 \\ 1 & 4 & 1 \\ 1 & -3 & 2 \end{bmatrix} \text{ and } B = \begin{bmatrix} 11 & -5 & -14 \\ -1 & -1 & 2 \\ -7 & 1 & 6 \end{bmatrix}$$

$$AB = \begin{bmatrix} 1 & -2 & 3 \\ 1 & 4 & 1 \\ 1 & -3 & 2 \end{bmatrix}\begin{bmatrix} 11 & -5 & -14 \\ -1 & -1 & 2 \\ -7 & 1 & 6 \end{bmatrix}$$

$$= \begin{bmatrix} 11+2-21 & -5+2+3 & -14-4+18 \\ 11-4-7 & -5-4+1 & -14+8+6 \\ 11+3-14 & -5+3+2 & -14-6+12 \end{bmatrix}$$

$$= \begin{bmatrix} -8 & 0 & 0 \\ 0 & -8 & 0 \\ 0 & 0 & -8 \end{bmatrix}$$

$\therefore AB = -8\,I_3$

$\Rightarrow -\frac{1}{8}(B) = A^{-1}$ ....(i)

The given system of equations can be written as,

$$\begin{bmatrix} 1 & -2 & 3 \\ 1 & 4 & 1 \\ 1 & -3 & 2 \end{bmatrix}\begin{bmatrix} x \\ y \\ z \end{bmatrix} = \begin{bmatrix} 6 \\ 12 \\ 1 \end{bmatrix}$$

Or, $AX = C$

Where, $X = \begin{bmatrix} x \\ y \\ z \end{bmatrix}$ and $C = \begin{bmatrix} 6 \\ 12 \\ 1 \end{bmatrix}$

As A^{-1} exists, the given system of equations has a unique solution

$\therefore X = A^{-1}C$

$\Rightarrow X = \left(-\frac{1}{8}B\right)C$ [Using (i)]

Solve the Following Type Questions

$$= -\frac{1}{8}\begin{bmatrix} 11 & -5 & -14 \\ -1 & -1 & 2 \\ -7 & 1 & 6 \end{bmatrix}\begin{bmatrix} 6 \\ 12 \\ 1 \end{bmatrix}$$

$$= -\frac{1}{8}\begin{bmatrix} 66-60-14 \\ -6-12+2 \\ -42+12+6 \end{bmatrix}$$

$$\begin{bmatrix} x \\ y \\ z \end{bmatrix} = -\frac{1}{8}\begin{bmatrix} -8 \\ -16 \\ -24 \end{bmatrix}$$

$$\begin{bmatrix} x \\ y \\ z \end{bmatrix} = \begin{bmatrix} 1 \\ 2 \\ 3 \end{bmatrix}$$

Hence the solution of equations is,

$x = 1, y = 2, z = 3.$ **Ans.**

5. *Given that:* $A = \begin{bmatrix} 1 & -1 & 0 \\ 2 & 3 & 4 \\ 0 & 1 & 2 \end{bmatrix}$ *and* $B = \begin{bmatrix} 2 & 2 & -4 \\ -4 & 2 & -4 \\ 2 & -1 & 5 \end{bmatrix}$

find AB.

Using this result, solve the following system of equation :

$x - y = 3,\ 2x + 3y + 4z = 17$ *and* $y + 2z = 7$

Sol.

We have, $A = \begin{bmatrix} 1 & -1 & 0 \\ 2 & 3 & 4 \\ 0 & 1 & 2 \end{bmatrix}$

and $B = \begin{bmatrix} 2 & 2 & -4 \\ -4 & 2 & -4 \\ 2 & -1 & 5 \end{bmatrix}$

Then

$$A\cdot B = \begin{bmatrix} 1\times2+(-1)(-4)+0\times2 & 1\times2+(-1)\times2+0\times-1 & 1\times(-4)+(-1)(-4)+0\times5 \\ 2\times2+3\times(-4)+4\times2 & 2\times2+3\times2+4\times(-1) & 2\times(-4)+3\times(-4)+4\times5 \\ 2\times0+1\times(-4)+2\times2 & 2\times0+1\times2+2\times(-1) & (-4)\times0+1\times(-4)+2\times5 \end{bmatrix}$$

$$\Rightarrow \quad AB = \begin{bmatrix} 6 & 0 & 0 \\ 0 & 6 & 0 \\ 0 & 0 & 6 \end{bmatrix} = 6\,I_3$$ **Ans.**

$$\Rightarrow \quad \frac{1}{6}AB = I$$

$$\Rightarrow \quad A^{-1} = \frac{1}{6}\cdot B$$

Now, writing the given system of equations in the matrix form AX = C as

$$\begin{bmatrix} 1 & -1 & 0 \\ 2 & 3 & 4 \\ 0 & 1 & 2 \end{bmatrix}\begin{bmatrix} x \\ y \\ z \end{bmatrix} = \begin{bmatrix} 3 \\ 17 \\ 7 \end{bmatrix}$$

Where, $A = \begin{bmatrix} 1 & -1 & 0 \\ 2 & 3 & 4 \\ 0 & 1 & 2 \end{bmatrix}, X = \begin{bmatrix} x \\ y \\ z \end{bmatrix}$

and $C = \begin{bmatrix} 3 \\ 17 \\ 7 \end{bmatrix}$

$$\Rightarrow \quad AX = C$$

$$\Rightarrow \quad X = A^{-1}C$$

$$\Rightarrow \quad X = \frac{1}{6}BC \left[\because A^{-1} = \frac{1}{6}B\right]$$

$$= \frac{1}{6}\begin{bmatrix} 2 & 2 & -4 \\ -4 & 2 & -4 \\ 2 & -1 & 5 \end{bmatrix}\begin{bmatrix} 3 \\ 17 \\ 7 \end{bmatrix}$$

$$= \frac{1}{6}\begin{bmatrix} 2\times3+2\times17+(-4)\times7 \\ (-4)\times3+2\times17+(-4)\times7 \\ 2\times3+(-1)\times17+5\times7 \end{bmatrix}$$

$$\begin{bmatrix} x \\ y \\ z \end{bmatrix} = \frac{1}{6}\begin{bmatrix} 12 \\ -6 \\ 24 \end{bmatrix} = \begin{bmatrix} 2 \\ -1 \\ 4 \end{bmatrix}$$

Hence, $x = 2, y = -1$ and $z = 4$. **Ans.**

6. *Solve the following linear equations by matrix method* $5x - 12y = -9,\ 7x - 6y = -8.$

Sol. Let $A = \begin{bmatrix} 5 & -12 \\ 7 & -6 \end{bmatrix}, X = \begin{bmatrix} x \\ y \end{bmatrix}$ and $B = \begin{bmatrix} -9 \\ -8 \end{bmatrix}$

Then, the equations can be written as $AX = B$.

Now, $|A| = \begin{vmatrix} 5 & -12 \\ 7 & -6 \end{vmatrix}$

$= -30 + 84 = 54 \neq 0$

∴ The system has a unique solution given by

$$X = A^{-1}B \quad \ldots(i)$$

$$A^{-1} = \frac{1}{|A|}(\text{adj } A) = \frac{1}{54}\begin{bmatrix} A_{11} & A_{21} \\ A_{12} & A_{22} \end{bmatrix}$$

where A_{ij} is the co-factor of a_{ij} in $|A|$ for all i, j in $1 \le i, j \le 2$.

$$= \frac{1}{54}\begin{bmatrix} -6 & 12 \\ -7 & 5 \end{bmatrix} \quad \ldots(ii)$$

Solve the Following Type Questions

∴ From (i) and (ii), we get

$$\begin{bmatrix} x \\ y \end{bmatrix} = \frac{1}{54}\begin{bmatrix} -6 & 12 \\ -7 & 5 \end{bmatrix}\begin{bmatrix} -9 \\ -8 \end{bmatrix}$$

$$= \frac{1}{54}\begin{bmatrix} 54-96 \\ 63-40 \end{bmatrix} = \frac{1}{54}\begin{bmatrix} -42 \\ 23 \end{bmatrix}$$

$$= \begin{bmatrix} \frac{-7}{9} \\ \frac{23}{54} \end{bmatrix}$$

Equating corresponding elements of equal matrices, we get

$x = \frac{-7}{9}$ and $y = \frac{23}{54}$ as the solution. **Ans.**

7. *Solve the following linear equations using the matrix method :*

$$x + y + z = 9$$
$$2x + 5y + 7z = 52$$
$$2x + y - z = 0$$

Sol. Given,

$$A = \begin{bmatrix} 1 & 1 & 1 \\ 2 & 5 & 7 \\ 2 & 1 & -1 \end{bmatrix}$$

$$\therefore |A| = 1(-5-7) - 1(-2-14) + 1(2-10)$$
$$= -12 + 16 - 8$$
$$= -4 \neq 0$$

∴ A^{-1} exists.

Now cofactors of A,

$$A_{11} = \begin{bmatrix} 5 & 7 \\ 1 & -1 \end{bmatrix} = -12$$

$$A_{12} = -\begin{bmatrix} 2 & 7 \\ 2 & -1 \end{bmatrix} = 16$$

$$A_{13} = \begin{bmatrix} 2 & 5 \\ 2 & 1 \end{bmatrix} = -8$$

$$A_{21} = -\begin{bmatrix} 1 & 1 \\ 1 & -1 \end{bmatrix} = 2$$

$$A_{22} = \begin{bmatrix} 1 & 1 \\ 2 & -1 \end{bmatrix} = -3$$

$$A_{23} = -\begin{bmatrix} 1 & 1 \\ 2 & 1 \end{bmatrix} = 1$$

$$A_{31} = \begin{bmatrix} 1 & 1 \\ 5 & 7 \end{bmatrix} = 2$$

$$A_{32} = -\begin{bmatrix} 1 & 1 \\ 2 & 7 \end{bmatrix} = -5$$

$$A_{33} = \begin{bmatrix} 1 & 1 \\ 2 & 5 \end{bmatrix} = 3$$

$$\therefore \text{Adj}(A) = \begin{bmatrix} -12 & 2 & 2 \\ 16 & -3 & -5 \\ -8 & 1 & 3 \end{bmatrix}$$

$$\therefore A^{-1} = \frac{1}{|A|}\text{adj}(A)$$

$$= -\frac{1}{4}\begin{bmatrix} -12 & 2 & 2 \\ 16 & -3 & -5 \\ -8 & 1 & 3 \end{bmatrix}$$

The given system of equations can be written as

$$AX = B$$

$$\begin{bmatrix} 1 & 1 & 1 \\ 2 & 5 & 7 \\ 2 & 1 & -1 \end{bmatrix}\begin{bmatrix} x \\ y \\ z \end{bmatrix} = \begin{bmatrix} 9 \\ 52 \\ 0 \end{bmatrix}$$

As $|A| \neq 0$, thus the system has a unique solution

$$\therefore X = A^{-1}B$$

$$= \frac{-1}{4}\begin{bmatrix} -12 & 2 & 2 \\ 16 & -3 & -5 \\ -8 & 1 & 3 \end{bmatrix}\begin{bmatrix} 9 \\ 52 \\ 0 \end{bmatrix}$$

$$= \frac{-1}{4}\begin{bmatrix} -108+104+0 \\ 144-156+0 \\ -72+52+0 \end{bmatrix}$$

$$= \frac{-1}{4}\begin{bmatrix} -4 \\ -12 \\ -20 \end{bmatrix} = \begin{bmatrix} 1 \\ 3 \\ 5 \end{bmatrix}$$

Equating corresponding elements of equal matrices, we get $x = 1, y = 3, z = 5$. **Ans.**

8. *Solve the following system of equations using matrices :*

$$x + y + z = 6$$
$$x - y + z = 2$$
$$2x + y - z = 1$$

Sol. Given,

$$A = \begin{bmatrix} 1 & 1 & 1 \\ 1 & -1 & 1 \\ 2 & 1 & -1 \end{bmatrix}$$

Solve the Following Type Questions

$\therefore \quad |A| = 1(1-1) - 1(-1-2) + 1(1+2)$

$= 3 + 3 = 6 \neq 0$

$\therefore A^{-1}$ exists.

Now cofactors of A,

$$A_{11} = \begin{bmatrix} -1 & 1 \\ 1 & -1 \end{bmatrix} = 1 - 1 = 0$$

$$A_{12} = -\begin{bmatrix} 1 & 1 \\ 2 & -1 \end{bmatrix} = -(-1-2) = 3$$

$$A_{13} = \begin{bmatrix} 1 & -1 \\ 2 & 1 \end{bmatrix} = 1 + 2 = 3$$

$$A_{21} = -\begin{bmatrix} 1 & 1 \\ 1 & -1 \end{bmatrix} = -(-1-1) = 2$$

$$A_{22} = \begin{bmatrix} 1 & 1 \\ 2 & -1 \end{bmatrix} = -1 - 2 = -3$$

$$A_{23} = -\begin{bmatrix} 1 & 1 \\ 2 & 1 \end{bmatrix} = -(1-2) = 1$$

$$A_{31} = \begin{bmatrix} 1 & 1 \\ -1 & 1 \end{bmatrix} = 1 + 1 = 2$$

$$A_{32} = -\begin{bmatrix} 1 & 1 \\ 1 & 1 \end{bmatrix} = (1-1) = 0$$

$$A_{33} = \begin{bmatrix} 1 & 1 \\ 1 & -1 \end{bmatrix} = -1 - 1 = -2$$

$$\therefore \quad \text{adj}(A) = \begin{bmatrix} 0 & 2 & 2 \\ 3 & -3 & 0 \\ 3 & 1 & -2 \end{bmatrix}$$

$$\therefore \quad A^{-1} = \frac{1}{|A|}\text{Adj}(A)$$

$$= \frac{1}{6}\begin{bmatrix} 0 & 2 & 2 \\ 3 & -3 & 0 \\ 3 & 1 & -2 \end{bmatrix}$$

The given system of equation can be written as

$$\begin{bmatrix} 1 & 1 & 1 \\ 1 & -1 & 1 \\ 2 & 1 & -1 \end{bmatrix}\begin{bmatrix} x \\ y \\ z \end{bmatrix} = \begin{bmatrix} 6 \\ 2 \\ 1 \end{bmatrix} \text{ or } AX = B$$

As $|A| \neq 0$, thus the system has a unique solution

$$\therefore \quad X = A^{-1}B$$

$$= \frac{1}{6}\begin{bmatrix} 0 & 2 & 2 \\ 3 & -3 & 0 \\ 3 & 1 & -2 \end{bmatrix}\begin{bmatrix} 6 \\ 2 \\ 1 \end{bmatrix}$$

$$= \frac{1}{6}\begin{bmatrix} 0+4+2 \\ 18-6+0 \\ 18+2-2 \end{bmatrix}$$

$$= \frac{1}{6}\begin{bmatrix} 6 \\ 12 \\ 18 \end{bmatrix} = \begin{bmatrix} 1 \\ 2 \\ 3 \end{bmatrix}$$

Equating corresponding elements of equal matrices, we get $x = 1, y = 2, z = 3$. **Ans.**

9. *Solve the following system of linear equations using matrix method :*

$3x + y + z = 1, 2x + 2z = 0, 5x + y + 2z = 2$

Sol. Writing the given equation in the matrix form AX = B as

$$\begin{bmatrix} 3 & 1 & 1 \\ 2 & 0 & 2 \\ 5 & 1 & 2 \end{bmatrix}\begin{bmatrix} x \\ y \\ z \end{bmatrix} = \begin{bmatrix} 1 \\ 0 \\ 2 \end{bmatrix}$$

where, $A = \begin{bmatrix} 3 & 1 & 1 \\ 2 & 0 & 2 \\ 5 & 1 & 2 \end{bmatrix}, X = \begin{bmatrix} x \\ y \\ z \end{bmatrix}$

and $B = \begin{bmatrix} 1 \\ 0 \\ 2 \end{bmatrix}$

$$\therefore \quad A^{-1} = \begin{bmatrix} 3 & 1 & 1 \\ 2 & 0 & 2 \\ 5 & 1 & 2 \end{bmatrix}$$

$$= 3(0-2) - 1(4-10) + 1(2-0) = 2 \neq 0$$

$\therefore$ A is non-singular.

$\therefore A^{-1}$ exists.

$\therefore$ The system has the unique solution

$$X = A^{-1}B.$$

$A_{11} = -2, A_{12} = 6, A_{13} = 2, A_{21} = -1, A_{22} = 1,$
$A_{23} = 2, A_{31} = 2, A_{32} = -4, A_{33} = -2.$

$$\therefore \quad \text{adj } A = \begin{bmatrix} A_{11} & A_{21} & A_{31} \\ A_{12} & A_{22} & A_{32} \\ A_{13} & A_{23} & A_{33} \end{bmatrix}$$

$$= \begin{bmatrix} -2 & -1 & 2 \\ 6 & 1 & -4 \\ 2 & 2 & -2 \end{bmatrix}$$

$$A^{-1} = \frac{\text{adj } A}{|A|} = \frac{1}{2}\begin{bmatrix} -2 & -1 & 2 \\ 6 & 1 & -4 \\ 2 & 2 & -2 \end{bmatrix}$$

$$= \begin{bmatrix} -1 & -1/2 & 1 \\ 3 & 1/2 & -2 \\ 1 & 1 & -1 \end{bmatrix}$$

Now, $AX = B \Rightarrow X = A^{-1}B$.

$$\begin{bmatrix} x \\ y \\ z \end{bmatrix} = \begin{bmatrix} -1 & -1/2 & 1 \\ 3 & 1/2 & -2 \\ 1 & 1 & -1 \end{bmatrix}\begin{bmatrix} 1 \\ 0 \\ 2 \end{bmatrix}$$

$$= \begin{bmatrix} 1 \\ -1 \\ -1 \end{bmatrix}$$

$\therefore \quad x = 1, y = -1, z = -1$ **Ans.**

10. *Solve the following linear equations using the matrix method :*

$$\frac{2}{x} + \frac{3}{y} + \frac{10}{z} = 4$$

$$\frac{4}{x} - \frac{6}{y} + \frac{5}{z} = 1$$

$$\frac{6}{x} + \frac{9}{y} - \frac{20}{z} = 2$$

Sol. Let $\frac{1}{x} = a$, $\frac{1}{y} = b$ and $\frac{1}{z} = c$, then the system of equations are,

$$2a + 3b + 10c = 4$$
$$4a - 6b + 5c = 1$$
$$6a + 9b - 20c = 2$$

So, $$A = \begin{bmatrix} 2 & 3 & 10 \\ 4 & -6 & 5 \\ 6 & 9 & -20 \end{bmatrix}$$

$\therefore \ |A| = 2(120 - 45) - 3(-80 - 30) + 10(36 + 36)$

$= 150 + 330 + 720$

$= 1200 \neq 0$

$\therefore \ A^{-1}$ exists.

Now cofactors of A,

$$A_{11} = \begin{bmatrix} -6 & 5 \\ 9 & -20 \end{bmatrix} = 120 - 45 = 75$$

$$A_{12} = -\begin{bmatrix} 4 & 5 \\ 6 & -20 \end{bmatrix} = -(-80 - 30) = 110$$

$$A_{13} = \begin{bmatrix} 4 & -6 \\ 6 & 9 \end{bmatrix} = 36 + 36 = 72$$

$$A_{21} = -\begin{bmatrix} 3 & 10 \\ 9 & -20 \end{bmatrix} = -(-60 - 90) = 150$$

$$A_{22} = \begin{bmatrix} 2 & 10 \\ 6 & -20 \end{bmatrix} = -40 - 60 = -100$$

$$A_{23} = -\begin{bmatrix} 2 & 3 \\ 6 & 9 \end{bmatrix} = -(18 - 18) = 0$$

$$A_{31} = \begin{bmatrix} 3 & 10 \\ -6 & 5 \end{bmatrix} = 15 + 60 = 75$$

$$A_{32} = -\begin{bmatrix} 2 & 10 \\ 4 & 5 \end{bmatrix} = -(10 - 40) = 30$$

$$A_{33} = \begin{bmatrix} 2 & 3 \\ 4 & -6 \end{bmatrix} = -12 - 12 = -24$$

$$\therefore \quad \text{adj}(A) = \begin{bmatrix} 75 & 150 & 75 \\ 110 & -100 & 30 \\ 72 & 0 & -24 \end{bmatrix}$$

$$\therefore \quad A^{-1} = \frac{1}{|A|}\text{adj}(A)$$

$$= \frac{1}{1200}\begin{bmatrix} 75 & 150 & 75 \\ 110 & -100 & 30 \\ 72 & 0 & -24 \end{bmatrix}$$

The given system of equations can be written as

$$AX = B$$

$$\begin{bmatrix} 2 & 3 & 10 \\ 4 & -6 & 5 \\ 6 & 9 & -20 \end{bmatrix}\begin{bmatrix} a \\ b \\ c \end{bmatrix} = \begin{bmatrix} 4 \\ 1 \\ 2 \end{bmatrix}$$

As $|A| \neq 0$, thus the system has a unique solution

$$\therefore \quad X = A^{-1}B$$

$$\Rightarrow \quad \begin{bmatrix} a \\ b \\ c \end{bmatrix} = \frac{1}{1200}\begin{bmatrix} 75 & 150 & 75 \\ 110 & -100 & 30 \\ 72 & 0 & -24 \end{bmatrix}\begin{bmatrix} 4 \\ 1 \\ 2 \end{bmatrix}$$

$$\Rightarrow \quad \begin{bmatrix} a \\ b \\ c \end{bmatrix} = \frac{1}{1200}\begin{bmatrix} 300 + 150 + 150 \\ 440 - 100 + 60 \\ 288 + 0 - 48 \end{bmatrix}$$

$$\Rightarrow \quad \begin{bmatrix} a \\ b \\ c \end{bmatrix} = \frac{1}{1200}\begin{bmatrix} 600 \\ 400 \\ 240 \end{bmatrix} = \begin{bmatrix} \frac{1}{2} \\ \frac{1}{3} \\ \frac{1}{5} \end{bmatrix}$$

Equating corresponding elements of equal matrices, we get, $a = \frac{1}{2}, b = \frac{1}{3}, c = \frac{1}{5}$

Since, $a = \frac{1}{x}, b = \frac{1}{y}, c = \frac{1}{z}$

Hence, the solution of the given system of equations is $x = 2$, $y = 3$ and $z = 5$ **Ans.**

Chapter 6. Differential Calculus

1. *If $y = \log_e (x+\sqrt{x^2+k^2})$, find $\frac{dy}{dx}$.*

Sol. Given,

$$y = \log_e (x+\sqrt{x^2+k^2})$$

Differentiating both sides w.r.t. x, we get

$$\frac{dy}{dx} = \frac{1}{x+\sqrt{x^2+k^2}}\left(1+\frac{2x}{2\sqrt{x^2+k^2}}\right)$$

$$= \frac{1}{x+\sqrt{x^2+k^2}}\left(\frac{\sqrt{x^2+k^2}+x}{\sqrt{x^2+k^2}}\right)$$

$$= \frac{1}{\sqrt{x^2+k^2}}$$ **Ans.**

2. *Differentiate : $\log\sqrt{\frac{1-\cos x}{1+\cos x}}$.*

Sol. Let

$$y = \log\sqrt{\frac{1-\cos x}{1+\cos x}}$$

$$= \log\sqrt{\frac{2\sin^2 x/2}{2\cos^2 x/2}}$$

$$= \log\sqrt{\frac{\sin^2 x/2}{\cos^2 x/2}}$$

$$= \log(\tan x/2)$$

Now, $$\frac{dy}{dx} = \frac{1}{\tan x/2}\,\frac{1}{2}\cdot\sec^2\frac{x}{2}$$

$$= \frac{1}{2}\frac{\cos x/2}{\sin x/2}\cdot\frac{1}{\cos^2 x/2}$$

$$= \frac{1}{2(\sin x/2)(\cos x/2)}$$

$$= \frac{1}{\sin x}$$

$$= \text{cosec } x.$$ **Ans.**

3. *Find $\frac{dy}{dx}$ when*

$$y = \frac{\tan^{-1}\sqrt{1+x^2}-1}{x}.$$

Sol. Given,

$$y = \tan^{-1}\frac{\sqrt{1+x^2}-1}{x}$$

Let $x = \tan\theta$

$$\therefore \quad y = \tan^{-1}\frac{\sqrt{1+\tan^2\theta}-1}{\tan\theta}$$

$$= \tan^{-1}\left(\frac{\sec\theta-1}{\tan\theta}\right)$$

$$= \tan^{-1}\left(\frac{1-\cos\theta}{\sin\theta}\right) \quad \left(\because \sec\theta = \frac{1}{\cos\theta}\right)$$

$$= \tan^{-1}\left[\frac{2\sin^2\frac{\theta}{2}}{2\sin\frac{\theta}{2}\cos\frac{\theta}{2}}\right]$$

$$= \tan^{-1}\left(\tan\frac{\theta}{2}\right) = \frac{\theta}{2} = \frac{1}{2}\tan^{-1} x$$

Differentiating w.r.t. x, we get

$$\therefore \quad \frac{dy}{dx} = \frac{1}{2(1+x^2)}$$ **Ans.**

4. *If $x = a\sin^3 t$ and $y = a\cos^3 t$, find $\frac{dy}{dx}$.*

Sol. Given,

$$x = a\sin^3 t$$

$$\therefore \quad \frac{dx}{dt} = 3a\sin^2 t.\cos t$$

Also, $$y = a\cos^3 t$$

$$\therefore \quad \frac{dy}{dt} = 3a\cos^2 t\,(-\sin t)$$

$$\frac{dy}{dx} = \frac{-3a\cos^2 t \cdot \sin t}{3a\sin^2 t \cdot \cos t} \left[\because \frac{dy}{dx} = \frac{dy/dt}{dx/dt}\right]$$

$$\therefore \quad \frac{dy}{dx} = -\cot t.$$ **Ans.**

5. *If* $x = \sqrt{a^{\sin^{-1} t}}$ *and* $y = \sqrt{a^{\cos^{-1} t}}$ *find* $\frac{dy}{dx}$.

Sol. Given,

$$x = \sqrt{a^{\sin^{-1} t}}$$

$$\frac{dx}{dt} = \frac{1}{2\sqrt{a^{\sin^{-1} t}}} a^{\sin^{-1} t}.\log_e a.\frac{(1)}{\sqrt{1-t^2}}$$

and we have, $y = \sqrt{a^{\cos^{-1} t}}$

$$\frac{dy}{dt} = \frac{1}{2\sqrt{a^{\cos^{-1} t}}} a^{\cos^{-1} t}.\log_e a.\frac{(-1)}{\sqrt{1-t^2}}$$

$$\therefore \quad \frac{dy}{dx} = \frac{dy}{dt}\Big/\frac{dx}{dt}$$

$$= \frac{1}{2\sqrt{a^{\cos^{-1} t}}}.a^{\cos^{-1} t}.\log_e a.\frac{(-1)}{\sqrt{1-t^2}}$$

$$\times 2\sqrt{a^{\sin^{-1}}}.\frac{1}{a^{\sin^{-1} t}}\frac{1}{\log_e a}.\sqrt{1-t^2}$$

$$= \frac{-a^{\cos^{-1} t}}{\sqrt{a^{\cos^{-1} t}}} \times \frac{\sqrt{a^{\sin^{-1} t}}}{a^{\sin^{-1} t}}$$

$$= \frac{-\sqrt{a^{\cos^{-1} t}}}{\sqrt{a^{\sin^{-1} t}}} = \frac{-y}{x}$$ **Ans.**

6. *Differentiate* $\cos x^3$ *w.r.t.* x^2.

Sol. Let $u = \cos x^3$ and $v = x^2$

Now, $\frac{du}{dx} = -\sin x^3 . 3x^2 = -3x^2 \sin x^3$

and $\frac{dv}{dx} = 2x$

$$\therefore \quad \frac{du}{dv} = \frac{du}{dx}\Big/\frac{dv}{dx}$$

$$= \frac{-3x^2 \sin x^3}{2x}$$

$$= \frac{-3}{2} x \sin x^3.$$ **Ans.**

Chapter 7. Applications of Derivatives

1. *Determine the point on the curve* $y = 3x^2 - 8x + 8$ *at which the normal is parallel to a line whose slope is 1/2.*

Sol. Given,

$$y = 3x^2 - 8x + 8 \quad \ldots(i)$$

then, $\frac{dy}{dx} = 6x - 8$

$\therefore$ Slope of normal $= \frac{1}{2}$ (given)

$$-\frac{1}{(dy/dx)} = \frac{1}{2}$$

$$-\frac{1}{(6x-8)} = \frac{1}{2}$$

$$-2 = 6x - 8$$

$$6x = 6$$

$$x = 1$$

Putting the value of x in equation (i), we get

$$y = (3)(1)^2 - 8(1) + 8$$

$$= 3 - 8 + 8 = 3$$

Hence, (1, 3) is the required point on the curve.

Ans.

2. *The sum of three positive numbers is 26. The second number is thrice as large as the first. If the sum of the square of these numbers is least, find the numbers.*

Sol. Let the three numbers be x, $3x$ and y.

$$\therefore \quad 4x + y = 26 \quad \text{(Given)}$$

$$\Rightarrow \quad y = 26 - 4x$$

Sum of squares $= S = x^2 + 9x^2 + y^2$

$$= 10x^2 + y^2$$

$$= 10x^2 + (26 - 4x)^2$$

To minimise S, $\frac{dS}{dx} = 0$ and $\frac{d^2S}{dx^2}$ is positive.

Differentiating S w.r.t. x, we get

$$\frac{dS}{dx} = 20x + 2(26 - 4x) \times (-4)$$

$$\Rightarrow \quad \frac{dS}{dx} = 20x - 208 + 32x$$

$$\Rightarrow \quad \frac{dS}{dx} = 52x - 208$$

Put $\frac{dS}{dx} = 0$

$$\Rightarrow \quad x = 4$$

Differentiating $\frac{dS}{dx}$ w.r.t. x, we get

$$\frac{d^2S}{dx^2} = 20 + 32 = 52 > 0$$

Hence, the numbers are 4, 12 and 10. **Ans.**

Solve the Following Type Questions

Chapter 9. Differential Equations

1. *Solve the differential equation :*

$$\text{cosec}^3 x\, dy - \text{cosec}\, y\, dx = 0.$$

Sol. Given, $\text{cosec}^3 x\, dy - \text{cosec}\, y\, dx = 0$

$$\Rightarrow \sin y\, dy = \sin^3 x\, dx$$

$$\Rightarrow \sin y\, dy = \frac{1}{4}[3 \sin x - \sin 3x]$$

$$[\because \sin 3x = 3 \sin x - 4 \sin^3 x]$$

Integrating both sides, we get

$$\Rightarrow \int \sin y\, dy = \frac{1}{4}\int [3 \sin x - \sin 3x]\, dx + c$$

Where, 'c' is a constant of integration.

$$\Rightarrow -\cos y = \frac{1}{4}\left[-3\cos x + \frac{\cos 3x}{3}\right] + c$$

$$\Rightarrow \cos y = \frac{1}{4}\left[3\cos x - \frac{\cos 3x}{3}\right] + c \quad \textbf{Ans.}$$

2. *Solve the differential equation :*

$$\log\left(\frac{dy}{dx}\right) = 2x - 3y$$

Sol. Given,

$$\log\left(\frac{dy}{dx}\right) = 2x - 3y$$

$$\Rightarrow \frac{dy}{dx} = e^{2x - 3y}$$

$$\Rightarrow \frac{dy}{dx} = \frac{e^{2x}}{e^{3y}}$$

$$\Rightarrow e^{3y}\, dy = e^{2x}\, dx$$

Integrating both sides, we get

$$\int e^{3y}\, dy = \int e^{2x}\, dx$$

$$\frac{e^{3y}}{3} = \frac{e^{2x}}{2} + c \quad \textbf{Ans.}$$

3. *Solve the differential equation :*

$$(x + 1)\, dy - 2xy\, dx = 0$$

Sol. Given, $(x + 1)\, dy - 2xy\, dx = 0$

$$\Rightarrow \frac{dy}{dx} = \frac{2xy}{x+1}$$

$$\Rightarrow \frac{1}{y}\, dy = \left(\frac{2x}{x+1}\right)dx$$

$$\Rightarrow \frac{1}{y}\, dy = \left(\frac{2x+2-2}{x+1}\right)dx$$

$$\Rightarrow \frac{1}{y}\, dy = \left[\frac{2(x+1)}{(x+1)} - \frac{2}{(x+1)}\right]dx$$

Integrating both sides, we get

$$\therefore \int \frac{1}{y}\, dy = \int 2\, dx - \int \frac{2}{(x+1)}\, dx$$

$$\log |y| = 2x - 2 \log |x + 1| + c$$

Where c is a constant.

$$\Rightarrow \log y + 2 \log (x + 1) = 2x + c$$

$$\Rightarrow \log [y (x + 1)^2] = 2x + c$$

$$\Rightarrow [y (x + 1)^2] = e^{2x + c}$$

$$\Rightarrow y = \frac{e^{2x+c}}{(x+1)^2} \quad \textbf{Ans.}$$

4. *Solve the following differential equation :*

$$x (x^2 - x^2y^2)\, dy + y (y^2 + x^2y^2)\, dx = 0.$$

Sol. Given,

$$\Rightarrow x (x^2 - x^2y^2)\, dy + y (y^2 + x^2y^2)\, dx = 0$$

$$\Rightarrow x^3 (1 - y^2)\, dy = -y^3 (1 + x^2)\, dx$$

Integrating both sides, we get

$$\Rightarrow \int \frac{1-y^2}{y^3}\, dy = -\int \frac{(1+x^2)}{x^3}\, dx$$

$$\Rightarrow \int \frac{1}{y^3} - \frac{1}{y}\, dy = -\int \frac{1}{x^3}\, dx - \int \frac{1}{x}\, dx$$

$$\Rightarrow \frac{y^{-3+1}}{-3+1} - \log y = -\frac{x^{-3+1}}{-3+1} - \log x + c$$

where 'c' is a constant

$$\Rightarrow -\frac{1}{2y^2} - \log y = \frac{1}{2x^2} - \log x + c$$

$$\Rightarrow c + \frac{1}{2x^2} + \frac{1}{2y^2} = \log x - \log y$$

$$\Rightarrow c + \frac{1}{2}\left[\frac{1}{x^2} + \frac{1}{y^2}\right] = \log \frac{x}{y}. \quad \textbf{Ans.}$$

5. *Solve the differential equation ;*

$$e^y [1 + x^2]\, dy - \frac{x}{y}\, dx = 0.$$

Sol. Given,

$$e^y [1 + x^2]\, dy - \frac{x}{y}\, dx = 0$$

Multiplying by 'y' in the given equation, we get

$$y \cdot e^y (1 + x^2) \cdot dy - x \cdot dx = 0$$

Dividing by $(1 + x^2)$, we get

$$y \cdot e^y \cdot dy - \frac{x}{1+x^2} \cdot dx = 0$$

$$\therefore y \cdot e^y \cdot dy = \frac{x}{1+x^2} \cdot dx$$

Integrating both sides, we get

$$\int y.e^y.dy = \int \frac{x}{1+x^2}\cdot dx$$

Integrating by parts,

$$y\int e^y dy - \int\left\{\frac{d}{dy}(y)\int e^y dy\right\}dy = \frac{1}{2}\int\frac{2x}{1+x^2}dx$$

$$y\cdot e^y - \int e^y dy = \frac{1}{2}\log(1+x^2)+c$$

where 'c' is constant.

$$y\cdot e^y - e^y = \frac{1}{2}\log(1+x^2)+c$$

$$e^y(y-1) = \log\sqrt{1+x^2}+c.$$

$\therefore \quad e^y(y-1) - \log\left(\sqrt{1+x^2}\right) = c$ **Ans.**

6. *Solve the differential equation :*

$(x^2 - yx^2)\,dy + (y^2 + xy^2)\,dx = 0.$

Sol. Given,

$$(x^2 - yx^2)\,dy + (y^2 + xy^2)\,dx = 0$$

$$\Rightarrow \quad x^2(1-y)\,dy + y^2(1+x)\,dx = 0$$

On dividing by x^2y^2 on both sides

$$\Rightarrow \quad \left(\frac{1-y}{y^2}\right).dy + \left(\frac{1+x}{x^2}\right).dx = 0$$

Integrating both sides, we get

$$\int\frac{1-y}{y^2}\cdot dy = -\int\frac{1+x}{x^2}\cdot dx$$

$$\Rightarrow \int\frac{1}{y^2}\cdot dy - \int\frac{1}{y}\cdot dy = -\int\frac{1}{x^2}\cdot dx - \int\frac{1}{x}\cdot dx$$

$$\Rightarrow \frac{y^{-2+1}}{-2+1} - \log y = \frac{-x^{-2+1}}{-2+1} - \log x + c$$

where 'c' is constant of integration.

$$\Rightarrow \quad \frac{y^{-1}}{-1} - \log y = \frac{x^{-1}}{1} - \log x + c$$

$$\Rightarrow \quad -\frac{1}{y} - \log y = \frac{1}{x} - \log x + c$$

$\Rightarrow \quad \left(\frac{1}{x}+\frac{1}{y}\right) - (\log x - \log y) + c = 0$ **Ans.**

7. *Solve the differential equation :*

$$\frac{dy}{dx} = e^{x-y} + x^2e^{-y}.$$

Sol. Given, $\dfrac{dy}{dx} = e^{x-y} + x^2 e^{-y}$

$$= \frac{e^x}{e^y} + \frac{x^2}{e^y} = \frac{1[e^x + x^2]}{e^y}$$

$$\Rightarrow \quad e^y\,dy = (e^x + x^2)\,dx$$

Integrating both sides, we get

$$\int e^y\,dy = \int (e^x + x^2)\,dx$$

$\Rightarrow \quad e^y = e^x + \dfrac{x^3}{3} + c.$ **Ans.**

8. *Solve the following differential equation :*

$$\frac{dy}{dx} - e^{y+x} = e^{x-y}.$$

Sol. Given,

$$\frac{dy}{dx} = e^{(x-y)} + e^{(x+y)}$$

$$\Rightarrow \quad \frac{dy}{dx} = e^x.(e^{-y}+e^y)$$

Integrating both sides, we get

$$\int\frac{dy}{e^{-y}+e^y} = \int e^x\,dx$$

$$\Rightarrow \quad \int\frac{e^y dy}{1+e^{2y}} = \int e^x\,dx$$

Putting $e^y = t$,

$$\therefore \quad e^y = \frac{dt}{dy}$$

$$\therefore \quad \int\frac{dt}{1+t^2} = e^x + c$$

where c is a constant

$$\Rightarrow \quad \tan^{-1} t = e^x + c$$

$\Rightarrow \quad \tan^{-1} e^y = e^x + c$ **Ans.**

9. *Solve the differential equation :*

$(xy^2 + x)\,dx + (x^2y + y)\,dy = 0.$

Sol. Given,

$$(xy^2 + x)\,dx + (x^2y + y)\,dy = 0$$

$$\Rightarrow \quad x(y^2+1)\,dx + y(x^2+1)\,dy = 0$$

$$\Rightarrow \quad x(y^2+1)\,dx = -y(x^2+1)\,dy$$

$$\Rightarrow \quad \frac{x}{x^2+1}dx = \frac{-y}{y^2+1}dy$$

Multiplying by 2 on both sides

$$\Rightarrow \quad \frac{2x}{x^2+1}dx = \frac{-2y}{y^2+1}dy$$

Integrating both sides, we get

$$\int\frac{2x}{x^2+1}dx = \int\frac{-2y\,dy}{y^2+1}$$

$$\Rightarrow \quad \log|x^2+1| = -\log|y^2+1| + c$$

$$\Rightarrow \quad \log|x^2+1| + \log|y^2+1| = c$$

where 'c' is a constant of integration

$$\Rightarrow \quad |x^2+1|\,|y^2+1| = e^c$$

$\Rightarrow \quad (x^2+1)(y^2+1) = \pm e^c$ **Ans.**

Solve the Following Type Questions

10. *Solve the differential equation :*

$$\frac{dy}{dx} = e^{x+y} + x^2 e^y.$$

Sol. Given,

$$\frac{dy}{dx} = e^{x+y} + x^2 e^y$$

$$= e^x \cdot e^y + x^2 \cdot e^y$$

$$\Rightarrow \quad \frac{dy}{dx} = e^y (e^x + x^2)$$

Integrating both sides, we get

$$\Rightarrow \quad \int \frac{dy}{e^y} = \int (e^x + x^2)\, dx$$

$$\Rightarrow \quad \int e^{-y} dy = \int e^x dx + \int x^2\, dx$$

$$\Rightarrow \quad -e^{-y} = e^x + \frac{x^3}{3} + c$$

where 'c' is a constant of integration

$$\Rightarrow \quad -\frac{1}{e^y} = e^x + \frac{x^3}{3} + c$$

$$\Rightarrow \quad e^{x+y} + \frac{e^y \cdot x^3}{3} + ce^y + 1 = 0.$$ **Ans.**

11. *Solve*

$$\frac{dy}{dx} = \frac{(1+\cos^2 x)\sin^2 y}{(1+\sin^2 y)\cos^2 x}.$$

Sol. Given, $\frac{dy}{dx} = \frac{(1+\cos^2 x)\sin^2 y}{(1+\sin^2 y)\cos^2 x}$

$$\int \frac{(1+\sin^2 y)}{\sin^2 y} dy = \int \frac{(1+\cos^2 x)}{\cos^2 x} dx$$

$$\int (\text{cosec}^2\, y + 1)\, dy = \int (\sec^2 x + 1)\, dx$$

$$\Rightarrow \quad -\cot y + y = \tan x + x + c$$

where 'c' is a constant integration.

$$\Rightarrow \quad y - x = \tan x + \cot y + c$$ **Ans.**

12. *Solve :*

$(x^2 - yx^2)\, dy + (y^2 + xy^2)\, dx = 0$

Sol. We have,

$$\Rightarrow \quad (x^2 - yx^2)\, dy + (y^2 + xy^2)\, dx = 0$$

$$\Rightarrow \quad (x^2 - yx^2)\, dy = -(y^2 + xy^2)\, dx$$

$$\Rightarrow \quad x^2 (1 - y)\, dy = -y^2 (1 + x)\, dx$$

$$\Rightarrow \quad \frac{(1-y)}{y^2} dy = -\frac{(1+x)}{x^2} dx$$

$$\Rightarrow \quad \frac{(1-y)}{y^2} . dy + \frac{(1+x)}{x^2} \cdot dx = 0$$

On integrating both sides, we get

$$\int \frac{(1-y)}{y^2} dy + \int \frac{(1+x)}{x^2} dx = c$$

where 'c' is a constant

$$\int \frac{1}{y^2} dy - \int \frac{1}{y} dy + \int \frac{1}{x^2} dx + \int \frac{1}{x} dx = c$$

$$-\frac{1}{y} - \log |y| - \frac{1}{x} + \log |x| = c$$

$$\log\left(\frac{x}{y}\right) - \frac{1}{x} - \frac{1}{y} = c$$

which is the required solution. **Ans.**

13. *Solve :*

$$\frac{dy}{dx} = 1 - xy + y - x$$

Sol. We have,

$$\frac{dy}{dx} = 1 - xy + y - x$$

$$\frac{dy}{dx} = 1 + y - x - xy$$

$$\frac{dy}{dx} = (1 + y) - x\,(1 + y)$$

$$\frac{dy}{dx} = (1 + y)\,(1 - x)$$

$$\frac{dy}{1+y} = (1 - x)\, dx$$

Integrating both sides, we get

$$\int \frac{1}{(1+y)} dy = \int (1-x)\, dx$$

$$\log (1 + y) = x - \frac{x^2}{2} + c$$

(where 'c' is constant of integration) **Ans.**

14. *Solve the following differential equation :*

$(3xy + y^2)\, dx + (x^2 + xy)\, dy = 0$

Sol. Given, $(3xy + y^2)\, dx + (x^2 + xy)\, dy = 0$

$$\Rightarrow \quad \frac{dy}{dx} = -\left(\frac{3xy + y^2}{x^2 + xy}\right)$$

(Homogeneous)

Put $y = vx$

$$\Rightarrow \quad \frac{dy}{dx} = v + x \cdot \frac{dv}{dx}$$

$$\Rightarrow \quad \frac{dv}{dx} \cdot x + v = -\left(\frac{3x.vx + v^2x^2}{x^2 + x.vx}\right)$$

$$\Rightarrow \quad \frac{dv}{dx}\cdot x = \frac{-3v-v^2}{1+v} - v$$

$$\Rightarrow \quad x\frac{dv}{dx} = \frac{-3v-v^2-v-v^2}{1+v}$$

$$\Rightarrow \quad x\frac{dv}{dx} = \frac{-2v^2-4v}{1+v}$$

Integrating both sides, we get

$$\int \frac{1+v}{2v^2+4v}\,dv = \int -\frac{1}{x}\,dx$$

$$\Rightarrow \quad \frac{1}{4}\int \frac{2+2v}{2v+v^2}\,dv = -\int \frac{1}{x}\,dx$$

$$\Rightarrow \quad \frac{1}{4}\log|v^2+2v| = -\log|x| + \log c$$

$$\Rightarrow \quad \frac{1}{4}\log\left(\frac{y^2}{x^2}+2\frac{y}{x}\right) = \log\frac{c}{x}$$

$$\Rightarrow \quad \log\left(\frac{y^2}{x^2}+\frac{2y}{x}\right) = 4\log\frac{c}{x}$$

$$\Rightarrow \quad \log\left(\frac{y^2}{x^2}+\frac{2y}{x}\right) = \log\frac{c^4}{x^4}$$

$$\Rightarrow \quad \frac{y^2}{x^2}+\frac{2y}{x} = \frac{c^4}{x^4}$$

$$\Rightarrow \quad x^2y^2 + 2x^3y = c^4$$ **Ans.**

(where 'c' is a constant)

15. *Solve the differential equation ;*

$$x\frac{dy}{dx} - y = \sqrt{x^2+y^2}.$$

Sol. The given differential equation can be expressed as,

$$x\frac{dy}{dx} - y = \sqrt{x^2+y^2}.$$

$$\Rightarrow \quad \frac{dy}{dx} = \frac{y}{x} + \frac{\sqrt{x^2+y^2}}{x}$$

$$\Rightarrow \quad \frac{dy}{dx} = \frac{y}{x} + \sqrt{1+\left(\frac{y}{x}\right)^2} \quad \text{...(i)}$$

Let, $\frac{y}{x} = v \Rightarrow y = vx$

Now differentiating both sides w.r.t. x, we get

$$\frac{dy}{dx} = v + x\,.\,\frac{dv}{dx}$$

Substituting this value in differential equation (i), we get

$$v + x\frac{dv}{dx} = v + \sqrt{1+v^2}$$

$$\Rightarrow \quad x\frac{dv}{dx} = \sqrt{1+v^2}$$

$$\Rightarrow \quad \frac{dx}{x} = \frac{dv}{\sqrt{1+v^2}}$$

Integrating both sides, we get

$$\int\frac{dx}{x} = \int\frac{dv}{\sqrt{1+v^2}}$$

$$\Rightarrow \quad \log x + \log c = \log(v + \sqrt{1+v^2})$$

where $\log c$ is a constant

$$= \log\left(\frac{y}{x} + \sqrt{1+\frac{y^2}{x^2}}\right)$$

$$= \log\left(\frac{y+\sqrt{x^2+y^2}}{x}\right)$$

$$= \log(y + \sqrt{x^2+y^2}) - \log x$$

$$\Rightarrow \quad \log(y + \sqrt{x^2+y^2}) = 2\log x + \log c$$

$$\Rightarrow \quad \log(y + \sqrt{x^2+y^2}) = \log cx^2$$

$$\Rightarrow \quad y + \sqrt{x^2+y^2} = cx^2$$ **Ans.**

16. *Solve the following differential equation.*

$$y - x\frac{dy}{dx} = x + y\frac{dy}{dx}, \textit{ when } y = 0 \textit{ and } x = 1.$$

Sol. Given, $y - x\frac{dy}{dx} = x + y\frac{dy}{dx}$, when $y = 0$ and $x = 1$

$$\Rightarrow \quad y - x = (x+y)\frac{dy}{dx}$$

$$\Rightarrow \quad \frac{dy}{dx} = \frac{y-x}{y+x} \quad \text{...(i)}$$

Put $y = vx$, we get

$$\frac{dy}{dx} = v + x\frac{dv}{dx}$$

Put the value in equation (i)

$$\Rightarrow \quad v + x\frac{dv}{dx} = \frac{vx-x}{vx+x} = \frac{x(v-1)}{x(v+1)}$$

$$\Rightarrow \quad v + x\frac{dv}{dx} = \frac{v-1}{v+1}$$

$$\Rightarrow \quad x\frac{dv}{dx} = \frac{v-1}{v+1} - \frac{v}{1}$$

$$\Rightarrow \quad x\frac{dv}{dx} = \frac{v-1-v^2-v}{v+1} = \frac{-(v^2+1)}{v+1}$$

$$\Rightarrow \quad \frac{v+1}{v^2+1}dv = -\frac{dx}{x}$$

Integrating both sides, we get

$$\int \frac{v+1}{v^2+1}dv + \int \frac{dx}{x} = 0$$

$$\Rightarrow \frac{1}{2}\int \frac{2v}{v^2+1}dv + \int \frac{1}{v^2+1}dv + \int \frac{dx}{x} = c$$

$$\Rightarrow \frac{1}{2}\log(v^2+1) + \tan^{-1}v + \log|x| = c$$

$$\Rightarrow \frac{1}{2}\log\left(\frac{y^2}{x^2}+1\right) + \tan^{-1}\left(\frac{y}{x}\right) + \log|x| = c$$

$$\left[\text{replacing } v = \frac{y}{x}\right]$$

Now, when $y = 0$ and $x = 1$

$$\Rightarrow c = \frac{1}{2}\log 1 + \tan^{-1}0 + \log 1$$

$$\Rightarrow c = 0 \qquad [\because \log 1 = 0]$$

$\therefore$ Required solution is

$$\frac{1}{2}\log\left(\frac{y^2}{x^2}+1\right) + \tan^{-1}\left(\frac{y}{x}\right) + \log|x| = 0$$

$$\Rightarrow \log\left(\frac{x^2+y^2}{x^2}\right) + 2\tan^{-1}\left(\frac{y}{x}\right) + 2\log x = 0$$

$$\Rightarrow \log\left(\frac{x^2+y^2}{x^2}\right) + \log x^2 + 2\tan^{-1}\left(\frac{y}{x}\right) = 0$$

$$\Rightarrow \log\left[\left(\frac{x^2+y^2}{x^2}\right)\cdot x^2\right] + 2\tan^{-1}\left(\frac{y}{x}\right) = 0$$

$$\Rightarrow \log(x^2+y^2) + 2\tan^{-1}\left(\frac{y}{x}\right) = 0 \quad \textbf{Ans.}$$

17. *Solve the differential equation :*

$$x(x-y)\,dy + y^2\,dx = 0.$$

Sol. Given, $x(x-y)\,dy + y^2\,dx = 0$

$$\Rightarrow \frac{dy}{dx} = \frac{-y^2}{x^2-xy} = \frac{y^2}{xy-x^2} \qquad \text{...(i)}$$

Let $y = vx \Rightarrow v = \frac{y}{x}$

$$\Rightarrow \frac{dy}{dx} = v + \frac{xdv}{dx}$$

Put the values in equation (i)

$$\Rightarrow v + \frac{xdv}{dx} = \frac{v^2x^2}{vx^2-x^2} = \frac{v^2}{v-1}$$

$$\therefore \frac{xdv}{dx} = \frac{v^2}{v-1} - v = \frac{v^2-v^2+v}{v-1}$$

$$\Rightarrow \frac{xdv}{dx} = \frac{v}{v-1}$$

$$\Rightarrow \frac{(v-1)}{v}dv = \frac{1}{x}dx$$

Integrating both sides, we get

$$\int\left(1-\frac{1}{v}\right)dv = \int \frac{1}{x}dx$$

$$\Rightarrow v - \log v = \log x + c$$

where 'c' is constant of integration.

$$\Rightarrow \frac{y}{x} - \log\frac{y}{x} = \log x + c$$

$$\Rightarrow \frac{y}{x} - \left[\log\frac{y}{x} + \log x\right] = c$$

$$\Rightarrow \frac{y}{x} - \log y = c$$

$$\Rightarrow \frac{y}{x} = \log y + c. \quad \textbf{Ans.}$$

18. *Solve the differential equation :*

$$2\frac{dy}{dx} = \frac{y}{x} + \frac{y^2}{x^2}.$$

Sol. Given, $\frac{2dy}{dx} = \frac{y}{x} + \frac{y^2}{x^2}$ ...(i)

Let $y = vx \Rightarrow \frac{dy}{dx} = v + x\frac{dv}{dx}$

Putting value in equation (i)

$$\Rightarrow 2\left(v + \frac{xdv}{dx}\right) = v + v^2$$

$$\Rightarrow 2x\frac{dv}{dx} = v^2 - v$$

Integrating both sides, we get

$$\int \frac{1}{v^2-v}dv = \frac{1}{2}\int \frac{dx}{x}$$

$$\Rightarrow \int \frac{1}{\left(v^2-v+\frac{1}{4}-\frac{1}{4}\right)}dv = \frac{1}{2}\int \frac{dx}{x}$$

$$\Rightarrow \int \frac{1.\,dv}{\left(v-\frac{1}{2}\right)^2 - \left(\frac{1}{2}\right)^2} = \frac{1}{2}\int \frac{dx}{x}$$

$$\Rightarrow \frac{1}{2\times\frac{1}{2}}\log\left|\frac{v-\frac{1}{2}-\frac{1}{2}}{v-\frac{1}{2}+\frac{1}{2}}\right| = \frac{1}{2}\log x$$

$$\Rightarrow \quad \log\frac{v-1}{v} = \frac{1}{2}\log x$$

$$\Rightarrow \quad \log\frac{y-x}{y} = \frac{1}{2}\log x + c$$

where 'c' is constant of integration. **Ans.**

19. *Solve the differential equation :*

$$sin^{-1}\left(\frac{dy}{dx}\right) = x + y.$$

Sol. Given,

$$\sin^{-1}\left(\frac{dy}{dx}\right) = x + y$$

$$\frac{dy}{dx} = \sin(x+y) \quad \ldots(i)$$

Put $v = x + y$, we get

$$\frac{dv}{dx} = 1 + \frac{dy}{dx}$$

$$\Rightarrow \quad \frac{dy}{dx} = \frac{dv}{dx} - 1$$

Putting the value of $x + y$ and $\frac{dy}{dx}$ in equation (i), we get

$$\frac{dv}{dx} - 1 = \sin v$$

$$\Rightarrow \quad \frac{dv}{dx} = \sin v + 1$$

$$\Rightarrow \quad dx = \frac{1}{1+\sin v}dv$$

On integrating both sides,

$$\int dx = \int \frac{1}{1+\sin v}dv$$

$$x = \int \frac{1}{1+\sin v} \times \frac{1-\sin v}{1-\sin v}dv$$

$$= \int \frac{1-\sin v}{1-\sin^2 v}dv$$

$$= \int \frac{1-\sin v}{\cos^2 v}dv$$

$$= \int (\sec^2 v - \sec v \tan v)\, dv$$

$$\therefore \quad x = \tan v - \sec v + c \quad [\text{Put } v = x + y]$$

$$x = \tan(x+y) - \sec(x+y) + c$$

where c is a constant of integration. **Ans.**

20. *Solve the differential equation :*

$$e^{x/y}\left(1-\frac{x}{y}\right) + (1+e^{x/y})\frac{dx}{dy} = 0 \text{ when } x = 0, y = 1$$

Sol. Given differential equation is,

$$e^{x/y}\left(1-\frac{x}{y}\right) + (1+e^{x/y})\frac{dx}{dy} = 0$$

$$\Rightarrow \quad \frac{dx}{dy} = \frac{-e^{x/y}(1-x/y)}{1+e^{x/y}} \quad \ldots(i)$$

Since (i) is of the form $\frac{dx}{dy} = g(x/y)$. It is a homogeneous differential equation.

Put $\quad x = vy$

$$\Rightarrow \quad \frac{dx}{dy} = v + y\frac{dv}{dy}$$

Substituting the values of x and $\frac{dx}{dy}$ in (i), we get

$$v + y\frac{dv}{dy} = \frac{-e^v(1-v)}{1+e^v}$$

$$\Rightarrow \quad y\frac{dv}{dy} = \frac{-e^v(1-v)}{1+e^v} - v$$

$$= \frac{-(v+e^v)}{1+e^v}$$

$$\Rightarrow \quad \frac{1+e^v}{v+e^v}dv = -\frac{dy}{y}$$

On integrating both sides,

$$\int \frac{1+e^v}{v+e^v}dv = -\int \frac{dy}{y}$$

$$\Rightarrow \quad \log|v+e^v| = -\log|y| + c$$

$$\Rightarrow \log|y(v+e^v)| = c$$

$$\Rightarrow \quad y\left(\frac{x}{y} + e^{x/y}\right) = \pm e^c$$

$$\Rightarrow \quad x + y\,e^{x/y} = A \quad \ldots(i)$$

Where, A is arbitrary constant.

When $x = 0, y = 1$, then A = 1

$$[\because 0 + 1.\, e^{0/1} = 0 + 1 \times 1 = 1]$$

Substituting this value of A in equation (i), we get

$$x + y\,e^{x/y} = 1,$$

which is the required particular solution. **Ans.**

21. *Solve the differential equation :*

$$cos^2 x \,.\, \frac{dy}{dx} + y = tan\ x.$$

Sol. Given, $\cos^2 x \cdot \frac{dy}{dx} + y = \tan x$

$$\Rightarrow \quad \frac{dy}{dx} + \frac{y}{\cos^2 x} = \frac{\tan x}{\cos^2 x}$$

$$\Rightarrow \quad \frac{dy}{dx} + y\sec^2 x = \tan x \cdot \sec^2 x \qquad ....(i)$$

This is a differential equation of the form

$$\frac{dy}{dx} + Py = Q$$

where P and Q both are function of x.

On comparing, $P = \sec^2 x$

$$\therefore \quad \int P\,.dx = \int \sec^2 x \,.\, dx = \tan x$$

Integrating factor $= e^{\int Pdx} = e^{\tan x}$

Multiplying both sides by integrating factor in equation (i), we get

$$e^{\tan x} \cdot \frac{dy}{dx} + y \cdot e^{\tan x} \cdot \sec^2 x = e^{\tan x} \cdot \tan x \sec^2 x$$

$$\Rightarrow \quad \frac{dy}{dx}(y.e^{\tan x}) = e^{\tan x} \cdot \tan x \cdot \sec^2 x$$

On integrating above equation, we get

$$\int \frac{d}{dx}(ye^{\tan x})\,dx = \int e^{\tan x} \tan x \sec^2 x\,dx$$

Substituting $\tan x = t$, we get

$$\frac{dt}{dx} = \sec^2 x$$

$$\Rightarrow \quad \sec^2 x\,dx = dt$$

$$\therefore \quad y\,e^t = \int e^t \,.\, t \,.\, dt$$

[on integrating by parts]

$$= t \,.\int e^t \,.\, dt - \int 1 \,.\, e^t \,.\, dt$$

$$= t \,.\, e^t - e^t + k$$

$$= e^t(t-1) + k$$

(where 'k' is contant)

$$\Rightarrow \quad y \,.\, e^{\tan x} = e^{\tan x}(\tan x - 1) + k. \qquad \textbf{Ans.}$$

22. *Solve* $(1-x^2)\frac{dy}{dx} - xy = x$

given $y = 2$, *when* $x = 0$.

Sol. Given,

$$(1-x^2)\frac{dy}{dx} - xy = x$$

$$\Rightarrow \quad \frac{dy}{dx} - \left(\frac{x}{1-x^2}\right)y = \frac{x}{1-x^2} \qquad ...(i)$$

This is a differental equation of the form

$$\frac{dy}{dx} + Py = Q$$

Here, $P = \dfrac{-x}{(1-x^2)}$

$$\therefore \text{ Integrating factor} = e^{\int Pdx} = e^{\int \frac{-x}{1-x^2}dx}$$

Here, $\quad e^{\int Pdx} = e^{\int \frac{-x}{1-x^2}dx}$

where $e^{\int Pdx}$ is the integrating factor

Let $\quad 1 - x^2 = t$

$$\therefore \quad \frac{dt}{dx} = -2x$$

then $e^{\int \frac{-x}{t} \cdot \frac{dt}{-2x}} = e^{\frac{1}{2}\int \frac{dt}{t}} = e^{\frac{1}{2}\log t}$

$$e^{\int Pdx} = e^{1/2 \log t} = e^{\log \sqrt{t}}$$

$$= \sqrt{t} = \sqrt{1-x^2}$$

Multiplying by I.F. on both sides, in equation (i) we get

$$\frac{dy}{dx} \cdot \sqrt{1-x^2} - \left(\frac{x}{1-x^2}\right)y\sqrt{1-x^2} = \frac{x}{1-x^2} \times \sqrt{1-x^2}$$

$$\therefore \quad \frac{d}{dx}[y \,.\, \sqrt{1-x^2}] = \frac{x}{\sqrt{1-x^2}}$$

Integrating both sides, we get

$$\int \frac{d}{dx}(y.\sqrt{1-x^2}).\,dx = \int \frac{x\,.\,dx}{\sqrt{1-x^2}}$$

$$\Rightarrow \quad y\,.\sqrt{1-x^2} = \frac{1}{2}\int \frac{2x\,.\,dx}{\sqrt{1-x^2}}$$

$$\Rightarrow \quad y.\sqrt{1-x^2} = -\frac{1}{2}\int \frac{-2x\,.\,dx}{\sqrt{1-x^2}} \qquad \begin{bmatrix} \text{Putting} \\ t = 1-x^2 \\ dt = -2x\,dx \end{bmatrix}$$

$$\therefore \quad \frac{-1}{2}\int \frac{-2x\,dx}{\sqrt{1-x^2}} = \frac{-1}{2}\int \frac{dt}{\sqrt{t}}$$

$$\Rightarrow \quad y.\sqrt{1-x^2} = -\frac{1}{2}\frac{t^{-1/2+1}}{\left(-\frac{1}{2}+1\right)} + c$$

$$\Rightarrow \quad y.\sqrt{1-x^2} = -t^{1/2} + c$$

$\Rightarrow \quad y.\sqrt{1-x^2} = -\sqrt{1-x^2} + c$

$\therefore \quad y\sqrt{1-x^2} = -\sqrt{1-x^2} + c$

$\Rightarrow \quad y\sqrt{1-x^2} + \sqrt{1-x^2} = c$ ...(ii)

where 'c' is constant of integration.

Putting $x = 0$, $y = 2$ in equation (ii) we get $c = 3$

$\therefore \quad y\sqrt{1-x^2} + \sqrt{1-x^2} = 3$ **Ans.**

23. *Solve the following differential equation :*

$ye^y\,dx = (y^3 + 2xe^y)\,dy$, given that $x = 0$, $y = 1$.

Sol. Given, $y\,e^y\,dx = (y^3 + 2x\,e^y)\,dy$

$\Rightarrow \quad y\,e^y \dfrac{dx}{dy} = y^3 + 2x\,e^y$

$\Rightarrow \quad \dfrac{dx}{dy} = \dfrac{2}{y} \cdot x + y^2 e^{-y}$

$\Rightarrow \quad \dfrac{dx}{dy} - \dfrac{2}{y} \cdot x = y^2 e^{-y}$ ...(i)

It is a linear differential equation in x.

$\therefore \quad P = -2/y$

$\therefore \quad \text{I.F.} = e^{\int -\frac{2}{y} dy}$

$= e^{-2\log|y|}$

$= e^{\log|y|^{-2}}$

$= |y^{-2}| = \dfrac{1}{y^2}$

Multiplying equation (i) by I.F

$\Rightarrow \quad \dfrac{1}{y^2}\dfrac{dx}{dy} - \dfrac{2}{y^3}x = e^{-y}$

$\Rightarrow \quad \dfrac{d}{dy}(x \cdot y^{-2}) = e^{-y}$

On integrating both sides

$\Rightarrow \int \dfrac{d}{dy}(x \cdot y^{-2}).\,dy = \int e^{-y}\,dy$

Therefore, general solution of (i) is

$x \cdot \dfrac{1}{y^2} = \int y^2 e^{-y} \cdot \dfrac{1}{y^2}\,dy + c$

$\Rightarrow \quad x \cdot \dfrac{1}{y^2} = \int e^{-y}\,dy + c$

$\Rightarrow \quad x \cdot \dfrac{1}{y^2} = \dfrac{e^{-y}}{-1} + c$ ...(ii)

Given that $x = 0$ when $y = 1$

$\Rightarrow \quad 0 = -e^{-1} + c$

$\Rightarrow \quad c = e^{-1}$

Substituting the value of c in (ii), the required particular solution of the given equation is

$\dfrac{x}{y^2} = -e^{-y} + e^{-1}$

$\Rightarrow \quad x = y^2(-e^{-y} + e^{-1}).$ **Ans.**

24. *Solve the differential equation :*

$\dfrac{dy}{dx} - 3y\cot x = \sin 2x$, given $y = 2$, when $x = \dfrac{\pi}{2}$.

Sol. Given,

$\dfrac{dy}{dx} - 3y\cot x = \sin 2x$

This is a linear differential equation of the form

$\dfrac{dy}{dx} + Py = Q$

where $P = -3\cot x$, $Q = \sin 2x$

$\therefore \quad \text{I.F.} = e^{\int P dx}$

$\Rightarrow \quad \text{I.F.} = e^{\int -3\cot x\,dx} = e^{-3\int \cot x\,dx}$

$= e^{-3\log\sin x} = e^{\log(\sin x)^{-3}}$

$= (\sin x)^{-3} = \dfrac{1}{\sin^3 x}$

$\therefore$ Its solution is $y \cdot (\text{I.F.}) = \int Q\,(\text{I.F.})\,dx$

$\Rightarrow \quad y \cdot \dfrac{1}{\sin^3 x} = \int \sin 2x \cdot \dfrac{1}{\sin^3 x}\,dx$

$\Rightarrow \quad \dfrac{y}{\sin^3 x} = \int \dfrac{2\sin x \cdot \cos x}{\sin^3 x}\,dx$

$\Rightarrow \quad \dfrac{y}{\sin^3 x} = 2\int \operatorname{cosec} x \,.\, \cot x \,.\, dx$

$\Rightarrow \quad \dfrac{y}{\sin^3 x} = -2\operatorname{cosec} x + c$

Putting $y = 2$ and $x = \pi/2$,

$\dfrac{2}{\sin^3 \pi/2} = -2\operatorname{cosec}(\pi/2) + c$

$\Rightarrow \quad 2 = -2 + c$

$\Rightarrow \quad c = 4$

$\therefore \quad \dfrac{y}{\sin^3 x} = -\dfrac{2}{\sin x} + 4$

$\Rightarrow \quad y = -2\sin^2 x + 4\sin^3 x$ **Ans.**

25. *Solve the differential equation :*

$$(1+y^2)\frac{dy}{dx} = \tan^{-1}(y) - x.$$

Sol. Given, $(1 + y^2)\dfrac{dy}{dx} = \tan^{-1}(y) - x$

$$\Rightarrow \quad \frac{dx}{dy} = \frac{\tan^{-1} y}{1+y^2} - \frac{x}{1+y^2}$$

$$\Rightarrow \quad \frac{dx}{dy} + \frac{1}{1+y^2}\cdot x = \frac{\tan^{-1} y}{1+y^2}$$

This is a linear differential equation of the form

$$\frac{dy}{dx} + Px = Q$$

$$\therefore \quad \text{Integrating factor} = e^{\int \frac{dy}{1+y^2}}$$

$$= e^{\tan^{-1} y}$$

Multiplying by I.F., we get

$$e^{\tan^{-1} y}.\frac{dx}{dy} + e^{\tan^{-1} y}.\frac{1}{1+y^2}\cdot x$$

$$= \frac{\tan^{-1} y}{1+y^2}.e^{\tan^{-1} y}$$

Integrating both sides, we get

$$e^{\tan^{-1} y}.x = \int e^{\tan^{-1} y}.\frac{\tan^{-1} y}{1+y^2}dy$$

Let $\tan^{-1} y = \theta$

$$\Rightarrow \quad y = \tan\theta$$

$$\Rightarrow \quad dy = \sec^2\theta\,.\,d\theta$$

$$\therefore \int e^{\tan^{-1} y}\,.\,\tan^{-1} y\,.\,\frac{1}{1+y^2}.dy$$

$$= \int e^{\theta}.\frac{\theta}{1+\tan^2\theta}.\sec^2\theta.d\theta$$

$$= \int e^{\theta}.\frac{\theta}{\sec^2\theta}.\sec^2\theta.d\theta$$

$$= \int e^{\theta}.\theta.d\theta$$

Integrating by parts

$$= \theta.\int e^{\theta}\,d\theta - \int 1.\,e^{\theta}\,.\,d\theta$$

$$= \theta\,.\,e^{\theta} - e^{\theta} + c$$

where c is constant of integration.

$$= \tan^{-1} y\,.\,e^{\tan^{-1} y} - e^{\tan^{-1} y} + c$$

$$= e^{\tan^{-1} y}(\tan^{-1} y - 1) + c$$

$$\Rightarrow e^{\tan^{-1} y}\cdot x = e^{\tan^{-1} y}(\tan^{-1} y - 1) + c.$$ **Ans.**

26. *Solve the differential equation*

$$y\,dx - (x + 2y^2)\,dy = 0.$$

Sol. Given,

$$y\,dx - (x + 2y^2)\,dy = 0$$

$$\Rightarrow \quad y\frac{dx}{dy} - x - 2y^2 = 0$$

$$\Rightarrow \quad \frac{dx}{dy} - \frac{x}{y} - 2y = 0$$

$$\Rightarrow \quad \frac{dx}{dy} - \frac{x}{y} = 2y \qquad ...(i)$$

This is a linear differential equation of the from

$$\frac{dy}{dx} + Px = Q$$

$$\therefore \text{I. F.} = e^{\int P dy} = e^{-\int \frac{1}{y}dy} = e^{-\log y} = e^{\log y^{-1}} = y^{-1} = \frac{1}{y}$$

Multiplying both sides of (i) by $\dfrac{1}{y}$, we get

$$\Rightarrow \quad \frac{1}{y}\left[\frac{dx}{dy} - \frac{x}{y}\right] = \frac{1}{y}\times 2y$$

Integrating both sides, we get

$$x.\frac{1}{y} = \int 2\,dy$$

$$\frac{x}{y} = 2y + c$$

where 'c' is a constant of integration

$$x = 2y^2 + cy,$$ **Ans.**

27. *Solve the differential equation $(y + \log x)\,dx - x\,dy = 0$, given that $y = 0$, when $x = 1$.*

Sol. Given, $(y + \log x)\,dx - x\,dy = 0$

$$\Rightarrow \quad \frac{dy}{dx} = \frac{y}{x} + \frac{\log x}{x}$$

$$\Rightarrow \quad \frac{dy}{dx} - \frac{y}{x} = \frac{\log x}{x},$$

This is linear differential equation of the form

$$\frac{dy}{dx} + Py = Q$$

Here, $P = -\frac{1}{x}, Q = \frac{\log x}{x}$

$\therefore$ I. F. $= e^{-\int \frac{1}{x} dx} = e^{-\log x} = \frac{1}{x}$

Solution of given equation will be given by

$\therefore y \cdot$ I. F. $= \int Q$ (I. F.) $dx + c$

Where c is constant of integration

$$\Rightarrow \quad y.\frac{1}{x} = \int \frac{\log x}{x}.\frac{1}{x}dx + c$$

On integrating by parts

$$\Rightarrow \quad \frac{y}{x} = \log x \int x^{-2}\,dx - \int\left[\frac{d}{dx}\log x \int x^{-2}dx\right]dx + c$$

$$\Rightarrow \quad \frac{y}{x} = \log x\left(-\frac{1}{x}\right) - \int \frac{1}{x}\left(-\frac{1}{x}\right)dx + c$$

$$\Rightarrow \quad \frac{y}{x} = \log x\left(-\frac{1}{x}\right) + \left(-\frac{1}{x}\right) + c$$

$$\Rightarrow \quad y = -\log_e x - 1 + cx$$

Given that, $y = 0$, when $x = 1$

$0 = -\log_e 1 - 1 + c$

$c = 1$ $\quad [\because \log_e 1 = 0]$

$\therefore \quad y = -\log_e x - 1 + x$ **Ans.**

28. *Solve the equation*

$$\sin x\frac{dy}{dx} - y = \cos^2 x \sin x \tan\frac{x}{2}.$$

Sol. Given,

$$\sin x\frac{dy}{dx} - y = \cos^2 x \sin x \tan\frac{x}{2}$$

$$\frac{dy}{dx} - \frac{1}{\sin x}y = \cos^2 x \tan\frac{x}{2} \quad ...(i)$$

$$\text{I.F.} = e^{\int -\frac{1}{\sin x}dx} = e^{-\int \operatorname{cosec} x\, dx}$$

$$= e^{\log|\operatorname{cosec} x + \cot x|}$$

$$= e^{\log\left|\frac{1}{\sin x} + \frac{\cos x}{\sin x}\right|}$$

$$= e^{\log\left|\frac{1+\cos x}{\sin x}\right|}$$

$$= e^{\log\left|\frac{2\cos^2 x/2}{2\sin x/2 \cos x/2}\right|}$$

$$= e^{\log|\cot x/2|}$$

$$= \cot x/2 \quad (\because e^{\log x} = x$$

Multiplying (i) by $\cot\frac{x}{2}$, we get.

$$\cot\frac{x}{2}\left(\frac{dy}{dx} - \frac{1}{\sin x}y\right) = \cos^2 x \tan\frac{x}{2}\cot\frac{x}{2} = \cos^2 x$$

Integrating both sides w.r.t. x, we get

$$y\cot\frac{x}{2} = \int \cos^2 x\,dx + c$$

$$= \int \frac{1}{2}(1 + \cos 2x)\,dx + c$$

$$\Rightarrow \quad y\cot\frac{x}{2} = \frac{1}{2}\left(x + \frac{\sin 2x}{2}\right) + c$$

$$\Rightarrow \quad y\cot\frac{x}{2} = \frac{x}{2} + \frac{\sin 2x}{4} + c,$$ where 'c' is a constant of integration which is required solution.

Ans.

29. *Solve the differential equation*

$$\tan x\frac{dy}{dx} + 2y = \sec x$$

Sol. Given, $\tan x\frac{dy}{dx} + 2y = \sec x$

$$\Rightarrow \quad \frac{dy}{dx} + \frac{2}{\tan x}y = \frac{1}{\sin x}$$

$$\therefore \quad P = \frac{2}{\tan x};\ Q = \frac{1}{\sin x}$$

$$\therefore \quad \text{I. F.} = e^{\int P\,dx} = e^{2\int \cot x\,dx} = e^{2\log \sin x}$$

$$= e^{\log_e \sin^2 x} = \sin^2 x$$

$\therefore \quad y$ (I. F.) $= \int Q$ (I. F) $dx + c$

where 'c' is constant of integration.

$$\Rightarrow \quad y.\sin^2 x = \int \frac{1}{\sin x}.\sin^2 x\,dx + c$$

$$\Rightarrow \quad (\sin^2 x)\,y = \int \sin x\,dx + c$$

$$\Rightarrow \quad (\sin^2 x)\,y = -\cos x + c$$

$\Rightarrow \quad (\sin^2 x)\,y + \cos x = c$ **Ans.**

30. *Solve the following differential equation :*
$x^2\,dy + (xy + y^2)\,dx = 0$, when $x = 1$ and $y = 1$

Solve the Following Type Questions

Sol. The given differential equation is,

$$x^2\,dy + (xy + y^2)\,dx = 0$$

$$\Rightarrow \quad \frac{dy}{dx} = -\frac{y(x+y)}{x^2}$$

which is homogeneous as each of the functions in the numerator and the denominator is a homogenous function of degree 2.

Putting $y = vx$ and $\frac{dy}{dx} = v + x\frac{dv}{dx}$, the given equation becomes

$$v + x\frac{dv}{dx} = \frac{-vx(x+vx)}{x^2}$$

$$\Rightarrow \quad x\frac{dv}{dx} = -v(1+v) - v$$

$$= -(2v + v^2)$$

$$= -v(v+2)$$

$$\Rightarrow \quad \frac{dv}{v(v+2)} = -\frac{1}{x}$$

$$\therefore \quad \int \frac{dv}{v(v+2)} = -\int \frac{dx}{x}$$

$$\int \frac{1}{2}\left(\frac{1}{v} - \frac{1}{v+2}\right)dv = -\int \frac{dx}{x}$$

$$\Rightarrow \quad \frac{1}{2}\log|v| - \frac{1}{2}|v+2|$$

$$= -\log|x| + \log c$$

$$\Rightarrow \quad \log\left|\frac{x\sqrt{v}}{\sqrt{v+2}}\right| = \log c$$

$$\Rightarrow \quad \frac{x\sqrt{v}}{\sqrt{v+2}} = c$$

$$\Rightarrow \quad \frac{x\sqrt{y}}{\sqrt{y+2x}} = c \qquad \left(\text{Putting } v = \frac{y}{x}\right)$$

Squaring both sides

$$\Rightarrow \quad x^2y = A(y + 2x) \qquad \ldots\text{(i)}$$

where $A = c^2$.

when $x = 1$ and $y = 1$, equation (i) becomes.

$$1 = A(1+2) \Rightarrow A = \frac{1}{3}$$

Putting the value of A in equation (i), we get

$$3x^2y = (y + 2x)$$ **Ans.**

31. *Solve the differential equation :*

$$x\frac{dy}{dx} + y = 3x^2 - 2$$

Sol. Given,

$$x\frac{dy}{dx} + y = 3x^2 - 2$$

$$\frac{dy}{dx} + \frac{1}{x}\cdot y = 3x - \frac{2}{x} \qquad \ldots\text{(i)}$$

Here, $P = \frac{1}{x}$ and $Q = 3x - \frac{2}{x}$

$$\text{I. F.} = e^{\int P\,dx} = e^{\int \frac{1}{x}dx} = e^{\log x} = x$$

Multiplying equation (i) by I. F.

$$x\frac{dy}{dx} + \frac{y}{x}\cdot x = \left(3x - \frac{2}{x}\right)\cdot x$$

$$x\frac{dy}{dx} + y = 3x^2 - 2$$

On integrating, we get

$$xy = \int (3x^2 - 2)\,dx + c$$

where c is a constant

$xy = x^3 - 2x + c$ which is the required solution. **Ans.**

Chapter 12. Three Dimensional Geometry

1. *Find the shortest distance between the lines* $\frac{x}{5} = \frac{y-2}{2} = \frac{z-3}{3}$ *and* $\frac{x+3}{5} = \frac{y-1}{2} = \frac{z+4}{3}$.

Sol. Given lines are

$$\frac{x}{5} = \frac{y-2}{2} = \frac{z-3}{3} \qquad \ldots\text{(i)}$$

$$\text{and} \quad \frac{x+3}{5} = \frac{y-1}{2} = \frac{z+4}{3} \qquad \ldots\text{(ii)}$$

Here, we see that the lines are parallel because their direction ratios are same.

Let N be the foot of perpendicular drawn from the point A (0, 2, 3) on the line (i) to the line (ii).

Any point on (ii) is N $(5k - 3, 2k + 1, 3k - 4)$

Direction ratios of

$$AN = 5k - 3,\ 2k + 1 - 2,\ 3k - 4 - 3$$

i.e., $5k - 3, 2k - 1, 3k - 7$

Direction ratios of (ii) are 5, 2, 3.

Here, AN is perpendicular to line (ii)

$\therefore 5(5k - 3) + 2(2k - 1) + 3(3k - 7) = 0$

$\Rightarrow \quad k = 1$

$\therefore$ N (2, 1, – 4)

$\therefore$ Shortest distance

$$AN = \sqrt{2^2 + (-1)^2 + (-7)^2}$$

$$\sqrt{2 + 1 + 49} = \sqrt{54}$$ **Ans.**

Chapter 13. Application of Integrals

1. *Find the area of the region bounded by the curves $x = 4y - y^2$ and the y-axis.*

Sol. Given curve is, $x = 4y - y^2$

i.e., $y^2 - 4y = -x$

$$y^2 - 4y + 4 = -x + 4$$

$$(y-2)^2 = -(x-4)$$

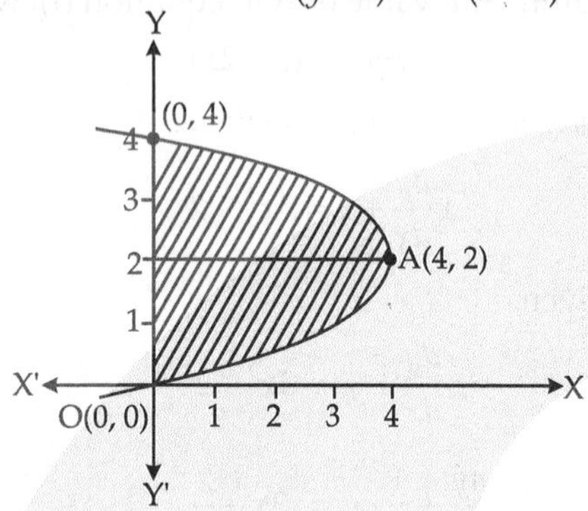

Which represents a parabola with vertex at A (4, 2).

The parabola meets y-axis *i.e.,* $x = 0$

$$4y - y^2 = 0$$

i.e., at $y = 0, 4$.

The area enclosed between the curve and the y-axis is

$$\int_0^4 x\,dy = \int_0^4 (4y - y^2)\,dy$$

$$= \left[4 \cdot \frac{y^2}{2} - \frac{y^3}{3}\right]_0^4$$

$$= \left[32 - \frac{64}{3}\right] - [0 - 0]$$

$$= \left|\frac{32}{3}\right| \text{ sq. units} = \frac{32}{3} \text{ sq. units} \quad \textbf{Ans.}$$

2. *Find the area of the region lying in the first quadrant bounded by the parabola $y^2 = 4x$, the x-axis and the ordinate $x = 4$.*

Sol. Given, $y^2 = 4x$

$\Rightarrow$ $y = \sqrt{4x}$ (in first quadrant)

Area under the curve $y^2 = 4x$ bounded by the x-axis, ordinate $x = 4$ is given as

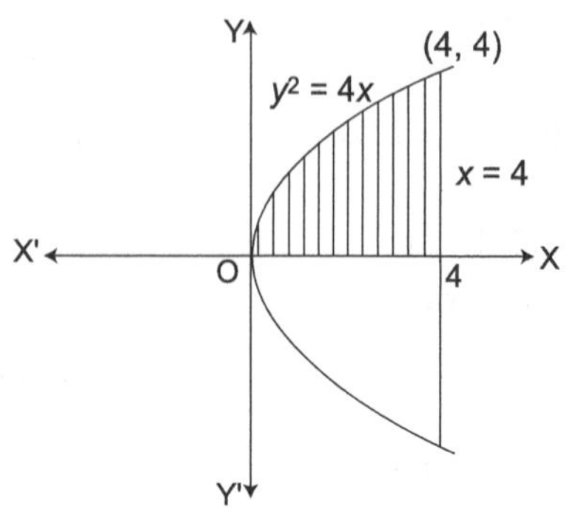

$$\text{Required Area} = \int_0^4 \sqrt{4x}\,dx$$

$$= \int_0^4 2\sqrt{x}\,dx$$

$$= \left[2 \times \frac{2}{3} x^{3/2}\right]_0^4 = \frac{4}{3} \times (4)^{3/2}$$

$$= \frac{32}{3} \text{ sq. units.} \quad \textbf{Ans.}$$

3. *Find the area bounded by the curve $y = \log_e x$, $y = 0$ and $x = e$.*

Sol. Given curve is $y = \log_e x$

$$\left[\begin{array}{l}\text{At } y = 0, 0 = \log_e x \Rightarrow x = 1 \\ \text{At } x = e, y = \log_e e \Rightarrow y = 1\end{array}\right.$$

Then, the area bounded by the curve $y = \log_e x$, $y = 0$ and $x = e$ is given as

$$\text{Required Area} = \int_1^e \log_e x\,dx$$

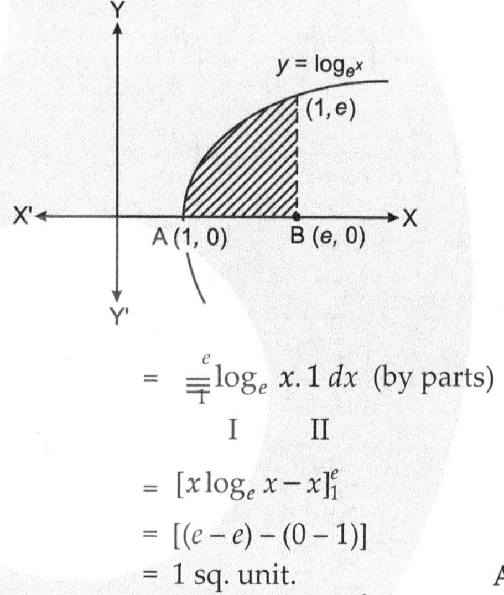

$$= \int_1^e \underset{\text{I}}{\log_e x} . \underset{\text{II}}{1}\,dx \text{ (by parts)}$$

$$= [x \log_e x - x]_1^e$$

$$= [(e - e) - (0 - 1)]$$

$$= 1 \text{ sq. unit.} \quad \textbf{Ans.}$$

4. *Draw a rough sketch of the curve $x^2 + y = 9$ and find the area enclosed by the curve, the x-axis and the lines $x + 1 = 0$ and $x - 2 = 0$.*

Sol. The given equation of the curve is,

$$x^2 + y = 9$$

$$\Rightarrow \quad y = 9 - x^2$$

$$\text{Required Area} = \int_{-1}^{2} y\,dx$$

$$= \int_{-1}^{2}(9-x^2)\,dx$$

$$= \left[9x-\frac{x^3}{3}\right]_{-1}^{2}$$

$$= \left[\left(18-\frac{8}{3}\right)-\left(-9+\frac{1}{3}\right)\right]$$

$= [27 - 3] = 24$ sq. units. **Ans.**

5. *Find the area enclosed by the curves $y = x^2$ and $y^2 = x$.*

Sol. Given curves are $y = x^2$ and $y^2 = x$ *i.e.*, $y = \sqrt{x}$ which represent the parabolas upward and right handed respectively having vertex (0, 0)

The points of intersection of the curves are O (0, 0) and A (1, 1)

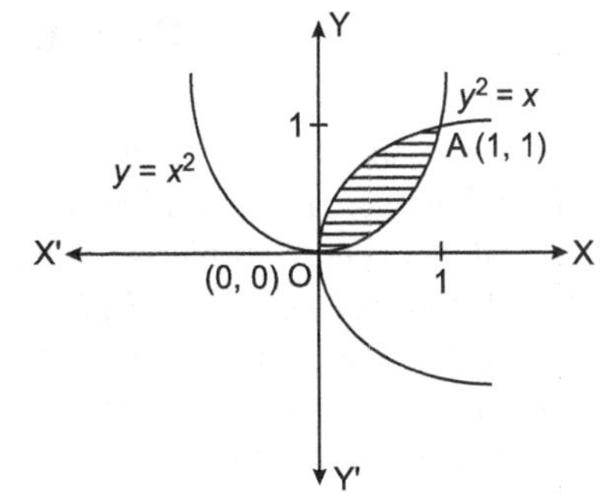

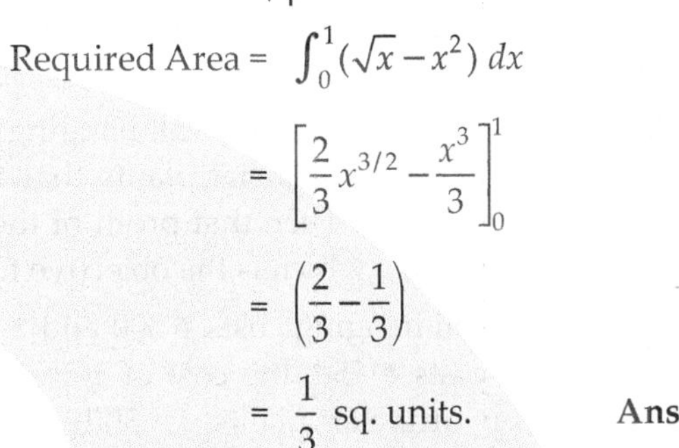

$$\therefore \quad \text{Required Area} = \int_{0}^{1}(\sqrt{x}-x^2)\,dx$$

$$= \left[\frac{2}{3}x^{3/2}-\frac{x^3}{3}\right]_{0}^{1}$$

$$= \left(\frac{2}{3}-\frac{1}{3}\right)$$

$= \dfrac{1}{3}$ sq. units. **Ans.**

Chapter 14. Applications of Calculus

1. *The marginal revenue function of a commodity is given by MR = $8 + 9x^2 - \frac{3}{2}x$, where x denotes the no. of units sold. Determine the demand function.*

Sol. We have,

$$MR = 8+9x^2-\frac{3}{2}x$$

We know that

$$MR = \frac{dR}{dx},$$

where R is the total revenue function

$$dR = MR\,dx$$

On integrating both sides, we get

$$\int dR = \int MR\,dx + c$$

where c is a constant of integration

$$R = \int\left(8+9x^2-\frac{3}{2}x\right)dx+c$$

$$R = 8x+3x^3-\frac{3}{4}x^2+c$$

Now, when $x = 0$, $R = 0 \Rightarrow c = 0$

$$\therefore \quad R = 8x+3x^3-\frac{3}{4}x^2$$

Hence, demand function is

$$P = \frac{R}{x}$$

$$\Rightarrow \quad P = 8+3x^2-\frac{3}{4}x.$$

$$= 3x^2-\frac{3}{4}x+8$$ **Ans.**

2. *A television manufacturer finds that the total cost for the production and marketing of x number of television sets is*

$C(x) = 300x^2 + 4200x + 13,500$.

Each product is sold for ₹ 8,400. Determine the breakeven points.

Sol. Given,

Total cost, $C(x) = 300x^2 + 4200x + 13500$

Total Revenue, $R(x) = Px$

[x is the no. of units sold]

$\therefore \quad R(x) = 8400x$

For breakeven points,

$$R(x) = C(x)$$

$$8400x = 300x^2 + 4200x + 13500$$

$$\Rightarrow 8400x - 300x^2 - 4200x - 13500 = 0$$

$$\Rightarrow -300x^2 + 4200x - 13500 = 0$$

$$\Rightarrow x^2 - 14x + 45 = 0$$

$$\Rightarrow x^2 - 9x - 5x + 45 = 0$$

$$\Rightarrow x(x-9) - 5(x-9) = 0$$

$$\Rightarrow (x-9)(x-5) = 0$$

$$\Rightarrow x = 5$$

or $\quad x = 9$

$\therefore$ For breakeven points $x = 5$ or 9 **Ans.**

Chapter 16. Linear Programming

1. *A shopkeeper deals in two items–wall hangings and artificial plants. He has ₹ 15,000 to invest and a space to store at most 80 pieces. A wall hanging costs him ₹ 300 and an artificial plant ₹ 150. He can sell a wall hanging at a profit of ₹ 50 and an artificial plant at a profit of ₹ 18. Assuming that he can sell all the items that he buys, formulate a linear programming problem in order to maximize his profit.*

Sol. Let x be the number of wall hangings and y be the number of artificial plants that the dealer buys and sells. Then that profit of the dealer is $Z = 50x + 18y$, which is the objective function.

As a wall hanging costs ₹ 300 and an artificial plant costs ₹ 150, the cost of x wall hangings and y artificial plants is $300x + 150y$. We are given that the dealer can invest at most ₹ 15,000. Hence, the investment constraint is,

$$300x + 150y \leq 15{,}000$$

i.e., $$2x + y \leq 100$$

As the dealer has space to store at most 80 pieces. We have another constraint (space constraint)

$$x + y \leq 80$$

Also the number of wall hangings and artificial plants can't be negative. Thus, we have the non-negativity constraints

$$x \geq 0,\ y \geq 0$$

Thus, the mathematical formulation of the linear programming problem is

Maximise $Z = 50x + 18y$

Subject to the constraints,

$$2x + y \leq 100,$$
$$x + y \leq 80$$
$$x \geq 0,\ y \geq 0$$ **Ans.**

2. *A manufacturer manufactures two types of tea-cups A and B. Three machines are needed for manufacturing the tea cups. The time in minutes required for manufacutring each cup on the machine is given below :*

Type of Cup	*Times in minutes*		
	Machine I	*Machine II*	*Machine III*
A	12	18	6
B	6	0	9

Each machine is available for a maximum of six hours per day. If the profit on each cup of type A is ₹ 1·50 and that on each cup of type B is ₹ 1·00, find the number of cups of each type that should be manufactured in a day to get maximum profit.

Sol. Let x be the number of A type tea cups and y be the number of B type tea cups.

The problem can be formulated as

Maximize the profit $P = 1{\cdot}50x + 1{\cdot}00y$

i.e., $$P = 1{\cdot}50x + y$$

Subject of the constraints

$$12x + 6y \leq 360 \text{ i.e., } 2x + y \leq 60$$
$$18x + 0{\cdot}y \leq 360 \text{ i.e., } x \leq 20$$
$$6x + 9y \leq 360 \text{ i.e., } 2x + 3y \leq 120$$

and $$x \geq 0,\ y \geq 0$$

(Non negative constriants)

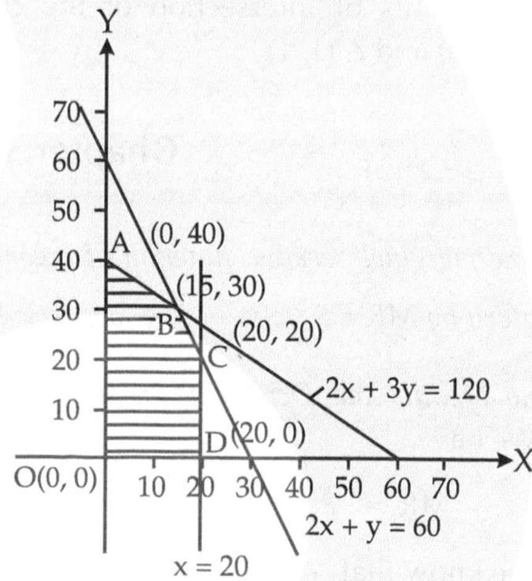

We draw the straight lines. $2x + y = 60$, $x = 20$, $2x + 3y = 120$, $x = 0$ and $y = 0$.

The shaded portion shows the feasible region. The point B and C are the points of intersection of lines $2x + y = 60$, with $2x + 3y = 120$ and $x = 20$ with $2x + y = 60$ respectively.

Thus, the 5 corner points of the feasible region are

O (0, 0), A (0, 40), B (15, 30), C (20, 20) and D (20, 0)

Corner Points	Objective function $Z = 1{\cdot}50\,x + y$
At O (0, 0)	$Z = 1{\cdot}50 \times 0 + 1 \times 0 = 0$
At A (0, 40)	$Z = 1{\cdot}50 \times 0 + 1 \times 40 = 40$
At B (15, 30)	$Z = 1{\cdot}50 \times 15 + 1 \times 30 = 52{\cdot}5$
At C (20, 20)	$Z = 1{\cdot}50 \times 20 + 1 \times 20 = 50$
At D (20, 0)	$Z = 1{\cdot}50 \times 20 + 1 \times 0 = 30$

Clearly the maximum profit is at $x = 15$ and $y = 30$.

Thus, the manufacturer should manufacture 15 cups of type A and 30 cups of type B to get maximum profit in a day. **Ans.**

3. *A farmer has a supply of chemical fertilizer of type A which contains 10% nitrogen and 6% phosphoric acid and of type B which contains 5% nitrogen and 10% phosphoric acid. After soil test, it is found that at least 7 kg of nitrogen and same quantity of phosphoric acid is required for a good crop. The fertilizer of type A costs ₹ 5.00 per kg and the type B costs ₹ 8.00 per kg. Using Linear programming, find how many kilograms of each type of the fertilizer should be bought to meet the requirement and for the cost to be minimum. Find the feasible region in the graph.*

Sol. Let x be the quantity of fertilizer of type A and y be the quantity of fertilizer of type B.

Now, x kg of fertilizer of type A contains 10% nitrogen and y kg of fertilizer of type B contains 5% nitrogen.

$$\frac{10}{100}x + \frac{5}{100}y \geq 7$$

Also, x kg of fertilizer of type A and y kg of fertilizer of type B contains 6% and 10% of phosphorus respectively.

$$\frac{6}{100}x + \frac{10}{100}y \geq 7$$

The cost of fertilizer of x kg of type A and y kg of type B will be ₹ $(5x + 8y)$

Hence the given L.P.P. is

$$\text{Minimize } Z = 5x + 8y$$

$$\frac{10}{100}x + \frac{5}{100}y \geq 7$$

or, $$2x + y \geq 140$$

$$\frac{6}{100}x + \frac{10}{100}y \geq 7$$

or $$3x + 5y \geq 350$$

$$x \geq 0, \quad y \geq 0$$

To solve this graphically draw the graphs of the equations.

$$2x + y = 140 \qquad \text{...(i)}$$

x	0	70
y	140	0

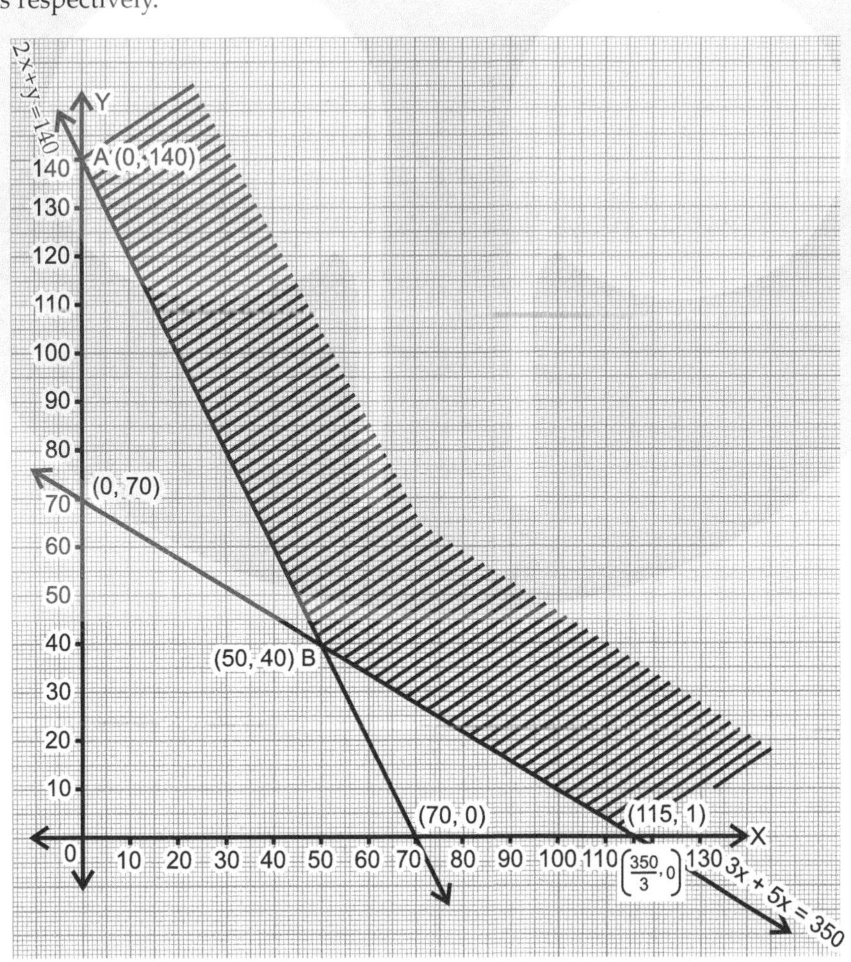

and $3x + 5y = 350$ ...(ii)

x	0	115	350/3
y	70	1	0

The intersection point of lines (i) & (ii) is 50 and 40.

Clearly the coordinates of the vertices of the feasible region are A (0, 140), B (50, 40), C (350/3, 0)

The values of the objective function at these corner points are given in the following table :

Corner Points	Objective Function $Z = 5x + 8y$
A (0, 140)	$Z = 0 + 8 \times 140 = 1120$
B (50, 40)	$Z = 5 \times 50 + 8 \times 40 = 570$
C (350/3, 0)	$Z = \frac{5 \times 350}{3} = \frac{1750}{3} = 583{\cdot}33$

From the above, we find that Z is minimum at (50, 40) *i.e.,* when $x = 50$ and $y = 40$.

Hence, the farmer should use 50 kg of fertilizer of type A and 40 kg of fertilizer of type B to meet the requirement and the minimum cost is ₹ 570. **Ans.**

Graph Based Questions

Set 9

Chapter 13. Application of Integrals

1. *Draw a rough sketch and find the area bounded by the curve $x^2 = y$ and $x + y = 2$.**

Sol. Given equation of curve is $x^2 = y$ and equation of line is $x + y = 2$

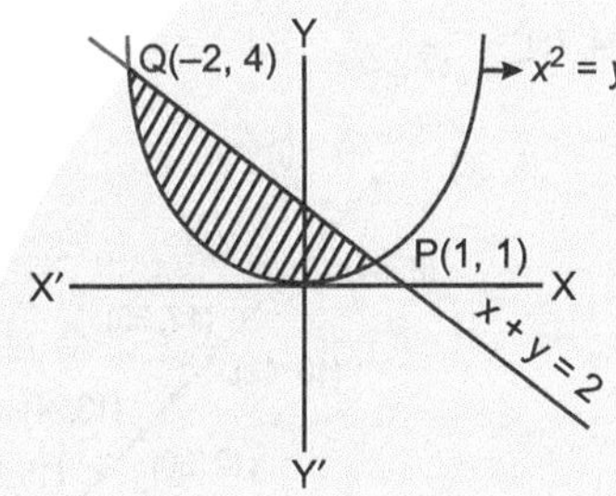

$\because \quad x^2 = y$ and $y = 2 - x$

$\therefore \quad x^2 = 2 - x$

$\Rightarrow \quad x^2 + x - 2 = 0$

$\Rightarrow \quad x^2 + 2x - x - 2 = 0$

$\Rightarrow x(x + 2) - (x + 2) = 0$

$\Rightarrow \quad (x + 2)(x - 1) = 0$

$x = -2$ and $x = 1$

when $x = -2$,

$y = 4$

(Putting $x = -2$ in curve $y = x^2$)

and when $x = 1$,

$y = 1$

(Putting $x = 1$ in curve $y = x^2$)

$\therefore$ The points are $(-2, 4)$ and $(1, 1)$.

$$\text{Required area} = \int_{-2}^{1}[(2 - x) - x^2]\,dx$$

$$= \left[2x - \frac{x^2}{2} - \frac{x^3}{3}\right]_{-2}^{1}$$

$$= \left[2 - \frac{1}{2} - \frac{1}{3}\right] - \left[-4 - 2 + \frac{8}{3}\right]$$

$$= \frac{7}{6} + \frac{10}{3}$$

$$= \frac{9}{2} \text{ sq. units.}$$ **Ans.**

2. *Draw a rough sketch of the curves $y^2 = x$ and $y^2 = 4 - 3x$ and find the area enclosed between them.**

Sol. Given equation of curves are,

$$y^2 = x \quad \text{...(i)}$$

and $$y^2 = 4 - 3x \quad \text{...(ii)}$$

These are the equations of parabola.

Solving equations (i) and (ii), we get

$x = 4 - 3x$

$\Rightarrow \quad 4x = 4$

$\Rightarrow \quad x = 1$

$\because \quad y^2 = 1$

$\Rightarrow \quad y = \pm 1$

So, $(1, 1)$ and $(1, -1)$ are the points of intersection of the two parabolas.

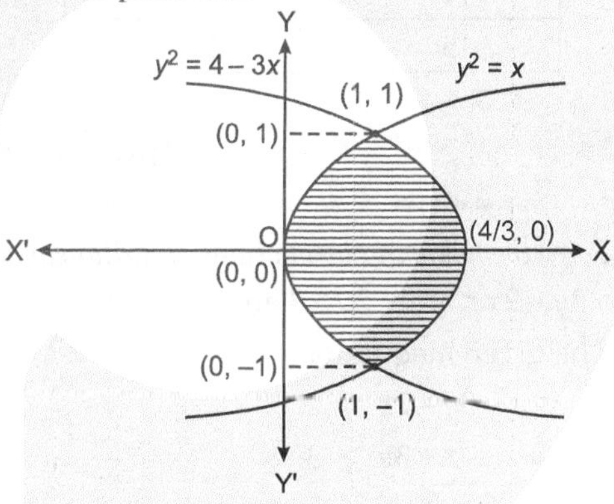

$$\because \text{Required area} = \int_{-1}^{1} x_{II}\,dy - \int_{-1}^{1} x_I\,dy$$

$$= \int_{-1}^{1} \frac{4 - y^2}{3}\,dy - \int_{-1}^{1} y^2\,dy \quad \left[\begin{array}{l} x_{II} = \dfrac{4 - y^2}{3} \\ x_I = y^2 \end{array}\right]$$

$$= \frac{1}{3}\left[4y - \frac{y^3}{3}\right]_{-1}^{1} - \left[\frac{y^3}{3}\right]_{-1}^{1}$$

$$= \frac{1}{3}\left[\left(4 \times 1 - \frac{1^3}{3}\right) - \left(4 \times (-1) - \frac{(-1)^3}{3}\right)\right]$$

$$- \left[\frac{1^3}{3} - \frac{(-1)^3}{3}\right]$$

* Frequently asked previous years Board Exam Questions.

$= \frac{1}{3}\left[4-\frac{1}{3}+4-\frac{1}{3}\right]-\left[\frac{1}{3}+\frac{1}{3}\right]$

$= \frac{1}{3}\left[8-\frac{2}{3}\right]-\frac{2}{3}$

$= \frac{1}{3}\times\frac{22}{3}-\frac{2}{3}$

$= \frac{22}{9}-\frac{2}{3}$

$= \frac{22-6}{9}$

$= \frac{16}{9}$ sq. units. **Ans.**

Chapter 15. Linear Regression

1. *From the following pairs of values of variables x and y, draw a scatter diagram and interpret the result.*

x	y
4	78
5	72
6	66
7	60
8	54
9	48
10	42
11	36
12	30
13	24
14	18
15	12

Sol. The scatter diagram of the given data is shown below :

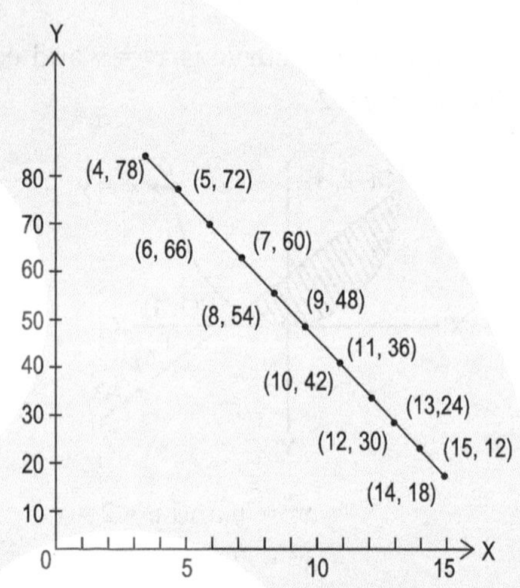

As all points lie on a straight line, there is a perfect negative correlation. **Ans.**

Chapter 16. Linear Programming

1. *Solve the following inequations simultaneously* $3y-2x<4$, $x+3y>3$ and $x+y\leq 5$.

Sol. The given inequations are :

$$3y-2x < 4 \quad \ldots(i)$$
$$x+3y > 3 \quad \ldots(ii)$$
and
$$x+y \leq 5 \quad \ldots(iii)$$

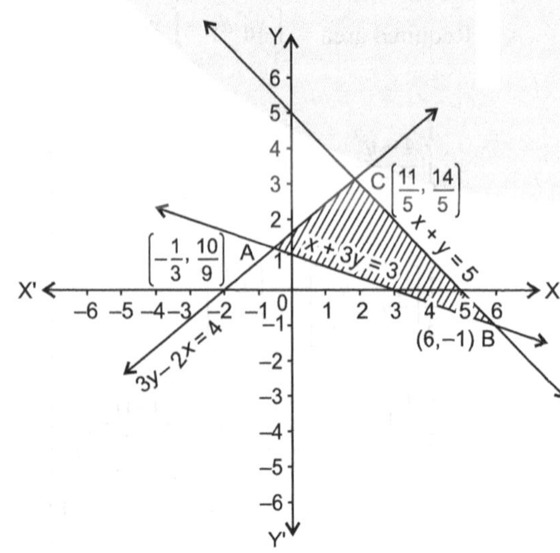

To draw the graph of $3y-2x<4$:

Draw the straight line $3y-2x=4$ which passes through the points $(-2, 0)$ and $\left(0,\frac{4}{3}\right)$.

The line divides the plane into two parts. Further as O (0, 0) satisfies the inequation $3y-2x<4$, ($\because 3.0-2.0=0<4$). Therefore, the graph consists of that part of the plane divided by the line $3y-2x=4$, which contains the origin.

Similarly, draw the graphs of other two inequations $x+3y>3$ and $x+y\leq 5$. Taking the equations $x+3y=3$ and $x+y=5$ which passes through (0, 1), (3, 0) and (0, 5), (5, 0) respectively.

The corner points are $A\left(-\frac{1}{3},\frac{10}{9}\right)$, $B(6,-1)$ and $C\left(\frac{11}{5},\frac{14}{5}\right)$ by solving inequations as equation.

Shade the common part of the graphs of all the three given inequations (i), (ii) and (iii).

Graph Based Questions

The solution set consists of all the points in the shaded part of the co-ordinate plane shown in the figure. The points on the line segment BC are included in the solution. **Ans.**

2. *Solve the following linear equations simultaneously*

$$2x + y - 3 \le 0,\ 2x + y - 6 > 0.$$

Sol. The adjoining figure shows the region corresponding to inequations $2x + y - 3 \le 0$ and $2x + y - 6 > 0$.

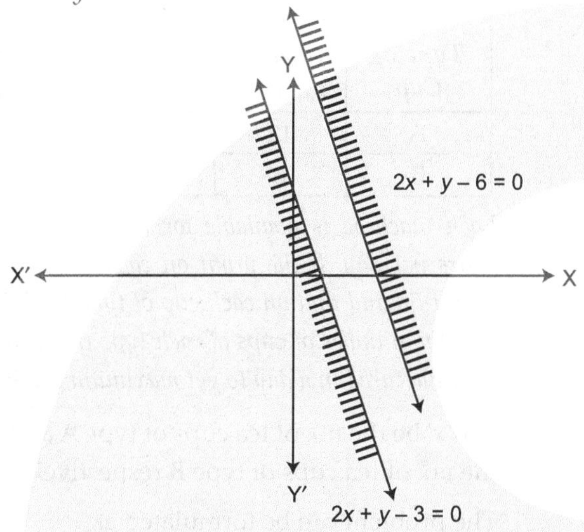

As the two lines are parallel, they never meet, and so the two shaded area never overlap. Hence, there is no solution of the given simultaneous linear equations. In other words, the solution set is empty. We also say that the given system of simultaneous linear inequations is infeasible. **Ans.**

3. *A company manufactures two types of toys A and B. A toy of type A requires 5 minutes for cutting and 10 minutes for assembling. A toy of type B requires 8 minutes for cutting and 8 minutes for assembling. There are 3 hours available for cutting and 4 hours available for assembling the toys in a day. The profit is ₹ 50 each on a toy of type A and ₹ 60 each on a toy of type B. How many toys of each type should the company manufacture in a day to maximize the profit ? Use linear programming to find the solution.*

Sol. The given data can be put in a tabular form as :

	Toy A	Toy B	Time in a day
Cutting time	5 min	8 min	180 min
Assembling time	10 min	8 min	240 min
Profit	₹50	₹60	
Assumed quantity	x	y	

Let the number of toys of type A and B be x and y respectively.

then, Profit function $Z = 50x + 60y$

$$5x + 8y \le 180$$

$$10x + 8y \le 240 \text{ or } 5x + 4y \le 120$$

$$x \ge 0, y \ge 0$$

Taking inequations as equations

$5x + 8y = 180$ $\qquad$ $5x + 4y = 120$

	A	B	E
x	0	36	12
y	22·5	0	15

	C	D	E
x	0	24	12
y	30	0	15

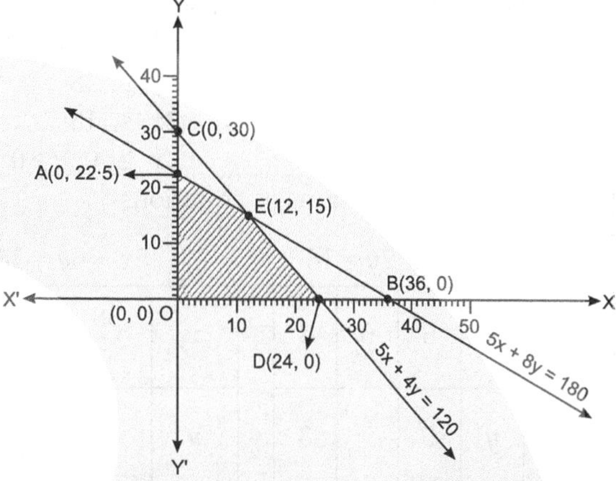

The vertices of the feasible region ODEAO are O (0, 0), D (24, 0), E (12, 15) and A (0, 22·5).

Corner points	Z = 50x + 60y
At O (0, 0)	Z = 0
At D (24, 0)	Z = 50 × 24 + 0 = 1200
At E (12, 15)	Z = 50 × 12 + 60 × 15 = 1500
At A (0, 22.5)	Z = 0 + 60 × 22·5 = 1350

Hence, the maximum profit is ₹ 1500 at E (12, 15).

Thus, the company should manufacture 12 toys of type A and 15 toys of type B in a day to maximize the profit. **Ans.**

4. *A mill owner buys two types of machines A and B for his mill. Machine A occupies 1,000 sqm of area and required 12 men to operate it; while machine B occupies 1,200 sqm of area and requires 8 men to operate it. The owner has 7,600 sqm of area available and 72 men to operate the machines. If machine A produces 50 units and machine B produces 40 units daily, how many machines of each type should he buy to maximise the daily output ? Use Linear Programming to find the solution.*

Sol. The data given in the problem are as under

	Machine A	Machine B	available
Area needed	1,000 sq.m	1,200 sq.m	7,600 sq.m
Labour force	12	8	72
Daily output	50 units	40 units	—

Let x and y be the number of machines A and B respectively then, Total output $Z = 50x + 40y$

subject to the constraints :

$$1{,}000x + 1{,}200y \le 7{,}600$$

or, $10x + 12y \le 76$

or, $5x + 6y \le 38$

$12x + 8y \le 72$

or, $3x + 2y \le 18$

$x \ge 0, y \ge 0$

The problem is to maximise

$$Z = 50x + 40y$$

subject to constraints, $5x + 6y \le 38$

$3x + 2y \le 18$

$x \ge 0, y \ge 0$

Taking inequations as equations

$3x + 2y = 18$

x	6	0	4
y	0	9	3

$5x + 6y = 38$

x	$\frac{38}{5}$	0	4
y	0	$\frac{19}{3}$	3

The graph of these inequations is as shown below :

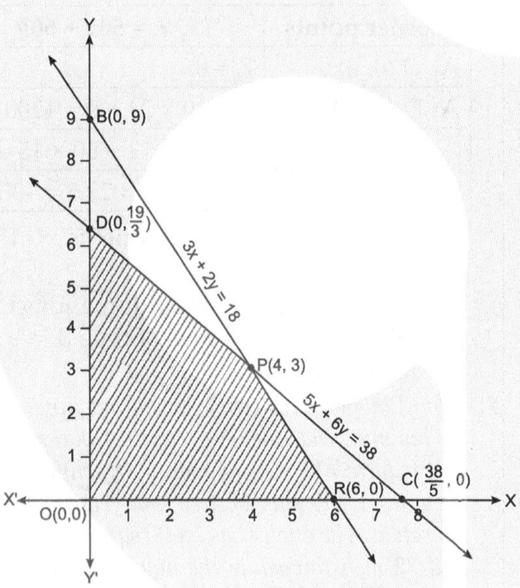

The vertices of the feasible region ORPDO are O (0, 0), R (6, 0), P (4, 3) and D $\left(0, \frac{19}{3}\right)$

Corner Points	Value of $Z = 50x + 40y$
At O (0, 0),	$Z = 0$
At R (6, 0),	$Z = 50 \times 6 + 0 = 300$
At P (4, 3),	$Z = 50 \times 4 + 40 \times 3 = 320$
At D $\left(0, \frac{19}{3}\right)$	$Z = 50 \times 0 + 40 \times \frac{19}{3}$
	$= \frac{760}{3} = 253{\cdot}33$

Thus, we see that Z is maximum at (4, 3).

∴ Number of machine A = 4

Number of machine B = 3. **Ans.**

5. *A manufacturer manufactures two types of tea-cups, A and B. Three machines are needed for manufacturing the tea cups. The time in minutes required for manufacturing each cup on the machines is given below :*

Type of Cup	Times in minutes		
	Machine I	Machine II	Machine III
A	12	18	6
B	6	0	9

Each machine is available for a maximum of six hours per day. If the profit on each cup of type A is ₹ 1·50 and that on each cup of type B is ₹ 1·00, find the number of cups of each type that should be manufactured in a day to get maximum profit.

Sol. Let 'x' be the no. of tea cups of type A and 'y' be the no. of tea cups of type B respectively.

The problem can be formulated as maximize the profit

$$Z = 1{\cdot}50x + 1{\cdot}00y$$

$$i.e., Z = 1{\cdot}50x + y$$

Subject to the constraints :

$12x + 6y \le 3600$ *i.e.,* $2x + y \le 600$

$18x + 0{\cdot}y \le 3600$ *i.e.,* $x \le 200$

$6x + 9y \le 3600$ *i.e.,* $2x + 3y \le 1200$

and $x \ge 0, y \ge 0$ (Non negative constraints)

We draw the straight lines, $2x + y = 600$, $x = 200$, $2x + 3y = 1200$, $x = 0$ and $y = 0$.

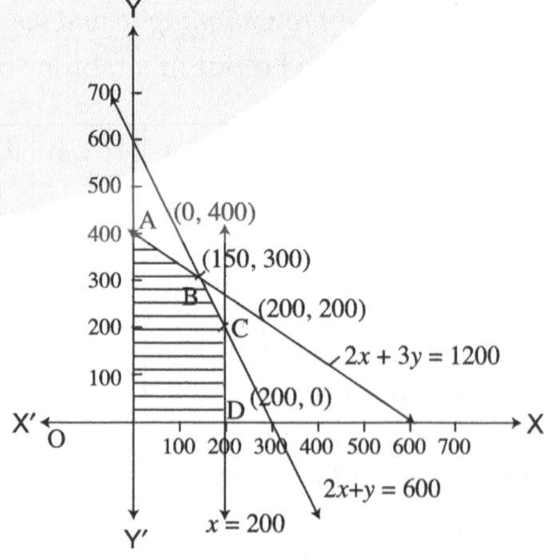

The shaded portion shows the feasible region. The points B and C are the points of intersection of lines $2x + y = 600$, with $2x + 3y = 1200$ and $x = 200$ with $2x + y = 600$ respectively.

Thus, the 5 corner points of the feasible region are O(0, 0), A(0, 400), B(150, 300), C(200, 200) and D (200, 0)

Corner Points	$Z = 1{\cdot}50\ x + y$
At O (0, 0)	Z = 0
At A (0, 400)	Z = 0 + 400 = 400
At B (150, 300)	Z = 1·50 × 150 + 300 = 525
At C (200, 200)	Z = 1·50 × 200 + 200 = 500
At D (200, 0)	Z = 1·50 × 200 + 0 = 300

Clearly, the maximum profit is at $x = 150$ and $y = 300$.

Thus, the manufacturer should manufacture 150 cups of type A and 300 cups of type B to get maximum profit in a day. **Ans.**

6. *A company produces two types of items, P and Q. Manufacturing of both items requires the metals gold and copper. Each unit of item P requires 3 gms of gold and 1 gm of copper while that of item Q requires 1 gm of gold and 2 gms of copper. The company has 9 gms of gold and 8 gms of copper in its store. If each unit of item P makes a profit of ₹ 50 and each unit of item Q makes a profit of ₹ 60, determine the number of units of each item that the company should produce to maximize profit. What is the maximum profit ?*

Sol. The given data can be put in a tabular form as :

Items	Total items present in store		Profit per unit
	9 gms	8 gms	
	Gold (gms)	Copper (gms)	
P	3	1	₹ 50
Q	1	2	₹ 60

Let x_1 = number of units of item P

x_2 = number of units of item Q.

Total profit maximize

$$Z = 50\,x_1 + 60\,x_2$$

$$3x_1 + x_2 \le 9$$

$$x_1 + 2x_2 \le 8$$

Such that, $x_1, x_2 \ge 0$

Now, plotting the inequations as equations

$$3x_1 + x_2 = 9 \quad \ldots\text{(i)}$$

and $$x_1 + 2x_2 = 8 \quad \ldots\text{(ii)}$$

x_1	0	3	2
x_2	9	0	3

x_1	0	8	2
x_2	4	0	3

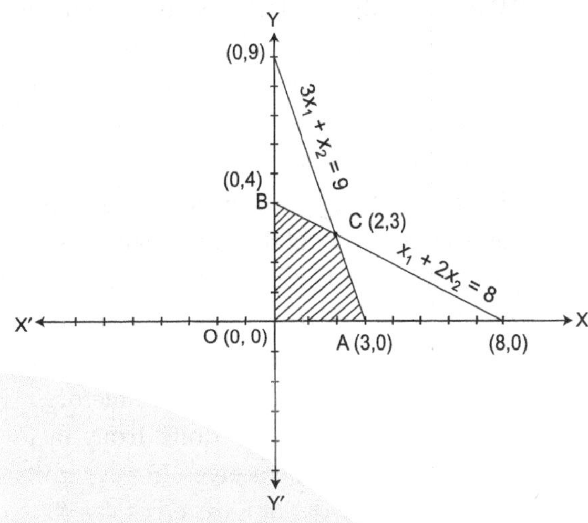

The corner points of the feasible region are O (0, 0), A (3, 0), B (0, 4) and C (2, 3)

Corner Points	$Z = 50\,x_1 + 60\,y_1$
At O (0, 0)	Z = 0
At A (3, 0)	Z = 50 × 3 + 60 × 0 = 150
At C (2, 3)	Z = 50 × 2 + 60 × 3 = 280
At B (0, 4)	Z = 50 × 0 + 60 × 4 = 240

∴ The company should produce 2 items of type P and 3 items of type Q.

∴ Maximum profit = ₹ 280 **Ans.**

7. *(Transportation Problem). A company has two factories located at P and Q and has three depots situated at A, B and C. The weekly requirement of the depots at A, B, C is respectively 5, 5 and 4 units, while the production capacity of the factories at P and Q are respectively 8 and 6 units. The cost of transportation per unit is given below :*

To / *From*	*Cost (In ₹/unit)*		
	A	B	C
P	16	10	15
Q	10	12	10

How many units should be transported from each factory to each depot in order that the transportation cost is minimum ?

Sol. The given data can be represented diagrammatically as

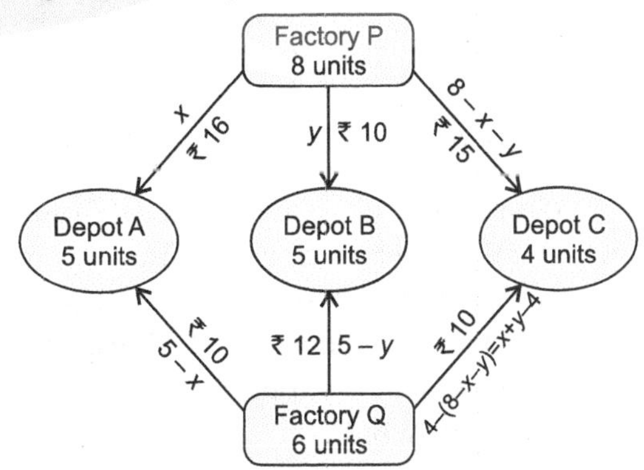

We note that total weekly production (of P and Q) = 8 + 6 = 14 units, and total weekly requirement at depots A, B, C = 5 + 5 + 4 = 14 units, so there is no mismatch between supply and demand.

Let factory P supply x units per week to depot A and y units to depot B, so that it supplies $(8 - x - y)$ units to depot C. Obviously, $0 \le x \le 5$, $0 \le y \le 5$, $0 \le 8 - x - y < 4$.

As depot A's requirement is 5 units and it receives x units from factory P, it must receive $(5 - x)$ units from factory Q. Similarly, depot B receives $(5 - y)$ units from factory Q and depot C receives $4 - (8 - x - y) = (x + y - 4)$ units from factory Q.

(As a cross check, quantity supplied from factory Q to depots A, B, C = $(5 - x) + (5 - y) + (x + y - 4)$ = 6 units = capacity of factory Q).

Thus, total transportation cost (in rupees).

$$= 16x + 10y + 15(8 - x - y) + 10(5 - x) + 12(5 - y) + 10(x + y - 4)$$

$$= x - 7y + 190$$

Hence, the given problem can be formulated as an linear programming program as

Find x and y which minimize

$$Z = x - 7y + 190$$

Subject to the constraints :

$$x + y \ge 4,$$
$$x + y \le 8,$$
$$x \ge 0,\ x \le 5$$
$$y \ge 0,\ y \le 5$$

The feasible region corresponding to these constraints is shown in the adjoining diagram as ABCDEF. We calculate the values of Z at the six corner points :

Corner Points	$Z = x - 7y + 190$
At A (4, 0)	$Z = 4 - 0 + 190 = 194$
At B (5, 0)	$Z = 5 - 0 + 190 = 195$
At C (5, 3)	$Z = 5 - 7 \times 3 + 190 = 174$
At D (3, 5)	$Z = 3 - 7 \times 5 + 190 = 158$
At E (0, 5)	$Z = 0 - 7 \times 5 + 190 = 155$
At F (0, 4)	$Z = 0 - 7 \times 4 + 190 = 162$

We see that Z is minimum (155) at point E (0, 5). Hence, $x = 0$, $y = 5$. Thus, for minimum transportation cost, factory P should supply 0, 5, 3 units to depots A, B, C respectively and factory Q should supply 5, 0, 1 units respectively to depots A, B, C.

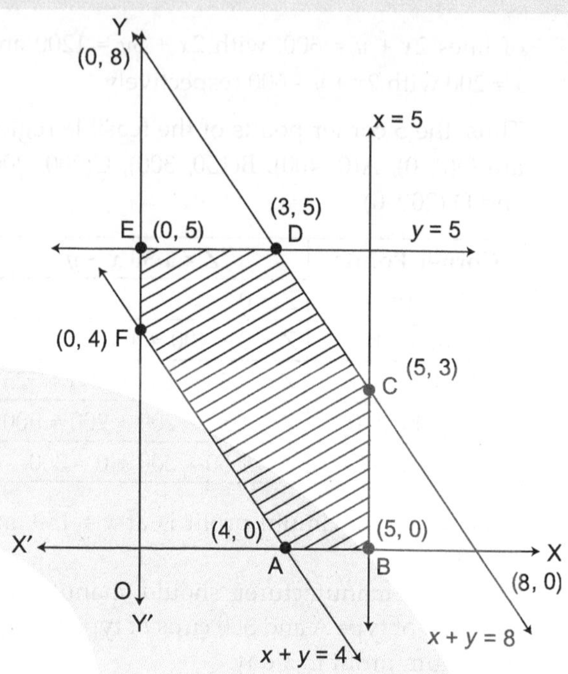

8. *A company manufactures two types of products A and B. Each unit of A requires 3 grams of nickel and 1 gram of chromium, while each unit of B requires 1 gram of nickel and 2 grams of chromium. The firm can produce 9 grams of nickel and 8 grams of chromium. The profit is ₹ 40 on each unit of product of type A and ₹ 50 on each unit of type B. How many units of each type should the company manufacture so as to earn maximum profit ? Use linear programming to find the solution.*

Sol.

Type of Product	Total items		Profit per unit
	9 gms Nickel (gms)	8 gms Chromium (gms)	
A	3	1	₹ 40
B	1	2	₹ 50

Let x = Number of units of type A

y = Number of units of type B.

Maximize $Z = 40x + 50y$

Subject to the constraints,

$$3x + y \le 9$$
$$x + 2y \le 8$$

and $\quad x, y \ge 0$

Consider the inequations as equations,

$$3x + y = 9$$

x	0	3	2
y	9	0	3

and $x + 2y = 8$

x	0	8	2
y	4	0	3

The solution set of this system is the shaded region in the diagram.

Now, we can determine the maximum value of Z by evaluating the value of Z at the four points (vertices) as shown below.

Vertices	$Z = 40x + 50y$
(0, 0)	Z = 40 × 0 + 50 × 0 = ₹ 0
(3, 0)	Z = 40 × 3 + 50 × 0 = ₹ 120
(0, 4)	Z = 40 × 0 + 50 × 4 = ₹ 200
(2, 3)	Z = 40 × 2 + 50 × 3 = ₹ 230

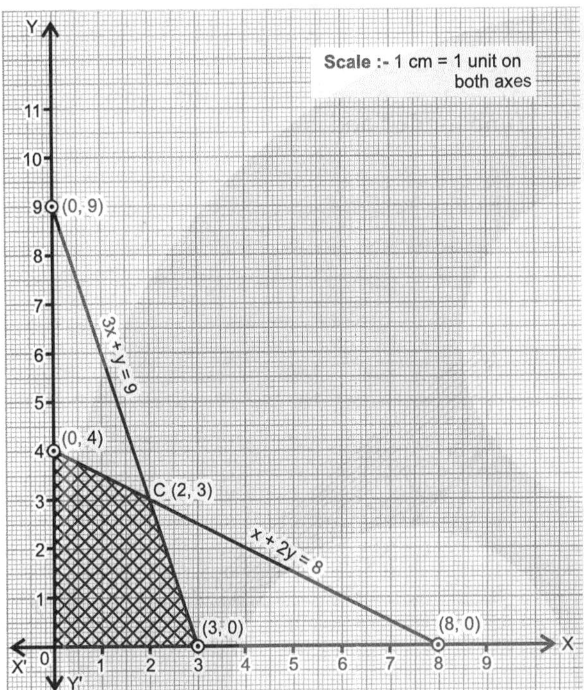

From graph,
Maximum profit, Z = ₹ 230

∴ Number of units of type A is 2 and number of units of type B is 3. **Ans.**

9. *A housewife wishes to mix together two kinds of food F_1 and F_2 in such a way that the mixture contains at least 10 units of vitamin A, 12 units of vitamin. B and 8 units of vitamin C. The vitamin contents of one kg of foods F_1 and F_2 are as below :*

	Vitamin A	*Vitamin B*	*Vitamin C*
Food F_1	1	2	3
Food F_2	2	2	1

One kg of food F_1 costs ₹ 6 and one kg of food F_2 costs ₹ 10. Formulate the above problem as a linear programming problem, and use iso-cost method to find the least cost of the mixture which will produce the diet.

Sol. Suppose x kg of food F_1 and y kg of food F_2, produce the required diet. Then the problem can be formulated as linear programming problem as :

$$\text{Minimize } Z = 6x + 10y,$$

Subject to the constraints :

$$x + 2y \geq 10 \quad ...(i)$$
$$2x + 2y \geq 12$$
$$\text{or} \quad x + y \geq 6 \quad ...(ii)$$
$$3x + y \geq 8 \quad ...(iii)$$
$$x \geq 0,\ y \geq 0$$

We draw the line $x + 2y = 10$, $x + y = 6$, $3x + y = 8$ in the adjoining graph and get the corner points as A (0, 8), B (1, 5), C (2, 4) and D (10, 0).

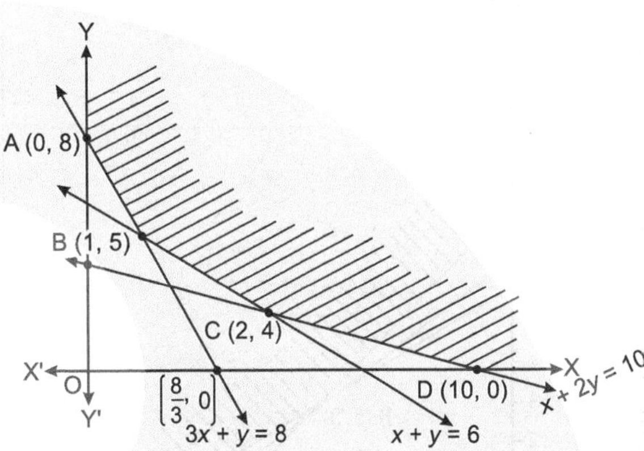

We use iso-cost method, seeing that in $Z = 6x + 10y$, the L.C.M. of coefficients 6 and 10 being 30, a convenient iso-cost line is $6x + 10y = 30$. We move this line parallel to itself away from origin, and observe that first passes through corner point C (2, 4) which give optimal solution.

∴ Minimum cost is 6 (2) + 10 (4) = ₹ 52. **Ans.**

10. *Two tailors P and Q earn ₹ 150 and ₹ 200 per day respectively. P can stitch 6 shirts and 4 trousers a day, while Q can stitch 10 shirts and 4 trousers per day. How many days should each work to produce at least 60 shirts and 32 trousers at minimum labour cost ?*

Sol. Let the tailor P work for x days and the tailor Q work for y days respectively.

$$Z = 150x + 200y$$

Here the problem can be formulated as an linear programming problem as follows

Minimize $Z = 150x + 200y$

Subject to the constraints :

$$6x + 10y \geq 60$$
$$\text{or,} \quad 3x + 5y \geq 30$$
$$4x + 4y \geq 32$$
$$\text{or,} \quad x + y \geq 8$$
$$\text{and} \quad x \geq 0, y \geq 0$$

Converting them into equations, we obtain the following equations :

$$\Rightarrow \quad 3x + 5y = 30,$$
$$x + y = 8$$

$$\Rightarrow \quad y = \frac{30-3x}{5}$$

$$\Rightarrow \quad y = 8 - x.$$

3x + 5y = 30			
x	0	10	5
y	6	0	3

x + y = 8			
x	0	8	5
y	8	0	3

The lines are shown on the graph paper and the feasible region (Unbounded convex) is the shaded region.

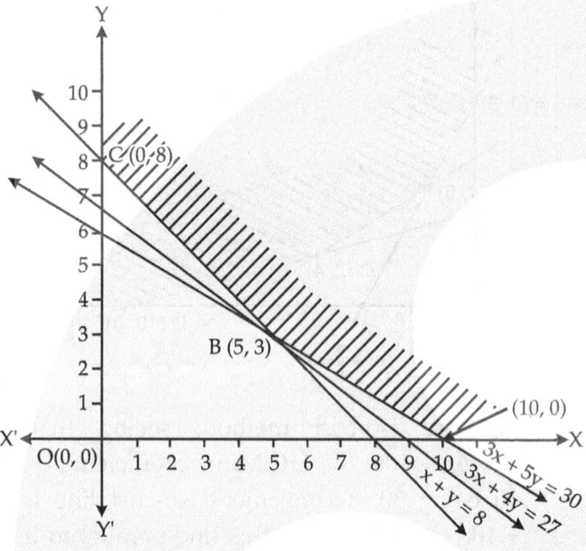

The corner points are

A (10, 0), B (5, 3) and C (0, 8)

Corner Points	$Z = 150\,x + 200\,y$
At A (10, 0),	$Z = 150 \times 10 + 0 = 1500$
At B (5, 3),	$Z = 150 \times 5 + 200 \times 3 = 1350$
At C (0, 8)	$Z = 0 + 200 \times 8 = 1600$

As the feasible region is unbounded, we draw the graph of the half-plane

$$150x + 200y < 1350$$

i.e., $\quad 3x + 4y < 27$

There is no point common with the feasible region, therefore, Z has minimum value. Minimum value of Z is ₹ 1350 and it occurs at the point B (5, 3).

Hence, the labour cost in ₹ 1350 when P works for 5 days and Q works for 3 days. **Ans.**

11. *A dietician wishes to mix two kinds of food X and Y in such a way that the mixture contains at least 10 units of vitamin A, 12 units of vitamin B and 8 units of vitamin C. The vitamin contents of one kg food is given below :*

Food	*Vitamin A*	*Vitamin B*	*Vitamin C*
X	1 unit	2 units	3 units
Y	2 units	2 units	1 unit

One kg of food X costs ₹ 24 and one kg of food Y costs ₹ 36. Using Linear Programming, find the least cost of the total mixture which will contain the required vitamins.

Sol. Let x be the number of units of food X and y be the number of units of food Y be mixed to obtain the desired diet. Then LPP of the given problem is :

Minimize $Z = 24x + 36y$

Subject to the constraints,

$$x + 2y \geq 10,$$
$$2x + 2y \geq 12 \text{ i.e., } x + y \geq 6$$
$$3x + y \geq 8$$
$$x \geq 0, y \geq 0$$

We draw the lines $x + 2y = 10$, $x + y = 6$, $3x + y = 8$ and obtain the feasible region (unbounded and convex) shown in the figure.

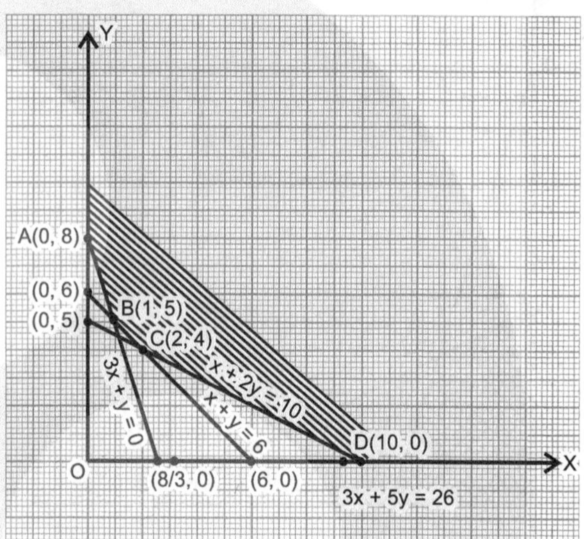

Thus corner points are A (0, 8), B (1, 5), C (2, 4) and D (10, 0).

The values of Z (in ₹) at these points are given in the following table :

Corner Points	Objective Function $Z = 24x + 36y$
A (0, 8)	$Z = 24 \times 0 + 36 \times 8 = 288$
B (1, 5)	$Z = 24 \times 1 + 36 \times 5 = 204$
C (2, 4)	$Z = 24 \times 2 + 36 \times 4 = 192$
D (10, 0)	$Z = 24 \times 10 + 36 \times 0 = 240$

As the feasible region is unbounded, we draw the graph of the half plane $24x + 36y < 192$ *i.e.,* $12x + 3y < 16$ and note that there is no point common with the feasible region. Therefore, Z has the minimum value and the minimum value is ₹ 192.

It occurs at C (2, 4).

i.e., when 2 kg of food X and 4 kg of food Y are mixed to get the desired diet. **Ans.**

12. *A carpenter has 90, 80 and 50 running feet respectively of teak wood, plywood and rosewood which is used to produce product A and product B. Each unit of product A requires 2, 1 and 1 running feet and each unit of product B requires 1, 2 and 1 running feet of teak wood, plywood and rosewood respectively. If product A is sold for ₹ 48 per unit and product B is sold for ₹ 40 per unit, how many units of product A and product B should be produced and sold by the carpenter, in order to obtain the maximum gross income ?*

*Formulate the above as a Linear Programming Problem and solve it, indicating clearly the feasible region in the graph.**

Sol. Let x units of product A and y unit of product B should be produced

Total running feet teak wood = 90

Total running feet plywood = 80

Total running feet rosewood = 50

For teak wood

$$2x + y \leq 90 \qquad \ldots(i)$$

For plywood

$$x + 2y \leq 80 \qquad \ldots(ii)$$

For rosewood

$$x + y \leq 50 \qquad \ldots(iii)$$

$$x \geq 0, y \geq 0$$

To maximise the gross income

Maximise $\quad Z = 48x + 40y$

From equation (i)

$$2x + y = 90$$

x	0	45
y	90	0

From equation (ii)

$$x + 2y = 80$$

x	0	80
y	40	0

From equation (iii)

$$x + y = 50$$

x	0	50
y	50	0

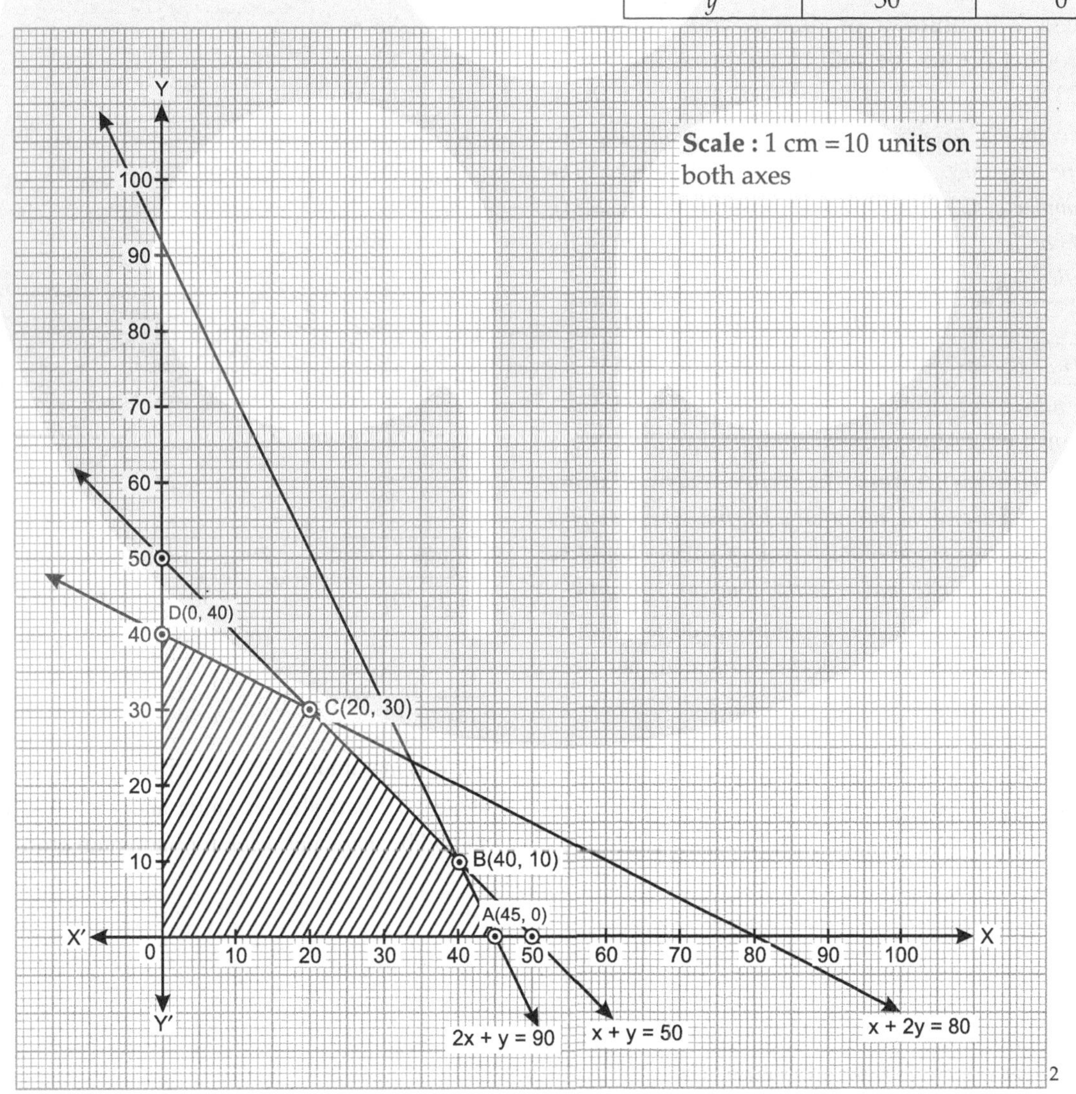

* Frequently asked previous years Board Exam Questions.

Corner points are A (45, 0), B (40, 10), C (20, 30) and D (0, 40)

For gross profit

$$Z = 48x + 40y$$

For point A,

$$Z = 48 \times 45 + 40 \times 0$$
$$= 2160 + 0$$
$$= ₹2160$$

For point B,

$$Z = 48 \times 40 + 40 \times 10$$
$$= 1920 + 400$$
$$= ₹2320$$

For point C,

$$Z = 48 \times 20 + 40 \times 30$$
$$= 960 + 1200$$
$$= ₹2160$$

For point D,

$$Z = 48 \times 0 + 40 \times 40$$
$$= ₹1600$$

$\therefore$ Z is maximum at B (40, 10).

Hence, 40 units of product A and 10 units of product B must be sold to get gross income of ₹2320. **Ans.**

13. *A company uses three machines to manufactue two types of shirts, half sleeves and full sleeves. The number of hours required per week on machine M_1, M_2 and M_3 for one shirt of each type is given in the following table :**

	M_1	M_2	M_3
Half sleeves	*1*	*2*	*8/5*
Full sleeves	*2*	*1*	*8/5*

None of the machines can be in operation for more than 40 hours per week. The profit on each half sleeve shirt is ₹ 1 and the profit on each full sleeve shirt is ₹ 1·50. How many of each type of shirts should be made per week to maximize the company's profit ?

Sol. Let the number of half sleeve shirts made per week be x and the number of full sleeve shirts made per week be y.

So, the objective function is,

Maximize $Z = x + 1.50y$

Subject to constraints,

$$x + 2y \le 40 \quad ...(i)$$
$$2x + y \le 40 \quad ...(ii)$$
$$\frac{8}{5}x + \frac{8}{5}y \le 40$$

or $x + y \le 25$ ...(iii)

and $x \ge 0$ and $y \ge 0$

From inequation (i), we have

$$x + 2y = 40$$

x	0	40
y	20	0

From inequation (ii), we have

$$2x + y = 40$$

x	0	20
y	40	0

From inequation (iii), we have

$$x + y = 25$$

x	0	25
y	25	0

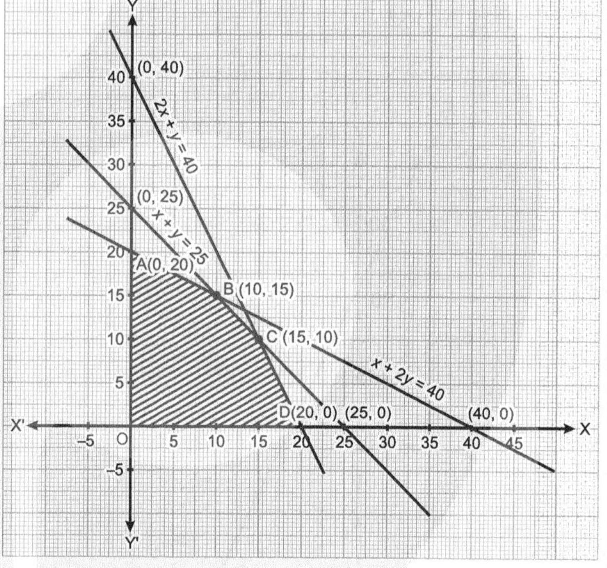

Points	$Z = x + 1.50y$
A (0, 20)	$0 + 1.50 \times 20 =$ ₹ 30
B (10, 15)	$10 + 1.50 \times 15 =$ ₹ 32.50
C (15, 10)	$15 + 1.50 \times 10 =$ ₹ 30
D (20, 0)	$20 + 1.50 \times 0 =$ ₹ 20

So, Z is maximum at B (10, 15).

Hence, 10 half sleeves shirts and 15 full sleeve shirts should be made per week for maximum profit. **Ans.**

* Frequently asked previous years Board Exam Questions.

Practice Exercises

Set 10

Chapter 1. Relations

1. Let R = {(a, a^3) : a is a prime number less than 5} be a relation. Find the range of R.
2. Show that the relation R on the set N of all natural numbers defined as R {$(x, y) : y = x + 5$ and $x < 4$} is transitive.
3. Show that the relation R defined by a R b iff b is divisible by $a, b \in$ N is reflexive and transitive.
4. Show that the relation R on the set A = {1, 2, 3} given by R = {(1, 2), (2, 1)} is symmetric but neither reflexive non-transitive.
5. Check whether the relation R on R defined by R = {$(a, b) : a \leq b^3$} is reflexive, symmetric or transitive.
6. Check whether the relation S defined by $l_1 S l_2$ iff $l_1 \perp l_2$, where l_1 and l_2 are straight lines in a plane is reflexive, symmetric or transitive.
7. Prove that R is an equivalence relation if R be a relation on the set A of ordered pairs of integers defined by (x, y) R (u, v) iff $xv = yu$.
8. Show that S is not an equivalence relation on R. If S be a relation on the set R of all real numbers defined by S = {$(a, b) \in R \times R : a^2 + b^2 = 1$}.

Answers

1. {8, 27}
5. Neither reflexive nor symmetric and transitive
6. Symmetric but not reflexive and transitive

Chapter 2. Functions

1. Find 5 * 7 if * be a binary operation on N given by $a * b$ = LCM (a, b) $\forall\ a, b \in$ N.
2. Find the identity element with respect to *, if the binary operation * on the set Z is defined by $a * b = a + b - 5$.
3. Show that the binary operation * defined by $a * b = ab + 1$ on Q is commutative.
4. Show that the logarithmic function $f : R_0^+ \to R$ given by $f(x) = \log_a x$, $a > 0$ is a bijection.
5. Show that the function f : N → N defined by

$$f(n) = \begin{cases} n+1, & \text{if } n \text{ is odd} \\ n-1, & \text{if } n \text{ is even} \end{cases}$$

is a bijection.
6. Show that the function f : R → R defined by $f(x) = 4x^3 + 7$, is a bijection.
7. Show that gof is a one-one function if f : A → B and g : B → C are one-one functions.
8. Let f : {1, 3, 4} → {1, 2, 5} and g : {1, 2, 5} → {1, 3} be given by f = {(1, 2), (3, 5), (4, 1)} and g = {(1, 3), (2, 3), (5, 1)}. Write down gof.
9. Let f : R → R and g : R → R be defined by $f(x) = x^2$ and $g(x) = x + 1$. Show that $fog \neq gof$.
10. If $f : R \to (0, 2)$ defined by $f(x) = \dfrac{e^x - e^{-x}}{e^x + e^{-x}} + 1$ is invertible. Find f^{-1}.
11. Show that the function $f : Q \to Q$ defined by $f(x) = 3x + 5$ is invertible. Also, find f^{-1}.

Answers

1. 35
2. 5
8. gof = {(1, 3), (3, 1), (4, 3)}
10. $f^{-1}(x) = \log_e \left(\dfrac{x}{2-x}\right)^{1/2}$
11. $f^{-1}(x) = \dfrac{x-5}{3}$

Chapter 3. Inverse Trigonometric Functions

Evaluate the following (Q. 1– Q. 3)

1. $\tan \frac{1}{2}\left[\cos^{-1}\frac{\sqrt{5}}{3}\right]$.

2. $\cos\left[\cos^{-1}\left(\frac{-\sqrt{3}}{2}\right)+\frac{\pi}{6}\right]$.

3. $\sin\left[\frac{\pi}{3}-\sin^{-1}\left(-\frac{1}{2}\right)\right]$.

4. Prove that $\sec^2(\tan^{-1}2)+\mathrm{cosec}^2(\cot^{-1}3)=15$.

5. If $\tan^{-1}x+\tan^{-1}y+\tan^{-1}z=\frac{\pi}{2}$, prove that :

   $xy+yz+zx=1$.

6. Prove that $\sin^{-1}\frac{12}{13}+\cos^{-1}\frac{4}{5}+\tan^{-1}\frac{63}{16}=\pi$

7. If $\cos^{-1}\frac{x}{2}+\cos^{-1}\frac{y}{3}=\alpha$, prove that :

   $9x^2-12xy\cos\alpha+4y^2=36\sin^2\alpha$.

Prove that (Q. 8 to Q. 13)

8. $\tan^{-1}\left[\frac{\sqrt{1+x^2}+\sqrt{1-x^2}}{\sqrt{1+x^2}-\sqrt{1-x^2}}\right]=\frac{\pi}{4}+\frac{1}{2}\cos^{-1}x^2$.

9. $\cos^{-1}\frac{4}{5}+\cot^{-1}\frac{5}{3}=\tan^{-1}\frac{27}{11}$

10. $\sin\left[\sin^{-1}\frac{1}{2}+\cos^{-1}\frac{3}{5}\right]=\frac{3+4\sqrt{3}}{10}$.

11. $\cos\left[\tan^{-1}\frac{15}{8}-\sin^{-1}\frac{7}{25}\right]=\frac{297}{425}$.

12. $\cot^{-1}\left[\frac{pq+1}{p-q}\right]+\cot^{-1}\left[\frac{qr+1}{q-r}\right]+\cot^{-1}\left[\frac{rp+1}{r-p}\right]=0$.

13. $\tan(2\tan^{-1}a)=2\tan(\tan^{-1}a+\tan^{-1}a^3)$

Solve the following equations :

14. $\tan^{-1}\frac{1}{2x+1}+\tan^{-1}\frac{1}{4x+1}=\tan^{-1}\frac{2}{x^2}$.

15. $\tan^{-1}(x+1)+\cot^{-1}(x-1)=\sin^{-1}\frac{4}{5}+\cot^{-1}\frac{3}{4}$.

16. $\sin^{-1}x+\sin^{-1}2x=\frac{\pi}{3}$.

17. $\tan^{-1}(x-1)+\tan^{-1}x+\tan^{-1}(x+1)=\tan^{-1}3x$.

Answers

1. $\frac{1}{2}(3-\sqrt{5})$ 2. -1

3. 1 14. $0, \frac{-2}{3}, 3$

15. $\pm 4\sqrt{\frac{3}{7}}$ 16. $\frac{1}{2}, \frac{-1}{2}, \frac{1}{2}\sqrt{\frac{3}{7}}, \frac{-1}{2}\sqrt{\frac{3}{7}}$

17. $x=0, \pm\frac{1}{2}$

Chapter 4. Determinants

1. Using properties of determinants evaluate

$$\begin{vmatrix} 1 & 374 & 1893 \\ 1 & 372 & 1892 \\ 1 & 371 & 1891 \end{vmatrix}.$$

2. Prove that

$$\begin{vmatrix} b+c & c+a & a+b \\ q+r & r+p & p+q \\ y+z & z+x & x+y \end{vmatrix}=2\begin{vmatrix} a & b & c \\ p & q & r \\ x & y & z \end{vmatrix}.$$

3. Using properties of determinants show that

$$\begin{vmatrix} 1 & a & a^2-bc \\ 1 & b & b^2-ac \\ 1 & c & c^2-ab \end{vmatrix}=0.$$

4. Evaluate $\begin{vmatrix} 42 & 6 & 1 \\ 28 & 4 & 7 \\ 14 & 2 & 3 \end{vmatrix}$ without expanding.

5. Evaluate $\begin{vmatrix} c-a & a-b & b-c \\ a-b & b-c & c-a \\ b-c & c-a & a-b \end{vmatrix}$.

6. Evaluate $\begin{vmatrix} 219 & 198 & 181 \\ 240 & 225 & 198 \\ 265 & 240 & 219 \end{vmatrix}$.

7. Solve $\begin{vmatrix} x+2 & 1 & -3 \\ 1 & x-3 & x-2 \\ -3 & -2 & 1 \end{vmatrix}=-8$

Practice Exercises

8. Prove that $\begin{vmatrix} a & b-c & c-a \\ a-c & b & c-a \\ a-b & b-a & c \end{vmatrix}$

$= (a+b-c)(b+c-a)(c+a-b)$.

9. Prove that $\begin{vmatrix} b^2+c^2 & a^2 & a^2 \\ b^2 & c^2+a^2 & b^2 \\ c^2 & c^2 & a^2+b^2 \end{vmatrix}$

$= \begin{vmatrix} a^2 & bc & ac+c^2 \\ a^2+ab & b^2 & ac \\ ab & b^2+bc & c^2 \end{vmatrix}$.

10. Prove that $\begin{vmatrix} 1 & a & bc \\ 1 & b & ca \\ 1 & c & ab \end{vmatrix} = \begin{vmatrix} 1 & 1 & 1 \\ a & b & c \\ a^2 & b^2 & c^2 \end{vmatrix}$.

11. Prove that : $\begin{vmatrix} x+9 & x & x \\ x & x+9 & x \\ x & x & x+9 \end{vmatrix} = 243(x+3)$

12. Solve for x

$$\begin{vmatrix} x^2 & x & 1 \\ 0 & 2 & 1 \\ 3 & 1 & 4 \end{vmatrix} = 28.$$

13. Show that

$$\begin{vmatrix} b & 1 & a \\ c & -a & 1 \\ 1 & -b & -c \end{vmatrix} = 1 + a^2 + b^2 + c^2.$$

14. Solve for x if

$$\begin{vmatrix} 3+x & 5 & 2 \\ 1 & 7+x & 6 \\ 2 & 5 & 3+x \end{vmatrix} = 0.$$

15. Express in factors $\begin{vmatrix} 1 & 1 & 1 \\ a & b & c \\ a^3 & b^3 & c^3 \end{vmatrix}$.

16. Show that

$$\begin{vmatrix} b+c & a & a \\ b & c+a & b \\ c & c & a+b \end{vmatrix} = 4abc.$$

17. Evaluate $\begin{vmatrix} 2 & 0 & 0 & 1 \\ -6 & -4 & 0 & 2 \\ -5 & -1 & -4 & 2 \\ -3 & 2 & 6 & -3 \end{vmatrix}$

18. Solve for x

$$\begin{vmatrix} x & -6 & -1 \\ 2 & -3x & x-3 \\ -3 & 2x & x+2 \end{vmatrix} = 0.$$

19. Show that

$$\begin{vmatrix} x-y-z & 2x & 2y \\ 2y & y-z-x & 2y \\ 2z & 2z & z-x-y \end{vmatrix} = (x+y+z)^3.$$

20. Using properties of determinants, solve for x.

$$\begin{vmatrix} x+a & b & c \\ c & x+b & a \\ a & b & x+c \end{vmatrix} = 0.$$

21. Using product of determinants. Prove that

$$\begin{vmatrix} 2bc-a^2 & c^2 & b^2 \\ c^2 & 2ac-b^2 & a^2 \\ b^2 & a^2 & 2ab-c^2 \end{vmatrix} = \begin{vmatrix} a & b & c \\ b & c & a \\ c & a & b \end{vmatrix}^2.$$

22. Without expanding prove that

$$\begin{vmatrix} a_1\alpha_1+b_1\beta_1 & a_1\alpha_2+b_1\beta_2 & a_1\alpha_3+b_2\beta_3 \\ a_2\alpha_1+b_2\beta_1 & a_2\alpha_2+b_2\beta_2 & a_2\alpha_3+b_2\beta_3 \\ a_3\alpha_1+b_1\beta_1 & a_3\alpha_3+b_3\beta_2 & a_3\alpha_2+b_3\beta_3 \end{vmatrix} = 0$$

23. Prove that

$$\begin{vmatrix} 1 & a & a^2 \\ \cos(p-d)x & \cos px & \cos(p+d)x \\ \sin(p-d)x & \sin px & \sin(p+d)x \end{vmatrix}$$

is independent of p.

24. Prove that for all values of θ

$$\begin{vmatrix} \sin\theta & \cos\theta & \sin 2\theta \\ \sin\left(\theta+\frac{2\pi}{3}\right) & \cos\left(\theta+\frac{2\pi}{3}\right) & \sin\left(2\theta+\frac{4\pi}{3}\right) \\ \sin\left(\theta-\frac{2\pi}{3}\right) & \cos\left(\theta-\frac{2\pi}{3}\right) & \sin\left(2\theta-\frac{4\pi}{3}\right) \end{vmatrix} = 0$$

25. Solve $\begin{vmatrix} \sin x & \cos x & \cos x \\ \cos x & \sin x & \cos x \\ \cos x & \cos x & \sin x \end{vmatrix} = 0$

in the interval $-\frac{\pi}{4} \le x \le \frac{\pi}{4}$.

Answers

1. 1 **4.** 0

5. 0 **6.** 0

7. $x = 2, 3$

12. $x = 2, \frac{-17}{7}$

14. 0, –1, –12

15. $(a - b)(b - c)(c - a)(a + b + c)$

17. 188

18. $x = 1, 2, -3$

20. $x = 0, -(a + b + c)$

25. $\frac{+\pi}{4}$

Chapter 5. Matrices

1. Construct a matrix of order 3 × 2 if

$$a_{ij} = \begin{Bmatrix} i-j+2 & i \neq j \\ (i+j)^2 & i = j \end{Bmatrix}$$

2. Find x and y if

$$x+y=\begin{bmatrix} 7 & 0 \\ 2 & 5 \end{bmatrix} \text{and } x-y=\begin{bmatrix} 3 & 0 \\ 0 & 3 \end{bmatrix}.$$

3. Find x and y if

$$2x+3y=\begin{bmatrix} 2 & 3 \\ 4 & 0 \end{bmatrix} \text{and } 3x+2y=\begin{bmatrix} -2 & 2 \\ 1 & -5 \end{bmatrix}.$$

4. If $A=\begin{bmatrix} 1 & -1 \\ 2 & 3 \end{bmatrix}, B=\begin{bmatrix} 2 & 1 \\ 1 & 0 \end{bmatrix}$, prove that

$$(A+B)^2 \neq A^2 + 2AB + B^2.$$

5. If $A=\begin{bmatrix} 3 & -2 \\ 4 & -2 \end{bmatrix}$, and $I=\begin{bmatrix} 1 & 0 \\ 0 & 1 \end{bmatrix}$, then find k so that $A^2 = kA - 2I$.

6. If $A=\begin{bmatrix} 1 & 1 \\ 0 & 1 \end{bmatrix}$, Prove that $A^n=\begin{bmatrix} 1 & n \\ 0 & 1 \end{bmatrix}$ for all positive n.

7. Find x and y such that

$$2\begin{bmatrix} x & y \\ -1 & 5 \end{bmatrix}=\begin{bmatrix} -2 & 0 \\ -1 & 5 \end{bmatrix}-\begin{bmatrix} 1 & 2 \\ 0 & 0 \end{bmatrix}.$$

8. If $(x\ y\ z) - (-5, 3, 0) = (-5\ \ 6\ 7)$, determine x, y and z.

9. If $\begin{bmatrix} 2 & -3 \\ 4 & 0 \end{bmatrix}-\begin{bmatrix} x_1 & x_2 \\ y_1 & y_2 \end{bmatrix}=\begin{bmatrix} -3 & 4 \\ 5 & -1 \end{bmatrix}$ determine x_1, x_2, y_1 and y_2.

10. Express in a single matrix

$$4\begin{bmatrix} 1 & 3 \\ 1 & -4 \end{bmatrix}-\frac{1}{2}\begin{bmatrix} 8 & 4 \\ 4 & 8 \end{bmatrix}.$$

11. Solve the equation

$$-2\left[X+\begin{bmatrix} 1 & 2 & 3 \\ 0 & 1 & 2 \\ 0 & 0 & 1 \end{bmatrix}\right]=3X+\begin{bmatrix} 1 & 0 & 0 \\ 0 & 0 & 0 \\ 0 & 0 & 1 \end{bmatrix}.$$

12. If $A=\begin{bmatrix} 1 & 1 & -1 \\ 2 & -3 & 4 \\ 3 & -2 & 3 \end{bmatrix}; B=\begin{bmatrix} -1 & -2 & -1 \\ 6 & 12 & 6 \\ 5 & 10 & 5 \end{bmatrix}$.

and $C=\begin{bmatrix} -1 & -1 & 1 \\ 2 & 2 & -2 \\ -3 & -3 & 3 \end{bmatrix}$, show that AB and CA are null matrix but $BA \neq 0$ and $AC \neq 0$.

13. Find a matrix A if

$$\begin{bmatrix} 0 & 2 \\ 1 & 1 \end{bmatrix}A\begin{bmatrix} 0 & 1 \\ 1 & 1 \end{bmatrix}=\begin{bmatrix} 1 & 0 \\ 1 & 1 \end{bmatrix}.$$

14. Given $A=\begin{bmatrix} 3 & -1 \\ 1 & 2 \end{bmatrix}, B=\begin{bmatrix} 3 \\ 1 \end{bmatrix}, C=\begin{bmatrix} 1 \\ -2 \end{bmatrix}$, find the matrix X, such that

$$AX = 3B + 2C.$$

15. Find the value of k, a non zero scalar if

$$2\begin{bmatrix} 1 & 2 & 3 \\ -1 & -3 & 2 \end{bmatrix}+k\begin{bmatrix} 1 & 0 & 2 \\ 3 & 4 & 5 \end{bmatrix}=\begin{bmatrix} 4 & 4 & 10 \\ 4 & 2 & 14 \end{bmatrix}.$$

16. Let $A=\begin{bmatrix} 0 & 1 \\ 0 & 0 \end{bmatrix}$, show that

$$(aI + bA)^n = a^n I + na^n - 1\, bA.$$

17. If $A=\begin{bmatrix} 3 & 5 \\ -4 & 2 \end{bmatrix}$, find $A^2 - 5A - 14I$.

18. Without using concept of inverse of matrix, find the matrix $\begin{bmatrix} x & y \\ z & u \end{bmatrix}$ such that

$$\begin{bmatrix} 5 & -7 \\ -2 & 3 \end{bmatrix}\begin{bmatrix} x & y \\ z & u \end{bmatrix}=\begin{bmatrix} -16 & -6 \\ 7 & 2 \end{bmatrix}.$$

19. Find the product of two matrices A and B where

$$A=\begin{bmatrix}-5 & 1 & 3\\ 7 & 1 & -5\\ 1 & -1 & 1\end{bmatrix}; B=\begin{bmatrix}1 & 1 & 2\\ 3 & 2 & 1\\ 2 & 1 & 3\end{bmatrix}$$

and use it to solve the equations $x + y + 2z = 1$; $3x + 2y + z = 7$; $2x + y + 3z = 2$.--

20. A man invests ₹ 50,000 into two types of bonds. The first bond pays 5% interest per year and the second bond pays 6% interest per year. Use matrix multiplication to determine how to divide ₹ 50,000 among two types of bonds so as to obtain an annual total interest of ₹ 2780.

Answers

1. $\begin{bmatrix}4 & 1\\ 3 & 16\\ 4 & 3\end{bmatrix}$

2. $x=\begin{bmatrix}5 & 0\\ 1 & 4\end{bmatrix}, y=\begin{bmatrix}2 & 0\\ 1 & 1\end{bmatrix}$

3. $x=\begin{bmatrix}-2 & 0\\ -1 & -3\end{bmatrix}, y=\begin{bmatrix}2 & 1\\ 2 & 2\end{bmatrix}$

4. $k=1$

5. $x=\frac{-3}{2}$, $y=-1$

6. $x=-10, y=9, z=7$

7. $x_1=5, x_2=-7, y_1=-1, y_2=1$

8. $\begin{bmatrix}0 & 10\\ 2 & -20\end{bmatrix}$

9. $\begin{bmatrix}-3/5 & -4/5 & -6/5\\ 0 & -2/5 & -4/5\\ 0 & 0 & -3/5\end{bmatrix}$

13. $\begin{bmatrix}-1 & -1\\ 0 & 1\end{bmatrix}$

14. $X=\begin{bmatrix}3\\ -2\end{bmatrix}$

15. $k=2$

17. – 40I

18. $\begin{bmatrix}1 & -4\\ 3 & -2\end{bmatrix}$

19. $4I, x=2, y=1, z=-1$

20. 22,000; 28,000

Chapter 6. Differential Calculus

(Q. 1.–Q. 6) : Differen;tiate y w.r.t. x :

1. $y=\frac{(x+1)(x-2)}{\sqrt{x}}$.

2. If $y=\sqrt{\frac{1-\cos x}{1+\cos x}}$, find $\frac{dy}{dx}$.

3. $y=\log\sqrt{\frac{1-\cos x}{1+\cos x}}$.

4. If $\sqrt{1-x^2}+\sqrt{1-y^2}=(x-y)$ then show that

$$\frac{dy}{dx}=\sqrt{\frac{1-y^2}{1-x^2}}.$$

5. $y=\cos^{-1}x$.

6. $y=\tan^{-1}\left[\frac{\sqrt{1+x^2}+\sqrt{1-x^2}}{\sqrt{1+x^2}-\sqrt{1-x^2}}\right]$.

7. Find $\frac{dy}{dx}$ if $y=\tan^{-1}\frac{\sqrt{1+x^2}-1}{x}$.

8. $y=\frac{2t}{1+t^2}$ and $x=\frac{1-t^2}{1+t^2}$.

9. Differentiate

$$\cos^{-1}\left(\frac{1-x^2}{1+x^2}\right) \text{ w.r.t. } \tan^{-1}\left(\frac{3x-x^3}{1-3x^2}\right)$$

10. If $y=e^{a^x}$, show that

$$\frac{d^2y}{dx^2}-2a\frac{dy}{dx}+(a^2+b^2)y=0.$$

11. If $y=(\sin^{-1}x)^2$, prove that

$$(1-x^2)\frac{d^2y}{dx^2}-x\frac{dy}{dx}=2.$$

12. If $y=\sin(\log x)$ prove that $x^2\frac{d^2y}{dx^2}+x\frac{dy}{dx}+y=0$.

13. If $y=A\cos nx+B\sin nx$, Prove that

$$y''+n^2y=0.$$

14. Find limit in the following cases :

(i) $\lim_{x\to-\infty} e^x$.

(ii) $\lim_{x\to\infty} \frac{\sum n^2}{n^3}$.

(iii) $\lim_{x\to a} \frac{(x+2)^{5/3}-(a+2)^{5/3}}{x-a}$.

(iv) $\lim_{x\to 0} \frac{x^3 \cot x}{1-\cos x}$.

15. Verify Rolle's theorem for $f(x) = x(x+3)e^{-x/2}$ in $[-3, 0]$.

16. Verify Lagrange's mean value theorem for $f(x) = \sin x - \sin 2x$ in $[0, \pi]$.

17. Verify Lagrange's mean value theorem for the following function

$$f(x) = 2x - x^2, \qquad 0 \le x \le 1$$

18. Check if Lagrange's mean value theorem is applicable to

$$f(x) = 4 - (6-x)^{2/3} \text{ in } [5, 7]$$

19. Find c of the Lagrange's mean value theorem for the function $f(x) = x(x-2)$ in the interval $[1, 2]$

Answers

1. $\frac{3}{2}\sqrt{x} - \frac{1}{2\sqrt{x}} + \frac{1}{x^{3/2}}$

2. $\sec^2 \frac{x}{2}$

3. $\operatorname{cosec} x$

5. $\frac{-1}{\sqrt{1-x^2}}$

6. $\frac{dy}{dx} = \frac{-x}{\sqrt{1-x^4}}$

7. $\frac{1}{2}\sec^2\frac{x}{2}$

8. $\frac{t^2-1}{2t}$

9. $\frac{2}{3}$

10. $\frac{1}{2(1+x^2)}$

14. (i) 0 (ii) $\frac{1}{3}$

(iii) $\frac{5}{3}(a+2)^{2/3}$ (iv) 2

19. $c = \frac{3}{2}$

Chapter 7. Applications of Derivatives

1. Find the point on the curve $y^2 = 4x$ which is nearest to the point $(2, -8)$.

2. A body is moving in straight line according to $S = t^3 - 6t^2 + 9t - 4$ where S is displacement in meters and t is time in seconds. Find the displacement and acceleration when velocity is 9 m/s.

3. Show that the function $f(x) = 2 - 3x + 3x^2 - x^3$ is decreasing on R.

4. Find the intervals in which the function $f(x) = x^3 - 6x^2 + 9x + 15$ is increasing and decreasing.

5. Show that $f(x) = \sin x$ is an increasing function on $(-\pi/2, \pi/2)$.

6. Prove that $f(x) = \log x$ do not have maxima or minima.

7. Find two positive numbers x and y such that $x + y = 60$ and xy^3 is maximum.

8. Show that the height of a closed cylinder of given volume and minimum surface area is equal to its diameter.

9. An open box with a square base is to be made out of quantity of card board whose axis is c^2 units, show that the maximum volume of the box is $\frac{c^3}{6\sqrt{3}}$ units.

10. Show that the volume of the greatest cylinder which can be inscribed in a cone of height 'h' and semi-vertical angle 30° is $\frac{4}{81}\pi h^3$.

Answers

1. $(4, -4)$

2. Displacement – 4 m and 0 m

Acceleration –12 m/sec² and 12 m/sec²

4. Increasing in $(-\infty, 1)$ and $(3, \infty)$ and decreasing in $(1, 3)$.

7. 15, 45.

Chapter 8. Integral Calculus

Evaluate the following (Q.1. – Q.22.)

1. $\int\left(\sin^2 x+\cos^2 x+\frac{x^3+2x}{\sqrt{x}}\right)dx.$

2. $\int x\,.\,(\log x)^2\,dx.$

3. $\int x\,.\,e^{2x}\cos x\,dx.$

4. $\int e^x\frac{(x\log x+1)}{x}dx.$

5. $\int\frac{dx}{\sin x+\tan x}.$

6. $\int(\log x)^2\,dx.$

7. $\int\frac{1}{1-\sin x}dx.$

8. $\int\frac{dx}{x\sqrt{x^6-1}}.$

9. $\int\left(\sqrt{1+\sin\frac{x}{2}}\right)dx.$

10. $\int\frac{10x^9+10^x\log 10}{10^x+x^{10}}dx.$

11. $\int\frac{\cot(\log x)}{x}dx.$

12. $\int\frac{\cot(x-\alpha)}{\cos x}dx.$

13. $\int(\log x)^2\sqrt{x}\,dx.$

14. $\int\frac{x+\sin x}{1+\cos x}dx.$

15. $\int\sin(\log x)\,dx.$

16. $\int e^{-x}\sin x\,dx.$

17. $\int e^{x/2}\left(\frac{2-\sin x}{1-\cos x}\right)dx.$

18. $\int\frac{x^4\,.\,dx}{(x+1)(x^2+1)}.$

19. $\int\sqrt{\tan x}\,dx.$

20. $\int\frac{x^2}{x^2-4}dx.$

21. $\int_0^{\pi/2}\frac{dx}{1+2\sin x+\cos x}$

22. $\int_{-1/2}^{1}\frac{dx}{\sqrt{1-x^2}}.$

Evaluate the following integrals as limit of sums.

23. $\int_1^3(2x+3)\,dx$

24. $\int_0^2(x^2+1)\,dx.$

25. $\int_0^1 e^{2-3x}\,dx.$

Answers

1. $x+\frac{2}{7}x^{7/2}+\frac{4}{3}x^{3/2}+c.$

2. $\frac{x^2}{2}(\log x)^2-\frac{x^2}{2}\log x+\frac{x^4}{4}+c.$

3. $e^{2x}\left[\frac{x}{\sqrt{5}}\cos\left(x-\tan^{-1}\frac{1}{2}\right)-\frac{1}{5}\cos\left(x-2\tan^{-1}\frac{1}{2}\right)\right]$

4. $e^x\log x+c.$

5. $\log(\operatorname{cosec}x-\cot x)+\frac{1}{4}\log\frac{1+\cos x}{1-\cos x}-\frac{1}{2(1+\cos x)}+c.$

6. $x(\log x)^2-2x\log x+2x+c.$

7. $\tan x+\sec x+c.$

8. $\frac{\sec^{-1}x^3}{3}+c.$

9. $4\left[\sin\frac{\pi}{4}-\cos\frac{x}{4}\right]+c.$

10. $\log[10^x+x^{10}]+c.$

11. $\log[\sin(\log x)]+c.$

12. $x\cos\alpha+\sin\alpha\,.\,\log(\sec x)+c.$

13. $x^{3/2}\left[\frac{2}{3}(\log x)^2-\frac{8}{9}\log(x)+\frac{16}{27}\right]+c.$

14. $x\tan\frac{x}{2}+c.$

15. $\frac{1}{2}[\sin x(\log x)-\cos(\log x)+c].$

16. $\frac{1}{13}e^{2x}(2\sin 3x-3\cos 3x)+c.$

17. $-2e^{x/2}\cot\frac{x}{2}+c.$

18. $-\log(x+1)+\frac{1}{2}\log(x^2+1)+c.$

19. $\frac{1}{\sqrt{2}}\tan^{-1}\left(\frac{\tan x-1}{\sqrt{2}\tan x}\right)+\frac{1}{2\sqrt{2}}\log\left|\frac{\tan x-\sqrt{2}\tan x+1}{\tan x+\sqrt{2}\tan x+1}\right|+c.$

20. $x+\log\left|\frac{x-2}{x+2}\right|+c.$

21. $\frac{1}{2}\log 3.$

22. $\frac{2\pi}{3}.$

23. 14

24. 14/3

25. $\frac{1}{3}(e^2-e^{-1}).$

Chapter 9. Differential Equations

Solve the following differential equations :

1. $\frac{dy}{dx}=\frac{x-2y+5}{2x+2y+1}.$

2. $3e^x \cdot \tan y \cdot dx+(1-e^x)\cdot \sec^2 y \cdot dy=0.$

3. $\frac{dy}{dx}=x(2\log x+1).$ when $y=0, x=2.$

4. $y-x\frac{dy}{dx}=a\left(y^2+\frac{dy}{dx}\right).$

5. $\frac{dy}{dx}=x^2(x-2).$ Given $y=2$, when $x=0.$

6. $\frac{dy}{dx}=e^{3x-2y}+x^2e^{-2y}.$

7. $\frac{dy}{dx}=\sin^{-1}x.$

8. $(x+1)\frac{dy}{dx}=e^{3x}(x+1)^2.$

9. $\frac{dy}{dx}=\frac{(1+x)y^2}{x^2(y-1)}.$

10. $(x-y-2)\,dx-(2x-2y-3)\,dy=0.$

11. $\frac{dy}{dx}+\frac{2x}{1+x^2}\cdot y=\frac{1}{(1+x^2)^2}.$

12. $\frac{dy}{dx}=\frac{y}{x}+\tan\frac{y}{x}.$

13. $\frac{dy}{dx}=(4x+y+1)^2.$

14. $(1+y^2)\,dx=(\tan^{-1}y-x)\,dy.$

15. $\frac{dy}{dx}+\frac{y}{x}=x^2.$ Given $y=1$, when $x=1.$

16. $(x+2y^2)\frac{dy}{dx}=y,\ y>0.$

Answers

1. $x^2-y^2-4xy+10x+2y=c_1.$

2. $3\log(e^x-1)=\log(\tan y)+c.$

3. $y=x^2\log x-4\log 2$

4. $\log y-\log(1-ay)=\log(x+a)+c.$

5. $y=\frac{x^4}{4}-\frac{2}{3}x^3.$

6. $\frac{e^{2y}}{2}=\frac{e^{3x}}{3}+\frac{x^3}{3}+c.$

7. $y=x\sin^{-1}x+\sqrt{1-x^2}+c.$

8. $\frac{y}{x+1}=\frac{e^{3x}}{3}+c.$

9. $\log y+\frac{1}{y}=\log x-\frac{1}{x}+c$

10. $(2y-x+4)+\log(x-y-1)=c.$

11. $y=\frac{\tan^{-1}x}{1+x^2}+c.$

12. $\log\sin\left(\frac{y}{x}\right)=\log x+c.$

13. $\frac{1}{2}\tan^{-1}\left(\frac{4x+y+1}{2}\right)=x+c.$

14. $e^{\tan^{-1}y}x=e^{\tan^{-1}y}(\tan^{-1}y-1)+c.$

15. $xy=\frac{x^4}{4}+\frac{3}{4}.$

16. $x=2y^2+cy.$

Chapter 10. Probability

1. In a single throw of two dice, what is the probability of getting a total of at least 10 ?
2. A coin is tossed 400 times. Out of which it shows head 220 times. Discuss whether the coin is unbiased or not.
3. In a single throw of dice, find the probability of getting a total of at most 9.
4. A pair of dice is thrown 4 times, if getting a even number is considered a success. Find the probability of 3 successes.
5. In a given race the odds in favour of four horses A, B, C and D are 1 : 3, 1 : 4, 1 : 5 and 1 : 6 respectively. Assume that a dead heat is impossible, find the chances that one of them wins the race.
6. On an average out of 12 games of chess played by A and B, A wins 6, B wins 4 and 2 games end in a tie. A and B plays a tournament of 3 games, calculate the probability that A and B wins alternate game, no game is tied up.
7. Bag 1 contains 2 white and 3 red balls and Bag 2 contains 4 white and 5 red balls. One ball is drawn at random from one of the bags and is found to be red. Find the probability that it was drawn from Bag 2.
8. A card from a pack of 52 cards is lost. From the remaining cards of pack, two cards are drawn and are found to be hearts. Find the probability of missing card to be a heart.
9. A bag contains 20 tickets with marked numbers 1 to 20. One ticket is drawn at random. Find the probability that it will be a multiple of 2 or 5.
10. In a lecture class 52% students cannot read what is written on board, 46% cannot hear and 32% can neither read nor hear. What percentage of students can read or hear on board ?
11. Assume that on an average one telephone out of 10 is busy. Six telephone numbers are randomly selected and called. Find the probability that four of them will be busy.
12. P is the probability that a man aged x will die in a year. Find the probability that out of n men $A_1, A_2, \ldots\ldots, A_n$ each aged x, A_1 will die in a year and be the first to die.
13. A coin is tossed four times. If X is the number of heads observed. Find the probability distribution of X.
14. Find mean, variance and Standard deviation of the number of heads in three tosses of a coin.
15. If the sum of mean and variance of a binomial distribution for 5 trials is 1·8, find the distribution.

Answers

1. $\frac{1}{6}$ 2. Perhaps biased. 3. $\frac{5}{6}$

4. $\frac{5}{16}$ 5. $\frac{319}{420}$ 6. $\frac{5}{36}$

7. $\frac{25}{32}$ 8. $\frac{11}{50}$ 9. 0·36

10. 66% 11. $\frac{243}{200000}$

12. $\frac{1}{n}\left[1-(1-p)^n\right]$

13.

0	1	2	3	4
1/16	1/3	3/8	1/4/	1/16

14. $\frac{3}{2}, \frac{3}{4}, \frac{\sqrt{3}}{2}$ 15. $\left(\frac{4}{5}+\frac{1}{5}\right)^5$

Chapter 11. Vectors

1. If sum of two unit vectors is a unit vector, show that the magnitude of their difference is $\sqrt{3}$.
2. If $\vec{a}, \vec{b}, \vec{c}$ represent sides of a triangle taken in order then show that $\vec{a}+\vec{b}+\vec{c}=0$.
3. Prove by vector method that in any triangle *ABC*,

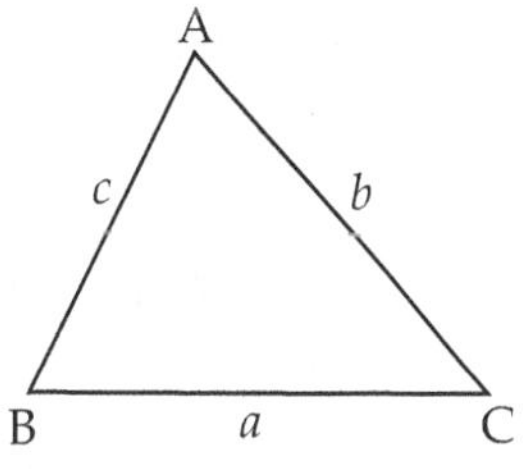

(i) $\cos A = \frac{b^2+c^2-a^2}{2bc}$

(ii) $a = b\cos C + c\cos B$.

4. If A, B, C and D, be any four points and E, F be the mid points of AC and BD respectively, prove that

$$\vec{AB}+\vec{CB}+\vec{CD}+\vec{AD}=4\vec{EF}$$

5. If two medians of a triangle are equal, show that triangle is isosceles.(By vector method)

6. In parallelogram $ABCD$, E is mid-point of AB. Using vectors find $\frac{AF}{FC}$.

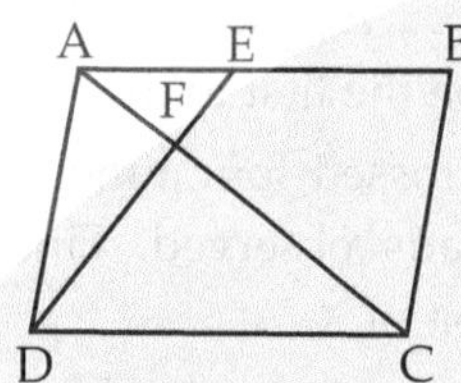

7. The vectors $2\hat{i}-\hat{j}+\hat{k}$, $\hat{i}-3\hat{j}-5\hat{k}$ and $3\hat{i}-4\hat{j}-4\hat{k}$ are the position vectors of the vertices of triangle ABC, prove that ABC is right angle triangle.

8. Given $\vec{a}=3i-j$ and $\vec{b}=2i+j-3k$ express $\vec{b}$ as $\vec{b_1}+\vec{b_2}$ such that $b_1 \parallel a$ and $b_2 \perp a$.

9. Prove that if the diagonals of a parallelogram intersect at right angle then all the sides of a parallelogram are equal.

10. If D is the mid-point of the side BC of a triangle ABC, then show $AB^2+AC^2=2(AD^2+BD^2)$.

11. If $\vec{a}$ and $\vec{b}$ are two mutually perpendicular vectors show that

$$(\vec{a}+\vec{b})^2=(\vec{a}-\vec{b})^2.$$

12. Let $\vec{a},\vec{b},\vec{c}$ be three vectors of magnitude 3, 4 and 5 respectively. If each one is per-pendicular to sum of the other two vectors, prove that

$$|\vec{a}+\vec{b}+\vec{c}|=5\sqrt{2}.$$

13. Prove the Pythagoras theorem by vector method.

14. Using vectors prove that

$\cos(A-B)=\cos A\cos B-\sin A.\sin B.$

15. If $\vec{A}=(1,1,1)$, $\vec{C}=(0,1,-1)$ are given vectors, then find a vector B satisfying the equations

$$\vec{A}\times\vec{B}=\vec{C} \text{ and } \vec{A}.\vec{B}=3.$$

16. Using vector method show that points $P(-2,1)$, $Q(-5,-1)$, $R(1,3)$ are collinear.

17. Let $\vec{A}=2i+k$, $\vec{B}=i+j+k$ and $\vec{C}=4i-3j+7k$. Find vector $\vec{R}$ which satisfies.

$$\vec{R}\times\vec{B}=\vec{C}\times\vec{B} \text{ and } \vec{R}.\vec{A}=0.$$

18. If the diagonals of parallelogram are equal. Prove by vector method that it is rectangle.

19. Find the area of parallelogram whose adjacent sides are determined by the vectors $\vec{a}=\hat{i}+2\hat{j}+3\hat{k}$ and $\vec{b}=-3\hat{i}-2\hat{j}+\hat{k}$.

20. Find area of parallelogram whose diagonals are $a=2\hat{i}+3\hat{j}+4\hat{k}$ and $b=\hat{i}+2\hat{j}-\hat{k}$

21. Find the value of λ which makes the vector $\hat{i}-\hat{j}+\hat{k}$, $2\hat{i}+\hat{j}-\hat{k}$ and $\lambda\hat{i}+\hat{j}+\lambda\hat{k}$ coplanar.

22. Show that the four points $A(4, 5, 1)$, $B(0,-1,-1)$, $C(3, 9, 4)$ and $D(-4, 4, 4)$ are coplanar.

23. Prove that

$$(\vec{P}-\vec{Q}).[(\vec{Q}-\vec{R})\times(\vec{R}-\vec{P})]=0.$$

24. If the volume of parallelopiped whose edges are $-12\hat{i}+\lambda\hat{k}, 3\hat{i}-\hat{k}, 2\hat{i}+\hat{j}-15\hat{k}$ is 546. Find λ.

25. Find a unit vector coplanar with $\hat{i}+\hat{j}+2\hat{k}$ and $\hat{i}+2\hat{j}+\hat{k}$ and perpendicular to $\hat{i}+\hat{j}+\hat{k}$.

Answers

6. $\frac{1}{2}$

15. $\frac{1}{3}[5\hat{i}+2\hat{j}+2\hat{k}]$

17. $-\hat{i}-8\hat{j}+2\hat{k}$

19. $3\sqrt{20}$

20. $\frac{\sqrt{158}}{2}$ square units.

21. $\lambda=-1$

24. -3 or 197

25. $\pm\frac{\hat{j}-\hat{k}}{\sqrt{2}}$.

Chapter 12. Three Dimensional Geometry

1. Find the angle between the lines whose direction cosines are given by $3l+m+5n=0$ and $6mn-2nl+5lm=0$.

2. Show that the following points are coplanar :

(a) $(-6,3,2)$, $(3,-2,4)$, $(5,7,3)$ and $(-13,17,-1)$

(b) $(0,4,3)$, $(1,1,-1)$, $(-1,-5,-3)$ and $(-2,-2,1)$.

Practice Exercises

3. Prove that the lines

$$\frac{x-2}{1}=\frac{y-4}{3}=\frac{z-6}{5}$$

and $\frac{x+1}{3}=\frac{y+3}{5}=\frac{z+5}{7}$ are coplanar.

4. Find the image of the point P (1, 6, 3) in the line $\frac{x}{1}=\frac{y-1}{2}=\frac{z-2}{3}$.

5. Find the foot of the perpendicular drawn from the point P (1, 6, 3) on the line $\frac{x}{1}=\frac{y-1}{2}=\frac{z-2}{3}$. Also find the equations and the length of perpendicular.

6. Find the shortest distance between two lines

$$\frac{x-3}{2}=\frac{y-8}{-1}=\frac{z-3}{1}$$

and $$\frac{x+3}{3}=\frac{y+7}{-2}=\frac{z-6}{-6}$$

Also find the equations of the lines of shortest distance.

7. Show that the two points (0, 4, 3), (–1, –5, –3), (–2, –2, 1) and (1, 1, –1) are coplanar and find the equation of common plane.

8. A plane meets the plane $x = 0$ when $x = 0$, $2y - 3z = 5$, and the plane $z = 0$, where $z = 0$, $7x + 4y = 10$. Find the equation to the plane.

9. Find the equations of the planes bisecting the angle between the planes $x + 2y - 2z + 6 = 0$ and $4x - y + 8z - 8 = 0$ and specify the one which bisects the acute angle.

10. Find the distance of the point (2, 3, 5) from xy plane.

11. Find the equation of the plane passing through points (2, 2, 1), (1, – 2, 3) and parallel to the line joining the points (2, 1, – 3), (– 1, 5, – 8).

12. Find the equation of the plane passing through the line of intersection of the planes $x + 2y + 3z - 5 = 0$ and $3x - 2y - z + 1 = 0$ and cutting off equal intercepts on OX and OZ axes.

13. Find the equation of the plane perpendicular to each of the planes $3x - y + z = 0$ and $x + 5y + 3z = 0$ at a distance of $\sqrt{6}$ from the origin.

14. Find the equation of the plane which bisects the line joining (5, – 2, 6) and (7, 2, 0) at right angle.

Answers

1. $\theta = \cos^{-1}\left(\frac{1}{6}\right)$.

4. (1, 0, 7).

5. (1, 3, 5), $x = 1$, $\frac{y-6}{3}=\frac{z-3}{-2}\sqrt{13}$.

6. $3\sqrt{30}$ units.

$$\frac{x-3}{2}=\frac{y-8}{6}=\frac{z-3}{-1}.$$

7. $9x - 5y + 6z + 2 = 0$.

8. $7x + 4y - 6z - 10 = 0$.

9. $7x + 5y + 2z + 10 = 0$ and acute angle bisector $x - 7y - 24z - 26 = 0$.

10. 5.

11. $12x - 11y - 16z + 14 = 0$.

12. $5x + 2y + 5z - 9 = 0$.

13. $x + y - 2z - 6 = 0$.

14. $x + 2y - 3z + 3 = 0$.

Practice Exercises

Chapter 13. Application of Integrals

1. Using definite integrals, find the area of the circle $x^2 + y^2 = a^2$.

2. Find the area of the parabola $y^2 = 4ax$ bounded by its latus rectum.

3. Find the area bounded by the curve $y^2 = 4x$ and the line $y = 3$ and y-axis.

4. Calculate the area under the curve $y = 2\sqrt{x}$ included between the lines $x = 0$ and $x = 1$.

5. Find the area between the curve $y = [x + 3]$, the x-axis and the ordinate $x = -6$ and $x = 0$.

6. Using integration, find the area of the triangle, whose vertices are (2, 5), (4, 7) and (6, 2).

7. Find the area of the region bounded by $y = |x - 1|$ and $y = 1$.

8. Find the area bounded by the lines $y = 4x + 5$, $y = 5 - x$ and $4y = x + 5$.

9. Indicate the region bounded by the curve $y = x \log x$ and $y = 2x - 2x^2$ and obtain the area enclosed by them.

10. Find the area bounded by the parabolas $y = 6x - x^2$ and $y = x^2 - x$.

Answers

1. πa^2 sq. units.
2. $\frac{8}{3}a^2$ sq. units.
3. $2\frac{1}{4}$ sq. units.
4. $\frac{4}{3}$ sq. units.
5. $4\frac{1}{2}$ sq. units.
6. 7 sq. units
7. 1 sq. units
8. $\frac{15}{2}$ sq. units
9. $\frac{7}{12}$ sq. units.
10. $\frac{125}{24}$ sq. units.

Chapter 14. Application of Calculus

1. Suppose the cost to produce some commodity is a linear function of output. Find cost as a function of output, if costs are ₹ 4,000 for 250 units and ₹ 5,000 for 350 units.
2. The total cost $C(x)$, associated with production and making of x units of an item is given by $C(x) = 0{\cdot}005x^3 - 0{\cdot}02x^2 + 30x + 5000$; find
   (i) the average cost function
   (ii) the average cost of output of 10 units
   (iii) the marginal cost function
   (iv) the marginal cost when 3 units are produced.
3. If the total cost function $C = 2x^2 - 3x + 8$, find the average cost function and marginal cost function, and marginal cost when 10 units are produced.
4. If $C = 2x\left(\frac{x+4}{x+1}\right) + 6$ is the total cost of production of x units of a certain product, show that the marginal cost falls continuously as the output x increases.
5. A manufacturer can sell x items of commodity of price ₹ $(330 - x)$ each. Find the revenue function. If the cost of producing x items is ₹ $(x^2 + 10x + 12)$, determine the profit function.
6. For the demand function $P = \frac{a}{x+b} - c$, where $ab > 0$, show that the marginal revenue decreases with the increase of x.
7. $C(x) = 5x + 350$ and $R(x) = 50x - x^2$ are respectively the total cost and total revenue functions of a company that produces and sells x units of a particular product. Find
   (i) the break-even values.
   (ii) the value of x that produce profit.
   (iii) the values of x that result in loss.
8. Suppose a manufacturer can sell x items per week at a price $P = 20 - 0{\cdot}01x$ rupees each when it cost, $y = 5x - 2000$ rupees to produce x item. Determine the number of items he should produce per week for maximum profit.
9. Verify for the cost functions,
   $$C(x) = ax\left(\frac{x+b}{x+c}\right) + d,\ a, b, c, d > 0$$
   that the average and marginal cost curves fall continuously with increasing output.
10. Find the relationship between the slopes of marginal revenue curve and the average revenue curve, for the demand function,
   $$P = a - bx.$$

Answers

1. $C = 10x + 1500$
2. (i) $AC = 0{\cdot}005x^2 - 0{\cdot}02x + 30 + \frac{5000}{x}$
   (ii) 530·3
   (iii) $MC = 0{\cdot}015x^2 - 0{\cdot}04x + 30$
   (iv) ₹ 30·015
3. (i) $AC = 2x + \frac{8}{x} - 3$
   (ii) $MC = 4x - 3$
   (iii) 37
5. $R(x) = 330x - x^2$, $P(x) = 320x - 12 - 2x^2$
7. (i) 10, 35
   (ii) $10 < x < 35$
   (iii) $x > 35$ or $x < 10$
8. 750 items.
9. The slope of MR curve is twice the slope of AR curve.

Practice Exercises

Chapter 15. Linear Regression

1. Find the equation of the regression line of y on x, if the observations (x, y) are the following (1, 4), (2, 8), (3, 2), (4, 12), (5, 10), (6, 4), (7, 6), (8, 6), (9, 18).
2. Use following data to find coefficient of correlation.

$$n = 10, \Sigma x = 55, \Sigma y = 4, \Sigma x^2 = 385, \Sigma y^2 = 192 \text{ and } (x + y)^2 = 967$$

3. Find (i) $\bar{x}$ and $\bar{y}$ (ii) b_{yx} and b_{xy}, (iii) $\rho\ (x, y)$ when the two regression lines are $3x + 12y = 19$, $9x + 3y = 46$.
4. The two lines of regression for a distribution (x, y) are $3x + 2y = 7$ and $x + 4y = 9$. Find the regression coefficient b_{yx} and b_{xy}.
5. For lines of regression $4x - 2y = 3$ and $2x - 3y = 5$. Find (i) b_{xy} and b_{yx} (ii) $\rho\ (x, y)$, (iii) y when $x = 3$.
6. The two lines of regression are $x + 3y = 11$ and $2x + y = 7$. Find the coefficient of correlation between x and y and estimate the value of x when $y = 4$.

Answers

1. regression equation of y on x is $18y = 15x + 65$.
2. $-0{\cdot}6812$
3. $\bar{x} = S,\ \bar{y} = \frac{1}{3},\ b_{xy} = \frac{-1}{3},\ b_{yx} = \frac{-1}{4},\ \rho\ (x, y) = -0{\cdot}29$
4. $b_{yx} = -\frac{1}{4},\ b_{xy} = \frac{-2}{3}$.
5. $b_{xy} = \frac{1}{2},\ b_{yx} = \frac{2}{3},\ \rho\ (x, y) = 0{\cdot}58,\ y = \frac{1}{3}$ when $x = 3$.
6. $r = -0{\cdot}408,\ x = 3/2$.

Chapter 16. Linear Programming

1. Determine the maximum and minimum values of the follwing functions and the values of x and y where they occur.

   (i) $f(x, y) = 3x + 5y$, vertices at (4, 8), (2, 4), (1, 1), (5, 2).

   (ii) $f(x, y) = x + 4y$, vertices at (0, 7), (0, 0), (6, 2), (5, 4).
2. Find the minimum and maximum values of the following functions and the values of x and y where they occur :

   (i) $Q = x + 3y$ subject to $x, y \geq 0$, $5x + 2y \leq 20$, $2y \geq x$.

   (ii) $f(x, y) = 10x + 12y$ subject to $x, y \geq 0$, $2x + 5y \geq 22$, $4x + 3y \geq 28$, $2x + 2y \leq 17$.
3. Solve graphically the linear programming problem to minimize the cost $c = 3x + 2y$ subject to the following constraints :

   $5x + y \geq 10,\ x + y \geq 6,\ x + 4y \geq 12,\ x \geq 0,\ y \geq 0.$

   Solve this problem using corner point method as well as by using iso-cost line method.
4. Maximize $Z = 3x + 4y$, if possible, subject to the constraints :

$$x - y \leq -1,\ -x + y \leq 0,\ x, y \geq 0.$$

5. A man has ₹ 1,500 for purchase of rice and wheat. A bag of rice and a bag of wheat cost ₹ 180 and ₹ 120, respectively. He has storage capacity of 10 bags only. He earns a profit of ₹ 11 and ₹ 9 per bag of rice and wheat respectively. Formulate an linear programming problem to maximize the profit and solve it.
6. A manufacturer makes two types of tea–cups, say, A and B. Three machines are needed for their manufacturing and the time (in minutes) required for each cup on the machines is given below :

Cup	Machines		
	I	II	III
A	12	18	6
B	6	0	9

   Each machine is available for a maximum of 6 hours per day. If the profit on each cup A is 75 paise and on each cup B is 50 paise, show that 15 tea-cups of type A and 30 of type B should be manufactured in a day to get maximum profit.
7. A factory owner purchases two types of machines, A and B, for his factory. The requirements and limitations for the machines are as follows :

Machine	Area occupied by the machine	Labour force for each machine	Daily output (in units)
A	1,000 sq.m	12 men	60
B	1,200 sq.m	8 men	40

He has an area of 9,000 sq. m available and 72 skilled men who operate the machines. How many machines of each type should he buy to maximize the daily output ?

8. A dietician wishes to mix two types of food in such a way that the vitamin contents of the mixture contain at least 8 units of vitamin A, and 10 units of vitamin C. Food I contains 2 units/kg of vitamin A and 1 unit/kg of vitamin C. Food II contains 1 unit/kg of vitamin A and 2 units/kg of vitamin C. It costs ₹ 5 per kg to purchase food I and ₹ 7 per kg to purchase food II. Determine the minimum cost of such a mixture.

9. A factory manufactures has two types of screws, A and B. Each type requiring the use of two machines–an automatic and a hand–operated. It take 4 minutes on the automatic and 6 minutes on the hand-operated machine to manutes a package of screws 'A'. While it takes 6 minutes on the automatic and 3 minutes on the hand-operated machine to manufacture a package of screws 'B'. Each machine is available for at most 4 hours on any day. The manufacturer can sell a package of screws 'A' at a profit of ₹ 7 and of screws 'B' at a profit of ₹ 10. Assuming that he can sell all the screws he can manufacture, how many package of each type should the factory owner produce in a day in order to maximize his profit ? Determine the maximum profit.

Answers

1. (i) Max. 52 at (4, 8), Min. 8 at (1, 1).

(ii) Max. 28 at (0, 7), Min. 0 at (0, 0).

2. (i) Max. 30 at (0, 10), Min 0 at (0, 0).

(ii) Max. 97 at $\left(\frac{5}{2}, 6\right)$, Min. $\frac{562}{7}$ at $\left(\frac{37}{7}, \frac{16}{7}\right)$

3. Min. 13 at (1, 5).

4. No solution exists.

5. 5 rice bags and 5 wheat bags, Max. profit = ₹ 100.

7. 4 type A, 3 type B or 6 type A, no machine of type B.

8. 2 kg food I, 4 kg food II. Min. cost = ₹ 38.

9. Max. profit ₹ 410 at (30, 20).

Practice Exercises

ISC Solved Paper 2020

(Maximum Marks : 100)
(Time allowed : Three hours)
*(Candidates are allowed additional 15 minutes for **only** reading the paper.*
They must NOT start writing during this time.)
This Question Paper consists of three sections A, B and C.
*Candidates are required to attempt all questions from **Section A** and all questions*
EITHER** from **Section B OR Section C.
***Section A :** Internal choice has been provided in three questions of four marks each and two questions of six marks each.*
***Section B :** Internal choice has been provided in two questions of four marks each.*
***Section C :** Internal choice has been provided in two questions of four marks each.*
All working, including rough work, should be done on the same sheet as, and adjacent to the rest of the answer.
The intended marks for questions or parts of questions are given in brackets [ ].
Mathematical tables and graph papers are provided.

SECTION – A (80 Marks)

Question 1. **[10 × 2]**

(i) *Determine whether the binary operation $*$ on R defined by $a * b = |a - b|$ is commutative. Also, find the value of $(-3) * 2$.*

(ii) *Prove that :*

$\tan^2(\sec^{-1} 2) + \cot^2(\text{cosec}^{-1} 3) = 11.$

(iii) *Without expanding at any stage, find the value of the determinant :*

$$\Delta = \begin{vmatrix} 20 & a & b+c \\ 20 & b & a+c \\ 20 & c & a+b \end{vmatrix}$$

(iv) *If* $\begin{pmatrix} 2 & 3 \\ 5 & 7 \end{pmatrix}\begin{pmatrix} 1 & -3 \\ -2 & 4 \end{pmatrix} = \begin{pmatrix} -4 & 6 \\ -9 & x \end{pmatrix}$, *find x.*

(v) *Find* $\frac{dy}{dx}$ *if* $x^3 + y^3 = 3axy$

(vi) *The edge of a variable cube is increasing at the rate of 10 cm/sec. How fast is the volume of the cube increasing when the edge is 5 cm long ?*

(vii) *Evaluate :* $\int_4^5 |x-5|\, dx$

(viii) *Form a differential equation of the family of the curves $y^2 = 4ax$.*

(ix) *A bag contains 5 white, 7 red and 4 black balls. If four balls are drawn one by one with replacement, what is the probability that none is white ?*

(x) *Let A and B be two events such that*

$$P(A) = \frac{1}{2},\ P(B) = p \text{ and } P(A \cup B) = \frac{3}{5}$$

find 'p' if A and B are independent events.

Solution 1.

(i) We have,

$$\begin{aligned} a * b &= |a - b| \\ &= |-(b-a)| \\ &= b - a \\ &= |b - a| \\ &= b * a \end{aligned}$$

$\Rightarrow \quad a * b = b * a$

So, $*$ is commutative on R.

Now, $\quad (-3) * 2 = |-3-2| = |-5| = 5$ **Ans.**

(ii) To prove : $\tan^2(\sec^{-1} 2) + \cot^2(\text{cosec}^{-1} 3) = 11$

Let $\quad \sec^{-1} 2 = \theta$

$\Rightarrow \quad \sec\theta = 2$

$\therefore \quad \tan^2\theta = \sec^2\theta - 1 = 2^2 - 1 = 3$

Also, let $\quad \text{cosec}^{-1} 3 = \alpha$

$\Rightarrow \quad \text{cosec}\,\alpha = 3$

$\therefore \quad \cot^2\alpha = \text{cosec}^2\alpha - 1 = 3^2 - 1 = 8$

Now, L.H.S. $= \tan^2(\sec^{-1} 2) + \cot^2(\text{cosec}^{-1} 3)$

$= \tan^2\theta + \cot^2\alpha$

$= 3 + 8 = 11 =$ R.H.S. **Hence Proved.**

(iii) Given :

$$\Delta = \begin{vmatrix} 20 & a & b+c \\ 20 & b & a+c \\ 20 & c & a+b \end{vmatrix}$$

Taking 20 common from C_1,

$$\Rightarrow \quad \Delta = 20\begin{vmatrix} 1 & a & b+c \\ 1 & b & a+c \\ 1 & c & a+b \end{vmatrix}$$

Applying $C_2 \to C_2 + C_3$,

$$\Delta = 20\begin{vmatrix} 1 & a+b+c & b+c \\ 1 & a+b+c & a+c \\ 1 & a+b+c & a+b \end{vmatrix}$$

Taking $(a+b+c)$ common from C_2,

$$\Delta = 20(a+b+c)\begin{vmatrix} 1 & 1 & b+c \\ 1 & 1 & a+c \\ 1 & 1 & a+b \end{vmatrix}$$

$\because \quad C_1 = C_2$

$\therefore \quad \Delta = 0$ **Ans.**

(iv) Given :

$$\begin{pmatrix} 2 & 3 \\ 5 & 7 \end{pmatrix}\begin{pmatrix} 1 & -3 \\ -2 & 4 \end{pmatrix} = \begin{pmatrix} -4 & 6 \\ -9 & x \end{pmatrix}$$

$$\Rightarrow \begin{pmatrix} 2\times1+3\times(-2) & 2\times(-3)+3\times4 \\ 5\times1+7\times(-2) & 5\times(-3)+7\times4 \end{pmatrix} = \begin{pmatrix} -4 & 6 \\ -9 & x \end{pmatrix}$$

$$\Rightarrow \begin{pmatrix} -4 & 6 \\ -9 & 13 \end{pmatrix} = \begin{pmatrix} -4 & 6 \\ -9 & x \end{pmatrix}$$

$\therefore \quad x = 13$ **Ans.**

(v) Given : $x^3 + y^3 = 3axy$

Differentiating w.r.t. x, we get

$$3x^2 + 3y^2\frac{dy}{dx} = 3a\left(1.y + x\frac{dy}{dx}\right)$$

$$\Rightarrow x^2 + y^2\frac{dy}{dx} = ay + ax\frac{dy}{dx}$$

$$\Rightarrow \frac{dy}{dx}(y^2 - ax) = ay - x^2$$

$$\Rightarrow \frac{dy}{dx} = \frac{ay - x^2}{y^2 - ax}$$ **Ans.**

(vi) Let the edge of a cube be x units.

$$\therefore \frac{dx}{dt} = 10 \text{ cm/sec} \qquad \text{[Given]}$$

Now, Volume of cube $(V) = x^3$

Differentiating w.r.t. t, we get

$$\frac{dV}{dt} = 3x^2\frac{dx}{dt}$$

$$\Rightarrow \frac{dV}{dt} = 3 \times 5^2 \times 10 \qquad [\because x = 5 \text{ cm}]$$

$$\Rightarrow \frac{dV}{dt} = 750$$

Hence, the volume of the cube is increasing at the rate 750 cm^3/sec. **Ans.**

(vii)

$$\int_4^5 |x-5|\,dx = \int_4^5 -(x-5)\,dx$$

$$= \int_4^5 (5-x)\,dx$$

$$= \left[5x - \frac{x^2}{2}\right]_4^5$$

$$= \left[5\times5 - \frac{5\times5}{2}\right] - \left[5\times4 - \frac{4\times4}{2}\right]$$

$$= \left[25 - \frac{25}{2}\right] - \left[20 - \frac{16}{2}\right]$$

$$= \frac{50-25}{2} - 12$$

$$= \frac{25}{2} - 12$$

$$= \frac{25-24}{2} = \frac{1}{2}$$ **Ans.**

(viii) Given : $y^2 = 4ax$

Differentiating both sides w.r.t. x, we get

$$2y\frac{dy}{dx} = 4a$$

$$\Rightarrow 2y\frac{dy}{dx} = \frac{y^2}{x} \qquad \left[\because 4a = \frac{y^2}{x}\right]$$

$$\Rightarrow 2\frac{dy}{dx} = \frac{y}{x}$$

$$\Rightarrow 2x\frac{dy}{dx} - y = 0,$$

which is the required differential equation. **Ans.**

(ix) Total number of balls = 5 + 7 + 4 = 16

Total number of non-white balls = 7 + 4 = 11

$\therefore$ Probability of drawing a non-white ball = $\frac{11}{16}$

Since, balls are drawn one by one with replacement, So, probability of drawing non-white balls 4 times

$$= \left(\frac{11}{16}\right)^4 = \frac{14641}{65536}$$ **Ans.**

(x) Given : $P(A) = \frac{1}{2}$, $P(B) = p$ and $P(A \cup B) = \frac{3}{5}$.

We know, $P(A \cup B) = P(A) + P(B) - P(A \cap B)$

$$\Rightarrow \frac{3}{5} = \frac{1}{2} + p - P(A \cap B)$$

$$\Rightarrow P(A \cap B) = \frac{1}{2} - \frac{3}{5} + p = p + \frac{1}{2} - \frac{3}{5}$$

$$= p + \frac{5-6}{10}$$

$$= p - \frac{1}{10}$$

Now, A and B are independent events.

$$\therefore P(A \cap B) = P(A)\cdot P(B)$$

$$\Rightarrow p - \frac{1}{10} = \frac{1}{2} \times p$$

$$\Rightarrow p - \frac{p}{2} = \frac{1}{10}$$

$$\Rightarrow \frac{p}{2} = \frac{1}{10}$$

$$\therefore p = \frac{2}{10} = \frac{1}{5}$$ **Ans.**

Comments of Examiners

1. (i) Many candidates failed to prove the given binary operation.
   (ii) Several candidates made errors while applying properties of inverse trigonometric functions, converting one inverse trigonometric function into another equivalent inverse trigonometric function and in simplification.
   (iii) In some cases, candidates expanded the determinant to solve it, though it was mentioned in the question - without expanding at any stage, find the value of the determinant.
   (iv) A few candidates made errors while finding the product of a matrix by a matrix.
   (v) Errors were made by some candidates while finding the derivative of implicit function. They also made errors in simplification.
   (vi) A few candidates made mistake in writing the final answer in the correct form.
   (vii) Many candidates were unable to identify the correct intervals of absolute function for the given limits.
   (viii) Most of the candidates differentiated the function correctly but did not eliminate the arbitrary constant. Some of them used the second-order derivative method for eliminating constant '*a*'.
   (ix) Many candidates made errors while finding the probability of drawing four balls one by one with replacement. This was due to a lack of knowledge of the basic idea of the probability of drawing balls with or without replacement.
   (x) Several candidates did not find the probability of mutually exclusive events and independent events by applying the concept of independent events.

Answering Tips

- Practice the concept of modulus function and properties of binary operations like closure, commutative, identity and inverse.
- Give adequate practice in the comprehension of well-defined mathematical operations.
- Learn and practice the conversion formulae for all inverse trigonometric functions with the help of diagrams by using trigonometric ratios and Pythagoras theorem.
- Clarify the difference between expanding and not expanding the determinant at any stage. Extensive practice in questions based on the properties of determinants.
- Practice the concept of the product of two matrices with suitable examples.
- Frequent practice in different types of functions to help in identifying and distinguishing between different types of functions.
- Practice the concept of differential coefficient as a rate measurer. Lay more emphasis on the rate of change of connected variables.
- Clarify the concept of an absolute function and the methods of identifying intervals where the function satisfies the given limits.
- Emphasise on framing the differential equation by differentiating and eliminating the arbitrary constants in the process of mathematical simplification.
- Clarify the concept of probability with and without replacement.
- Emphasise on mutually exclusive events and independent events.

Question 2. **[4]**

If the function $f: R \to R$ be defined as

$$f(x) = \frac{3x+4}{5x-7}, \left(x \neq \frac{7}{5}\right) \text{ and } g: R \to R \text{ be defined as}$$

$$g(x) = \frac{7x+4}{5x-3}, \left(x \neq \frac{3}{5}\right) \text{ show that } (g \circ f)(x) = (f \circ g)(x).$$

Solution 2.

Given : $f(x) = \dfrac{3x+4}{5x-7}, x \neq \dfrac{7}{5}$ and $g(x) = \dfrac{7x+4}{5x-3}, 3x \neq \dfrac{3}{5}$

Now, $(g \circ f)(x) = g\{f(x)\}$

$$= g\left(\frac{3x+4}{5x-7}\right)$$

$$= \frac{7\left(\frac{3x+4}{5x-7}\right)+4}{5\left(\frac{3x+4}{5x-7}\right)-3}$$

$$= \frac{\frac{21x+28+20x-28}{5x-7}}{\frac{15x+20-15x+21}{5x-7}}$$

$$= \frac{41x}{41} = x$$

Also, $(f \circ g)(x) = f\{g(x)\}$

$$= f\left(\frac{7x+4}{5x-3}\right)$$

$$= \frac{3\left(\frac{7x+4}{5x-3}\right)+4}{5\left(\frac{7x+4}{5x-3}\right)-7}$$

$$= \frac{\frac{21x+12+20x-12}{5x-3}}{\frac{35x+20-35x+21}{5x-3}} = \frac{41x}{41} = x$$

$\because \quad (g\,o\,f)(x) = (f\,o\,g)(x) = x$

Thus, $(g\,o\,f)(x) = (f\,o\,g)(x)$ **Hence Proved.**

Comments of Examiners

2. Most of the candidates solved this question correctly. However, in some cases, the correct concept was applied, but the simplification was not done correctly.

Answering Tips

- Adequate practice needs to be given on fundamental operations in algebraic expressions and mathematical simplification.

Question 3. [4]

(a) *If* $\cos^{-1}\frac{x}{2} + \cos^{-1}\frac{y}{3} = \theta$, *then prove that*

$9x^2 - 12xy\cos\theta + 4y^2 = 36\sin^2\theta$

OR

(b) *Evaluate :* $\cos(2\cos^{-1}x + \sin^{-1}x)$ *at* $x = \frac{1}{5}$.

Solution 3.

(a) Given : $\cos^{-1}\frac{x}{2} + \cos^{-1}\frac{y}{3} = \theta$

$$\Rightarrow \cos^{-1}\left[\frac{x}{2}\times\frac{y}{3} - \sqrt{1-\frac{x^2}{4}}.\sqrt{1-\frac{y^2}{9}}\right] = \theta$$

$$\left[\because \cos^{-1}a + \cos^{-1}b = \cos^{-1}\left(ab - \sqrt{1-a^2}.\sqrt{1-b^2}\right)\right]$$

$$\Rightarrow \frac{xy}{6} - \sqrt{1-\frac{x^2}{4}}.\sqrt{1-\frac{y^2}{9}} = \cos\theta$$

$$\Rightarrow \frac{xy}{6} - \cos\theta = \sqrt{1-\frac{x^2}{4}}.\sqrt{1-\frac{y^2}{9}}$$

Squaring both sides, we get

$$\Rightarrow \left(\frac{xy}{6} - \cos\theta\right)^2 = \left(1-\frac{x^2}{4}\right)\left(1-\frac{y^2}{9}\right)$$

$$\Rightarrow \frac{x^2y^2}{36} + \cos^2\theta - \frac{2xy}{6}\cos\theta = 1 - \frac{x^2}{4} - \frac{y^2}{9} + \frac{x^2y^2}{36}$$

$$\Rightarrow \frac{x^2}{4} + \frac{y^2}{9} - \frac{xy\cos\theta}{3} = 1 - \cos^2\theta$$

$$\Rightarrow \frac{9x^2 + 4y^2 - 12xy\cos\theta}{36} = \sin^2\theta$$

$$\Rightarrow \quad 9x^2 + 4y^2 - 12xy\cos\theta = 36\sin^2\theta$$

$$\Rightarrow \quad 9x^2 - 12xy\cos\theta + 4y^2 = 36\sin^2\theta$$

Hence Proved.

OR

(b) $\cos(2\cos^{-1}x + \sin^{-1}x) = \cos[\cos^{-1}x + \cos^{-1}x + \sin^{-1}x]$

$= \cos[\cos^{-1}x + (\cos^{-1}x + \sin^{-1}x)]$

$$= \cos\left[\cos^{-1}x + \frac{\pi}{2}\right]$$

$$= \cos\left(\frac{\pi}{2} + \cos^{-1}x\right)$$

$$= -\sin(\cos^{-1}x) \qquad \left[\because \cos\left(\frac{\pi}{2}+\theta\right) = -\sin\theta\right]$$

$$= -\sin\left(\sin^{-1}\sqrt{1-x^2}\right) \qquad \left[\because \cos^{-1}x = \sin^{-1}\sqrt{1-x^2}\right]$$

$$= -\sqrt{1-x^2}$$

$$= -\sqrt{1-\left(\frac{1}{5}\right)^2} \qquad \left[\because x = \frac{1}{5}\text{(Given)}\right]$$

$$= -\sqrt{1-\frac{1}{25}} = -\sqrt{\frac{25-1}{25}}$$

$$= -\sqrt{\frac{24}{25}} = -\frac{2\sqrt{6}}{5}$$

Ans.

Comments of Examiners

3. Maximum number of candidates applied the formula of inverse trigonometric functions correctly but could not simplify the same.

Answering Tips

- Practice the formula of all inverse trigonometric functions for sum/difference for two or more terms and give adequate practice in solving questions to get the required result.

Question 4. **[4]**

Using properties of determinants, show that

$$\begin{vmatrix} x & p & q \\ p & x & q \\ q & q & x \end{vmatrix} = (x-p)(x^2+px-2q^2)$$

Solution 4.

To prove :

$$\begin{vmatrix} x & p & q \\ p & x & q \\ q & q & x \end{vmatrix} = (x-p)(x^2+px-2q^2)$$

Let $\Delta = \text{LHS} = \begin{vmatrix} x & p & q \\ p & x & q \\ q & q & x \end{vmatrix}$

Applying $C_1 \to C_1 - C_2$, we get

$$\Delta = \begin{vmatrix} x-p & p & q \\ p-x & x & q \\ q-q & q & x \end{vmatrix}$$

Taking $(x-p)$ as common from C_1, we get

$$\Delta = (x-p)\begin{vmatrix} 1 & p & q \\ -1 & x & q \\ 0 & q & x \end{vmatrix}$$

Now, expanding along R_1, we get

$$\Delta = (x-p)[1(x^2-q^2)-p(-x-0)+q(-q-0)]$$
$$= (x-p)[x^2-q^2+px-q^2]$$
$$= (x-p)(x^2+px-2q^2) = \text{RHS}.$$

Hence Proved.

Comments of Examiners

4. Only some candidates attempted to solve this question by expanding the determinant without applying the properties. A few candidates applied the properties correctly but made mathematical simplification errors.

Answering Tips

- Should hold a clear understanding of the properties of determinants with the help of examples to avoid mathematical simplification errors.

Question 5. **[4]**

Verify Rolle's theorem for the function, $f(x) = -1 + \cos x$ in the interval $[0, 2\pi]$.

Solution 5.

Given : $f(x) = -1 + \cos x$ in $[0, 2\pi]$

1. Since, $\cos x$ is continuous in $[0, 2\pi]$, so, $f(x)$ is a continuous function in $[0, 2\pi]$.
2. $f(x)$ is differentiable on $(0, 2\pi)$
3. $\because$ $f(0) = -1 + \cos 0 = -1 + 1 = 0$

   and $f(2\pi) = -1 + \cos(2\pi)$
   $= -1 + \cos(2\pi - 0)$
   $= -1 + \cos 0$
   $= -1 + 1 = 0$

$\therefore$ $f(0) = f(2\pi)$

Thus, $f(x)$ satisfies all conditions of Rolle's theorem.

So, there exists a value of x, say $c \in (0, 2\pi)$ such that $f'(c) = 0$

Now, $f(x) = -1 + \cos x$

$\therefore$ $f'(x) = 0 - \sin x$

$\Rightarrow$ $f'(c) = 0 - \sin c$

$\because$ $f'(c) = 0$

$\therefore$ $\sin c = 0$

or $\sin c = 0$

$\therefore$ $c = 0, \pi, 2\pi, ...$

Thus, $c = \pi \in (0, 2\pi)$, such that $f'(c) = 0$

Hence, Rolle's theorem is verified. **Hence Proved.**

Comments of Examiners

5. Majority of the candidates committed errors while identifying open and closed intervals in the process of applying the properties of mean value theorems. They also made errors while finding the value of 'x' in the open interval when $f'(x) = 0$.

Answering Tips

- Understand the concept of open and closed intervals and their significance with the help of examples.
- A clear understanding of that continuity is in the closed interval [ ] and differentiability is in the open interval ( ).

Question 6. **[4]**

If $y = e^{m\sin^{-1}x}$, prove that

$$(1-x^2)\frac{d^2y}{dx^2} - x\frac{dy}{dx} = m^2y$$

Solution 6.

Given : $y = e^{m \sin^{-1} x}$

Differentiating w.r.t. x, we get

$$\frac{dy}{dx} = e^{m \sin^{-1} x} \frac{d}{dx} (m \sin^{-1} x)$$

$$\Rightarrow \quad \frac{dy}{dx} = y \times \frac{m}{\sqrt{1-x^2}} \quad \ldots(i)$$

or $$\sqrt{1-x^2} \frac{dy}{dx} = m y$$

Squaring both LHS and RHS,

$$(1-x^2)\left(\frac{dy}{dx}\right)^2 = m^2y^2$$

Differentiating w.r.t. x,

$$\Rightarrow \quad (1-x^2) \times 2\left(\frac{dy}{dx}\right)\left(\frac{d^2y}{dx^2}\right) + \left(\frac{dy}{dx}\right)^2 (-2x) = m^2 \sqrt{2y} \frac{dy}{dx}$$

$$\Rightarrow \quad (1-x^2) \frac{d^2y}{dx^2} + \frac{dy}{dx}(-x) = m^2y$$

$$\Rightarrow \quad (1-x^2) \frac{d^2y}{dx^2} - x \frac{dy}{dx} = m^2y$$ **Hence Proved.**

Comments of Examiners

6. Most of the candidates made errors while differentiating second time and further simplification.

Answering Tips

- Practice successive differentiation and their applications by solving numerable.
- Practice needs to be on how to solve a variety of problems based on successive differentiation.

Question 7. [4]

(a) *The equation of tangent at (2, 3) on the curve $y^2 = px^3 + q$ is $y = 4x - 7$. Find the values of 'p' and 'q'.*

OR

(b) *Using L' Hospital's rule, evaluate :*

$$\lim_{x \to 0} \frac{xe^x - \log(1+x)}{x^2}$$

Solution 7.

(a) Given : Equation of curve is,

$$y^2 = px^3 + q$$

Differentiating both sides w.r.t. x, we get

$$2y \frac{dy}{dx} = 3px^2$$

$$\Rightarrow \quad \frac{dy}{dx} = \frac{3px^2}{2y}$$

$$\therefore \quad \left(\frac{dy}{dx}\right)_{(2,3)} = \frac{3p \times 2^2}{2 \times 3}$$

or $$m = \left(\frac{dy}{dx}\right)_{(2,3)} = \frac{12p}{6} = 2p$$

Since, $y = 4x - 7$ is the tangent to the curve at point (2, 3).

So, on comparing with $y = mx + c$, we get

$$m = 4$$

Now, $2p = 4$

$$\Rightarrow \quad p = \frac{4}{2} = 2$$

Since, point (2, 3) lies on the curve,

$$\therefore \quad 3^2 = p \times 2^3 + q$$

$$\Rightarrow \quad 9 = 2 \times 8 + q \quad [\because p = 2]$$

$$\Rightarrow \quad 9 - 16 = q$$

$$\Rightarrow \quad q = -7$$

Hence, $p = 2$ and $q = -7$ **Ans.**

OR

(b) $$\lim_{x \to 0} \frac{xe^x - \log(1+x)}{x^2} \quad \left[\frac{0}{0} \text{form}\right]$$

Differentiating w.r.t. x, on numerator and denominator

$$= \lim_{x \to 0} \frac{1.e^x + xe^x - \frac{1}{1+x}}{2x} \quad \left[\frac{0}{0} \text{form}\right]$$

Again, differentiating w.r.t. x, on numerator and denominator,

$$= \lim_{x \to 0} \frac{e^x + 1e^x + xe^x + \frac{1}{(1+x)^2}}{2}$$

Using limits, $x = 0$

$$\frac{e^0 + e^0 + 0.e^0 + \frac{1}{(1+0)^2}}{2} = \frac{1+1+1}{2} = \frac{3}{2}$$ **Ans.**

Comments of Examiners

7. (a) Some candidates made mathematical simplification errors while finding the values of 'p' and 'q', though the application of the concept was correct.

(b) Only a few candidates made mistakes while applying L' Hospital's rule the second time. Simplification errors were once again made by some candidates.

Answering Tips

- Thorough Practice of geometrical interpretation of derivative of a function and the concept of tangent and normal.
- A clear understanding of the condition of perpendicular and parallelism with the help of different examples.
- Practice the concept and application of L' Hospital's rule.

Question 8. **[4]**

(a) *Evaluate :* $\int \frac{dx}{\sqrt{5x-4x^2}}$

OR

(b) *Evaluate :* $\int \sin^3 x \cos^4 x\, dx$

Solution 8.

(a) $\int \frac{dx}{\sqrt{5x-4x^2}} = \frac{1}{2}\int \frac{dx}{\sqrt{\frac{5}{4}x - x^2}}$

$$= \frac{1}{2}\int \frac{dx}{\sqrt{\frac{5}{4}x - x^2 + \left(\frac{5}{8}\right)^2 - \left(\frac{5}{8}\right)^2}}$$

$$= \frac{1}{2}\int \frac{dx}{\sqrt{\left(\frac{5}{8}\right)^2 - \left(x^2 - \frac{5}{4}x + \left(\frac{5}{8}\right)^2\right)}}$$

$$= \frac{1}{2}\int \frac{dx}{\sqrt{\left(\frac{5}{8}\right)^2 - \left(x - \frac{5}{8}\right)^2}}$$

$$= \frac{1}{2}\sin^{-1}\left(\frac{x - \frac{5}{8}}{\frac{5}{8}}\right) + c$$

$$\left[\because \int \frac{dx}{\sqrt{a^2 - x^2}} = \sin^{-1}\frac{x}{a} + c\right]$$

$$= \frac{1}{2}\sin^{-1}\left(\frac{8x-5}{5}\right) + c$$ **Ans.**

OR

(b) Let $I = \int \sin^3 x \cos^4 x\, dx$

$= \int \sin^2 x . \cos^4 x\, (\sin x\, dx)$

$= \int (1 - \cos^2 x) \cos^4 x\, (\sin x\, dx)$

Let $\cos x = t$

$\Rightarrow$ $-\sin x\, dx = dt$

or $\sin x\, dx = -dt$

$\therefore$ $I = \int (1 - t^2)\, t^4\, (-dt)$

$= -\int (t^4 - t^6)\, dt$

$= \int (t^6 - t^4)\, dt$

$= \frac{t^7}{7} - \frac{t^5}{5} + c$

$= \frac{\cos^7 x}{7} - \frac{\cos^5 x}{5} + c$ **Ans.**

Comments of Examiners

8. (a) Most of the candidates made mistakes while writing the Quadratic expression in the form of a difference of two perfect squares which lead to further simplification mistakes.
   (b) This question was not answered correctly by most candidates. They committed errors in integration by substitution and simplification of Trigonometric functions.

Answering Tips

- Practice on solving different types of special integrals.
- Practice the method of integration by substitution and give adequate practice in solving trigonometric functions by integration by substitution.

Question 9. **[4]**

Solve the differential equation

$$(1 + x^2)\frac{dy}{dx} = 4x^2 - 2xy$$

Solution 9.

Given differential equation is,

$$(1 + x^2)\frac{dy}{dx} = 4x^2 - 2xy$$

$$\Rightarrow \quad \frac{dy}{dx} = \frac{4x^2}{1+x^2} - \frac{2xy}{1+x^2}$$

or $\frac{dy}{dx} + \frac{2xy}{1+x^2} = \frac{4x^2}{1+x^2}$

This is of the form $\frac{dy}{dx} + Py = Q$

where, $P = \frac{2x}{1+x^2}$ and $Q = \frac{4x^2}{1+x^2}$

Thus, the solution of given differential equation is,

$$y \times \text{I.F.} = \int (Q \times \text{I.F.})\, dx + c$$

where, $\text{I.F.} = e^{\int P dx}$

$$= e^{\int \frac{2x}{1+x^2} dx}$$

$$= e^{\log(1+x^2)} \qquad \left[\because \int \frac{f'(x)}{f(x)} dx = \log[f(x)] + c\right]$$

$$= 1 + x^2 \qquad [\because e^{\log(a)} = a]$$

$\therefore \quad y \times (1+x^2) = \int \left[\frac{4x^2}{1+x^2} \times (1+x^2)\right] dx + c$

$\Rightarrow \quad (1+x^2)\, y = \int 4x^2\, dx + c$

$\Rightarrow \quad (1+x^2)\, y = \frac{4x^3}{3} + c$ **Ans.**

Comments of Examiners

9. Most of the candidates solved the differential equation correctly but made errors while integrating a term at the end of the solution. A few candidates did not write the constant 'C' at the end of the solution.
In a few cases, candidates were not able to identify the differential equation and incorrectly noted it as a homogeneous equation.

Answering Tips

- Give adequate practice in identifying and solving different types of differential equations.
- Understand the importance of a constant in the solution of a differential equation.

Question 10. [4]

Three persons A, B and C shoot to hit a target. Their probabilities of hitting the target are $\frac{5}{6}, \frac{4}{5}$ *and* $\frac{3}{4}$ *respectively. Find the probability that :*

(i) *Exactly two persons hit the target.*

(ii) *At least one person hits the target.*

Solution 10.

Given : $P(A) = \frac{5}{6}$, $P(B) = \frac{4}{5}$ and $P(C) = \frac{3}{4}$

So, $P(\bar{A}) = 1 - P(A) = 1 - \frac{5}{6} = \frac{1}{6}$

$P(\bar{B}) = 1 - P(B) = 1 - \frac{4}{5} = \frac{1}{5}$

$P(\bar{C}) = 1 - P(C) = 1 - \frac{3}{4} = \frac{1}{4}$

(i) Probability (exactly two persons hit the target)

$$= P(A) \cdot P(B) \cdot P(\bar{C}) + P(A) \cdot P(\bar{B}) \cdot P(C) + P(\bar{A}) \cdot P(B) \cdot P(C)$$

$$= \frac{5}{6} \times \frac{4}{5} \times \frac{1}{4} + \frac{5}{6} \times \frac{1}{5} \times \frac{3}{4} + \frac{1}{6} \times \frac{4}{5} \times \frac{3}{4}$$

$$= \frac{1}{6} + \frac{1}{8} + \frac{1}{10}$$

$$= \frac{20+15+12}{120} = \frac{47}{120}$$ **Ans.**

(ii) Probability (at least one person hits the target)

$= 1 -$ No one will hit the target

$$= 1 - P(\bar{A}\bar{B}\bar{C})$$

$$= 1 - \left(\frac{1}{6} \times \frac{1}{5} \times \frac{1}{4}\right)$$

$$= 1 - \frac{1}{120}$$

$$= \frac{120-1}{120}$$

$$= \frac{119}{120}$$ **Ans.**

Comments of Examiners

10. Many candidates could not find the probabilities correctly under the given conditions.

Answering Tips

- Clear understanding and practice of the concept related to independent events.

Question 11. [6]

Solve the following system of linear equations using matrices:

$$x - 2y = 10,\ 2x - y - z = 8,\ -2y + z = 7$$

Solution 11.

Given equations are,

$$x - 2y = 10$$
$$2x - y - z = 8$$
and
$$-2y + z = 7$$

The given equations can be arranged in matrix form as,

$$\begin{bmatrix} 1 & -2 & 0 \\ 2 & -1 & -1 \\ 0 & -2 & 1 \end{bmatrix}\begin{bmatrix} x \\ y \\ z \end{bmatrix} = \begin{bmatrix} 10 \\ 8 \\ 7 \end{bmatrix}$$

Let $A = \begin{bmatrix} 1 & -2 & 0 \\ 2 & -1 & -1 \\ 0 & -2 & 1 \end{bmatrix}$, $X = \begin{bmatrix} x \\ y \\ z \end{bmatrix}$ and $B = \begin{bmatrix} 10 \\ 8 \\ 7 \end{bmatrix}$

$\therefore$ $AX = B$ ...(i)

Now, $|A| = \begin{vmatrix} 1 & -2 & 0 \\ 2 & -1 & -1 \\ 0 & -2 & 1 \end{vmatrix}$

$$= 1(-1-2) + 2(2+0) + 0(-4+0)$$

$$= -3 + 4 = 1 \neq 0$$

So, A is invertible.

Now, the minors of the elements of $|A|$ are

$M_{11} = -3, M_{12} = 2, M_{13} = -4$
$M_{21} = -2, M_{22} = 1, M_{23} = -2$
$M_{31} = 2, M_{32} = -1, M_{33} = 3$

Co-factors of the elements of $|A|$ are

$A_{11} = -3, A_{12} = -2, A_{13} = -4$
$A_{21} = 2, A_{22} = 1, A_{23} = 2$
$A_{31} = 2, A_{32} = 1, A_{33} = 3$

We know, $\text{adj. } A = \begin{bmatrix} A_{11} & A_{12} & A_{13} \\ A_{21} & A_{22} & A_{23} \\ A_{31} & A_{32} & A_{33} \end{bmatrix}^T$

$\therefore$ $\text{adj. } A = \begin{bmatrix} -3 & -2 & -4 \\ 2 & 1 & 2 \\ 2 & 1 & 3 \end{bmatrix}^T$

$$= \begin{bmatrix} -3 & 2 & 2 \\ -2 & 1 & 1 \\ -4 & 2 & 3 \end{bmatrix}$$

$\therefore$ $A^{-1} = \frac{1}{|A|}(\text{adj. } A)$

$$= 1.\begin{bmatrix} -3 & 2 & 2 \\ -2 & 1 & 1 \\ -4 & 2 & 3 \end{bmatrix}$$

From (i), we have

$AX = B$

$\Rightarrow$ $A^{-1}AX = A^{-1}B$ [Multiplying A^{-1} on both sides]

$\Rightarrow$ $IX = A^{-1}B$ $[AA^{-1} = I]$

$\Rightarrow$ $X = A^{-1}B$

$\Rightarrow$ $\begin{bmatrix} x \\ y \\ z \end{bmatrix} = \begin{bmatrix} -3 & 2 & 2 \\ -2 & 1 & 1 \\ -4 & 2 & 3 \end{bmatrix}\begin{bmatrix} 10 \\ 8 \\ 7 \end{bmatrix}$

$\Rightarrow$ $\begin{bmatrix} x \\ y \\ x \end{bmatrix} = \begin{bmatrix} -30+16+14 \\ -20+8+7 \\ -40+16+21 \end{bmatrix}$

$\Rightarrow$ $\begin{bmatrix} x \\ y \\ z \end{bmatrix} = \begin{bmatrix} 0 \\ -5 \\ -3 \end{bmatrix}$

$\therefore$ $x = 0, y = -5$ and $z = -3$. **Ans.**

Comments of Examiners

11. Some of the candidates made simplification errors in the process of finding the values of variables x, y and z. A few candidates made errors while calculating cofactors and inverse of a matrix.

Answering Tips

- Given adequate practice in computing cofactors, the inverse of a given matrix and in finding the solution of a given system of equations in the matrix form.

Question 12. [6]

(a) *Show that the radius of a closed right circular cylinder of given surface area and maximum volume is equal to half of its height.*

OR

(b) *Prove that the area of right-angled triangle of given hypotenuse is maximum when the triangle is isosceles.*

Answer 12.

(a) Let S be the given surface area of a closed right circular cylinder whose radius is r and height is h.

$\therefore$ $S = 2\pi rh + 2\pi r^2$

$\Rightarrow$ $h = \frac{S - 2\pi r^2}{2\pi r}$ ...(i)

Now, volume of cylinder (V) $= \pi r^2 h$

$\therefore \quad V = \pi r^2 \left(\frac{S - 2\pi r^2}{2\pi r}\right)$ [From equation (i)]

$\Rightarrow \quad V = \frac{Sr - 2\pi r^3}{2}$

$\therefore \quad \frac{dV}{dr} = \frac{S - 6\pi r^2}{2}$ [differentiating w.r.t. r.] ...(ii)

and $\frac{d^2V}{dr^2} = \frac{-12\pi r}{2} = -6\pi r$ [Again diff. w.r.t. r.] ...(iii)

For maximum and minimum value,

Put $\frac{dV}{dr} = 0$ [from eq. (ii)]

$\Rightarrow \quad \frac{S - 6\pi r^2}{2} = 0$

$\Rightarrow \quad S = 6\pi r^2$

Putting the value of S in equation (i), we get

$$h = \frac{6\pi r^2 - 2\pi r^2}{2\pi r} = \frac{4\pi r^2}{2\pi r} = 2r$$

or $r = \frac{h}{2}$

and $\left[\frac{d^2V}{dr^2}\right]_{r=h/2} = -6\pi \times \frac{h}{2}$ [from eq. (iii)]

$= -3\pi h < 0$

[$\because$ h cannot be negative]

So, volume is maximum when $r = \frac{h}{2}$.

Hence Proved.

OR

(b) Let h be the hypotenuse of the right-angled triangle and x be its altitude.

So, base $= \sqrt{h^2 - x^2}$ [using Pythagoras theorem]

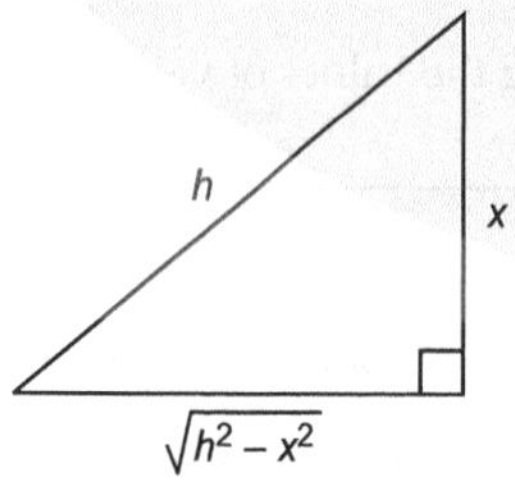

Now, area of triangle (A) $= \frac{1}{2} \times \sqrt{h^2 - x^2} \times x$

$$\therefore \quad \frac{dA}{dx} = \frac{1}{2}\left[x\left\{\frac{1}{2}(h^2 - x^2)^{-1/2} \times (-2x)\right\} + \sqrt{h^2 - x^2}\right]$$

$$= \frac{1}{2}\left[\frac{-x^2}{\sqrt{h^2 - x^2}} + \sqrt{h^2 - x^2}\right]$$

$$= \frac{1}{2}\left[\frac{-x^2 + h^2 - x^2}{\sqrt{h^2 - x^2}}\right]$$

$$= \frac{1}{2}\left[\frac{h^2 - 2x^2}{\sqrt{h^2 - x^2}}\right]$$

and
$$\frac{d^2A}{dx^2} = \frac{1}{2}\left[\frac{\sqrt{h^2 - x^2}.(-4x) - (h^2 - 2x^2)\left\{\frac{1}{2}(h^2 - x^2)^{-1/2}(-2x)\right\}}{\left(\sqrt{h^2 - x^2}\right)^2}\right]$$

$$= \frac{1}{2}\left[\frac{-4x\left(\sqrt{h^2 - x^2}\right) + \frac{(h^2 - 2x^2)x}{\sqrt{h^2 - x^2}}}{(h^2 - x^2)}\right]$$

$$= \frac{1}{2}\left[\frac{-4x(h^2 - x^2) + (h^2 - 2x^2)x}{(h^2 - x^2).\sqrt{h^2 - x^2}}\right]$$

$$= \frac{1}{2}\left[\frac{-4xh^2 + 4x^3 + h^2x - 2x^3}{(h^2 - x^2)^{3/2}}\right]$$

$$= \frac{1}{2}\left[\frac{2x^3 - 3xh^2}{(h^2 - x^2)^{3/2}}\right]$$

For maximum and minimum value,

Put $\frac{dA}{dx} = 0$

$\Rightarrow \quad \frac{1}{2}\left[\frac{h^2 - 2x^2}{\sqrt{h^2 - x^2}}\right] = 0$

$\Rightarrow \quad h^2 - 2x^2 = 0$

$\Rightarrow \quad x = \frac{h}{\sqrt{2}}$

$$\therefore \quad \left(\frac{d^2A}{dx^2}\right)_{x=\frac{h}{\sqrt{2}}} = \frac{1}{2}\left[\frac{2 \times \frac{h^3}{2\sqrt{2}} - 3 \times \frac{h}{\sqrt{2}} \times h^2}{\left(h^2 - \frac{h^2}{2}\right)^{3/2}}\right]$$

$$= \frac{1}{2}\left[\frac{h^3 - 3h^3}{\left(\frac{h^2}{2}\right)^{3/2}\sqrt{2}}\right]$$

$$= \frac{-2h^3}{2\times\frac{h^3}{2\sqrt{2}}\times\sqrt{2}}$$

$$= -2 < 0$$

Thus, A is maximum at $x = \frac{h}{\sqrt{2}}$

Now, $\quad \text{Base} = \sqrt{h^2 - x^2}$

$$= \sqrt{h^2 - \frac{h^2}{2}} = \sqrt{\frac{h^2}{2}} = \frac{h}{\sqrt{2}}$$

and $\quad \text{Altitude} = x = \frac{h}{\sqrt{2}}$

$$\text{Base} = \sqrt{h^2 - x^2}$$

$$= \sqrt{h^2 - \left(\frac{h}{\sqrt{2}}\right)^2} = \frac{h}{\sqrt{2}}$$

Since, $\quad \text{Base} = \text{Altitude} = \frac{h}{\sqrt{2}}$

Hence, the triangle is isosceles. **Hence Proved.**

Comments of Examiners

12. (a) Several candidates did not apply the correct formula for the surface area of a closed right circular cylinder to express the volume of the cylinder as a function to be maximised and get the same in terms of one variable. Some could not complete the second-order derivative test for maximisation/ minimisation. A few candidates found the value of the variable but could not prove that the radius of the cylinder is equal to half of the height of the cylinder.

(b) Many candidates made errors while equating the first derivative to zero and could not complete the solution. Some could not complete the second-order derivative test for maximisation/ minimisation. A few candidates found the value of the variable but could not prove that the triangle is isosceles.

Answering Tips

- Revision of mensuration formulae of 2-dimensional and 3-dimensional figures/objects are always needed.
- Understand the method of identifying the objective function, re-writing in terms of one variable and applying the concept of maxima/minima (1st derivative test and 2nd derivative test).
- Proper clarification of the method of forming a function in terms of one.

Question 13. **[6]**

(a) *Evaluate :*

$$\int tan^{-1}\sqrt{\frac{1-x}{1+x}}dx$$

OR

(b) *Evaluate :* $\int\frac{2x+7}{x^2-x-2}dx$

Answer 13.

(a) Let $I = \int \tan^{-1}\sqrt{\frac{1-x}{1+x}}dx$

Let $\quad x = \cos t$

$\Rightarrow \quad dx = -\sin t\, dt$

$\therefore \quad I = \int \tan^{-1}\sqrt{\frac{1-\cos t}{1+\cos t}}\,(-\sin t\, dt)$

$$= \int \tan^{-1}\sqrt{\frac{2\sin^2\frac{t}{2}}{2\cos^2\frac{t}{2}}}\,(-\sin t)\, dt$$

$$= \int \tan^{-1}\left(\tan\frac{t}{2}\right)(-\sin t)\, dt$$

$$= -\int\frac{t}{2}\sin t\, dt$$

$$= -\frac{1}{2}\int t\sin t\, dt$$

$$= -\frac{1}{2}\left[t(-\cos t) - \int 1.(-\cos t)\, dt\right]$$

[Integrating by parts]

$$= \frac{1}{2}t\cos t - \frac{1}{2}\sin t + c$$

$$= \frac{1}{2}t\cos t - \frac{1}{2}\sqrt{1-\cos^2 t} + c$$

$$= \frac{1}{2}x\cos^{-1}x - \frac{1}{2}\sqrt{1-x^2} + c \qquad \textbf{Ans.}$$

OR

(b) $\int\frac{2x+7}{x^2-x-2}dx = \int\frac{2x+7}{x^2-2x+x-2}dx$

$$= \int\frac{2x+7}{(x-2)(x+1)}dx$$

Let $\frac{2x+7}{(x-2)(x+1)} = \frac{A}{x-2} + \frac{B}{x+1}$

$$\Rightarrow \quad \frac{2x+7}{(x-2)(x+1)} = \frac{A(x+1)+B(x-2)}{(x-2)(x+1)}$$

$$\Rightarrow \quad 2x+7 = (A+B)x + A - 2B$$

Comparing both sides, we get

$$A+B = 2 \text{ and } A-2B = 7$$

Solving these equations for A and B, we get

$$A = \frac{11}{3} \text{ and } B = -\frac{5}{3}$$

Now, $\int \frac{2x+7}{(x-2)(x+1)}dx = \frac{11}{3}\int \frac{1}{(x-2)}dx - \frac{5}{3}\int \frac{1}{x+1}dx$

$$= \frac{11}{3}\log|x-2| - \frac{5}{3}\log|x+1| + c$$ **Ans.**

Comments of Examiners

13. (a) Most candidates attempted this problem to the extent of simplifying the inverse trigonometric function by taking correct substitution but made errors while applying the concept of integration by parts to integrate the function.

(b) Majority of the candidates applied the concept of standard integrals in the right manner but made simplification errors.

Answering Tips

- Comprehensively understanding the concepts of Integration by substitution and integration by parts.
- Clarify the concept of integration by a partial fraction to the students thoroughly providing a variety of examples.

Question 14. **[6]**

The probability that a bulb produced in a factory will fuse after 150 days of use is 0·05.

Find the probability that out of 5 such bulbs:

(i) *None will fuse after 150 days of use.*

(ii) *Not more than one will fuse after 150 days of use.*

(iii) *More than one will fuse after 150 days of use.*

(iv) *At least one will fuse after 150 days of use.*

Solution 14.

Let p be the probability that the bulb will fuse after 150 days of use.

[Since it is a case of binomial distribution]

So, $\quad p = 0.05 = \frac{5}{100} = \frac{1}{20}$

$\therefore \quad q = 1 - p = 1 - \frac{1}{20} = \frac{19}{20}$

and $\quad n = 5$

We know, $\quad P(x = r) = {}^nC_r p^r q^{n-r}$

(i) $P(x=0) = {}^5C_0\left(\frac{1}{20}\right)^0\left(\frac{19}{20}\right)^5$

$$= 1 \times 1 \times \left(\frac{19}{20}\right)^5 = \left(\frac{19}{20}\right)^5$$ **Ans.**

(ii) $P(x \le 1) = P(x=0) + P(x=1)$

$$= \left(\frac{19}{20}\right)^5 + {}^5C_1\left(\frac{1}{20}\right)^1\left(\frac{19}{20}\right)^4$$

[Putting value of P from (i)]

$$= \left(\frac{19}{20}\right)^5 + 5 \times \frac{1}{20} \times \left(\frac{19}{20}\right)^4$$

$$= \left(\frac{19}{20}\right)^4\left(\frac{19}{20} + \frac{1}{4}\right)$$

$$= \left(\frac{19}{20}\right)^4\left(\frac{19+5}{20}\right)$$

$$= \left(\frac{19}{20}\right)^4 \frac{24}{20}$$

$$= \frac{6}{5}\left(\frac{19}{20}\right)^4$$ **Ans.**

(iii) $P(x > 1) = 1 - P(x \le 1)$

$$= 1 - \frac{6}{5}\left(\frac{19}{20}\right)^4$$ **Ans.**

(iv) $P(x \ge 1) = 1 - P(x = 0)$

$$= 1 - \left(\frac{19}{20}\right)^5$$

[Putting value of P from (i)]

$$= 1 - \left(\frac{19}{20}\right)^5$$ **Ans.**

Comments of Examiners

14. Most candidates who attempted this question committed simplification errors while finding the probability values of success and failure. These incorrect values led the incorrect solutions. Some candidates made errors while calculating the probability in the case of bulbs, not more than one will fuse after 150 days of use.

Answering Tips

- Understand the concepts of probability distribution and their properties.
- Ample practice needs to be on different types of probability distribution problems.

SECTION – B (20 Marks)

Question 15. **[3×2]**

(a) *Write a vector of magnitude of 18 units in the direction of the vector* $\hat{i}-2\hat{j}-2\hat{k}$.

(b) *Find the angle between the two lines :*

$$\frac{x+1}{2}=\frac{y-2}{5}=\frac{z+3}{4} \text{ and } \frac{x-1}{5}=\frac{y+2}{2}=\frac{z-1}{-5}$$

(c) *Find the equation of the plane passing through the point (2, –3, 1) and perpendicular to the line joining the points (4, 5, 0) and (1, –2, 4).*

Solution 15.

(a) Let $\vec{a} = \hat{i}-2\hat{j}-2\hat{k}$

Then, $|\vec{a}| = \sqrt{1^2+(-2)^2+(-2)^2}$

$= \sqrt{1+4+4} = \sqrt{9} = 3$

$\therefore$ Unit vector, $\hat{a} = \frac{\vec{a}}{|\vec{a}|}$

$= \frac{1}{3}\hat{i} - \frac{2}{3}\hat{j} - \frac{2}{3}\hat{k}$

Hence, the required vector of magnitude 18 units is $18\hat{a}$.

$\therefore$ Required vector $= \frac{18}{3}\hat{i}-\frac{36}{3}\hat{j}-\frac{36}{3}\hat{k}$

$= 6\hat{i}-12\hat{j}-12\hat{k}$ **Ans.**

(b) Given lines are,

$$\frac{x+1}{2}=\frac{y-2}{5}=\frac{z+3}{4} \quad \text{...(i)}$$

and

$$\frac{x-1}{5}=\frac{y+2}{2}=\frac{z-1}{-5} \quad \text{...(ii)}$$

$\therefore$ Direction ratios of the (i), $\langle a_1, b_1, c_1\rangle = \langle 2, 5, 4\rangle$ and, direction ratios of line (ii), $\langle a_2, b_2, c_2\rangle = \langle 5, 2, -5\rangle$.

The angle between two lines is given as,

$$\cos\theta = \frac{a_1a_2+b_1b_2+c_1c_2}{\sqrt{a_1^2+b_1^2+c_1^2}.\sqrt{a_2^2+b_2^2+c_2^2}}$$

$$= \frac{2\times5+5\times2+4\times(-5)}{\sqrt{2^2+5^2+4^2}.\sqrt{5^2+2^2+(-5)^2}}$$

$$= \frac{10+10-20}{\sqrt{4+25+16}.\sqrt{25+4+25}}$$

$$= \frac{0}{\sqrt{45}.\sqrt{54}}$$

$\Rightarrow \cos\theta = 0$

$\therefore \theta = \frac{\pi}{2} = 90°$

Hence, the angle between the two lines is $\frac{\pi}{2}$. **Ans.**

(c) The general equation of a plane is,

$$a(x-x_1)+b(y-y_1)+c(z-z_1)=0$$

So, the equation of the plane passing through (2, – 3, 1) is,

$$a(x-2)+b(y+3)+c(z-1)=0 \quad \text{...(i)}$$

The direction ratios of the line joining the points (4, 5, 0) and (1, –2, 4) are (1 – 4, –2 – 5, 4 – 0) *i.e.,* (–3, – 7, 4).

Since, the plane is perpendicular to the line whose direction ratios are – 3, – 7 and 4.

So, direction ratios of the normal to the plane are – 3, –7 and 4.

Putting these values in equation (i), we get

$$-3(x-2)+(-7)(y+3)+4(z-1)=0$$

$$-3x+6-7y-21+4z-4=0$$

$\Rightarrow$ $$-3x-7y+4z-19=0$$

or $$3x+7y-4z+19=0$$

which is the required equation of plane. **Ans.**

Comments of Examiners

15. (a) Majority of the candidates answered this question correctly. However, a few candidates applied the formula of a unit vector for the given vector incorrectly.
 (b) In some cases, candidates applied the formula for the angle between the two lines correctly but committed errors in the process of simplification.
 (c) Most candidates attempted this part well, barring a few who applied the correct concept but made simplification errors at the end.

Answering Tips

- Practice the concept of unit vector and magnitude of a vector, dot, and cross product of two vectors and their properties with examples.
- Thoroughly understand the concept of a plane and direction ratios normal to the plane and finding the equation of a plane satisfying different conditions.
- Give adequate practice in solving problems based on the above concepts.

Question 16. [4]

(a) *Prove that*

$$\vec{a}.[(\vec{b}+\vec{c})\times(\vec{a}+3\vec{b}+4\vec{c})]=[\vec{a}\ \vec{b}\ \vec{c}]$$

OR

(b) *Using vectors, find the area of the triangle whose vertices are:*

A (3, –1, 2), B (1, –1, –3) and C (4, –3, 1)

Solution 16.

(a) Consider,

$$\text{L.H.S.} = \vec{a}.[(\vec{b}+\vec{c})\times(\vec{a}+3\vec{b}+4\vec{c})]$$

$$= \vec{a}.[\vec{b}\times\vec{a}+3(\vec{b}\times\vec{b})+4(\vec{b}\times\vec{c})+\vec{c}\times\vec{a}+3(\vec{c}\times\vec{b})+4(\vec{c}\times\vec{c})]$$

$$= \vec{a}.[\vec{b}\times\vec{a}+0+4(\vec{b}\times\vec{c})+\vec{c}\times\vec{a}+3(\vec{c}\times\vec{b})+0]$$

$$= \vec{a}.(\vec{b}\times\vec{a})+4[\vec{a}.(\vec{b}\times\vec{c})]+\vec{a}.(\vec{c}\times\vec{a})+3[\vec{a}.(\vec{c}\times\vec{b})]$$

$$= 0+4[\vec{a}\ \vec{b}\ \vec{c}]+0+3[\vec{a}\ \vec{c}\ \vec{b}]$$

$$= 4[\vec{a}\ \vec{b}\ \vec{c}]-3[\vec{a}\ \vec{b}\ \vec{c}]$$

$$= [\vec{a}\ \vec{b}\ \vec{c}] = \text{R.H.S.}$$ **Hence Proved.**

OR

(b) Given : Vertices of ΔABC are A (3, –1, 2), B (1, –1, –3) and C (4, –3, 1).

A(3, –1, 2)

B(1, –1, –3) C(4, –3, 1)

So, Position vector of A $= (3\hat{i}-\hat{j}+2\hat{k})$

Position vector of B $= (\hat{i}-\hat{j}-3\hat{k})$

Position vector of C $= (4\hat{i}-3\hat{j}+\hat{k})$

$\therefore$ $\vec{AB}$ = (p.v. of B) – (p.v. of A)

$= (\hat{i}-\hat{j}-3\hat{k})-(3\hat{i}-\hat{j}+2\hat{k})$

$= -2\hat{i}+0-5\hat{k}$

and $\vec{AC}$ = (p.v. of C) – (p.v. of A)

$= (4\hat{i}-3\hat{j}+\hat{k})-(3\hat{i}-\hat{j}+2\hat{k})$

$= \hat{i}-2\hat{j}-\hat{k}$

We know, area of $\Delta ABC = \frac{1}{2}|\vec{AB}\times\vec{AC}|$

Now, $\vec{AB}\times\vec{AC} = \begin{vmatrix} \hat{i} & \hat{j} & \hat{k} \\ -2 & 0 & -5 \\ 1 & -2 & -1 \end{vmatrix}$

$= (0-10)\hat{i}-(2+5)\hat{j}+(4-0)\hat{k}$

$= -10\hat{i}-7\hat{j}+4\hat{k}$

$\therefore$ $|\vec{AB}\times\vec{AC}| = \sqrt{(-10)^2+(-7)^2+(4)^2}$

$= \sqrt{100+49+16}$

$= \sqrt{165}$

So, Area of $\Delta ABC = \frac{1}{2}|\vec{AB}\times\vec{AC}|$

$= \frac{1}{2}\times\sqrt{165}$

$= \frac{\sqrt{165}}{2}$ sq. units **Ans.**

Comments of Examiners

16. (a) A few candidates made simplification errors in applying dot and cross product while expanding the left-hand side of the equation.
(b) Many candidates made errors in the process of finding any two sides of the triangle and some applied incorrect formula to find the area of the triangle.

Answering Tips

- Comprehensively understanding of the area of a triangle by using position vectors and applications based on the concept.
- Give ample practice in solving problems to avoid simplification errors to the maximum extent.

Question 17. **[4]**

(a) *Find the image of the point (3, –2, 1) in the plane $3x - y + 4z = 2$*

OR

(b) *Determine the equation of the line passing through the point (–1, 3, –2) and perpendicular to the lines :*

$$\frac{x}{1} = \frac{y}{2} = \frac{z}{3} \text{ and } \frac{x+2}{-3} = \frac{y-1}{2} = \frac{z+1}{5}$$

Solution 17.

(a) Given: Equation of plane is,

$$3x - y + 4z = 2 \quad \text{...(i)}$$

Let Q (x_1, y_1, z_1) be the image of the point P (3, –2, 1) in the given plane.

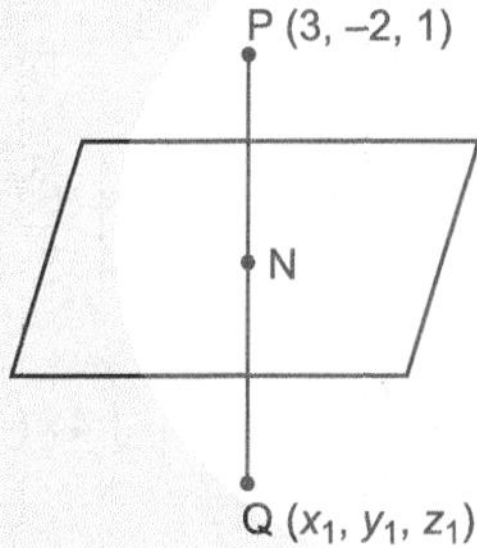

The equation of a line passing through (3, –2, 1) and perpendicular to the given plane is,

$$\frac{x-3}{3} = \frac{y+2}{-1} = \frac{z-1}{4} = k \text{ (say)}$$

So, the coordinates of a general point on this line are $(3k + 3, -k - 2, 4k + 1)$.

If N is the foot of the perpendicular from P to the given plane, then N lies on the plane.

Let the coordinates of N be $(3k + 3, -k - 2, 4k + 1)$.

$\therefore 3(3k + 3) - (-k - 2) + 4(4k + 1) - 2 = 0$ [Putting N in eq. (i)]

$\Rightarrow \quad 9k + 9 + k + 2 + 16k + 4 - 2 = 0$

$\Rightarrow \quad 26k + 13 = 0$

$\Rightarrow \quad k = -\frac{13}{26}$

$= -\frac{1}{2}$

So, coordinates of N $= \left[3\times\left(-\frac{1}{2}\right)+3, -\left(-\frac{1}{2}\right)-2, 4\times\left(-\frac{1}{2}\right)+1\right]$

$= \left(\frac{3}{2}, -\frac{3}{2}, -1\right)$

Since, N is the mid-point of PQ,

$\therefore$ By mid-point formula,

$\frac{3+x_1}{2} = \frac{3}{2}$

$\Rightarrow \quad x_1 = 3 - 3 = 0$

Also, $\frac{-2+y_1}{2} = -\frac{3}{2}$

$\Rightarrow \quad y_1 = -3 + 2 = -1$

and $\frac{1+z_1}{2} = -1$

$\Rightarrow \quad 1 + z_1 = -2$

$\Rightarrow \quad z_1 = -2 - 1 = -3$

Hence, the required image of the point P is Q(0, –1, –3). **Ans.**

OR

(b) Let the equation of required line be,

$$\frac{x-x_1}{a} = \frac{y-y_1}{b} = \frac{z-z_1}{c}$$

$\because$ The line passes through (–1, 3, –2).

$$\therefore \quad \frac{x+1}{a} = \frac{y-3}{b} = \frac{z+2}{c} \quad \text{...(i)}$$

Now, given lines are

$$\frac{x}{1} = \frac{y}{2} = \frac{z}{3}$$

and $$\frac{x+2}{-3} = \frac{y-1}{2} = \frac{z+1}{5}$$

Since, the required line is perpendicular to these two lines

$\therefore \quad a \times 1 + b \times 2 + 3 \times c = 0$

$\Rightarrow \quad a + 2b + 3c = 0 \quad \text{...(ii)}$

and $a \times (-3) + b \times 2 + c \times 5 = 0$

$\Rightarrow \quad -3a + 2b + 5c = 0 \quad \text{...(iii)}$

Solving equations (ii) and (iii) using cross-multiplication method,

$$\frac{a}{10-6} = \frac{b}{-9-5} = \frac{c}{2+6}$$

$$\Rightarrow \quad \frac{a}{4} = \frac{b}{-14} = \frac{c}{8} = k \text{ (say)}$$

$$\Rightarrow \quad a = 4k, b = -14k \text{ and } c = 8k.$$

Putting these values in equation (i), we get

$$\frac{x+1}{4k} = \frac{y-3}{-14k} = \frac{z+2}{8k}$$

$$\Rightarrow \quad \frac{x+1}{2} = \frac{y-3}{-7} = \frac{z+2}{4},$$

which is the required equation of line. **Ans.**

Comments of Examiners

17. (a) Many candidates found the incorrect value of the image of the point in the given plane. A few candidates applied irrelevant method to solve the problem.

(b) Most of the candidates did not apply the concept of perpendicular lines and the condition for the same correctly. Some candidates made calculation errors while applying the cross-multiplication rule to find the direction ratios.

Answering Tips

- Stress on concept building by giving adequate problems on three-dimensional geometry number for practice.
- Show students the different methods of finding the equation of a straight line by applying different types of conditions.

Question 18. **[6]**

Draw a rough sketch of the curves $y^2 = x$ and $y^2 = 4 - 3x$ and find the area enclosed between them.

Solution 18.

Given equation of curves are,

$$y^2 = x \quad \text{...(i)}$$

and

$$y^2 = 4 - 3x \quad \text{...(ii)}$$

These are the equations of parabola.

Solving equations (i) and (ii), we get

$$x = 4 - 3x$$

$$\Rightarrow \quad 4x = 4$$

$$\Rightarrow \quad x = 1$$

$$\therefore \quad y^2 = 1$$

$$\Rightarrow \quad y = \pm 1$$

So, (1, 1) and (1, –1) are the points of intersection of the two parabolas.

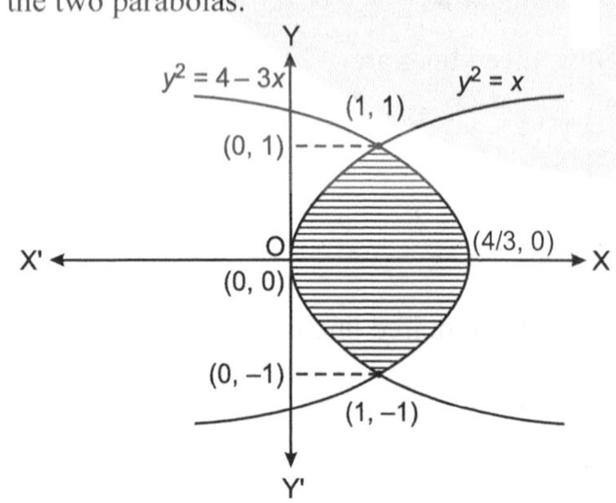

$$\therefore \text{ Required area} = \int_{-1}^{1} x_{II}\, dy - \int_{-1}^{1} x_I dy$$

$$= \int_{-1}^{1} \frac{4-y^2}{3} dy - \int_{-1}^{1} y^2 dy \qquad \left[\begin{array}{l} x_{II} = \frac{4-y^2}{3} \\ x_I = y^2 \end{array}\right]$$

$$= \frac{1}{3}\left[4y - \frac{y^3}{3}\right]_{-1}^{1} - \left[\frac{y^3}{3}\right]_{-1}^{1}$$

$$= \frac{1}{3}\left[\left(4\times1 - \frac{1^3}{3}\right) - \left(4\times(-1) - \frac{(-1)^3}{3}\right)\right] - \left[\frac{1^3}{3} - \frac{(-1)^3}{3}\right]$$

$$= \frac{1}{3}\left[4 - \frac{1}{3} + 4 - \frac{1}{3}\right] - \left[\frac{1}{3} + \frac{1}{3}\right]$$

$$= \frac{1}{3}\left[8 - \frac{2}{3}\right] - \frac{2}{3}$$

$$= \frac{1}{3} \times \frac{22}{3} - \frac{2}{3}$$

$$= \frac{22}{9} - \frac{2}{3}$$

$$= \frac{22-6}{9}$$

$$= \frac{16}{9} \text{ sq. units.}$$ **Ans.**

Comments of Examiners

18. Only a few candidates attempted this question well. Many candidates were not able to associate the equations of the curves as a parabola and made mistakes while sketching the graph. The concept of symmetry was not used by a few candidates and hence, they could not find the common portion and limits correctly.

Answering Tips

- Give sufficient practice in sketching various types of curves and interpretation of various graphs.
- Practice the method of identifying the limits of a definite integral and the area bounded by the region for the given curves and X-axis, etc.

SECTION – C (20 Marks)

Question 19. [3 × 2]

(a) *The selling price of a commodity is fixed at ₹ 60 and its cost function is C (x) = 35x + 250*

(i) *Determine the profit function.*

(ii) *Find the break even points.*

(b) *The revenue function is given by R(x) = 100x – x² – x³. Find*

(i) *The demand function.*

(ii) *Marginal revenue function.*

(c) *For the lines of regression 4x – 2y = 4 and 2x – 3y + 6 = 0, find the mean of 'x' and the mean of 'y'.*

Solution 19.

(a) Given : $C(x) = 35x + 250$

and $S(x) = 60x$

(i) $P(x) = S(x) - C(x)$

$= 60x - 35x - 250$

$= 25x - 250$ **Ans.**

(ii) For break even points,

$P(x) = 0$

$\Rightarrow 25x - 250 = 0$ [Putting P(x) from (i)]

$\Rightarrow x = \dfrac{250}{25} = 10$

$\therefore x = 10$ **Ans.**

(b) Given : $R(x) = 100x - x^2 - x^3$

(i) We know, $R(x) = P(x) \times x$

$\therefore P(x) = \dfrac{R(x)}{x}$

$= \dfrac{100x - x^2 - x^3}{x}$

$= 100 - x - x^2$ **Ans.**

(ii) We know M.R. $= \dfrac{dR}{dx}$

$= \dfrac{d}{dx}(100x - x^2 - x^3)$

$= 100 - 2x - 3x^2$ **Ans.**

(c) Given : Lines of regression are,

$4x - 2y = 4$ ...(i)

and $2x - 3y + 6 = 0$

$\Rightarrow 2x - 3y = -6$ ...(ii)

Multiplying equation (ii) by 2 and then subtracting equation (i) from it,

$$\begin{array}{r} 4x - 6y = -12 \\ 4x - 2y = 4 \\ - \quad + \quad - \\ \hline -4y = -16 \end{array}$$

$\Rightarrow y = \dfrac{16}{4} = 4$

Putting $y = 4$ in equation (i), we get

$4x - 2 \times 4 = 4$

$\Rightarrow 4x = 4 + 8 = 12$

$\Rightarrow x = \dfrac{12}{4} = 3$

$\therefore \bar{x} = 3$ and $\bar{y} = 4$ **Ans.**

Comments of Examiners

19. (a) This question was well attempted by most candidates. A few candidates, however, made calculation errors in simplification.

(b) Most of the candidates attempted this part correctly, but a few candidates made mistakes while calculating the demand function and marginal revenue function.

(c) Most of the candidates solved this question by correctly identifying the mean of variables 'x' and 'y'. In some cases, candidates applied the concept of solving regression equations correctly but again made simplification errors.

Answering Tips

- Understand all technical terms and formulae of the chapter on cost function along with examples.
- Practice the concept of the mean of x and y by solving the two equations simultaneously.
- Try to maintain the highest level of accuracy while solving problems from this topic.

Question 20. **[4]**

(a) *The correlation coefficient between x and y is 0.6. If the variance of x is 225, the variance of y is 400, mean of x is 10 and mean of y is 20, find*

(i) *the equations of two regression lines.*

(ii) *the expected value of y when x = 2.*

OR

(b) *Find the regression coefficients* b_{yx}, b_{xy} *and correlation coefficient 'r' for the following data :*

(2, 8), (6, 8), (4, 5), (7, 6), (5, 2).

Solution 20.

(a) Given : $r = 0{\cdot}6$, $\bar{x} = 10$, $\bar{y} = 20$, var. $(x) = (\sigma_x^2)$ $= 225$ and var. $(y) = \sigma_y^2 = 400$.

(i) $\sigma_x = \sqrt{225} = 15$ and $\sigma_y = \sqrt{400} = 20$

$$b_{yx} = r\frac{\sigma_y}{\sigma_x} = 0{\cdot}6 \times \frac{20}{15} = \frac{4}{5}$$

and $$b_{xy} = r\frac{\sigma_x}{\sigma_y} = 0{\cdot}6 \times \frac{15}{20} = \frac{9}{20}$$

So, line of regression of y on x is

$$y - \bar{y} = b_{yx}(x - \bar{x})$$

$$\Rightarrow \quad y - 20 = \frac{4}{5}(x - 10)$$

$$\Rightarrow \quad 5y - 100 = 4x - 40$$

$$\Rightarrow \quad 4x - 5y + 60 = 0 \qquad \text{...(i)}$$

Line of regression of x on y is

$$x - \bar{x} = b_{xy}(y - \bar{y})$$

$$\Rightarrow \quad x - 10 = \frac{9}{20}(y - 20)$$

$$\Rightarrow \quad 20x - 200 = 9y - 180$$

$$\Rightarrow \quad 20x - 9y - 20 = 0$$ **Ans.**

(ii) At $x = 2$

$$4 \times 2 - 5y + 60 = 0 \qquad \text{[Putting } x = 2 \text{ in eq. (i)]}$$

$$\Rightarrow \quad 5y = 68$$

$$\therefore \quad y = \frac{68}{5}$$ **Ans.**

OR

(b)

x	x^2	y	y^2	xy
2	4	8	64	16
6	36	8	64	48
4	16	5	25	20
7	49	6	36	42
5	25	2	4	10
$\Sigma x = 24$	$\Sigma x^2 = 130$	$\Sigma y = 29$	$\Sigma y^2 = 193$	$\Sigma xy = 136$

$$b_{yx} = \frac{\Sigma xy - \frac{1}{n}\Sigma x \Sigma y}{\Sigma x^2 - \frac{1}{n}(\Sigma x)^2}$$

$$= \frac{136 - \frac{1}{5}(24)(29)}{130 - \frac{1}{5}(24)^2}$$

$$= \frac{\frac{136 \times 5 - 696}{5}}{\frac{130 \times 5 - 576}{5}}$$

$$= \frac{680 - 696}{650 - 576}$$

$$= \frac{-16}{74} = -0.22$$

$$b_{xy} = \frac{\Sigma xy - \frac{1}{n}\Sigma x \Sigma y}{\Sigma y^2 - \frac{1}{n}(\Sigma y)^2}$$

$$= \frac{136 - \frac{1}{5}(24)(29)}{193 - \frac{1}{5}(29)^2}$$

$$= \frac{136 \times 5 - 696}{193 \times 5 - 841}$$

$$= \frac{680 - 696}{965 - 841}$$

$$= \frac{-16}{124} = -0.13$$

$$r = \pm\sqrt{b_{yx}\, b_{xy}} = \pm\sqrt{(-0.22)(-0.13)}$$

$$= \pm\sqrt{0.0286} = \pm 0.17$$

Since, r has same sign as regression coefficients

$\therefore$ $r(x, y) = -0.17$

$\therefore$ $b_{yx} = -0.22$, $b_{xy} = -0.13$

and $r = -0.17$ **Ans.**

Comments of Examiners

20. (a) Many candidates applied the incorrect formulae for calculation of byx and bxy, thereby made errors in finding the regression equations which led to the incorrect expected value of 'y' when $x = 2$.

(b) Many candidates wrote incorrect formulas for regression coefficients which caused errors in finding the regression coefficients hence led to the incorrect correlation coefficient.

Answering Tips

- Learn The formulae for calculating regression coefficients *byx* and *bxy* correctly.
- Practice multiple problems based on regression.

Question 21. **[4]**

(a) *The marginal cost of the production of the commodity is 30 + 2x, it is known that fixed costs are ₹ 200, find*

(i) *The total cost.*

(ii) *The cost of increasing output from 100 to 200 units.*

OR

(b) *The total cost function of a firm is given by* $C(x) = \frac{1}{3}x^3 - 5x^2 + 30x - 15$ *where the selling price per unit is given as ₹ 6. Find for what value of x will the profit be maximum.*

Solution 21.

(a) Given : M.C. = 30 + 2x

(i) We know, $\text{M.C.} = \frac{dC}{dx}$

$$\Rightarrow \frac{dC}{dx} = 30 + 2x$$

Intergrating both sides w.r.t. x

$$\int \frac{dC}{dx}\,dx = \int (30+2x)dx$$

$$\Rightarrow C = 30x + \frac{2x^2}{2} + k$$

$$\Rightarrow C = 30x + x^2 + k$$

It is given that,

When $x = 0$, C = 200

$\therefore$ 200 = 0 + 0 + k

$\Rightarrow$ k = 200

$\therefore$ Total cost C (x) = 30x + x^2 + 200. **Ans.**

(ii) Cost of increasing output from 100 to 200

$$= \int_{100}^{200} (30+2x)dx$$

$$= \left[30x + x^2\right]_{100}^{200}$$

$$= [30 \times 200 + (200)^2] - [30 \times 100 + (100)^2]$$

$$= [6{,}000 + 40{,}000] - [3{,}000 + 10{,}000]$$

$$= 46{,}000 - 13{,}000$$

= ₹ 33,000 **Ans.**

OR

(b) Given : $C(x) = \frac{1}{3}x^3 - 5x^2 + 30x - 15$

[$\because$ P = 6 $\therefore$ S(x) = 6x]

$$\therefore \quad P(x) = S(x) - C(x)$$

$$= 6x - \frac{1}{3}x^3 + 5x^2 - 30x + 15$$

$$= 5x^2 - \frac{1}{3}x^3 - 24x + 15$$

Now, $$\frac{dP}{dx} = 10x - \frac{3x^2}{3} - 24$$

$$= 10x - x^2 - 24$$

For maximum and minimum profit,

Put $$\frac{dP}{dx} = 0$$

$$\Rightarrow 10x - x^2 - 24 = 0$$

$$\Rightarrow x^2 - 10x + 24 = 0$$

$$\Rightarrow (x-6)(x-4) = 0$$

$$\Rightarrow x = 6 \text{ or } x = 4$$

Now, $$\frac{d^2P}{dx^2} = 10 - 2x$$

$$\therefore \left(\frac{d^2P}{dx^2}\right)_{x=6} = 10 - 2 \times 6 = -2 < 0$$

So, at x = 6, P(x) is maximum.

and $$\left(\frac{d^2P}{dx^2}\right)_{x=4} = 10 - 2 \times 4 = 2 > 0$$

So, at x = 4, P(x) is minimum.

Hence, at x = 6, profit is maximum. **Ans.**

Comments of Examiners

21. (a) Several candidates integrated MC(x) to find C(x) without adding a constant to C(x). This led to an incorrect solution. Some candidates committed simplification errors in the process of applying limits to find an increase in the output cost.

(b) Many candidates made errors while finding the profit function. Some made errors in the process of finding the value of 'x' at which the profit was maximum while applying the concept of maxima and minima. In a few cases, candidates maximised the cost function in place of the profit function as well as they failed to apply the second-order derivative for maximisation.

Answering Tips

- Practice with the help of examples, the concept and interpretation of all Marginal cost/revenue functions and application of calculus in Commerce and Economics.
- The concepts of application of derivatives regarding increasing/decreasing functions and maxima/minima in the context of functions like Cost function, Marginal Cost and Average Cost, etc.
- Practice the application of integration in Commerce and Economics supported with an appropriate number of examples.

Question 22. **[6]**

A company uses three machines to manufactue two types of shirts, half sleeves and full sleeves. The number of hours required per week on machine M_1, M_2 and M_3 for one shirt of each type is given in the following table :

	M_1	M_2	M_3
Half sleeves	*1*	*2*	*8/5*
Full sleeves	*2*	*1*	*8/5*

None of the machines can be in operation for more than 40 hours per week. The profit on each half sleeve shirt is ₹ 1 and the profit on each full sleeve shirt is ₹ 1·50. How many of each type of shirts should be made per week to maximize the company's profit ?

Solution 22.

Let the number of half sleeve shirts made per week be x and the number of full sleeve shirts made per week be y.

So, the objective function is,

$$\text{Maximize } Z = x + 1.50y$$

Subject to constraints,

$$x + 2y \leq 40 \quad \text{...(i)}$$

$$2x + y \leq 40 \quad \text{...(ii)}$$

$$\frac{8}{5}x + \frac{8}{5}y \leq 40$$

or $$x + y \leq 25 \quad \text{...(iii)}$$

and $$x \geq 0 \text{ and } y \geq 0$$

From inequation (i), we have

$$x + 2y = 40$$

x	0	40
y	20	0

From inequation (ii), we have

$$2x + y = 40$$

x	0	20
y	40	0

From inequation (iii), we have

$$x + y = 25$$

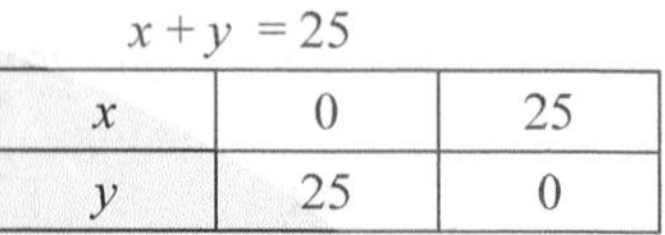

x	0	25
y	25	0

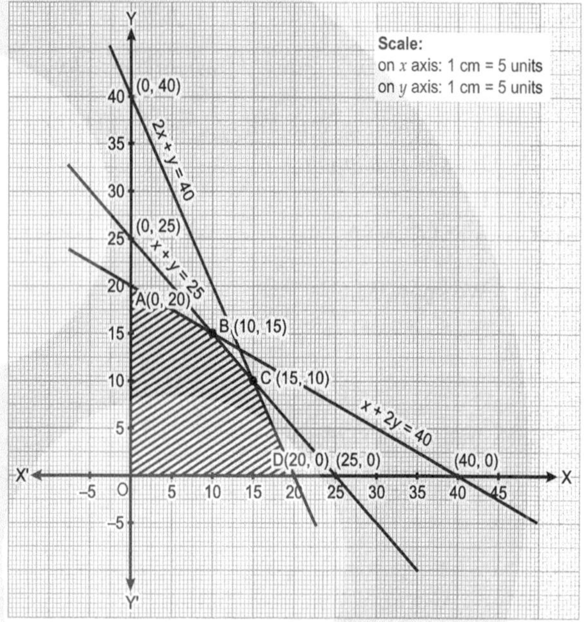

Points	$Z = x + 1.50y$
A (0, 20)	$0 + 1.50 \times 20 =$ ₹ 30
B (10, 15)	$10 + 1.50 \times 15 =$ ₹ 32.50
C (15, 10)	$15 + 1.50 \times 10 =$ ₹ 30
D (20, 0)	$20 + 1.50 \times 0 =$ ₹ 20
O (0, 0)	0

So, Z is maximum at B (10, 15).

Hence, 10 half sleeves shirts and 15 full sleeve shirts should be made per week for maximum profit. **Ans.**

Comments of Examiners

22. Many candidates could not express the given constraints in the form of linear inequalities correctly. Several candidates did not write non-negative constraints $x \geq 0$ and $y \geq 0$.

Answering Tips

- Revise the concept of solving linear inequalities.
- Understand and practice the method of identifying the constraints and forming the corresponding linear inequalities and sketching the graph of the same to identify the feasible region for the objective function.